ANNALS OF
THE NEW YORK ACADEMY
OF SCIENCES

Volume 899

EDITORIAL STAFF

Executive Editor
BARBARA M. GOLDMAN

Managing Editor
JUSTINE CULLINAN

Associate Editor
JOYCE HITCHCOCK

The New York Academy of Sciences
2 East 63rd Street
New York, New York 10021

THE NEW YORK ACADEMY OF SCIENCES
(Founded in 1817)

BOARD OF GOVERNORS, September 15, 1999–September 15, 2000

BILL GREEN, *Chairman of the Board*
TORSTEN WIESEL, *Vice Chairman of the Board*
RODNEY W. NICHOLS, *President and CEO* [ex officio]

Honorary Life Governors
WILLIAM T. GOLDEN   JOSHUA LEDERBERG

JOHN T. MORGAN, *Treasurer*

Governors

D. ALLAN BROMLEY	LAWRENCE B. BUTTENWIESER	PRAVEEN CHAUDHARI
JOHN H. GIBBONS	RONALD L. GRAHAM	HENRY M. GREENBERG
ROBERT G. LAHITA	MARTIN L. LEIBOWITZ	JACQUELINE LEO
WILLIAM J. McDONOUGH	KATHLEEN P. MULLINIX	JOHN F. NIBLACK
SANDRA PANEM	RICHARD RAVITCH	RICHARD A. RIFKIND
	SARA LEE SCHUPF	JAMES H. SIMONS

ELEANOR BAUM, *Past Chairman of the Board*

HELENE L. KAPLAN, *Counsel* [ex officio]   PETER H. KOHN, *V.P. & Secretary* [ex officio]

REACTIVE OXYGEN SPECIES
FROM RADIATION TO MOLECULAR BIOLOGY
A Festschrift in Honor of Daniel L. Gilbert

ANNALS OF THE NEW YORK ACADEMY OF SCIENCES
Volume 899

REACTIVE OXYGEN SPECIES
FROM RADIATION TO MOLECULAR BIOLOGY
A Festschrift in Honor of Daniel L. Gilbert

Edited by Chuang Chin Chiueh

The New York Academy of Sciences
New York, New York
2000

Copyright © 2000 by the New York Academy of Sciences. All rights reserved. Under the provisions of the United States Copyright Act of 1976, individual readers of the Annals are permitted to make fair use of the material in them for teaching and research. Permission is granted to quote from the Annals provided that the customary acknowledgment is made of the source. Material in the Annals may be republished only by permission of the Academy. Address inquiries to the Permissions Department (editorial@nyas.org) at the New York Academy of Sciences.

Copying fees: For each copy of an article made beyond the free copying permitted under Section 107 or 108 of the 1976 Copyright Act, a fee should be paid through the Copyright Clearance Center, Inc., 222 Rosewood Drive, Danvers, MA 01923 (www.copyright.com).

∞ The paper used in this publication meets the minimum requirements of American National Standard for Information Sciences—Permanence of Paper for Printed Library Materials. ANSI Z39.48-1984.

Library of Congress Cataloging-in-Publication Data

Reactive oxygen species : from radiation to molecular biology : a festschrift in honor of Daniel L. Gilbert / editor, Chuang Chin Chiueh.
    p. cm. — (Annals of the New York Academy of Sciences, v. 899)
    Includes bibliographical references and index.
    ISBN 1-57331-238-X (cloth : alk. paper) — ISBN 1-57331-239-8 (pbk : alk. paper)
    1. Active oxygen—Physiological effect—Congresses. 2. Active oxygen—Pathophysiology—Congresses. 3. Gilbert, Daniel L. I. Chiueh, Chuang C. II. Gilbert, Daniel L. III. Series
    Q11 .N5 vol. 899
    [QP535.O1]
    500 s—dc21
    [612.2'2]
                                                                     99-086493
                                                                        CIP

K-M Research/PCP
Printed in the United States of America
ISBN 1-57331-238-X (cloth)
ISBN 1-57331-239-8 (paper)
ISSN 0077-8923

ANNALS OF THE NEW YORK ACADEMY OF SCIENCES

Volume 899

Reactive Oxygen Species

From Radiation to Molecular Biology[a]

A Festschrift in Honor of Daniel L. Gilbert

Editor
CHUANG CHIN CHIUEH

CONTENTS

Editor's Remarks. *By* CHUANG CHIN CHIUEH.	ix
Part I. Introduction	
Fifty Years of Radical Ideas. *By* DANIEL L. GILBERT.	1
Human Limits for Hypoxia: The Physiological Challenge of Climbing Mt. Everest. *By* JOHN B. WEST.	15
Part II. Radiation	
Radiation, Radicals, and Images. *By* JAMES B. MITCHELL, ANGELO RUSSO, PERIANNAN KUPPUSAMY, AND MURALI C. KRISHNA.	28
Radioprotection by Antioxidants. *By* JOSEPH F. WEISS AND MICHAEL R. LANDAUER.	44
Ionizing Radiation Potentiates the Induction of Nitric Oxide Synthase by Interferon-γ and/or Lipopolysaccharide in Murine Macrophage Cell Lines: Role of Tumor Necrosis Factor-α. *By* LESLIE C. MCKINNEY, ELIZABETH M. AQUILLA, DEBORAH COFFIN, DAVID A. WINK, AND YORAM VODOVOTZ.	61
Part III. Molecular Biology	
Genetic Responses to Free Radicals: Homeostasis and Gene Control. *By* BEATRIZ GONZÁLEZ-FLECHA AND BRUCE DEMPLE.	69
Mapping Oxidative DNA Damage and Mechanisms of Repair. *By* STEVEN A. AKMAN, TIMOTHY R. O'CONNOR, AND HENRY RODRIGUEZ.	88
Post-Transcriptional Regulation of Lung Antioxidant Enzyme Gene Expression. *By* LINDA BIADASZ CLERCH.	103

[a]This volume is the result of a symposium held in honor of **Daniel L. Gilbert** on July 2, 1998 at the National Institutes of Health in Bethesda, Maryland, hosted by the NIH ROS Interest Group and the NIH Laboratories and co-hosted by the Oxygen Club.

Part IV. Biochemistry

Oxidative Stress, Mitochondrial Respiration, and Parkinson's Disease.
By GERALD COHEN 112

Regulation of Mitochondrial Respiration by Oxygen and Nitric Oxide.
By ALBERTO BOVERIS, LIDIA E. COSTA, JUAN J. PODEROSO,
MARIA C. CARRERAS, AND ENRIQUE CADENAS 121

Free Radicals and Antioxidants in the Year 2000. A Historical Look to the Future.
By JOHN M. C. GUTTERIDGE AND BARRY HALLIWELL 136

Reflections on the Role of the Thiol Group in Biology. *By* NIELS HAUGAARD. . . . 148

Differential Regulation of MAP Kinase Signaling by Pro- and Antioxidant Biothiols. *By* YUICHIRO J. SUZUKI, SUSAN S. SHI, REGINA M. DAY, AND JEFFREY B. BLUMBERG 159

Enzyme-Like Activity of Glycated Cross-Linked Proteins in Free Radical Generation. *By* MOON B. YIM, SA-OUK KANG, AND P. BOON CHOCK 168

Molecular Markers of Oxidative Stress Vulnerability. *By* INGEBORG HANBAUER AND MARZIA SCORTEGAGNA 182

Protein Oxidation. *By* EARL R. STADTMAN AND RODNEY L. LEVINE 191

Mechanisms of Cell Death Governed by the Balance between Nitrosative and Oxidative Stress. *By* MICHAEL GRAHAM ESPEY, KATRINA M. MIRANDA, MARTIN FEELISCH, JON FUKUTO, MATHEW B. GRISHAM, MICHAEL P. VITEK, AND DAVID A. WINK 209

Nitrone Inhibition of Age-Associated Oxidative Damage. *By* ROBERT A. FLOYD AND KENNETH HENSLEY 222

Part V. Nervous System

Effects of Atypical Antioxidative Agents, S-nitrosoglutathione, and Manganese on Brain Lipid Peroxidation Induced by Iron Leaking from Tissue Disruption. *By* PEKKA RAUHALA AND CHUANG C. CHIUEH. 238

Effect of MAO-B Inhibitors on MPP^+ Toxicity *in Vivo*. *By* RUEY-MEEI WU, RONG-CHI CHEN, AND CHUANG C. CHIUEH 255

Neuroprotective Strategies in Parkinson's Disease Using the Models of 6-Hydroxydopamine and MPTP. *By* EDNA GRÜNBLATT, SILVIA MANDEL, AND MOUSSA B.H. YOUDIM 262

Neuroprotective Antioxidants from Marijuana. *By* A.J. HAMPSON, M. GRIMALDI, M. LOLIC, D. WINK, R. ROSENTHAL, AND J. AXELROD 274

A Positive-Feedback Model for the Loss of Acetylcholine in Alzheimer's Disease. *By* GERALD EHRENSTEIN, ZYGMUNT GALDZICKI, AND G. DAVID LANGE. . . . 283

Microglial Contribution to Oxidative Stress in Alzheimer's Disease. *By* CAROL A. COLTON, OLGA N. CHERNYSHEV, DANIEL L. GILBERT, AND MICHAEL P. VITEK 292

Part VI. Nutrition

Antioxidant Status and Human Health: Use of Cyclic Voltammetry for the Evaluation of the Antioxidant Capacity of Plasma and of Edible Plants. *By* SHLOMIT CHEVION AND MORDECHAI CHEVION 308

Antioxidants in Nutrition. *By* SLOBODAN V. JOVANOVIC AND MICHAEL G. SIMIC 326

Part VII. Cancer and Other Diseases

Free Radical Intermediates in Sonodynamic Therapy. *By* VLADIMÍR MIŠÍK AND PETER RIESZ 335

Glucose Deprivation-Induced Oxidative Stress in Human Tumor Cells: A Fundamental Defect in Metabolism? *By* DOUGLAS R. SPITZ, JULIA E. SIM, LISA A. RIDNOUR, SANDRA S. GALOFORO, AND YONG J. LEE 349

Cytomegalovirus Gene Regulation by Reactive Oxygen Species: Agents in Atherosclerosis. *By* EDITH SPEIR 363

Reactive Species in Sickle Cell Disease. *By* MUTAY ASLAN, DENYSE THORNLEY-BROWN, AND BRUCE A. FREEMAN 375

Invited Poster Papers

Dopamine Stimulates Astrocytic C6-D2L Cells via Tyrosine Kinase and p38 MAPK Activation. *By* YONGQUAN LUO AND GEORGE S. ROTH 392

Selenium Biochemistry: Mammalian Selenoenzymes. *By* THRESSA C. STADTMAN 399

Cytoprotective Properties of Nisoldipine and Amlodipine against Oxidative Endothelial Cell Injury. *By* I. TONG MAK, JINGYUN ZHANG, AND WILLIAM B. WEGLICKI 403

The Origin of Dinitrosyl-Iron Complex in Endothelial Cells. *By* ANDREI M. KOMAROV, I. TONG MAK, AND WILLIAM B. WEGLICKI..... 407

Neutrophil-Endothelial Cell Interactions: Inverse Correlation between Nitric Oxide and Superoxide Anions. *By* JAYASREE NATH AND SANTHANAM KAUSALYA 411

Daniel L. Gilbert: Explorer of Life. *By* CLAIRE GILBERT AND RAYMOND L. GILBERT 415

Index of Contributors 425

The New York Academy of Sciences believes it has a responsibility to provide an open forum for discussion of scientific questions. The positions taken by the participants in the reported conferences are their own and not necessarily those of the Academy. The Academy has no intent to influence legislation by providing such forums.

Reactive Oxygen Species: From Radiation to Molecular Biology

Editor's Remarks

CHUANG CHIN CHIUEH[a]

Unit on Neurodegeneration & Neuroprotection, National Institute of Mental Health, National Institutes of Health, Bethesda, Maryland 20892-1264, USA

This special issue is based on the Festschrift held on July 2, 1998 at the National Institutes of Health (NIH) honoring Dr. Daniel L. Gilbert who has dedicated a half century of his professional career in fostering free radical research (see the first chapter). The idea of organizing this Festschrift originated from a conversation with Professor Carol Colton (Georgetown University) and Dr. Liana Harvath (Food and Drug Administration) on October 2, 1997. This motion was immediately endorsed and supported by the Oxygen Club, the NIH Reactive Oxygen Species Interest Group, and the NIH Special Events Office, Ms. Hilda Madine. We also took this occasion to celebrate the birthdays of Dr. Gilbert and Dr. James Mitchell, two NIH scientists with an avid interest in Free Radical Research.

For this Festschrift, Dr. Gilbert invited his colleagues to present their research findings on reactive oxygen species and to discuss the following areas where oxyradicals play significant roles: radiation, molecular biology, biochemistry, aging (nervous system and cardiovascular system), cancer, and nutrition. It was Dr. Gilbert's idea to further the celebration by the publication of symposium talks and enlist further contributions from other scientists. We sincerely appreciate the enthusiastic support of the *Annals of the New York Academy of Sciences* for providing not only the excellent editorial assistance by Ms. Joyce Hitchcock but also the whole volume of 899 as an early issue of the year 2000. Some of the book chapters are devoted to pertinent and timely ideas about free radicals, and are not limited to reactive oxygen species only but also included reactive nitrogen species, reactive lipid species, reactive thiyl species, and reactive carbon species.

Dr. Gilbert believed that free radicals exist in our body and could have important implications in pathogenesis and in the treatments of diseases (Chiueh, C.C., C. Colton & D.L. Gilbert. 1994. Ann. N. Y. Acad. Sci. **738**: 1–471). For example, nitric oxide is one of the pollutants generated by cars but this radical is also generated in endothelial cells of our body as the endothelium-derived relaxing factor (EDRF, Furchgott, R.F., M.T. Khan & D. Jothianandan. 1987. Federation Proceedings **46**(3): 385; Ignarro, L.J. *et al.* 1987. Federation Proceedings **46**(3): 644; Palmer, R.M.J., A.G. Ferrige & S. Moncada. 1987. Nature **327**: 524–526). Recognition of the importance of free radicals in biology and medicine has been enhanced by the awarding of the 1998 Nobel Prize in Physiology and Medicine to Professors Robert F. Furchgott,

[a]e-mail: chiueh@helix.nih.gov

Louis J. Ignarro, and Farid Murad for their work on nitric oxide. Numerous biochemical pathways are now known to utilize nitric oxide including the reaction of nitric oxide with glutathionyl radicals, converting these radicals to a more potent antioxidant than glutathione and thus enhancing cellular antioxidative defenses against peroxynitrite, cysteine proteases and HIV-1 protease (Rauhala, P., A.M.-Y. Lin & C.C. Chiueh. 1998. FASEB J. **12:** 165–173; Chiueh, C.C. & P. Rauhala. 1999. Free Rad. Res. **31:** 641–650; Hawkins, V., Q. Shen & C.C. Chiueh. 1999. J. Biomed. Sci. **6:** 433–438).

Dr. James Mitchell and I contributed to the design of the front cover of this book using a modified version of the logo of the Oxygen Club of the Greater Washington Area (FIG. 1). We also highlighted the logo and Dr. Gilbert's name with gold color, to symbolize his role as a shining radical scientist. The Oxygen Club succeeded the Oxygen Forum organized by Dr. Terry Hoffeld and Dr. Michael Trush (1981–1985) at NIH. Dr. Michael Simic (National Bureau of Standards), Professor William Weglicki (George Washington University) and Dr. Gilbert organized the first monthly seminar of the Oxygen Club on May 4, 1987 at the NIH Clinical Center Lipsett Amphitheater. The seminar was sometimes held at the Armed Force Radiation Research Institute in the Naval Medical Center. The Oxygen Club is now registered as a non-profit organization for promoting science and education in free radical research (Employer Identification Number: 52-1624142). The Oxygen Club also collaborates with the NIH Reactive Oxygen Species Interest Group in the organization of seminars, symposia, workshops, and the annual spring dinner meeting, as described on the Club's web site (http://www.nih.gov/sigs/rosig/ or http://rsb.info.nih.gov/o2-club/).

Dr. Gilbert's courage has been an inspiration to all who know him. During World War II he suffered from a severe foot injury in combat and was hospitalized for six months for which he is still receiving disability compensation. He has also valiantly fought a brain tumor from 1984 and is now wheelchair bound. But like a true street fighter, he has continued to work at NIH one day a week. He and Professor Carol

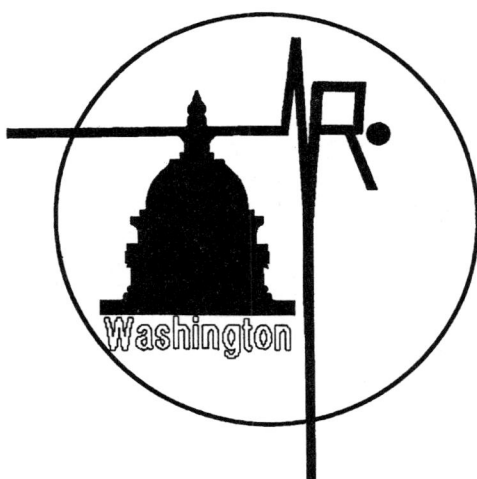

FIGURE 1. The logo of the Oxygen Club of the Greater Washington Area (designed by Dr. James Mitchell). <http://rsb.info.nih.gov/02-club/>

Colton have just published their new book on reactive oxygen species (see the first chapter). Despite his physical disability, Dr. Gilbert is still very active in attending seminars, conferences and meetings. In fact, based on the nomination of the Oxygen Club President, Professor Linda Clerch (Georgetown University) at the 1998 annual meeting of the Oxygen Society in Washington, D.C., Dr. Gilbert was presented with a lifetime achievement award for his longstanding work in promoting free radical research. Accordingly, we salute Dr. Gilbert for his vision, courage, and leadership as a shining radical researcher who made life easier for junior scientists to devote their professional career to this critical research field in biology and medicine.

The Gilberts, Daniel, Claire and Raymond, are frequent travelers. Their adventures have reached every corner of the earth's five continents. They are generously sharing their beautiful memories imprinted in old and new photos appearing in the last chapter of this Festschrift book. Finally, this book is dedicated by all of us to their courageous and loving husband and father, Dr. Daniel L. Gilbert.

Fifty Years of Radical Ideas

DANIEL L. GILBERT[a]

Unit on Reactive Oxygen Species, BNP, NINDS, National Institutes of Health, Bethesda, Maryland 20892-4156, USA

ABSTRACT: My role in the free radical theory of oxygen toxicity is discussed. Rebeca Gerschman and I published several papers on this subject. This sparked my interest in geochemistry and I developed the idea that oxygen was the best qualified biological potential energy source for the following reasons: great abundance, easily accessible, possession of a high thermodynamic potential, and its slow reaction rate. Ionization radiation can be viewed as a catalyst for reactive oxygen species since a killing dose imparts an infinitesimal small amount of energy. Next, Carol A. Colton and I showed that in the mammalian brain that stimulated microglia produce the superoxide radical anion and its implications in Alzheimer's disease is discussed. More recently, I have become interested in the role of sulfhydryl groups in transcription factors.

PRELUDE

"Physiology was Porter's Religion," a Barger quote[1] in 1974 about the person who founded the American Journal of Physiology in 1898. That is also my feelings. In the autumn of 1948, I entered the Department of Physiology at the State University of Iowa and asked Dr. Harry Hines, the chairman if I could start my research. He told me that I should wait until the fall of 1949, since I just started graduate studies; but I insisted. He finally gave in and allowed me to measure blood flow in dystrophic muscles of rats. He was using timed intra-peritoneal injections of ^{32}P to measure blood flow.[2,3] I did the radioactivity measurements in the laboratory of Dr. Titus Evans, the first editor of *Radiation Research*.[4,5] I was given the wrong statistical formula to calculate my results, so I did the radical approach and learned the statistics by myself. I couldn't find the tables that I wanted, so I calculated the Student (Gossett) statistical table with a hand calculator.[6] Today, one can do this with computer software.

REBECA GERSCHMAN

Dra. Rebeca Gerschman (FIG. 1) gave a seminar on oxygen toxicity in 1953 and theorized that it was due to free radicals and that the free radicals were the cause of radiation damage. I transferred to the University of Rochester in 1950 and took the radiation courses instead of repeating the medical physiology and biochemistry courses. The University of Rochester was selected to study the biological effects of radiation for the World War II Manhattan Project. I think that I was the only one at

Address for correspondence: e-mail: dangil@helix.nih.gov

FIGURE 1. Dra. Rebeca Gerschman, my first collaborator.

the seminar who knew what a free radical was. Immediately after the seminar, I was so impressed that we began a collaboration. The result was the 1954 Science pape,[7] in which it was stated "Free radical formation is also expected in normal oxidative metabolism." The univalent reduction of oxygen (FIG. 2) results in the formation of two free radicals plus hydrogen peroxide. Today we refer to these species as reactive oxygen species or ROS.

Dr. Wallace O. Fenn, my thesis advisor, encouraged me to pursue the oxygen toxicity studies with Rebeca; Rahn mentioned that Fenn liked those who showed "pluck, patience, and perseverance..."[8] However, my Ph.D. thesis was on muscle

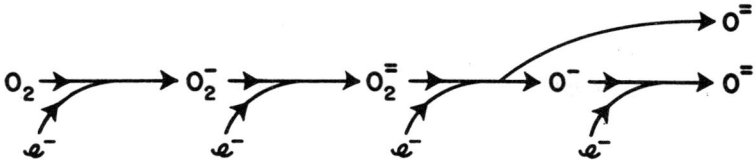

FIGURE 2. The univalent reduction of oxygen. Notice that when one electron is added, the superoxide radical anion, which is a free radical anion. When two electrons are added, hydrogen peroxide, which is not a free radical, is formed. When another electron is added, the very reactive hydroxyl radical is formed.

calcium in which one of my conclusions was that there was a calcium pump pushing the calcium out of the cell. I know of no other report about a calcium pump before this time.[9]

We continued our many papers on the effects of high oxygen tensions on DNA[10] to survival of mice as a function of oxygen pressure (FIG. 3).[11] As Rebeca Gerschman wrote "It is plausible that a continuous small 'slipping' in the [antioxidant] defense could be a factor contributing to aging and death..."[2]

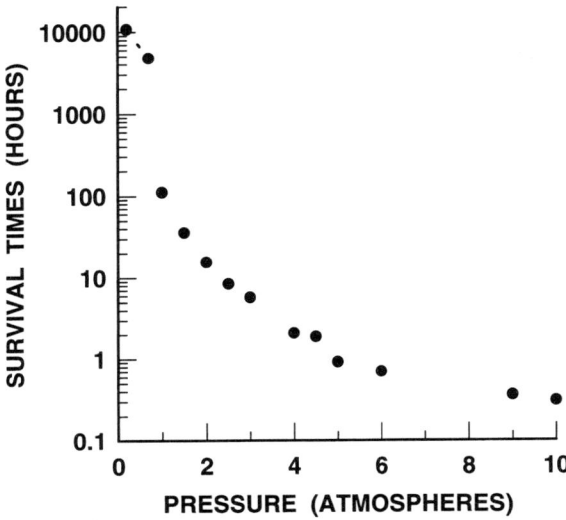

FIGURE 3. Survival of female mice as a function of oxygen pressure. From 1 to 10 atmospheres, the relationship between pressure and survival time is: $\log T = \log a - b*\log P$ where T is the survival time in hours, $a = 102$ hours, $b = 2.73$, and P is the oxygen pressure.

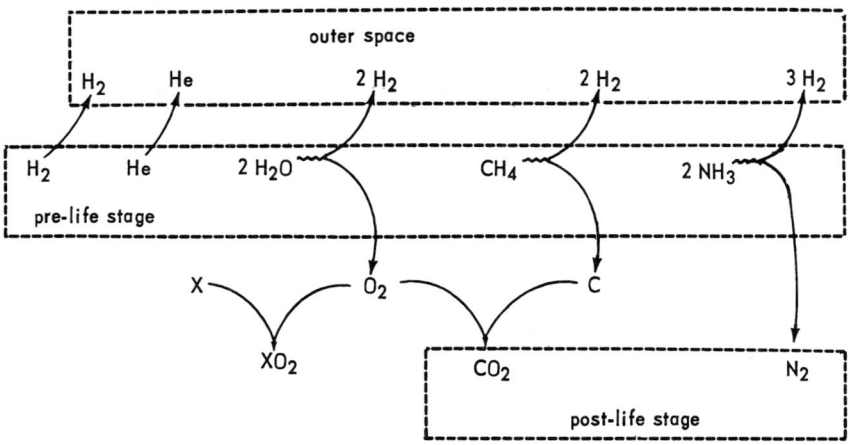

FIGURE 4. Evolution of atmospheres. The universe consists of a predominance of hydrogen, and hence it would be expected that a substantial amount of hydrogen be bound to the other abundant elements: oxygen to form water, carbon to form methane, and nitrogen to form ammonia. This early atmosphere might be lost early in the evolution of the planet giving rise to a secondary partially oxidized atmosphere.

GEOCHEMISTRY

Why was oxygen toxic? I tried to answer this question in my 1960 paper.[13] FIGURE 4 illustrates some of my ideas about planetary atmospheres evolving from a pre-life stage to a post-life stage. Over 90% of the universe is composed of hydro-

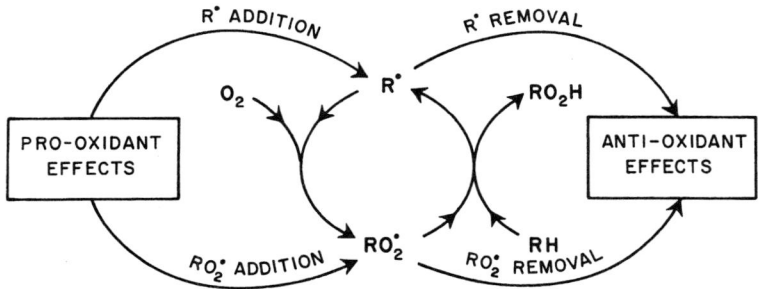

FIGURE 5. Chain reaction. Increasing the free radicals is a pro-oxidant effect and decreasing them is an antioxidant effect. An example is the effect of nitric oxide (a known free radical) reacting with the superoxide radical anion producing the very toxic peroxynitrite anion. The nitric oxide is an antioxidant when it removes the superoxide radical anion and a pro-oxidant when it forms the peroxynitrite anion.

gen, about 7% is composed of chemically inert helium, and the remaining 0.1% of the universe is composed of the rest of the elements. Since there is such an abundance of hydrogen, one expects that hydrogen to be combined with the other elements. Oxygen is the third most abundant element in the universe and is the best qualified biological potential energy source, not only because of its great abundance, but also because it is easily accessible. It is a gas under ordinary circumstances. There are two other reasons: its high thermodynamic potential and its slow reaction rate. Only fluoride, chloride, ozone have higher potentials, but they react rapidly.

Because of their very nature, antioxidants under the appropriate conditions become prooxidants.[14] FIGURE 5 illustrates a chain reaction. Increasing free radicals is a pro-oxidant effect, whereas decreasing free radicals is an antioxidant effect.

FIGURE 6. "The first reader to use [the] Madison Building facilities [in the Library of Congress] was Daniel L. Gilbert of the National Institutes of Health, seen here in the Geography and Map Reading Room on Monday morning, March 17, examining a 1590 manuscript map of a mountain in Peru." *From* Library of Congress Information Bulletin, Vol. 39, No. 16, April 18, 1980, p. 132. Actually, I didn't find the Pariacaca mountain at this time, but I did find it later in the Geography and Map Reading Room.

Since nitric oxide is a free radical, it can combine with other free radicals and remove them; this action is an antioxidant effect. When it combines with other free radicals, it can possibly become a prooxidant. Ionizing radiation is a catalyst for oxygen toxicity. A human can be killed by 5 grays of radiation, which is equivalent to an absorbed radiation of 5 joules/kg or 0.0012 Kcal/kg. The killing dose of radiation raises the temperature of only 0.0012°C. Radiation in the presence of oxygen produces oxidizing free radicals.[15]

LOW AND HIGH EFFECTS OF OXYGEN

When I was asked to be the editor of a book on oxygen,[16] I accepted with great delight. The first chapter[17] was going to be about a few sentences about the discovery of oxygen, but it developed into something more substantial. I wanted to present the role of oxygen in hypoxia to hyperoxia. Father Jose Acosta climbed a pass by Pariacaca and was the first one to describe altitude sickness (hypoxia) in 1590. In 1980, I (FIG. 6) was the first reader in the new Madison Building of the Library of

FIGURE 7. The twin peaks of Pariacaca, or as it is presently known, Tullujuto. Photo taken by Dr. Claire Gilbert in an airplane we rented specifically to view this mountain.

Congress. Later in 1990, we rented an airplane just to view Pariacaca.[18] The mountain pass by Pariacaca reaches an altitude of 16,000 ft., the same elevation as the summit of Mt. Blanc. This mountain, more recently known as Tullujuto (FIG. 7), reaches an altitude of over 19,500 ft.

NEGATIVE SURFACE CHARGE

Since biological membranes possess a negative surface charge,[19] the surface is more acid than the bulk phase. Thus, there is more $HO_2\cdot$ at the surface than in the bulk phase. The pK of $HO_2\cdot/O_2^-$ is 4.8. Since the $HO_2\cdot$ species is more reactive than the O_2^- species, the membranes are more susceptible to the deleterious effects of ROS.

FIGURE 8. Dr. Carol A. Colton, my most recent major collaborator. Photo taken by Dr. Claire Gilbert.

INVERTEBRATE SYNAPSES AND THE RAT HIPPOCAMPUS SLICE

For the past fifteen years, Dr. Carol A. Colton (FIG. 8) has collaborated with me. Together with her late husband, Dr. Joel S. Colton, we have studied the effects of ROS on invertebrate synapses. First, we showed that hydrogen peroxide decreases the function of the lobster neuromuscular junction.[20] Next we demonstrated the deleterious effects of O_2^- on the squid giant synapse:[21] we also found that the squid giant synapse could function for over a half hour without any oxygen.[22] FIGURE 9

FIGURE 9. Photograph of a lobster grabbing a squid taken by Dr. Claire Gilbert. We have performed experiments on both these invertebrates.

shows a lobster grabbing a squid, the two invertebrate species that we have used in our experiments. We were also the first to provide evidence that hydrogen peroxide inhibits long-term-potentiation in the rat hippocampus slice.[23]

ROS SOURCES IN MAMMALS

We also demonstrated that in mammals that activated microglia are sources of the superoxide radical anion.[24] Microglia release ROS by two mechanisms; the oxygen burst membrane mechanism mediated by NADPH (FIG. 10) and by mitochondria. When nitric oxide was demonstrated to be the endothelial relaxing factor (EDRF), we decided to experiment with both nitric oxide and the superoxide radical anion production in rat microglia.[25] FIGURE 11 shows the time course of O_2^- (less than 1 h) and NO· (1 day) production. The conclusion of this type of experiment is that even though the potentially damaging peroxynitrite can be formed very rapidly when these two species react with each other, they certainly don't get it from microglia. Others have shown that nitric oxide can be good and bad.[26] For example, the antioxidant reduced glutathione (GSH) is about a hundred times more effective as an antioxidant when it is combined with nitric oxide to form S-nitrosoglutathione (GSNO)[27] against the iron induced prooxidant state of rat *in vivo* nigrostrial dopamine neurons.

ALZHEIMER'S DISEASE

Our experiments also even addressed the question, what is the relationship between Alzheimer's disease and ROS? Internal polymerization (lipofuscin) does not

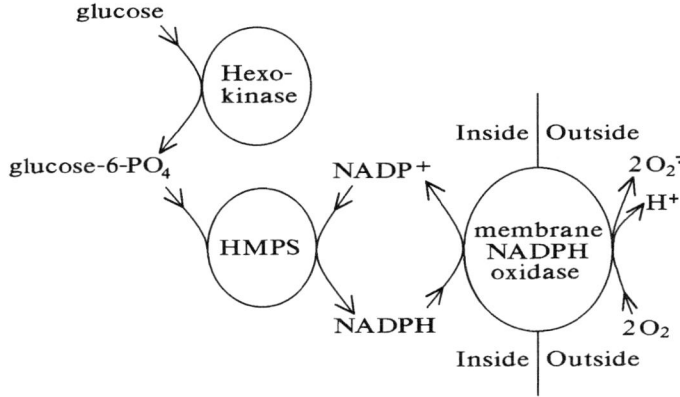

HMPS = Hexose monophosphate shunt

FIGURE 10. Release of ROS from activated microglia during an oxygen burst.

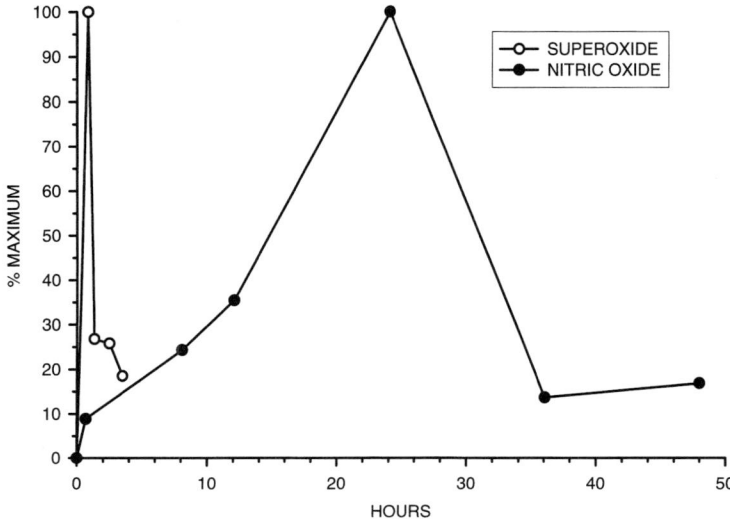

FIGURE 11. Time course of the superoxide radical anion production is much faster (less than 1 h) compared to the much slower nitric oxide production (1 day).

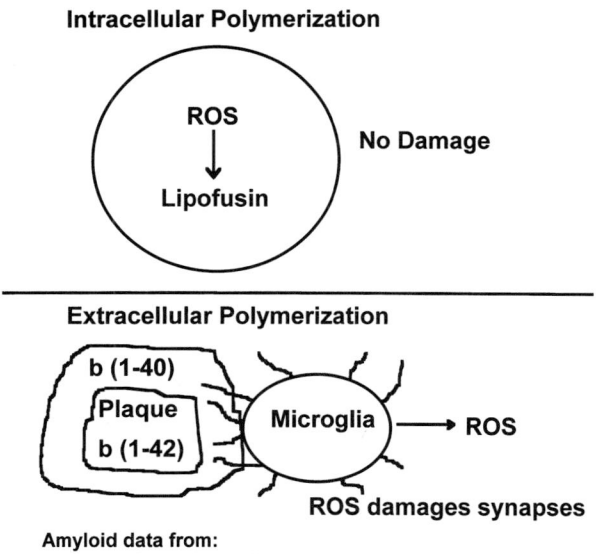

FIGURE 12. Internal polymerization compared to external polymerization: the role of microglia.

generally cause disease; on the other hand, external polymerization can cause disease, as shown by Kelly.[28] It seems that the beta-sheet structure gives Alzheimer's disease plaques unusual chemical stability. Formic acid is the only chemical which can dissolve Alzheimer's plaques. It has been shown that $A\beta(1-42)$ is more effective than $A\beta(1-40)$ in causing the initial seeding of the plaque.[29] Soto and his group[30,31] have shown that inserting a proline residue inhibits the formation of the beta-sheet structure in plaques. External polymerization activate microglia[32] to release ROS due to frustrated phagocytosis[33,34] (FIG. 12).

ROS ACT AS SECOND MESSENGERS

Damped ROS oscillation graph (FIG. 13) is described.[35,36] ROS induce the formation of antioxidant enzymes, which in turn reduce ROS; this takes time. Although sulfur is not so plentiful as hydrogen, carbon, or nitrogen, its role in transcription factors as a redox regulator is extremely important, permitting the toxic ROS to act as second messengers.

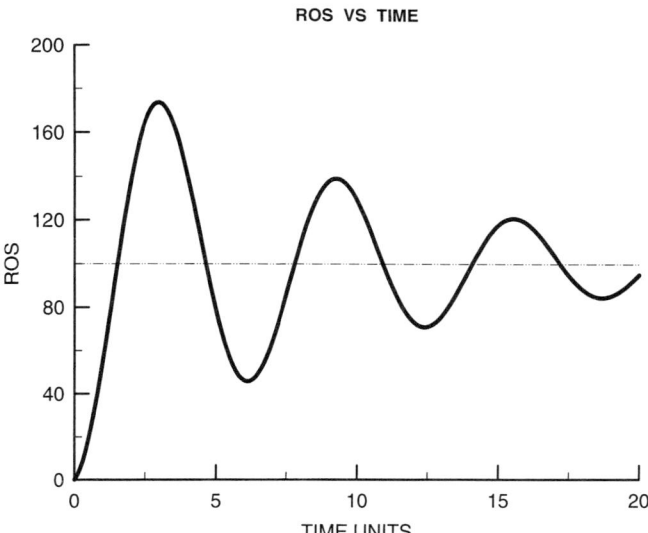

FIGURE 13. Damped oscillation graph. When these redox sensitive transcription factors are activated by ROS, they signal the DNA to produce antioxidant mechanisms to combat the ROS. However, this process takes time and during this interval, the ROS continues to increase. Then, the ROS decreases as the antioxidant defenses increase. After that, the transcription are no longer activated, which results in an increase in ROS. This cycle continues until the ROS is stabilized. If not for these transcription factors, the organism could not exist as the oxygen increased over a period of time.

FIGURE 14. Dr. Claire Gilbert sitting at the highest spot in the Inca city of Machu Picchu, Peru. This place is known as the *intihuatana*, or "the place to which the sun is tied." Photo taken by author.

PRESENT ACTIVITIES

The book Carol A. Colton and I edited demonstrates that I am still active.[37] However, FIGURE 14 shows my Goddess, Dr. Claire Gilbert, sitting on the *intihuatana* or "the place to which the sun is tied,"[38] which is the highest location in Machu Picchu, Peru.

REFERENCES

1. BARGER, A.C. 1974. William Townsend Porter. *In* Dictionary of American Biography: Supplement Four. 1946–1950. J.A. Garraty & E.T. James, Eds. :675–677. Charles Scribner's Sons. New York.
2. GILBERT, D.L., C.D. JANNEY & H.M. HINES. 1950. Circulatory transfer of P^{32} to skeletal muscles under various experimental conditions. Am. J. Physiol. **163:** 575–579.
3. GILBERT, D.L. 1999. From the breath of life to reactive oxygen species, Chapter 1. *In* Reactive Oxygen Species in Living Organisms: An Interdisciplinary Approach. D.L. Gilbert & C.A. Colton, Eds. :3–31. Kluwer Academic/Plenum Pub. New York.
4. BARR, N.F. 1972. Titus C. Evans. Managing Editor 1954–1972. Rad. Res. **50:** iii.

5. EVANS, T.C. 1972. Editorial. Fifty volumes of Radiation Research. Rad. Res. **50:** v–xvi.
6. GILBERT, D.L. 1951. Levels of significance—A direct table of the two-tail probabilities. Math. Tables and Other Aids to Comp. **5:** 85–86.
7. GERSCHMAN, R., D.L. GILBERT, S.W. NYE, P. DWYER & W.O. FENN. 1954. Oxygen poisoning and x-irradiation: a mechanism in common. Science **119:** 623–626.
8. RAHN, H. 1979. Wallace Osgood Fenn: August 23, 1893–September 20, 1971. Biograph. Memoirs, Nat. Acad. Sci. **50:** 141–173.
9. GILBERT, D.L. & W.O. FENN. 1957. Calcium equilibrium in muscle. J. Gen. Physiol. **40:** 393–408.
10. GILBERT, D.L., R. GERSCHMAN, J. COHEN & W. SHERWOOD. 1957. The influence of high oxygen pressures on the viscosity of solutions of sodium desoxyribonucleic acid and of sodium alginate. J. Am. Chem. Soc. **79:** 5677–5680.
11. GERSCHMAN, R., D.L. GILBERT & D. CACCAMISE. 1958. Effect of various substances on survival times of mice exposed to different high oxygen tensions. Am. J. Physiol. **192:** 563–571.
12. GERSCHMAN, R. 1959. Oxygen effects in biological systems. Symp. Spec. Lect. XXI Int. Cong. Physiol. Sciences, pp. 222–226.
13. GILBERT, D.L. 1960. Speculation on the relationship between organic and atmospheric evolution. Perspect. Biol. Med. **4:** 58–71.
14. GILBERT, D.L. 1963. The role of pro-oxidants and antioxidants in oxygen toxicity. Rad. Research Suppl. **3:** 44–53.
15. GILBERT, D.L. 1965. Ninth Bowditch Lecture. Atmosphere and oxygen. The Physiologist **8:** 9–34.
16. GILBERT, D.L., Ed. 1981. Oxygen and Living Processes: An Interdisciplinary Approach. Springer-Verlag. New York.
17. GILBERT, D.L. 1981. Perspective on the history of oxygen and life. *In* Oxygen and Living Processes: An Interdisciplinary Approach. D.L. Gilbert, Ed. :1–43. Springer-Verlag. New York.
18. GILBERT, D.L. 1991. The Pariacaca or Tullujuto story: political realism? Respir. Physiol. **86:** 147–157.
19. GILBERT, D.L. & G. EHRENSTEIN. 1984. Membrane surface charge. Curr. Topics Membr. Transport **22:** 407–421.
20. COLTON, C.A., J.S. COLTON & D.L. GILBERT. 1986. Changes in synaptic transmission produced by hydrogen peroxide. J. Free Rad. Biol. Med. **2:** 141–148.
21. COLTON, C., J. YAO, Y. GROSSMAN & D. GILBERT. 1991. The effect of xanthine/xanthine oxidase generated reactive oxygen species on synaptic transmission. Free Rad. Res. Commun. **14:** 385–393.
22. COLTON, C.A., J.S. COLTON & D.L. GILBERT. 1992. Oxygen dependency of synaptic transmission at the squid giant *Loligo pealei* synapse. Comp. Physiol. Biochem. **102A:** 279–283.
23. COLTON, C.A., L. FAGNI & D. GILBERT. 1989. The action of hydrogen peroxide on paired pulse and long-term potentiation in the hippocampus. Free Rad. Biol. Med. **7:** 3–8.
24. COLTON, C.A. & D.L. GILBERT. 1987. Production of superoxide anions by a CNS macrophage, the microglia. FEB Lett. **223:** 284–288.
25. COLTON, C.A., O. CHERNYSHEV, J. SNELL & D.L. GILBERT. 1994. Induction of superoxide anion and nitric oxide production in cultured microglia. Ann. N.Y. Acad, Sci. **738:** 54–63.
26. CHIUEH, C.C., D.L. GILBERT & C.A. COLTON, Eds. 1994. Neurobiology of NO· and ·OH. Ann. N.Y. Acad. Sci. **738:** 1–471.
27. RAUHALA, P., A.M. LIN & C.C. CHIUEH, 1999, Neuroprotection by S-nitrosoglutathione of brain dopamine neurons from oxidative stress. FASEB J. **12:** 165–173.
28. KELLY, J.W. 1996. Alternative conformations of amyloidogenic proteins govern their behavior. Curr. Opin. Struct. Biol. **6:** 11–17.
29. JARRETT, J.T., E.P. BERGER & P.T. LANSBURY, JR. 1993. The carboxy terminus of the beta amyloid is critical for the seeding of amyloid formation: Implications for the pathogenesis of Alzheimer's disease. Biochemistry **32:** 4693–4697.

30. SOTO, C., M.S. KINDY, M. BAUMANN & B. FRANGIONE. 1996. Inhibition of Alzheimer's amyloidosis by peptides that prevent β-sheet conformation. Biochem. Biophys. Res. Commun. **226:** 672–680.
31. SOTO, C., E.M. SIGURDSSON, L. MORELLI, R.A. KUMAR, E.M. CASTAÑO & B. FRANGIONE. 1998. β-sheet breaker peptides inhibit fibrillogenesis in a rat brain model of amydoidosis: implications for Alzheimer's therapy. Nature Med. **4:** 822–826.
32. COLTON, C.A., O.N. CHERNYSHEV, D.L. GILBERT & M.P. VITEK. 2000. Microglial contribution to oxidative stress in Alzheimer's disease. Ann. N.Y. Acad. Sci. **899:** this volume.
33. MOSSMAN, B.T. & A. CHURG. 1998. Mechanisms in the pathogenesis of asbestosis and silicosis. Am. J. Respir. Care Med. **157:** 1666–1680.
34. GILBERT, D.L., C.A. COLTON & O. CHERNYSHEV. 1998. Plaques: are they the cause of Alzheimer's disease? 15. The First Regional Meeting on Medical Sciences: The Role of Free Radicals in Health and Disease. Jerusalem, Israel and Amman, Jordan.
35. GILBERT, D.L. 1998. Why do so many transcription factors contain sulfur? Abstract 122. Free Rad. Biol. Med. **25** (Suppl. 1)**:** S50.
36. GILBERT, D.L. 1999. Evidence that life originated in a sulfur environment. ISSOL '99 Book of Program & Abstracts 69 (Abstr. P1.18).
37. GILBERT, D.L. & C.A. COLTON, Eds. 1999. Reactive Oxygen Species in Biological Systems: An Interdisciplinary Approach. Kluwer Acad./Plenum Pub. New York.
38. BINGHAM, H. 1963. Lost City of the Incas: The Story of Machu Picchu and Its Builders. Atheneum. New York.

Human Limits for Hypoxia

The Physiological Challenge of Climbing Mt. Everest

JOHN B. WEST[a]

*Department of Medicine, University of California San Diego,
La Jolla, California 92093-0623, USA*

ABSTRACT: Climbing Mt. Everest without supplementary oxygen presents a fascinating physiological challenge because, at the summit, humans are very near the limit of tolerance to hypoxia. It was not until 1978 that the feat was accomplished, and this was after many unsuccessful attempts over a period of more than 50 years, and several physiological studies that suggested that it would be impossible. An analysis shows that the critical factors for reaching the summit are the enormous hyperventilation which is necessary to maintain the alveolar P_{O_2} at viable levels, the fact that the barometric pressure is substantially higher than predicted by the Standard Atmosphere, and the severe respiratory alkalosis that assists loading of oxygen by the blood in the lung. Even so the maximal oxygen consumption on the summit is extremely low with the result that climbers are critically vulnerable to unexpected setbacks such as changes in the weather.

It is a great pleasure to contribute to this Festschrift in honor of Dan Gilbert, whom I have known for many years through the History Section of the American Physiological Society. It also gives me an opportunity to acknowledge his scholarship in the history of high-altitude physiology which I unashamedly put to good use in a recent book.[1] This is a convenient place also to point to two superb historical chapters by Dan on the history of the discovery of oxygen,[2] and the evolution of the composition of the Earth's atmosphere.[3]

I have not worked on the biology of oxygen radicals but I have a strong interest in another facet of oxygen biology, namely severe oxygen deprivation, especially that associated with extreme altitude. It is a remarkable fact that the oxygen partial pressure at the summit of Mt. Everest (altitude 8848 m, 29,028 ft) is right at the limit of human tolerance to hypoxia. The story of human attempts to reach the summit, first with supplementary oxygen, and then without, has many fascinating physiological overtones.

A good place to start is the gloomy prediction made by Thomas W. Hinchliff, president of the (British) Alpine Club in 1876, when he looked at the Chilean Andes from Santiago.

> Lover of mountains as I am, and familiar with such summits as those of Mont Blanc, Monte Rosa and other Alpine heights, I could not repress a strange feeling as I looked at Tupungato and Aconcagua, and reflected that endless successions of men must in all

---

[a]Address for correspondence: John B. West, M.D., Ph.D., UCSD Department of Medicine 0623A, 9500 Gilman Drive, La Jolla, CA 92093-0623. Voice: 858-534-4192; fax: 858-534-4812.
e-mail: jwest@ucsd.edu

probability be forever debarred from their lofty crests.... Those who, like Major Godwin Austen, have had all the advantages of experience and acclimatization to aid them in attacks upon the higher Himalayas, agree that 21,500 ft. [6553 m] is near the limit at which man ceases to be capable of slightest further exertion (ref.4, pp. 90–91).

Two years later, the eminent French physiologist Paul Bert published his monumental book *La Pression Barométrique*[5] in which he described experiment designed to determine what was the critical injurious factor at high altitude. He exposed animals to low barometric pressures on the one hand, and to low concentrations of oxygen on the other, and showed convincingly that neither by itself was the important factor, but rather the product of the two. We now call this the partial pressure of oxygen.

In 1909, the Italian aristocrat, the Duke of the Abruzzi, reached the remarkable altitude of 7500 m in the Karakorum mountains. This ascent astonished climbers and physiologists alike. Indeed, J.S. Haldane from Oxford argued that it would be impossible for human beings to remain alive at this immense altitude unless the lung actively secreted oxygen into the blood. To test this, Haldane organized the 1911 Anglo-American Pikes Peak expedition during which four men from Oxford and Yale Universities spent 5 weeks on the summit of Pikes Peak (altitude 4300 m) and made many important observations on the process of acclimatization to high altitude.[6] Mabel FitzGerald was also invited to join the expedition but, it is said, was not permitted to stay with the four men in the Summit House because she was unchaperoned. Instead she was sent off on her own, accompanied only by a mule, to collect alveolar gas samples from the miners in Colorado, and her resulting data at intermediate altitudes are still frequently cited.[7]

Haldane believed in oxygen secretion until his death in 1936, and the second edition of his book on respiration[8] has a whole chapter devoted to the topic. Although perhaps the notion seems quaint to us today, it was not unreasonable then, because as Haldane pointed out, the swim bladder of the fish often contains a very high concentration of oxygen, and it is a diverticulum of the alimentary canal as is the lung. However, convincing evidence against oxygen secretion was obtained by August and Marie Krogh among others,[9] and we now recognize that all gases move across the pulmonary blood-gas barrier by passive diffusion.

In 1920, A.M. Kellas carried out a remarkable study of the physiological problems of climbing Mt. Everest, but his work is almost unknown.[10] Kellas was a

TABLE 1. Comparisons of Kellas' values and predictions with currently accepted values on physiology at extreme altitudes on Mt. Everest

	Kellas Value	Current value
Summit barometric pressure (mmHg)	251	250–253
Alveolar P_{O_2} (mmHg)	23.6	35
Arterial S_{O_2} (%)	42	70
Arterial pH	7.4	7.7
$\dot{V}_{O_2 max}$ near summit (ml/min)	970	1070
Maximum climbing rate near summit (ft/hour)	300–350	330

lecturer in chemistry at the Middlesex Hospital Medical School in London and carried out a large series of expeditions in the Himalayas, mainly climbing alone with only a few sherpas. In 1920, he wrote a fascinating manuscript that was never published, but which contains remarkable predictions on human physiology at the summit of Mt. Everest. As can be seen from TABLE 1, some of Kellas' predictions were extraordinarily accurate though some were not. However the importance of Kellas' work was not so much the actual values he obtained, but the fact that he was perceptive enough to ask the important questions. In particular, he recognized that the barometric pressure on the summit was crucial as was the alveolar P_{O_2} and arterial oxygen saturation. Although he seriously underestimated the extent of the hyperventilation, and therefore got erroneously low values for the alveolar P_{O_2} and arterial O_2 saturation, his predictions for maximal oxygen consumption and climbing rate near the summit were remarkably accurate. Kellas concluded his study as follows:

> Mt. Everest could be ascended by a man of excellent physical and mental constitution in first rate training, without adventitious aids [supplementary oxygen] if the physical difficulties of the mountain are not too great, and with the use of oxygen even if the mountain can be classified as difficult from the climbing point of view.

It took another 58 years to prove that the first part of this was true! Dramatically, Kellas died just as the first expedition to Mt. Everest in 1921 had its first view of Everest.

In 1924 during the third Everest expedition, E.F. Norton got to within 300 m of the summit without using supplementary oxygen. In fact as FIGURE 1 shows, this altitude was not surpassed until the mountain was climbed in 1953 when oxygen was

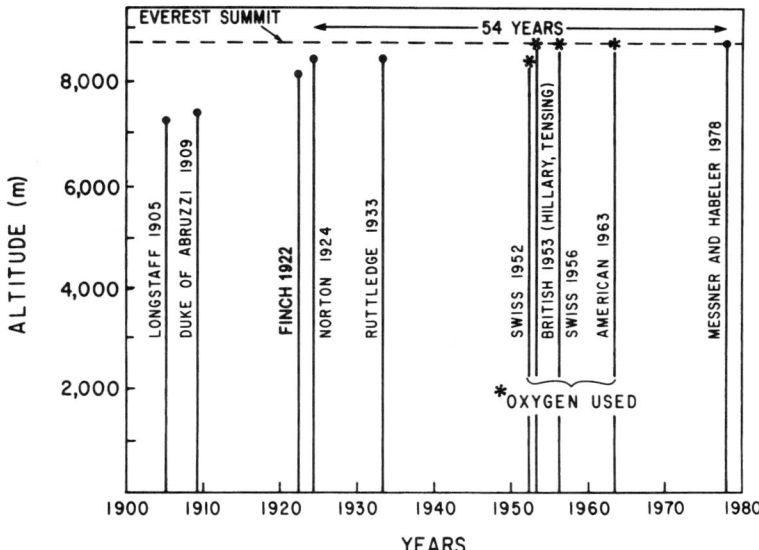

FIGURE 1. Highest altitudes attained by climbers during this century. Note that as early as 1924, Norton ascended to within 300 m of the summit of Mt. Everest. However without supplementary oxygen, the last 300 m took 54 years.

used, and it was not until 54 years later that the first "oxygenless" ascent was made. The fact that the last 300 m took 54 years is a graphic indication of how close the summit is to the limit of human tolerance to hypoxia.

Over the course of the next 30 years, a number of physiologists tackled the question of whether it would be possible for humans to reach the highest point on Earth. One of these was the Italian physiologist, Rudolfo Margaria, who carried out a series of studies in a low-pressure chamber. As FIGURE 2 shows, he found that at a barometric pressure of 300 mmHg which corresponds to an altitude below the summit, the maximum human power fell to zero.[11] He therefore argued that Everest could never be climbed without supplementary oxygen.

Later, Margaria collaborated with the British physiologist, Joseph Barcroft, from Cambridge, and the two of them went to Oxford to use the low-pressure chamber there to determine whether the mountain could be climbed if 100% oxygen was breathed. They predicted that it could, but Barcroft calculated the weight of the oxygen cylinders that it would be necessary for a climber to carry, and concluded that it would be impracticable to reach the summit in this way. An interesting sidelight on these studies was that the subjects measured their work capacity by stepping on and off a box in the low-pressure chamber while breathing 100% oxygen. They developed severe pain in the legs which they were at a loss to understand. In retrospect they were clearly suffering from "bends" caused by nitrogen bubbles in the legs and this was perhaps the first recorded instance of decompression sickness in humans at low barometric pressures, although it was not recognized as such.

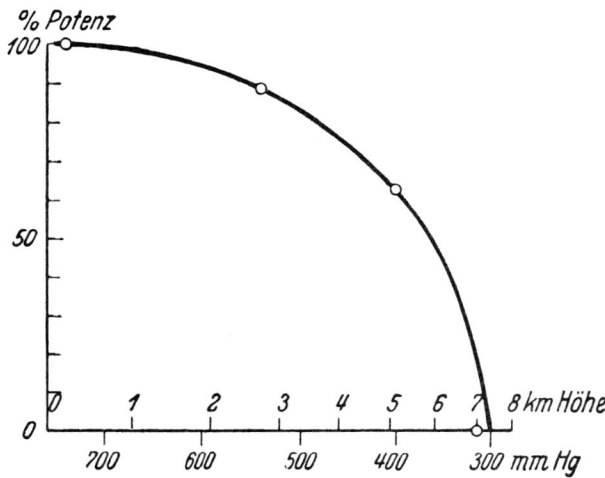

FIGURE 2. Margaria's study of the decline in power (*Potenz*) or work rate as barometric pressure was reduced in a low-pressure chamber. The horizontal axis shows the barometric pressure and also height (*Höhe*). Note that the power output fell to zero at a barometric pressure of about 300 which corresponds to an altitude below the summit of Mt. Everest. This study therefore predicted that it would not be possible to reach the summit without supplementary oxygen. From Margaria.[11]

A few years later, Yandell Henderson from Yale University made another series of calculations using data from well-acclimatized subjects at high altitude. As FIGURE 3 shows, he found that the maximal oxygen consumption rapidly fell as the altitude of Mt. Everest was approached. Henderson's conclusion from these data was that near the summit "the rate of ascent must approach zero: in other words, a minimum of progress in an unlimited amount of time." He therefore argued that it would never be possible to ascend Mt. Everest without supplementary oxygen.[12]

At this time World War II intervened and all thoughts of climbing Mt. Everest were shelved. However in 1952 the Swiss were given permission to attempt Mt. Everest, and they very nearly succeeded in the spring. It is interesting that their attempt failed primarily because they paid insufficient attention to two physiological factors. On the one hand, the oxygen equipment was hopelessly inadequate. The equipment had been designed so that the climbers could breathe oxygen at rest, but not during climbing. However the oxygen stores of the body are so small that any advantage of breathing oxygen at rest is immediately lost once climbing begins.

In addition, the Swiss greatly underestimated the need for adequate hydration at extreme altitude. Two climbers, Lambert and Tenzing, put in their final camp at 8400 m but, because of a logistical mix-up, they found themselves without a stove. As a result, they became desperately thirsty and acutely dehydrated. Lambert wrote: "We were overtaken by a consuming thirst, which we could not appease. There was nothing to drink. An empty tin gave us an idea: a fragment of ice and the candle-

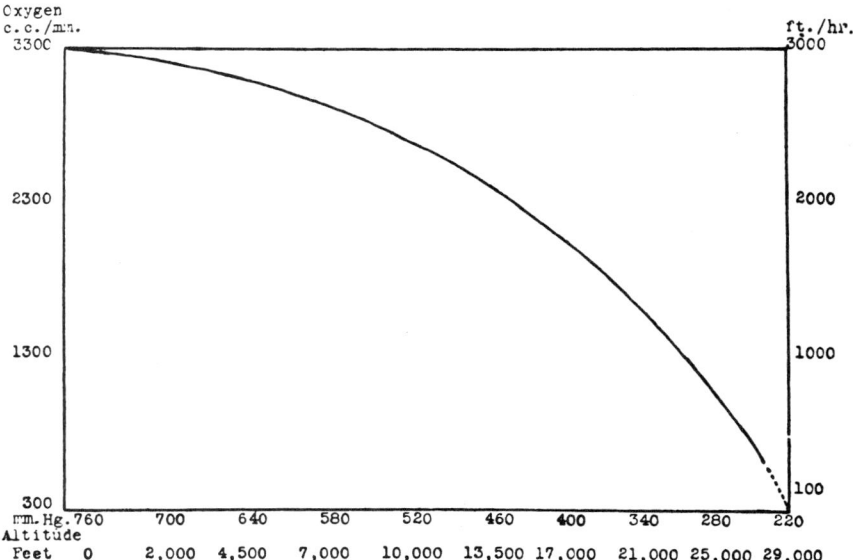

FIGURE 3. Yandell Henderson's study of maximal oxygen uptake plotted against altitude based on measurements made in acclimatized subjects. His conclusion from these data was that near the Everest summit "the rate of ascent must approach zero." Thus he argued that the mountain could not be climbed without supplementary oxygen. From Henderson.[12]

flame produced a little lukewarm water."[13] Contrast this with Edmund Hillary's account during the first successful ascent of Everest one year later when, at their final camp at almost the same altitude as the Swiss, the evening meal included "pint after pint of hot lemon drink crammed with sugar" followed by "great mugs of 'lemonade'."[14] Of course Tenzing who accompanied Lambert on the Swiss expedition summited with Hillary 12 months later.

The reasons why the British expedition succeeded where the Swiss had failed had a lot to do with the British physiologist L.C.G.E. Pugh. He carried out an extensive study of the physiological problems at extreme altitude during the spring of 1952 on Cho Oyu, not far from Everest. He wrote an extensive report which included studies of hydration, oxygen equipment, food, clothing, etc. In particular, the oxygen equipment that evolved from these studies was critical. John Hunt, the leader of the successful 1953 expedition stated:

> Among the numerous items in our inventory, I would single out oxygen for special mention. Many of our material aids were of great importance; only this, in my opinion, was vital to success.... But for oxygen, without the much-improved equipment which we were given, we should certainly not have got to the top. (ref. 15, p. 228).

The success of the 1953 expedition showed that Everest could be climbed if supplementary oxygen was used, but whether an "oxygenless" ascent was possible remained moot. In 1960–1961, Hillary and Pugh organized the so-called Silver Hut expedition to further clarify the physiological problems at extreme altitude, and I was lucky enough to be invited to take part. In fact this was my introduction to high-altitude physiology and medicine. I was one of a group of physiologists who wintered at an altitude of 5800 m (19,000 ft), and in the following spring, an attempt was made to climb Makalu (altitude 8481 m) without supplementary oxygen. This was unsuccessful mainly because one of the climbers became very ill a little below the summit. We took a bicycle ergometer to the Makalu Col (altitude 7440 m) and the resulting measurements of maximal oxygen consumption remain the highest in the field to date.[16] FIGURE 4 shows the results obtained, and it can be seen that extrapolation of the line from the highest altitude to the barometric pressure of the Everest summit again made an "oxygenless" ascent seem improbable. In fact the maximal oxygen consumption at the altitude of the summit was apparently equal to the basal oxygen uptake leaving no oxygen available for physical work.

The final proof that humans can reach the summit of Mt. Everest without supplementary oxygen was provided by Reinhold Messner and Peter Habeler in the spring of 1978. Messner described the experience in lyrical terms.

> On reaching the top, I sit down and let my legs dangle into space.... Now, after the hours of torment... I have nothing more to do than breathe, a great peace floods my whole being. I breathe like someone who has run the race of his life and knows that he may now rest for ever.... In my state of spiritual abstraction, I no longer belong to myself and to my eyesight. I am nothing more than a single, narrow, gasping lung, floating over the mists and the summits (ref. 17, p. 180).

Of course this moving statement was written some time after the expedition and the accounts of both climbers indicate that both men were in desperate straits when they reached the summit. However this was an epochal event in the history of high-altitude physiology. Two years later Messner summited again without supplementary oxygen, this time climbing alone from the north side.

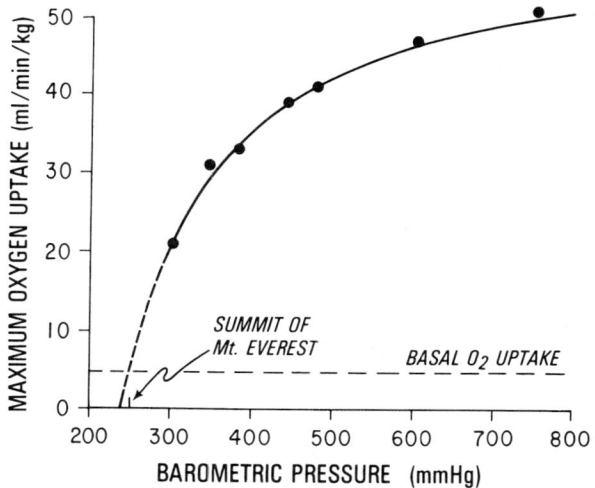

FIGURE 4. Maximal oxygen uptake in acclimatized subjects plotted against barometric pressure using the data from the Silver Hut expedition.[16] Note that the maximum oxygen uptake at the summit was predicted to be the same as the basal oxygen uptake indicating that no work would be possible. Also note that $\dot{V}_{O_2max}$ near the summit is exquisitely sensitive to barometric pressure. From West and Wagner.[19]

The extraordinary achievements of Messner and Habeler renewed interest in the physiological challenge presented by these extreme altitudes, and we decided to organize a special physiological expedition to attempt to obtain measurements on the summit. This was the 1981 American Medical Research Expedition to Everest (AMREE), and the audacious plan was to obtain a series of measurements at four sites on the mountain.[18] The first was the Base Camp at an altitude of 5400 m (17,700 ft). Then over 10 tons of equipment was carried through the Khumbu ice fall, and a laboratory camp was set up at an altitude of 6300 m (20,700 ft) in the Western Cwm. We also planned to obtain a few measurements at the highest Camp 5 situated at 8050 m (26,400 ft). Most enterprising of all was the hope of obtaining a few measurements on the summit itself (8848 m; 29,028 ft). Of course this was very ambitious because many expeditions attempt to climb Mt. Everest without success. In fact we looked back at the six expeditions immediately prior to our own and not one of them succeeded in reaching the summit. Success very much depends on weather; if the weather is bad, summiting is out of the question. Also climbers deteriorate rapidly above about 7000 m and expeditions frequently do not have fit enough people for the last summit push. However we were lucky and obtained a few measurements on the summit. FIGURE 5 shows Dr. Christopher Pizzo sitting on the summit of Mt. Everest taking alveolar gas samples on May 24, 1981.

In an extensive theoretical analysis carried out prior to the expedition[19] it became clear that there were three measurements that were critical to understanding how people could reach the summit. The first was the barometric pressure because this

FIGURE 5. Dr. Christopher Pizzo taking alveolar gas samples while sitting on the summit of Mt. Everest on October 24, 1981.

determines the inspired P_{O_2}, that is the starting point of the oxygen cascade from the atmosphere to the mitochondria. The barometric pressure had never been measured directly on the Everest summit previously. We obtained a large series of measurements at the Base Camp (5400 m) and Camp 2 (6300 m), six measurements at Camp 5 (8050 m) and one measurement on the summit itself.[20] The value was 253 mmHg which was a little higher than most of us had expected, and considerably higher than the pressure predicted from the Standard Atmosphere.[21] The measurements at the Base Camp, Camp 5 and summit, where the altitudes are accurately known, enabled us to draw a barometric pressure-altitude relationship for extreme altitudes on Mt. Everest (FIG. 6). We also got out weather balloon data which gave predicted barometric pressures at an altitude of 8848 m from balloons launched in New Delhi which has the same latitude as Mt. Everest. These data showed that the predicted barometric pressure on the summit varied substantially throughout the year with the highest pressures in the summer and the lowest in the winter. The difference of about 10 mmHg means that the maximal oxygen consumption in midwinter is appreciably lower than in May and October when the mountain is normally climbed and is one reason why an "oxygenless" ascent has never made in midwinter.

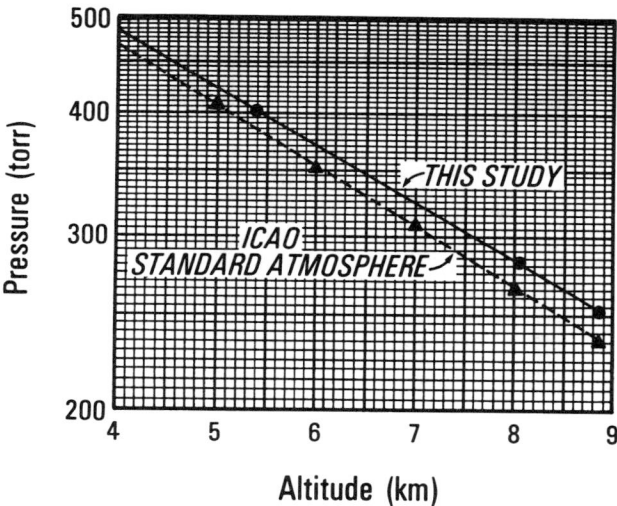

FIGURE 6. The *upper line* shows the barometric pressure-altitude relationship measured during the 1981 American Medical Research Expedition to Everest. The three data points are for altitudes that are accurately known, the right lower point being the Everest summit. The *lower broken line* shows the relationship for the Standard Atmosphere.[21] From West et al.[20]

The second critical physiological factor for climbing to the summit was the extent of the hyperventilation. It was clear from the theoretical analysis that the alveolar Po_2 could only be kept at viable levels if the alveolar Pco_2 was enormously reduced by hyperventilation. We decided that the best way to measure this was to take alveolar gas samples on the summit, return them to UCSD, and thus determine the alveolar Pco_2 and Po_2.[22] The results showed that the alveolar Pco_2 fell approximately linearly with increasing altitude and that, on the summit, its value was between 7 and 8 mmHg. Since the normal value at sea level is 40 mmHg, this extremely low value indicates enormous hyperventilation.

When the alveolar Po_2 and Pco_2 were plotted on an oxygen-carbon dioxide diagram, an interesting result emerged (FIG. 7). The Po_2 falls as one goes from sea level (top right on the graph) to the summit of Mt. Everest (bottom left) because of the reduction in Po_2 in the inspired air. The Pco_2 falls because of increasing hyperventilation. FIGURE 7 shows that when a certain altitude is reached (about 7000 m), there is no further fall in the alveolar Po_2. In fact this is defended against the fall of Po_2 in the inspired air by the process of extreme hyperventilation. It is clear that a climber needs a reasonably good hypoxic ventilatory response in order to mount this degree of hyperventilation. Other measurements on the expedition showed that climbers who were unfortunate to have a low hypoxic ventilatory response generally did badly at extreme altitudes.

Although the alveolar Po_2 was kept up reasonably well at a value of about 35 mmHg by the extreme hyperventilation, an important variable is the arterial Po_2.

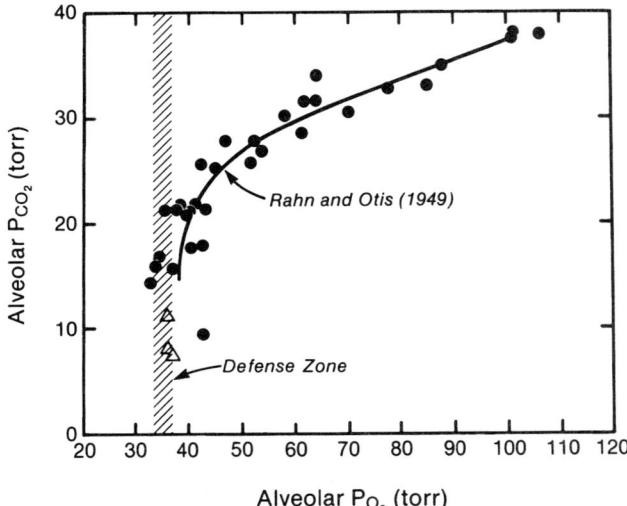

FIGURE 7. Oxygen-carbon dioxide diagram showing the alveolar gas values collated by Rahn and Otis (25) on acclimatized subjects at high altitude. The sea level values are *top right*. In addition, the *triangles* at lower left show the mean values from AMREE at altitudes of 8050 m, 8400 m, and the summit. Note that after a certain altitude has been exceeded (about 7000 m) the alveolar P_{O_2} is defended at a value of about 35 mmHg by the process of extreme hyperventilation. Modified from West et al.[22]

We were not able to take arterial blood gas samples on the summit; there is a limit to what can be done in this hostile environment. However we were able to calculate the arterial P_{O_2} by carrying out a Bohr integration along the pulmonary capillary as the oxygen was being loaded. We did this using the alveolar gas and venous blood data that had been collected. The results showed that the P_{O_2} in the pulmonary capillary blood increased very slowly along the capillary and reached a value of only about 28 mmHg at the end of the capillary. In other words there was a P_{O_2} difference of about 7 mmHg between alveolar gas and end-capillary blood, a hallmark of diffusion limitation of oxygen transfer. Clearly the lung has not evolved for these extraordinarily hypoxic conditions. In a subsequent simulated ascent of Mt. Everest in a low-pressure chamber, Operation Everest II, it was possible to take arterial samples at the barometric pressure equivalent to the summit, and the value was 30 mmHg which was in good agreement.[23]

The extraordinarily low alveolar P_{CO_2} on the summit immediately raises the question of the acid-base status of the climber. To obtain information on this we took venous blood samples on two climbers in the morning after their summit climb and thus measured the base excess of the blood. Parenthetically, there was strong evidence that base excess was changing very slowly at these great altitudes. If we assume that the arterial P_{CO_2} is equal to the alveolar value, the Henderson-Hasselbalch equation gives an arterial pH of between 7.7 and 7.8. In other words the climbers had an extremely severe respiratory alkalosis. TABLE 2 summarizes the alveolar gas and

TABLE 2. Alveolar gas and estimated arterial blood values on the summit of Mt. Everest as determined by the American Medical Research Expedition to Everest[a]

Altitude (m)	Barometric Pressure (torr)	Inspired P_{O_2} (torr)	Alveolar P_{O_2} (torr)	Arterial P_{O_2} (torr)	Arterial P_{CO_2} (torr)	pH	S_{aO_2}
8848 (summit)	253	43	35	28	7.5	>7.7	70
Sea level	760	149	100	95	40	7.4	97

[a]From West et al.[22]

estimated arterial blood values as best we can determine them at the present time. When I have an opportunity of showing these data to physicians who treat patients in the intensive care setting, I often ask them whether they have ever seen patients with such deranged blood gases as these. So far no one has come forward to claim this!

The extreme respiratory alkalosis has another interesting consequence. It greatly increases the oxygen affinity of the hemoglobin, which is a characteristic found in many high-altitude mammals such as the llama and vicuña.[24] Indeed it is found frequently throughout the animal kingdom when animals are required to tolerate severe hypoxia. Interestingly a whole series of strategies have been developed to increase the oxygen affinity of hemoglobin in the presence of oxygen deprivation (TABLE 3), the best known example to physicians being the high oxygen affinity of fetal hemoglobin. The environment of the human fetus is extremely hypoxic, the P_{O_2} of the descending aorta being approximately 25 mmHg. It is remarkable that a climber at extreme altitude is able to develop a similar increase in oxygen affinity of his hemoglobin by the mechanism of extreme respiratory alkalosis.

We saw earlier that many studies had predicted that it would be impossible to reach the summit of Mt. Everest without supplementary oxygen. Therefore we were particularly careful to obtain good measurements of maximal oxygen consumption ($\dot{V}_{O_2\text{max}}$) for the summit conditions. However of course it is not possible to put a bicycle ergometer on the summit. Instead we took climbers who were extremely well acclimatized to very high altitude, and had them pedal the bicycle ergometer in the Camp 2 laboratory at 6300 m while they were breathing 14% oxygen. Thus they

TABLE 3. Strategies for increasing oxygen affinity of hemoglobin in hypoxia

Strategy	Subject/Animal
Different sequence in globin chain	Human fetus, bar-headed goose, toad-fish
Decrease in 2,3-DPG	Fetus of dog, horse, pig
Decrease in ATP	Trout, eel
Different Hb, small Bohr effect	Tadpole
Mutant Hb (Andrew–Minneapolis)	Family in Minnesota
Respiratory alkalosis	Climber at extreme altitude

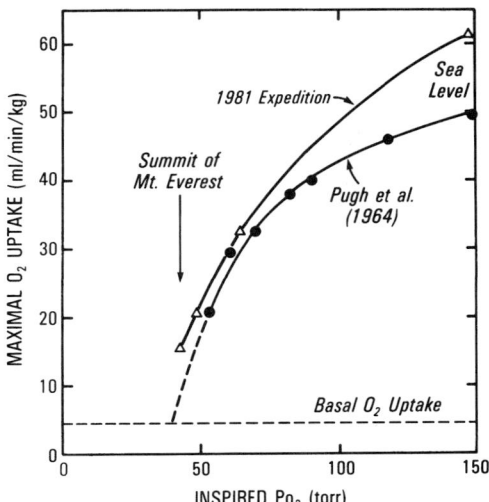

FIGURE 8. Comparison of the relationship between maximal oxygen uptake and inspired P_{O_2} for AMREE (*top line*) and the Silver Hut expedition (*bottom line*). Note that the lines are close at extreme altitudes but the small leftward displacement of the AMREE line explains how Messner and Habeler were able to reach the summit without supplementary oxygen. (AMREE data from West et al.[26])

were exposed to the double insult of very high altitude and a low inspired oxygen concentration. This procedure gave them the same inspired P_{O_2} as on the summit, that is 43 mmHg, and we found that the maximal oxygen consumption was about 1 liter of oxygen per minute. This is a miserable $\dot{V}_{O_2 max}$ being equivalent to walking slowly on the level, but apparently is just sufficient to allow a climber to reach the summit. In fact the climbing rates of Messner and Habeler near the summit in 1978 fit well with this measured maximal oxygen consumption.

Finally, where did we and others go wrong? How is it that previous predictions indicated that the $\dot{V}_{O_2 max}$ on the summit was essentially zero. FIGURE 8 shows a comparison of the results obtained on AMREE with those found on the Silver Hut expedition of 1960–1961. It can be seen that the shapes of the lines relating the $\dot{V}_{O_2 max}$ to inspired P_{O_2} were very similar, but the AMREE data were slightly left shifted. This is the explanation of how it is just possible for human beings to reach the highest point on Earth. If some evolutionary biologist can think of a reason for this, it would be very interesting to know about it.

REFERENCES

1. WEST, J.B. 1998. High Life: A History of High Altitude Physiology and Medicine. Oxford University Press. New York.
2. GILBERT, D.L. 1981. Perspective on the history of oxygen and life. *In* Oxygen and Living Processes. D.L. Gilbert, Ed. Springer-Verlag. New York.

3. GILBERT, D.L. 1996. Evolutionary aspects of atmospheric oxygen and organisms. *In* Handbook of Physiology, Section 4: Environmental Physiology. M.J. Fregly & C.M. Blatteis, Eds. Oxford University Press. New York.
4. HINCHLIFF, T.W. 1876. Over the Sea and Far Away. Longmans Green. London.
5. BERT, P. 1878. La Pression Barométrique. Masson. Paris.
6. DOUGLAS, C.G., J.S. HALDANE, Y. HENDERSON & E.C. SCHNEIDER. 1913. Physiological observations made on Pikes Peak, Colorado, with special reference to adaptation to low barometric pressures. Phil. Trans. R. Soc. Lond. Ser. B **203**: 185–381.
7. FITZGERALD, M.P. 1913. The changes in the breathing and the blood of various altitudes. Phil. Trans. R. Soc. Lond. Ser. B **203**: 351–371.
8. HALDANE, J.S. & J.G. PRIESTLEY. 1935. Respiration. Oxford University Press (Clarendon). London and New York.
9. KROGH, A. 1910. On the mechanism of the gas-exchange in the lungs. Skand. Arch. Physiol. **23**: 248–278.
10. WEST, J.B. 1987. Alexander M. Kellas and the physiological challenge of Mt. Everest. J. Appl. Physiol. **63**: 3–11.
11. MARGARIA, R. 1930. Die Arbeitsfahgkeit des Menschen bei vermindertem Luftdruck. Arbeitsphysiologie **2**: 261–272.
12. HENDERSON, Y. 1939. The last thousand feet on Everest. Nature **143**: 921–923.
13. DITTERT, R., G. CHEVALLEY & R. LAMBERT. 1954. Forerunners to Everest. George Allen and Unwin. London, p. 151.
14. HILLARY, E.P. 1955. High Adventure. Hodder & Stoughton. London, pp. 192–197.
15. HUNT, J. 1953. The Ascent of Everest. Hodder & Stoughton. London.
16. PUGH, L.G.C.E., M.B. GILL, S. LAHIRI, J.S. MILLEDGE, M.P. WARD & J.B. WEST. 1964. Muscular exercise at great altitudes. J. Appl. Physiol. **19**: 431–440.
17. MESSNER, R. 1979. Everest: Expedition to the Ultimate. Kaye & Ward. London.
18. WEST, J.B. 1984. Human physiology at extreme altitudes on Mount Everest. Science **223**: 784–788.
19. WEST, J.B. & P. WAGNER. 1980. Predicted gas exchange on the summit of Mt. Everest. Respir. Physiol. **42**: 1–16.
20. WEST, J.B., S. LAHIRI, K.H. MARET, R.M. PETERS, JR. & C.J. PIZZO. 1983. Barometric pressures at extreme altitudes on Mt. Everest: physiological significance. J. Appl. Physiol. **54**: 1188–1194.
21. ICAO 1964. Manual of the ICAO Standard Atmosphere. Int. Civil Aviation Org. Montreal,Quebec.
22. WEST, J.B., P.H. HACKETT, K.H. MARET, J.S. MILLEDGE, R.M. PETERS, JR., C.J. PIZZO & R.M. WINSLOW. 1983. Pulmonary gas exchange on the summit of Mt. Everest. J. Appl. Physiol. **55**: 678–687.
23. SUTTON, J.R., J.T. REEVES, P.D. WAGNER, B.M. GROVES, A. CYMERMAN, M.K. MALCONIAN, P.B. ROCK, P.M. YOUNG, S.D. WALTER & C.S. HOUSTON. 1988. Operation Everest II: oxygen transport during exercise at extreme simulated altitude. J. Appl. Physiol. **64**: 1309–1321.
24. HALL, F.G. 1937. Adaptations of mammals to high altitude. J.Mammol. **18**: 469–472.
25. RAHN, H. & A.B. OTIS. 1949. Man's respiratory response during and after acclimatization to high altitude. Am. J. Physiol. **157**: 445–462.
26. WEST, J.B., S.J. BOYER, D.J. GRABER, P.H. HACKETT, K.H. MARET, J.S. MILLEDGE, R.M. PETERS, JR., C.J. PIZZO, M. SAMAJA, F.H. SARNQUIST, R.B. SCHOENE & R.M. WINSLOW. 1983. Maximal exercise at extreme altitudes on Mount Everest. J. Appl. Physiol. **55**: 688–698.

Radiation, Radicals, and Images

JAMES B. MITCHELL,[a,c] ANGELO RUSSO,[a] PERIANNAN KUPPUSAMY,[b] AND MURALI C. KRISHNA[a]

[a]*Radiation Biology Branch, Division of Clinical Sciences, National Cancer Institute, Bethesda, Maryland 20892, USA*

[b]*EPR Center, Johns Hopkins University, School of Medicine, Baltimore, Maryland, USA*

ABSTRACT: Nitroxide stable free radicals exhibit varied chemical and biological properties. Their biological applications have been greatly expanded over the past few years. Not only have they been shown to exhibit potent antioxidant and radioprotective properties, but also they can serve as *in vivo* functional imaging probes that non-invasively report on the oxygen status and redox properties of tissue, which may have utility in clinical biomedical research.

The distinguished scientist who is honored and recognized by this special issue, Dr. Daniel L. Gilbert, was firmly rooted in the radiation sciences. Along with his colleague, Dr. Rebeca Gerschman, Dr. Gilbert provided evidence that the toxicity induced by radiation and oxygen poisoning share a common pathway, namely, oxidizing free radicals.[1] Not only was this an important mechanistic observation that has fostered a wealth of research in the area of oxidative stress, but also it has served to forge a solid relationship between radiation biologists and other basic scientists interested in free radicals and their role in biology and medicine. Dan Gilbert did much to promote this relationship through his work and commitment to the Washington Area Oxygen Club, where for many years he provided a forum for all scientists interested in free radicals to present and share their research.

INTRODUCTION

After the discovery of x-rays by Wilhem Röntgen,[2] the field of radiation biology was established rather quickly because within weeks after the discovery there were reports of tissue toxicity among physicians and physicists who for many hours each day were in direct contact with x-ray producing discharge tubes.[3] Toxicities included hair loss and erythema which in some instances culminated in severe burns and desquamation. These early observations of radiation-mediated tissue damage quickly led to the idea of using x-rays for therapeutic purposes, namely the treatment of cancer.[4] Not knowing the mechanism of radiation-induced toxicity did not inhibit its widespread use for both diagnostic and therapeutic purposes in the early 1900s.

[c]Address for correspondence: James B. Mitchell, Ph.D., Radiation Biology Branch, National Cancer Institute, Bldg. 10, Room B3-B69, Bethesda, MD 20892. Voice: 301-496-7511; Fax: 301-480-2238.
  e-mail: jbm@helix.nih.gov

Radiation continues to be widely used in cancer treatment. The reasons that radiation is not always successful in eradicating cancer are multiple; however, a major factor stems from a similar response of normal and cancerous tissue to radiation exposure. Because normal tissue (within the radiation field) is damaged by radiation treatment, the optimal, tumoricidal radiation dose can not always be delivered to the tumor. Ideally, what is needed in radiation therapy is an agent that could selectively protect normal tissues against radiation damage. If such an agent were available, more radiation could be delivered to the tumor and the increase in effective treatment hopefully would translate into enhanced tumor control and eradication. Indeed, several chemical agents have been developed over the years that provide radioprotection both at the cellular and whole animal level.[5,6] Because ionizing radiation produces highly reactive free radicals, the timing of the administration of a chemical radioprotector and the radiation exposure is crucial as shown in FIGURE 1.

The biological consequences of exposure to ionizing radiation are mediated by a series of physical, chemical, biochemical, and cellular responses initiated after the deposition of the radiation in the medium. Some of these stages are represented in FIGURE 1. Radiation produces ionization through the generation of secondary electrons which cause secondary ionizations. The time scale of the initial steps of energy deposition and bond scission is in the order of 10^{-13} seconds. The energy associated with ionizing radiation is significantly greater than the bond energies of many molecules and can cause homolytic bond scission. Since water is the main constituent of

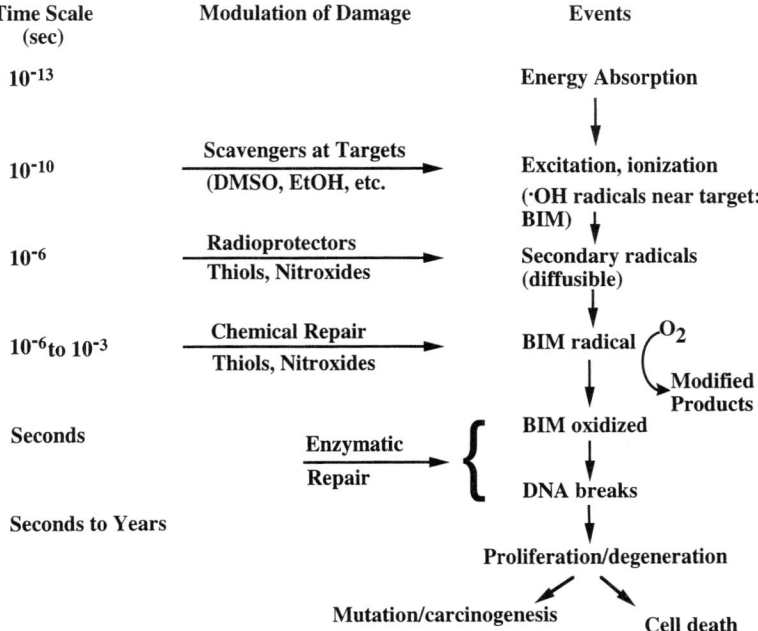

FIGURE 1. Time scales of events initiated by deposition of ionizing radiation in biologic matter.

cellular matter, ionization of water to produce secondary species with high reactivity and short life times (10^{-10}–10^{-9} s) such as the ·OH radical, aquated electrons, or H atoms, would be expected to mediate the chemical reactions which damage biologically important molecules (BIMs).

$$H_2O \rightarrow H\cdot + \cdot OH + e_{aq}\cdot + H^+ \quad (1)$$
(Primary radiolysis products)

$$\cdot OH + XH \rightarrow X\cdot + H_2O \quad (2)$$
(Secondary radicals, life time in microseconds)

Antioxidants that scavenge the primary products of radiolysis can decrease the concentration of these highly reactive species and also the generation of secondary radicals at critical targets. The cellular membrane, proteins/enzymes, and DNA are examples of biologically important molecules (BIMs) that are possible critical targets in radiation-induced cytotoxicity. However, DNA is thought be the crucial target which when damaged can result in both cell death and genetic alterations.[3]

$$X\cdot + BIM\text{-}H \rightarrow XH + BIM\cdot \quad (3)$$
(Radiation induced lesion on DNA)

Hence, eliminating primary radiolysis products close to chromatin/DNA or species with intermediate reactivity (X·) which can diffuse to DNA by using radical scavengers (ScH) should provide protection to DNA.

$$\cdot OH + ScH \rightarrow Sc\cdot + H_2O \quad (4)$$

$$X\cdot + ScH \rightarrow Sc\cdot + XH \quad (5)$$

The scavenger radical, when sufficiently unreactive, would no longer cause any biologic damage. In the absence of such scavengers, cellular oxygen can react with the radical on the target (BIM·) and "fix" the damage by forming peroxyl radical.

However, some scavengers, in addition to scavenging reactive species and inhibiting damage, can also restore damaged target molecules by processes called chemical repair.

$$BIM\cdot + ScH \rightarrow BIM\text{-}H + Sc\cdot \quad (6)$$
(chemical repair)

Most of the chemical radioprotectors tested so far contain sulfhydryl groups which are converted to disulfides by radical scavenging or chemical repair. Based on a thorough structure-activity relationship studies, thiols with the ability to scavenge reactive species close to DNA were found to be most effective radioprotectors.[7–9] The thiyl radicals were not reactive enough to cause further damage while the disulfides are ineffective in providing radioprotection and require active processes, such as thiol reductases, to restore their radioprotective properties. Aminothiols such as WR-1065 are effective in protecting against radiation induced damage based on its ability to accumulate in the vicinity of DNA at sufficiently high concentrations.[10] The precursor of WR-1065, namely WR-2721 is being evaluated and used clinically.[11,12]

Evidence strongly suggests that radiation-induced cytotoxicity results from radiolytically generated reactive species.[3] The evidence is based, in part, on the observed protective effects of aminothiols, which can scavenge reactive species. Recent studies have suggested that a diradical intermediate with a life time in the order of 1 µs is involved in the formation of DNA double strand breaks. Reducing agents which can repair such species might provide radioprotection. However, thiol compounds react with both primary species with high reactivity and short life time (10^{-9} s), as well as the secondary species with intermediate reactivity and life time (1 µs) with similar efficiencies. The scavenging ability of the thiols might provide protection by scavenging oxidants generated close to the critical cellular target. In order to gain more information on the role of the reactive species involved in the formation of DNA double strand breaks, it is necessary to use reducing agents having differing proclivities to react with radicals produced by radiation. An added requirement would be that these agents accumulate to similar extent at all the sites of radical scavenging and also the oxidation products be not significantly reactive.

Stable nitroxide free radicals and their one-electron reduced products, namely the hydroxylamines are a new class of recycling antioxidants (FIG. 2). By undergoing one-electron transfer reactions, nitroxides are reduced to the corresponding hydroxylamines or oxidized to the corresponding oxoammonium cation species.[13] Therefore, once administered *in vivo*, all three forms can exist. The nitroxide/oxoammonium cation pair constitutes an efficient redox couple and mimics the enzymic action of superoxide dismutase (SOD) in a pH-dependent manner[14] and also confers catalase-like action to heme proteins such as myoglobin, cytochrome C etc.[15] The nitroxide radical, though chemically stable, can participate in radical-radical recombination reactions with a variety of free radicals possessing a wide range of reactivities. The hydroxylamine on the other hand, can function as a classic

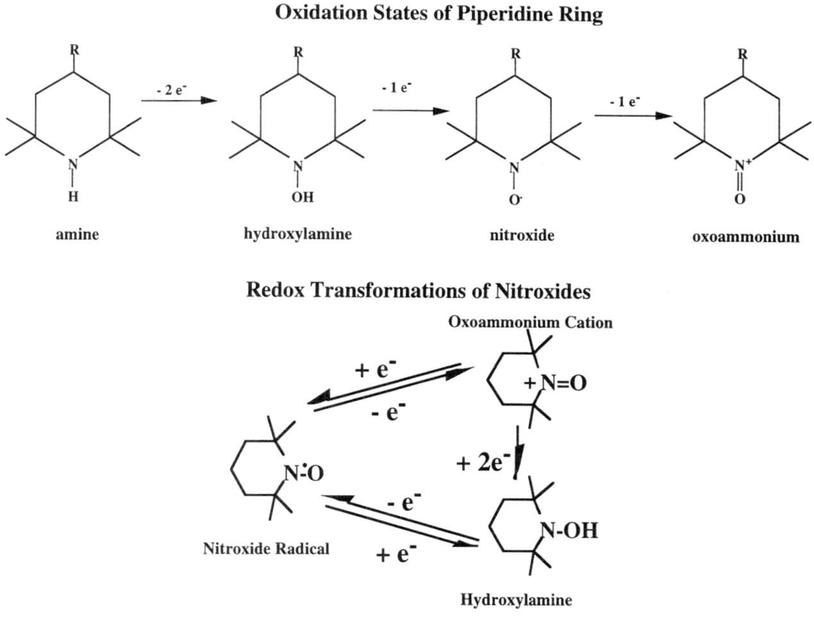

FIGURE 2. The various oxidation states of nitroxides and interconversion of nitroxide between other oxidation states.

antioxidant such as thiols, ascorbate, etc. by donating the H-atom. Its reaction efficiencies depend on the species with which it interacts. With highly reactive species such as OH radicals, the hydroxylamine is an efficient scavenger; whereas, with species of moderate oxidation potential, hydrogen atom donation by the hydroxylamine proceeds slowly. Not unexpectedly, accumulation of the nitroxide or hydroxylamine at critical sites necessary for radioprotection is influenced significantly by the ring substituents. Based on differences in the inherent reactivity of the nitroxide and hydroxylamine, studies designed to investigate the species that facilitates the formation of radiation induced DNA double strand breaks are feasible.

NITROXIDES AS RADIOPROTECTORS

In Vitro *Studies*

Nitroxide stable free radicals have been used as spin labels for biophysical studies and as contrast agents for magnetic resonance imaging.[16–19] Nitroxides are known to react with a variety of biological oxidants including free radicals.[20–23] The observation that nitroxides react with oxyradicals[24] and protect against superoxide and hydrogen peroxide cytotoxicity[25] led us to investigate nitroxides as radiation protectors (FIG. 3). We discovered that nitroxides protect cells and animals against the lethal effects of ionizing radiation.[26–28] This was a particularly interesting finding in that a free radical species (nitroxide) could protect against damage imposed by radiation which inflicts damage through free radical production. Chinese hamster V79

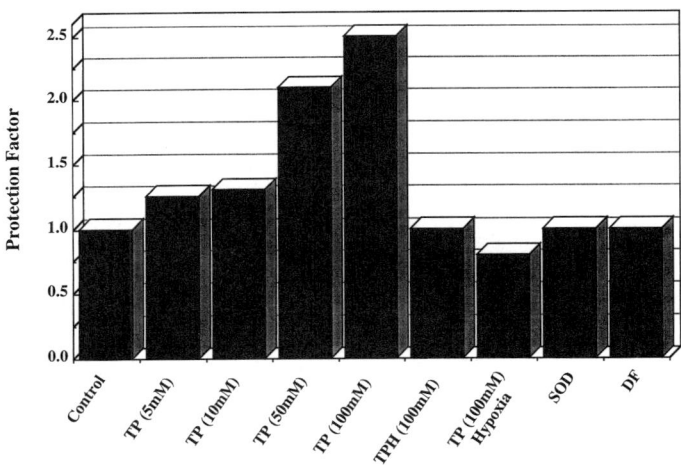

FIGURE 3. Protection of Chinese hamster V79 cells treated with Tempol (TP) at concentrations of 5, 10, 50 and 100 mM (aerobic and hypoxic), Tempol-H (TPH) at 100 mM, SOD (100 µg/ml), Desferal (500 µM) and exposed to x-rays. Protection factors were calculated at the 10% survival by dividing the radiation dose of control cells by the radiation dose of treated cells. (Data adapted from Mitchell *et al.*,[26] with permission.)

cells in the presence of the nitroxide, Tempol, were protected against the aerobic lethal damage induced by ionizing radiation in a concentration dependent manner as shown in FIGURE 3.[26] From the data presented in FIGURE 3 and subsequent studies, several observations regarding the effects of Tempol on the radiation response can be made.[26] First, Tempol-mediated radioprotection of V79 cells was concentration dependent and required mM concentrations to exert significant radioprotection. Second, the hydroxylamine form of Tempol, did not provide radioprotection. Third, Tempol modestly radiosensitized cells irradiated under hypoxic conditions, in agreement with previous studies.[29–31] Fourth, neither exogenous SOD nor Desferal (DF) altered the radiation response. This last point suggests that it is unlikely (although it cannot be ruled out) that the SOD mimetic action of nitroxides is responsible for aerobic radioprotection. Previous reports suggested that SOD protected bone marrow progenitor cells against radiation cytotoxicity *in vitro* and *in vivo* [32–34]; however, not all authors have been able to reproduce these findings.[35] We have repeatedly demonstrated in tissue culture studies (*in vitro*) that exogenously applied SOD does not afford aerobic radioprotection. Likewise, it is unlikely that Tempol radioprotection results from its effect upon metal reduction since the metal chelator, DF, does not provide *in vitro* radioprotection. Lastly, nitroxides must be present during irradiation to provide radioprotection. Pretreating cells with Tempol and rinsing it away prior to radiation does not provide protection nor does adding Tempol immediately after irradiation. This latter point suggests that the protection afforded by Tempol must be through interaction with radiation induced free radical species which have very short life times (μs range).

Radiation exposure can result in the formation of carbon-centered radicals on BIM (in the case of radiation, the most likely critical intracellular target is the DNA) as shown in Eq. 3. As proposed in Eqs. 4 and 5, nitroxides could scavenge radiation-induced reactive species. Likewise, Eqs. 8a and 8b (see below) demonstrate that nitroxides have the capability to restore a carbon-centered radical on BIM by donating an electron (BIM-H). It is possible that a combination of Eqs. 7, 8 and 9 operate to provide radioprotection.

Radiation-induced, unrepaired, DNA double strand breaks and subsequent production of chromosome aberrations have been closely linked with radiation-induced cell killing.[36–38] A recent study [39] showed a direct correlation between Tempol-mediated *in vitro* radioprotection (cell survival) and reduced DNA double strand breaks. Johnstone et al. showed that Tempol significantly reduced the frequency of radiation-induced chromosome aberrations in human peripheral blood lymphocytes.[40] Both of these studies strongly suggest that Tempol-mediated radioprotection results in a reduction in DNA damage.

The structural requirements for nitroxide to function as an effective radioprotector were determined by systematic screening of various nitroxides in an *in vitro* radiobiologic assay. The effect of ring size, substituents, and the ring oxidation state of the test compounds in providing protection to mammalian cells exposed to oxidative damage induced by exposure to ionizing radiation under aerobic conditions were evaluated by monitoring the clonogenic viability of Chinese hamster lung fibroblast cells. The results are shown in FIGURE 4A. Nitroxides of three different ring types were studied, namely the five-membered saturated pyrrolidine ring (A), five-membered unsaturated pyrroline ring, and the six-membered saturated ring piperidine.

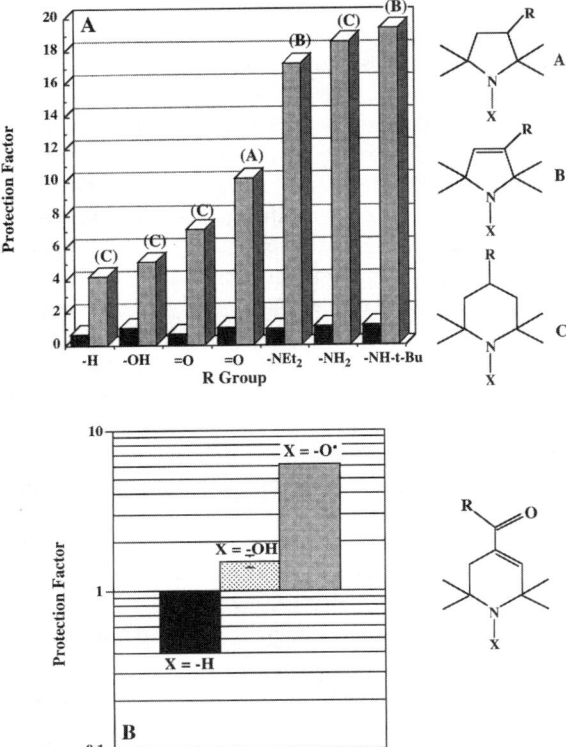

FIGURE 4. A: Dependence of ring type and ring substitutes of nitroxides on the aerobic radioprotection of Chinese hamster V79 cells. Cells were exposed to a 12 Gy radiation dose and protection factors were calculated by the ratio obtained by dividing the surviving fraction of cells treated with the test compound by that of untreated cells. The ring types are shown on right and substituents are indicated below: Columns in black represent hydroxylamine, columns in gray represent nitroxides. A = pyrrolidine; B = pyrroline; C = piperdine. X= –OH, hydroxylamine or X = –O·, nitroxide. **B:** Dependence of the oxidation state of the nitroxide on modifying the aerobic radiation response of Chinese hamster V79 cells. X = H secondary amine, X= –OH, hydroxylamine, X = –O·, nitroxide.

The ring oxidation states tested were the free radical form nitroxide (X = –O·) and its corresponding reduced form, the hydroxylamine (X = –OH). The following observations can be made from this study:

(a) The nitroxides afford radioprotection, the hydroxylamines provide minimal radioprotection.

(b) The ring size or saturation do not have any influence on the observed radioprotection.

(c) The substituents on the ring have significant influence on the radioprotective effects. Positively charged nitroxides have enhanced radioprotective effects compared to neutral or negatively charged nitroxides. These observations support the following conclusions:

(i) Though the hydroxylamines and nitroxides function as radical scavengers and antioxidants, only nitroxides exhibit radioprotective effects. Therefore, the chemical species causing radiation induced cytotoxicity should have intermediate reactivity and life-time (in microseconds).

(ii) Like aminothiols, nitroxides with positively charged substituents provide enhanced radioprotection compared to neutral or negatively charged nitroxides by accumulating at sufficient concentrations at sites of damage; i.e., DNA and prevent/repair the damage.

To study the influence of the oxidation state of the ring in modifying the radiobiologic effects on V79 cells, the clonogenic viability of V79 cells were tested using a fixed concentration of the agent with similar ring and substituent but varying in the oxidation state. Three oxidation states were tested for a given ring, the nitroxide ($X = O\cdot$), the one-electron reduced hydroxylamine ($X = -OH$), and the fully reduced secondary amine ($X = -H$) The results are shown in FIGURE 4B. In agreement with the results shown in FIGURES 3 and 4A, the nitroxide was effective in providing radioprotection; whereas, the corresponding hydroxylamine exhibited minimal protective effects. It should be noted that both are chemically capable of scavenging free radicals, though with differing reactivities. The data support the hypothesis that radiation-induced cytotoxicity is mediated (at least in part) by chemical species with life-times in the order of microseconds, which are efficiently scavenged by nitroxides compared to the corresponding hydroxylamines. The following reactions might be underlying the protective effects of nitroxides (RRNO) and the hydroxylamines (RRNOH).

Electron Transfer

$$X\cdot + H^+ \rightarrow XH + RRNO^+ \quad (7a)$$

$$X\cdot + RRNOH \rightarrow XH + RRNO\cdot \quad (7b)$$

Chemical Repair

$$BIM\cdot + H^+ + RRNO\cdot \rightarrow BIM\text{-}H + RRNO^+ \quad (8a)$$

$$BIM\cdot + H^+ + RRNOH \rightarrow BIM\text{-}H + RRNO\cdot \quad (8b)$$

Nitroxides and hydroxylamines can participate in reactions involving electron transfer and scavenge radicals or provide chemical repair of the target molecules by H-atom donation. However, the rate of reaction for nitroxides are expected to be higher than the corresponding hydroxylamines. In addition to the above mentioned class of reactions, nitroxides can also participate in radical-recombination reactions and provide additional detoxification pathways.

Radical-Recombination

$$X\cdot + RRNO\cdot \rightarrow RRNO\text{-}X \quad (9)$$

$$BIM\cdot + RRNO\cdot \rightarrow RRNO\text{-}BIM \quad (10)$$

While the latter reaction modifies the target molecule, it can be subjected to enzymatic repair processes.

The fully reduced amine on the other hand exhibits modest sensitizing effect, presumably mediated by generating additional reactive species,

$$X\cdot + RRNH \rightarrow RRN\cdot \quad (11)$$

where RRN· possesses sufficient reactivity to mediate further damage. In addition, fixing of damage on targets cannot be excluded.

$$BIM\cdot + RRN\cdot \rightarrow RRN\text{-}BIM \quad (12)$$

Cellular studies suggest that both nitroxides and hydroxylamines can participate in detoxification reactions as antioxidants, the relative efficiency of nitroxide in depleting radicals in the vicinity of the target (DNA) and/or repairing the damaged sites (Eqs. 7a and 8a) compared to that of the hydroxylamines (Eqs. 7b and 8b) makes nitroxides better radioprotectors. In addition, radical-radical recombination reactions (Eqs. 9 and 10) provide additional protective capabilities to nitroxides. The secondary amines on the other hand can form reactive nitrogen species that can damage target molecules (Eqs. 11 and 12).

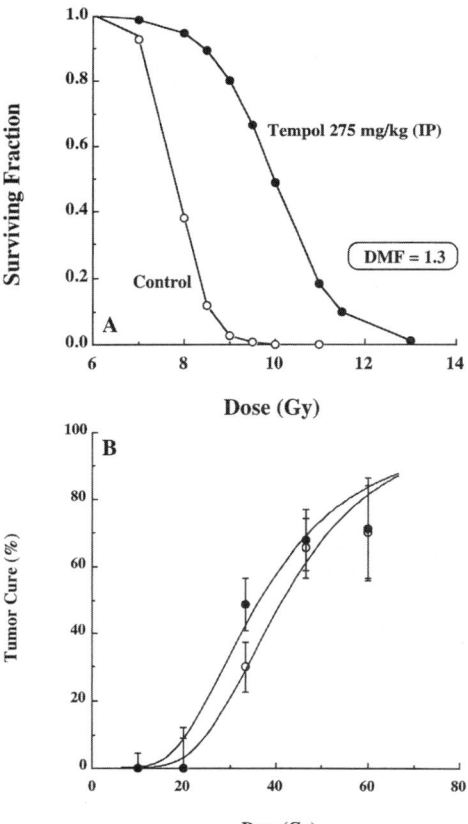

FIGURE 5. *In vivo* studies of Tempol in C3H mice. **A:** Survival of mice 30 days after whole body radiation. Control mice received saline IP and treated mice received Tempol 275 mg/kg IP 10 minutes prior to radiation. The dose modification factor (DMF) is 1.3. (Data from Hahn *et al.*,[27] with permission.) **B:** Radiation tumor control curves for animals treated in the absence or presence of Tempol. *Closed circles*: Tempol-treated mice; *open circles*: PBS-treated mice. There was no statistical difference between the two curves ($p = 0.54$). The calculated $TCD_{50/30}$ values for Tempol-treated and PBS-treated mice were 36.7 and 41.8 Gy, respectively. There was no statistical difference between these values ($p = 0.32$). The error bars represent one standard error above and below the value shown. (Data adapted from Hahn *et al.*,[44] with permission.)

In Vivo *Studies*

Because nitroxides provided radioprotection in *in vitro* experiments, we next explored their utility *in vivo*. The radioprotective effects of Tempol were studied in C3H mice.[27] Since pharmacology studies showed that Tempol is rapidly reduced *in vivo* to the hydroxylamine (a form of the nitroxide that does not provide *in vitro* radioprotection as discussed above), animals were injected with Tempol and received whole body irradiation within 10 min after injection as shown in FIGURE 5A. Tempol provided a dose modification factor of 1.3 at the $LD_{50/30}$ level. Since radiation-induced lethality to mice in 30 days results from cytotoxicity to the bone marrow, Tempol provided protection to bone marrow stem cells. In the course of subsequent studies it was noted that *in vivo* administration of Tempol resulted in a prompt drop in mean arterial blood pressure.[41] A drop in blood pressure could possibly lower oxygen tension in various tissue compartments (steel effect) such as the bone marrow producing hypoxia.[42] Thus, the *in vivo* radioprotection afforded by Tempol could be due to hypoxia-mediated radioprotection.[43] However, a different nitroxide (3-carbamoyl-PROXYL, a five membered ring nitroxide) which did not drop blood pressure *in vivo* also provided radioprotection.[41] Thus, nitroxide radioprotection appears to result from chemical interactions with damaging free radicals produced by radiation as outlined above.

The ability to selectively protect normal tissues in cancer patients receiving radiation treatment would be most advantageous. If selective protection of normal tissues were possible, higher radiation doses could be delivered to the tumor accompanied with higher local control rates. The key, however, is *selective* normal tissue protection because if a systemic radioprotector also protects the tumor, no advantage would be realized. Therefore, we next studied whether Tempol would also protect against local irradiation delivered to a tumor[44] as shown in FIGURE 5B. The administration of Tempol to tumor bearing mice at the same concentration and timing as shown in FIGURE 5A resulted in no protection of tumor. To identify a mechanism for the apparent differential protection of the hematopoietic system (see above) and tumor tissue, pharmacologic studies were carried out. The percentage of oxidized compound (nitroxide) was approximately two-fold greater in the bone marrow compartment, compared to RIF-1 tumor at the time of irradiation. Greater bioreduction of Tempol occurs in RIF-1 tumor. Such a result may provide at least a partial explanation for the absence of tumor radioprotection, since it previously had been demonstrated that the oxidized form of Tempol is the active radioprotector (see above). These preliminary data imply that a potential difference exists between normal and tumor tissues with respect to bio-reduction.

NITROXIDES AS FUNCTIONAL IMAGING PROBES

To further study the differential bio-reduction of nitroxides in normal versus tumor tissue we turned to the possibility of imaging free radicals *in vivo*. Nitroxides are electron paramagnetic resonance (EPR) detectable and the one-electron reduced product, the hydroxylamine is diamagnetic and hence EPR-silent. Nitroxides, when administered *in vivo*, are converted by cellular redox processes to the hydroxylamine, while the hydroxylamine is converted back to the nitroxide. The ratio of the nitroxide to the hydroxylamine depends on several parameters, such as pO_2 and the redox status of the tissue being examined (FIG. 6). Swartz and co-workers have used this principle to obtain metabolic information in tissue using *in vivo* EPR spectroscopy on live animals.[45] Such information can be spatially encoded using magnetic field gradients and a spatial image of differences in tissue metabolism can be obtained based on the differences in metabolism of nitroxide.

Since solid tumors are known to differ in redox status, as well as oxygen status, compared to normal tissue,[46,47] EPR imaging studies using nitroxides as redox sensitive spin probes might non-invasively provide valuable metabolic information. Issues that have been addressed recently are: 1) whether nitroxides could be detected in tumor tissue, and 2) whether differential reduction rates of nitroxide probes could be discerned in tumor versus normal tissues.[44,48] EPR imaging (EPRI) experiments have been performed using an EPR spectrometer operating at 1.2 GHz corresponding to a magnetic field of 40 mT.[49] A specially built bridged-loop surface resonator was used.[50] The open structure of the resonator is ideal for localized study of metabolic activity in large objects. With this imaging system, a cylindrical volume of 10 mm diameter and 5 mm depth could be probed.

Mice bearing ~1 cm diameter tumors were anesthetized and the tail vein cannulated with a heparin-filled 30 gauge catheter for 3-carbamoyl-PROXYL (3-CP) infusion (160 mg/kg). Either the right leg with tumor or the left leg with normal tissue (muscle, skin) was utilized for the imaging studies. The presence of nitroxide in normal and tumor tissue was readily detected by using EPRI. A two dimensional spatial image of the distribution of 3-CP in normal muscle and RIF-1 tumor as a function of time is shown in FIGURE 7. The panels in the top row show the clearance of the nitroxide in normal muscle as a function of time and the corresponding images in the bottom row show clearance from tumors. The data from the images indicate that the rate of clearance of nitroxide in tumors is faster than in normal tissue. This is consistent with previous observations that tumors provide a strong reducing environment when compared to normal tissue and as such results in faster reduction of the nitroxide.[44,48] These observations agree with earlier studies that suggest hypoxic cells within tumors reduce nitroxides more efficiently than well oxygenated normal

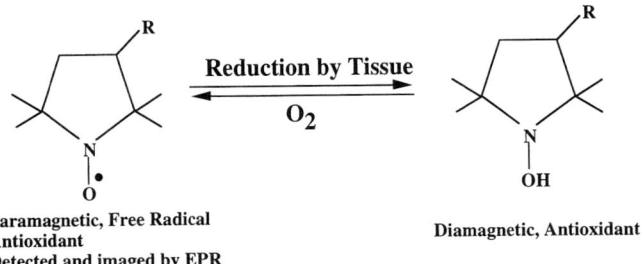

FIGURE 6. Schematic description of the dependence of tissue redox status on the oxidation states of the nitroxide.

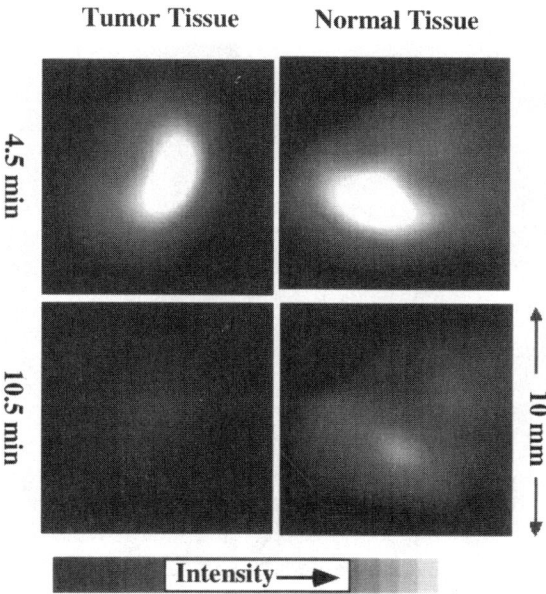

FIGURE 7. Spatially resolved clearance of nitroxide in normal and tumor tissue. Following a tail vein infusion of 160 mg/kg of 3-CP, a series of two-dimensional images of the nitroxide from normal muscle (*top*) and tumor (*bottom*) were measured using L-band EPR imaging instrumentation. The nitroxide in normal tissue persisted for longer than 16 min, while in tumor it was cleared within 10 min of infusion. The white areas represent maximum uptake of the nitroxide probe. Image acquisition parameters: projections, 16; gradient, 15 G/cm; acquisition time, 1.5 min. (Data adapted from Kuppsamy *et al.*,[48] with permission)

tissue.[51] Estimates of oxygen concentration in the tissues shown in FIGURE 7 using EPRI/oximetry indeed confirmed that the tumor tissue was much lower in oxygen concentration than normal tissue.[48] A three dimensional image of the normal tissue and tumor after nitroxide administration provides information on the physical architecture of the tissue based on the nitroxide distribution. FIGURE 8 (*top*) shows the nitroxide distribution in 0.3 mm adjacent slices in normal tissue (leg muscle) and the bottom panels show the corresponding slices from RIF-1 and SCCVII tumors. It is clear from these spatial images that the nitroxide distributes differently in normal and tumor tissue, which may reflect differences in the vasculature of the microenvironment associated with tumors.

Collectively, the studies shown in FIGURES 7 and 8 establish the feasibility of detecting nitroxides in tumors and clearly show major differences in nitroxide distribution and metabolism between normal and tumor tissue. As the technique evolves and becomes more sensitive, EPRI may play a useful role in advancing functional imaging in clinical medicine.

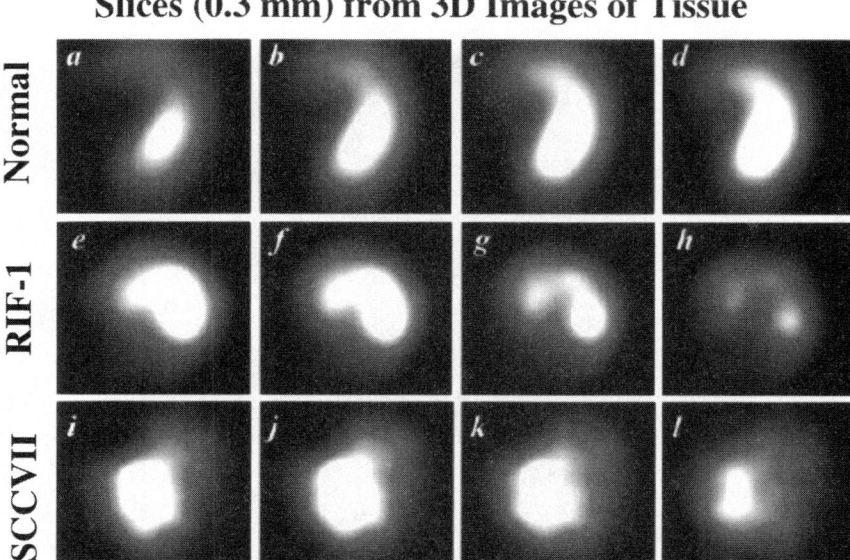

FIGURE 8. Visualization of tumor heterogeneity monitored by nitroxide uptake. Following a tail vein infusion of 160 mg/kg of 3-CP, three-dimensional EPR images of the nitroxide in normal muscle and tumor were measured. A few selected adjacent slices of 0.3 mm thickness obtained from the 3D image of normal (**A–D**) and tumor (**E–F**) tissue are shown. The white areas represent maximum uptake of the nitroxide probe. The images show some differences between the two tissues in terms of anatomy and physical architecture. The tumor tissue shows significant heterogeneity of nitroxide uptake when compared to the normal tissue. Image acquisition parameters: projections, 100; gradient, 15 G/cm; acquisition time, 10 min. (Data adapted from Kuppsamy *et al.*,[48] with permission.)

REFERENCES

1. GERSCHMAN, R., D.L. GILBERT, S.W. NYE, P. DWYER & W.O. FENN. 1954. Oxygen poisoning and X-irradiation: a mechanism in common. Science **119**: 623–626.
2. RONTGEN, W.C. 1895. Uber eine neue Art von Strahlen. Sitzungs-Berichte Phys.-med. Gesellschaft **9**: 132–141.
3. VON SONNTAG, C. 1987. The Chemical Basis of Radiation Biology. Taylor and Francis, London.
4. GRUBBE, E.H. 1933. Priority in the therapeutic use of X-rays. Radiology **21**: 156–162.
5. ALEXANDER, P. & A. CHARLESBY. 1954. Physico-chemical methods of protection aganist ionizing radiations. Radiobiology Symposium 49. Butterworth, London.
6. BUMP, E.A. & K. MALAKER. 1998. Radioprotectors. Chemical, Biological, and Clinical Perspectives. CRC Press. Boca Raton, FL.
7. ZHENG, S., G.L. NEWTON, G. GONICK, R.C. FAHEY & J.F. WARD. 1988. Radioprotection of DNA by thiols: relationship between the net charge on a thiol and its ability to protect DNA. Radiat. Res. **114**: 11–27.

8. AGUILERA, J.A., G.L. NEWTON, R.C. FAHEY & J.F. WARD. 1992. Thiol uptake by Chinese hamster V79 cells and aerobic radioprotection as a function of net charge on the thiol. Radiat. Res. **130:** 194–204.
9. NEWTON, G.L., J.A. AGUILERA, J.F. WARD & R.C. FAHEY. 1996. Binding of radioprotective thiols and disulfides in Chinese hamster V79 cell nuclei. Radiat. Res. **146:** 298–305.
10. FAHEY, R.C. 1988. Protection of DNA by thiols. Pharmacol. Ther. **39:** 101–108.
11. LIU, T., Y. LIU, S. HE, Z. ZHANG & M.M. KLIGERMAN. 1992. Use of radiation with or withoutWR-2721 in advanced rectal cancer. Cancer **69:** 2820–2825.
12. KLIGERMAN, M.M., T. LIU, Y. LIU, B. SCHEFFLER, S. HE & Z. ZHANG. 1992. Interim analysis of a randomized trail of radiation therapy of rectal cancer with/without Wr-2721. Int. J. Radiat. Oncol. Biol. Phys. **22:** 799–802.
13. KRISHNA, M.C., D.A. GRAHAME, A. SAMUNI, J.B. MITCHELL & A. RUSSO. 1992. Oxoammonium cation intermediate in the nitroxide-catalyzed dismutation of superoxide. Proc. Natl. Acad. Sci. USA **89:** 5537–5541.
14. KRISHNA, M.C., A. RUSSO, J.B. MITCHELL, S. GOLDSTEIN, H. DAFNI & A. SAMUNI. 1996. Do nitroxides antioxidants act as scavengers of superoxide or as SOD mimics? J. Biol. Chem. **271:** 26026–26031.
15. KRISHNA, M.C., A. SAMUNI, J. TAIRA, S. GOLDSTEIN, J.B. MITCHELL & A. RUSSO. 1996. Stimulation by nitroxides of catalase-like activity of hemeproteins. J. Biol. Chem. **271:** 26018–26025.
16. BENNETT, H.F., H.M. SWARTZ, R.D. BROWN III & S.H. KOENIG. 1987. Modification of relaxation of lipid protons by molecular oxygen and nitroxides. Invest. Radiol. **22:** 502–507.
17. MCCONNELL, H.M. 1965. Spin Labelling: Theory and Applications. pp. 525–560.
18. SWARTZ, H.M. 1983. Interactions between cells and nitroxides and their implications for their uses as biophysical probes and as metabolically responsive contrast agents for in vivo NMR. Bull. Mag. Res. **8:** 172–175.
19. GALLEZ, G., R. DEMEURE, R. DEBUYST, D. LEONARD, F. DEJEHET & P. DUMONT. 1992. Evaluation of Nonionic nitroxyl lipids as potential organ-specific contrast agents for magnetic resonance imaging. Magn. Reson. Med. **10:** 445-455.
20. BELKIN, S., R.J. MEHLHORN, K. HIDEG, O. HANKOVSKY & L. PACKER. 1987. Reduction and destruction rates of nitroxide spin probes. Arch. Biochem. Biophys. **256:** 232–243.
21. CHATEAUNEUF, J., J. LUSZTYK & K.U. INGOLD. 1988. Absolute rate constants for the reactions of some carbon-centered radicals with 2,2,6,6-tetramethylpiperidine-N-oxyl. J. Org. Chem. **53:** 1629–1632.
22. MEHLHORN, R.J. & L. PACKER. 1984. Electron paramagnetic resonance spin destruction methods for radical detection. Methods in Enzymology **105:** 215–220.
23. NILSSON, U.A., L.I. OLSSON, G. CARLIN & A.C. BYLUND-FELLENIUS. 1989. Inhibition of lipid peroxidation by spin labels. Relationships between structure and function. J. Biol. Chem. **264:** 11131–11135.
24. SAMUNI, A., C.M. KRISHNA, P. RIESZ, E. FINKELSTEIN & A. RUSSO. 1989. Superoxide reaction with nitroxide spin-adducts. Free Radic. Biol. Med. **6:** 141–148.
25. MITCHELL, J.B., A. SAMUNI, M.C. KRISHNA, W.G. DEGRAFF, M.S. AHN, U. SAMUNI & A. RUSSO. 1990. Biologically active metal-independent superoxide dismutase mimics. Biochemistry **29:** 2802–2807.
26. MITCHELL, J.B., W. DEGRAFF, D. KAUFMAN, M.C. KRISHNA, A. SAMUNI, E. FINKELSTEIN, M.S. AHN, S.M. HAHN, J. GAMSON & A. RUSSO. 1991. Inhibition of oxygen-dependent radiation-induced damage by the nitroxide superoxide dismutase mimic, Tempol. Arch. Biochem. Biophys. **289:** 62–70.
27. HAHN, S.M., Z. TOCHNER, C.M. KRISHNA, J. GLASS, L. WILSON, A. SAMUNI, M. SPRAGUE, D. VENZON, E. GLATSTEIN, J.B. MITCHELL & A. RUSSO. 1992. Tempol, a stable free radical, is a novel murine radiation protector. Cancer Res. **52:** 1750–1753.
28. HAHN, S.M., L. WILSON, C.M. KRISHNA, J. LIEBMANN, W. DEGRAFF, J. GAMSON, A. SAMUNI, D. VENZON & J.B. MITCHELL. 1992. Identification of nitroxide radioprotectors. Radiat. Res. **132:** 87–93.

29. MILLAR, B.C., E.M. FIELDEN & C.E. SMITHEN. 1977. Polyfunctional radiosensitzers III, effect of the biradical (Ro-03-6061) in combination with other radiosensitizers on the survival of hypoxic V-79 cells. Radiat. Res. **69:** 489–499.
30. MILLAR, B.C., E.M. FIELDEN & C.E. SMITHEN. 1978. Polyfunctional radiosensitizers IV. The effect of contact time and temperature on sensitization of hypoxic Chinese hamster cells in vitro by bifunctional nitroxyl compounds. Br. J. Cancer **37:** 73–79.
31. MILLAR, B.C., T.C. JENKINS, E.M. FIELDEN & S. JINKS. 1983. Polyfunctional radiosensitizers. VI. Dexamethasone inhibits shoulder modification by uncharged nitroxyl biradicals in mammalian cells irradiated in vitro. Radiat. Res. **96:** 160–172.
32. PETKAU, A., W.S. CHELACK, S.D. PLESKACH, B.E. MEEKER & C.M. BRADY. 1975. Radioprotection of mice by superoxide dismutase. Biochem. Biophys. Res. Commun. **65:** 886–893.
33. PETKAU, A. & W.S. CHELACK. 1984. Radioprotection by superoxide dismutase of macrophage progenitor cells from mouse bone marrow. Biochem. Biophys. Res. Commun. **119:** 1089–1095.
34. PETKAU, A. 1987. Role of superoxide dismutase in modification of radiation injury. Br. J. Cancer Suppl. **8:** 87–95.
35. ABE, M., T. NISHIDAI, Y. YUKAWA, M. TAKAHASHI, K. ONO, M. HIRAOKA & N. RI. 1981. Studies on the radioprotective effects of superoxide dismutase in mice. Int. J. Radiat. Oncol. Biol. Phys. **7:** 205–209.
36. PUCK, T.T. 1958. Action of radiation on mammalian cells: III. Relationships between reproductive death and induction of chromosome anomalies by X-irradiation of euploid human cells in vitro. Proc. Natl. Acad. Sci. USA **44:** 772–280.
37. CARRANO, A.V. 1973. Chromosome aberrations and radiation-induced cell death: II. Predicted and observed cell survival. Mutat. Res. **17:** 355–366.
38. BEDFORD, J.S., J.B. MITCHELL, H.G. GRIGGS & M.A. BENDER. 1978. Radiation-induced cellular reproductive death and chromosome aberrations. Radiat. Res. **76:** 573–586.
39. DEGRAFF, W.G., M.C. KRISHNA, D. KAUFMAN & J.B. MITCHELL. 1992. Nitroxide-mediated protection against x-ray- and neocarzinostatin-induced DNA damage. Free Radic. Biol. Med. **13:** 479–487.
40. JOHNSTONE, P.A.S., W.G. DEGRAFF & J.B. MITCHELL. 1995. Protection of radiation-induced chromosomal aberrations by the nitroxide Tempol. Cancer **75:** 2323–2327.
41. HAHN, S.M., A.M. DELUCA, D. COFFIN, M.C. KRISHNA & J.B. MITCHELL. 1998. In vivo radioprotection and effects on blood pressure of the stable free radical nitroxides. Int. J. Radiat. Oncol. Biol. Phys. **42:** 839–842.
42. ALLALUNIS-TURNER, M.J., T.L. WALDEN & C. SAWICH. 1989. Induction of marrow hypoxia by radioprotective agents. Radiat. Res. **118:** 581–586.
43. HALL, E.J. 1994. The oxygen effect and reoxygenation. *In* Radiobiology for the Radiologist. J.B. Lippincott Co. Philadelphia, PA, pp. 133–152.
44. HAHN, S.M., F.J. SULLIVAN, A.M. DELUCA, M.C. KRISHNA, N. WERSTO, D. VENZON, A. RUSSO & J.B. MITCHELL. 1997. Evaluation of tempol radioprotection in a murine tumor model. Free Radic. Biol. Med. **22:** 1211–1216.
45. BACIC, G., M.J. NILGES, R.L. MAGIN, T. WALCZAK & H.M. SWARTZ. 1989. In vivo localized spectroscopy reflecting metabolism. Magn. Reson. Med. **10:** 266–272.
46. THOMLINSON, R.H. & L.H. GRAY. 1955. The histological structure of some human lung cancers and the possible implications for radiotherapy. Br. J. Cancer **9:** 539–549.
47. HOCKEL, M., K. SCHLENGER, B. ARAL, M. MITZE, U. SCHAFFER & P. VAUPEL. 1996. Association between tumor hypoxia and malignant progression in advanced cancer of the uterine cervix. Cancer Res. **56:** 4509–4515.
48. KUPPUSAMY, P., M. AFEWORKI, R.A. SHANKAR, D. COFFIN, M.C. KRISHNA, S.M. HAHN, J.B. MITCHELL & J.L. ZWEIER. 1998. In vivo electron paramagnetic resonance imaging of tumor heterogeneity and oxygenation in a murine model. Cancer Res. **58:** 1562–1568.

49. KUPPUSAMY, P., M. CHZHAN, K. VIJ, M. SHTEYNBUK, D.J. LEFER, E. GIANNELLA & J.L. ZWEIER. 1994. Three-dimensional spectral-spatial EPR imaging of free radicals in the heart: a technique for imaging tissue metabolism and oxygenation. Proc. Natl. Acad. Sci. USA **91:** 3388–3392.
50. WALCZAK, T. & H.M. SWARTZ. 1989. A 1GHz in vivo ESR spectrometer with a surface probe. Physica Medica **5:** 195–202.
51. SWARTZ, H.M. 1990. Principles of the metabolism of nitroxides and their implications for spin trapping. Free Radic. Res. Commun. **9:** 399–405.

Radioprotection by Antioxidants[a]

JOSEPH F. WEISS[b,c] AND MICHAEL R. LANDAUER[d]

[b]Office of International Health Programs, U.S. Department of Energy, EH-63/270CC, 19901 Germantown Road, Germantown, Maryland 20874, USA

[d]Radiation Casualty Management Team, Armed Forces Radiobiology Research Institute, Bethesda, Maryland 20889, USA

ABSTRACT: The role of reactive oxygen species in ionizing radiation injury and the potential of antioxidants to reduce these deleterious effects have been studied in animal models for more than 50 years. This review focuses on the radioprotective efficacy and the toxicity in mice of phosphorothioates such as WR-2721 and WR-151327, other thiols, and examples of radioprotective antioxidants from other classes of agents. Naturally occurring antioxidants, such as vitamin E and selenium, are less effective radioprotectors than synthetic thiols but may provide a longer window of protection against lethality and other effects of low dose, low-dose rate exposures. Many natural antioxidants have antimutagenic properties that need further examination with respect to long-term radiation effects. Modulation of endogenous antioxidants, such as superoxide dismutase, may be useful in specific radiotherapy protocols. Other drugs, such as nimodipine, propranolol, and methylxanthines, have antioxidant properties in addition to their primary pharmacological activity and may have utility as radioprotectors when administered alone or in combination with phosphorothioates.

INTRODUCTION
RADIOPROTECTORS, RADIATION INJURY, AND OXYGEN EFFECTS

The first *in vivo* studies on protection by antioxidants against ionizing radiation were performed a half century ago. Patt *et al.* reported in 1949 that the thiol amino acid, cysteine, protected rats from a lethal dose of X-rays.[1] Discussions soon began concerning the actions of ionizing radiation on cells and how radioprotective chemicals might help elucidate the mechanisms of interaction of radiation and molecules of biological importance. Gerschman and her co-workers[2] hypothesized that both radiation injury and oxygen poisoning occur through the formation of reactive oxygen species. They demonstrated that antioxidants such as cysteine, glutathione, β-mercaptoethylamine (cysteamine, MEA), propyl gallate, and nordihydroguiaretic acid, that had been shown at that time to protect mice against the lethal effects of radiation,

---

[a]Views presented in this paper are those of the authors. Many of the reviewed animal studies were done at the Armed Forces Radiobiology Research Institute (AFRRI), Bethesda, Maryland, and were conducted according to the principles enunciated in the *Guide for the Care and Use of Laboratory Animals* prepared by the Institute of Laboratory Animal Resources, National Research Council. The contributions of the AFRRI staff, who participated in these studies, are gratefully acknowledged.

[c]Address for correspondence: Voice: 301- 903-1846; fax: 301- 903-1413.
e-mail: joseph.weiss@eh.doe.gov

could also increase the survival time of mice exposed to high oxygen tensions. Another interesting observation was that high oxygen pressures were more toxic to mice simultaneously or previously exposed (up to 2 h) to high-dose irradiation. Interest in the interrelationship between oxygen effects and the acute and late effects of radiation exposure, including therapeutic effects, remains strong. Similarly, studies on the modulation of radiation-induced reactive oxygen species by antioxidants and parallel studies on other pathologies involving free radicals remain an active area of biomedical research.

COMPARISON OF EFFICACY AND TOXICITY OF RADIOPROTECTIVE AGENTS

This review of radioprotection by antioxidants focuses on comparative radioprotection and toxicity studies in mice by using the most effective phosphorothioate agents, initially developed by the Walter Reed Army Institute of Research (WRAIR) and designated as WR compounds,[3] and other classes of protectors. A number of reviews and books on radioprotection that discuss classes of protectors and mechanisms of action have been previously cited.[4] The review by Murray and McBride[5] discusses in greater detail the mechanisms of action of many of the agents covered in this paper. This review will not cover in detail the nitroxide class of protectors, which is discussed in another chapter of this volume.[6] A discrete classification of protectors is often difficult. For example, some synthetic protectors or pharmacological agents are often similar or identical to endogenous or "naturally occurring" protectors, e.g., cysteamine and dithiolthiones. In addition, both chemical and biological protectors may lead to similar results, but through different pathways.

Various procedures have been used to compare the efficacy of protective agents in experimental animals and the most reliable has been to determine the dose reduction factor or the dose modifying factor (DMF). DMFs are determined by irradiating mice with and without administered agents at a range of radiation doses and comparing the endpoint of interest. For example, the DMF for 30-day survival ($LD_{50/30}$ drug-treated ÷ $LD_{50/30}$ vehicle-treated) quantifies protection of the hematopoietic system.[7] With the sufficient loss of hematopoietic stem cells, death follows due to infection, hemorrhage, and anemia. The gastrointestinal syndrome in mice can be assessed by determining survival at 6 or 7 days (GI death) after comparatively high doses of whole-body irradiation.[7,8] The most informative and useful preclinical studies relate protective effects to the drug's toxicity in the same animal model. For example, drugs should not be administered at doses greater than one-fourth or one-half the dose that causes death in 10% of the mice of a specific strain.[7] Of further use in comparing agents is the determination of protection based on doses that cause other measurable toxicities, such as "behavioral toxicity" or "performance decrement."[9,10] Alterations in the locomotor activity of rodents has been widely used to assess the behavioral toxicity of a variety of agents including radioprotectors[10,11] as well as the effects of ionizing radiation alone.[9]

THIOLS AND PHOSPHOROTHIOATES

From the earliest days of the nuclear era, the potential applications of radioprotective chemicals in the event of nuclear accidents, as well as the possibility of protecting normal tissues but not tumors during radiotherapy, were considered. In the 1950s, through a program supported by the Atomic Energy Commission, aminoethylisothiourea was developed along with an increased understanding of sulfur-containing radioprotectors. For almost three decades, WRAIR supported an Antiradiation Drug Development Program in which more than 4,000 compounds were synthesized and screened in mice.[3] The most significant contribution of the WRAIR initiatives was the development of WR-2721 (S-2-(3-aminopropylamino)ethylphosphorothioic acid) and many related phosphorothioates, including WR-151327 (S-3-(3-methylaminopropylamino)propylphosphorothioic acid) and WR-3689 (S-2-(3-methylaminopropylamino)ethylphosphorothioic acid).[3,7] During this period, WR-2721 (Ethyol®, amifostine, ethiofos, gammaphos) was introduced into cancer clinical trials to study protection against normal tissue damage caused by radiotherapy and various chemotherapeutic agents.[12] The phosphorylated compounds serve as prodrugs for the active free aminothiols, e.g., WR-1065 (2-(3-aminopropylamino)ethanethiol) from WR-2721 and WR-151326 (3-(3-methylaminopropylamino)propanethiol) from WR-151327, and their corresponding disulfides formed *in vivo*. Phosphorothioates and other aminothiols, which are usually administered shortly before irradiation, have been hypothesized to act by one or by a combination of effects: scavenging of radiation-induced free radicals before their reaction with biomolecules; induced hypoxia; formation of mixed disulfides; scavenging of metals; repair of DNA through hydrogen donation to carbon-centered radicals; and genome stabilization. Thiols may have further effects at the physiological and immunological levels that may not be readily predicted by considering only free-radical events.

Different sulfhydryl compounds exhibit a range of radioprotective effects, indicating the importance of structural elements other than the thiol moiety. TABLE 1 provides a comparison of the toxicity and radioprotective effects for the representative thiols WR-2721, cysteamine (MEA), diethyldithiocarbamate (DDC), and N-acetyl-L-cysteine (NAC).[13] With respect to comparative toxicity, NAC is one of the least toxic thiols. Its lethal toxicity is comparable to glutathione, for which it is a prodrug, and it has negligible behavioral toxicity. On the other hand, DDC was the most behaviorally toxic of the thiols tested in mice. The lethal toxicity of DDC, however, is much less than another dithiol, dithiothreitol, which is too toxic for drug use. The most protective compound, even taking into consideration its toxicity, was WR-2721. A more extensive comparison of protection against γ-irradiation by the thiols administered at equitoxic doses (one-fourth LD_{10}) indicated that the DMF for 30-day survival for WR-2721 (200 mg/kg) was 1.8. Equitoxic doses of other thiols, such as NAC, DDC, or mercaptopropionylglycine (MPG), only provided a DMF of 1.1. WR-3689 and WR-151327 provided protection similar to WR-2721 when administered to mice at one-fourth the LD_{10} either intraperitoneally (IP) or orally (PO). The DMFs for IP treatment were 1.6–1.8, and the DMFs for PO treatment were 1.2–1.3. Phosphorothioates, in contrast to other thiols, also provide moderate protection against high-LET (linear energy transfer) exposure such as neutrons.[8] Presumably,

TABLE 1. Comparison of toxicity and radioprotective effects of sulfhydryl compounds in male CD2F1 mice

Treatment	WR-2721	MEA	DDC	NAC	saline
Toxicity indices: acute lethality and behavioral effects					
$\frac{1}{2} LD_{10}$ (mg/kg)[a]	400	200	800	1000	—
ED_{50} (mg/kg)[b]	271	103	224	>1000	—
Behavioral toxicity index[c]	3.0	3.9	7.1	<2	—
Radioprotection at lethal and sublethal radiation exposures					
30-day survival (%)[d]	100	62	62	25	0
Cell-mediated immunity (%)[e]	74	66	47	58	31

[a] Maximum tolerated dose = one-half the LD_{10}.
[b] ED_{50} = effective dose that disrupted horizontal locomotor activity by 50% from vehicle control levels at 1 h postinjection.
[c] LD_{10} divided by ED_{50}. The higher the ratio, the greater the behavioral toxicity.
[d] Protection against radiation-induced lethality. Thirty-day survival evaluated by treating mice IP with equitoxic doses ($\frac{1}{2} LD_{10}$) of drugs before exposure to 13 Gy cobalt-60 radiation (1 Gy/min).
[e] Percent of nonirradiated control: 24-h delayed-type hypersensitivity response in drug-treated ($\frac{1}{2} LD_{10}$) vs. saline-treated mice irradiated at 7 Gy cobalt-60 (0.4 Gy/min).

it is more difficult to protect against the direct effects on DNA of high-LET radiation in contrast to the predominately indirect effects of low-LET radiation (X, γ), which are mediated by reactive oxygen species.

We have shown that a number of thiols also protect against radiation-induced immunosuppression (TABLE 1).[13,14] At sublethal radiation doses in mice, suppression of delayed-type hypersensitivity (DTH) provides a measure of the overall effect on cellular immunity.[15] Statistical analysis indicated that protection against radiation-induced immunosuppression, as measured by the 24-h DTH response, was greatest for WR-2721 and MEA. NAC followed by DDC were also protective in this assay. The least protection (data not shown) was provided by MPG and 2-mercaptoethane sulfonic acid (MESNA). There is substantial clinical experience in the use of thiols with relatively lower toxicity (DDC, MPG, NAC) for treating various conditions, and continued study of their potential for specific cell protection against lower doses of radiation is warranted.

The complexity of the various effects of individual thiol agents must always be taken into consideration. DDC, for example, may have dose-dependent effects, including immunostimulation at very low doses[16] and inhibition of CuZnSOD at high doses.[17] On the other hand, it has been proposed that DDC can mimic glutathione peroxidase activity.[18] Therefore, the ability of DDC to sensitize tumor tissue and protect normal tissues against radiation in the same animal[19] may reflect differential glutathione and antioxidant enzyme levels in tumor vs. normal tissues. MPG, in contrast to other thiols that appear to be effective only when administered before radiation, has a small protective effect also when administered after radiation exposure.[20]

A number of thiols that protect against radiation-induced lethality were reported to have antimutagenic[20] and anticarcinogenic[21] effects. When administered to mice either before or after radiation exposure, even very low doses of phosphorothioates were found to protect against mutagenesis at the hypoxanthine-guanine phosphoribosyl transferase locus.[22] Anticarcinogenic effects have been demonstrated when high doses of phosphorothioates are administered to mice before radiation exposure.[22-24] Current studies on phosphorothioates are directed not only toward investigating the ability of protectors to protect against cell killing, but also on how they influence mutagenesis, transformation, and carcinogenesis.[22] The structural similarity of the metabolites of WR-compounds (thiols and disulfides) to endogenous polyamines, which are also antioxidants, suggests that they may interact with DNA similarly and that they may influence DNA protection, repair, and synthetic processes.[22,25,26] Detailed investigations on the diverse preirradiation and postirradiation activities of WR-compounds, functioning as both radical scavengers and pro-drugs of disulfides that interact with polyamine systems, signal an important new stage in radioprotector development.[26-29] Because of the dose-limiting toxicities of phosphorothioates and disulfides,[4,30,31] it will be important also to determine structural-toxicity relationships that can lead to less toxic radioprotective compounds.

NATURALLY OCCURRING, NUTRITIONAL, AND RELATED ANTIOXIDANTS

In general, naturally occurring radioprotectors protect only against lower doses or lower dose-rates of radiation exposure in comparison to synthetic thiol agents. DMFs lower than 1.3 for 30-day survival are generally reported. However, natural antioxidants may also provide benefits of low toxicity and possible increased benefits from longer use including antimutagenic effects. They often exhibit a long window of protection, i.e., they provide some protection when administered hours before or after radiation exposure. Exogenous administration of antioxidants, such as glutathione, superoxide dismutase (SOD), antioxidant vitamins (A, C, and E), the disulfide lipoic acid, as well as substances that mimic or induce activity of endogenous antioxidant systems (e.g., selenium, zinc and copper salts, and metal complexes), have shown protection against hematopoietic syndrome death.[5,8,32-40] Natural antioxidants such as vitamin E may also protect the intestine from physiological damage, but in contrast to synthetic thiols, do not protect rodents against GI death after high dose radiation.[8] Hormone-related antioxidants, including melatonin[41] and dehydroepiandrosterone (DHEA) and its metabolite 5-androstenediol,[42] also have radioprotective properties.

Besides the more common nutrient antioxidants, a large number of plants contain antioxidant phytochemicals that have been reported to be radioprotective in various model systems. These include green tea (polyphenols), Chinese herbal medicines, Ayurvedic preparations, cruciferous vegetables (e.g., cabbage and broccoli), dithiolthiones, *Panax ginseng*, *Eleutherococcus senticosus* or Shigoka extract, *Gingko biloba* extract (flavone glycosides and terpene lactones), milk thistle (silymarin), curcumin, garlic (allicin), and lycopene. Most of these botanical or alternative medicines would be considered dietary supplements in the United States and would not be subject to strict regulation by the Food and Drug Administration. In Germany, there is a regulatory body (German Commission E) that evaluates the safety and efficacy of herbs.[43] There are major problems in evaluating studies of radioprotection by these compounds because they are often administered as complex extracts, and the results of animal experiments are difficult to compare since they are rarely reported in the form of DMFs. Notable exceptions are recent studies on the flavonoids orientin and vicenin extracted from *Ocimum sanctum*. These compounds provided significant protection against chromosome aberrations and lethality when administered to mice at nontoxic doses before radiation exposure.[44,45]

The lower degree of protection against lethality provided by most natural antioxidants when compared to chemical agents may be related to their modulation of later reactions, e.g., interaction of radiation-induced radicals of biomolecules with reactive oxygen species evolved during normal cellular processes.[32] Some natural antioxidants can protect against lethality, even when administered to mice after radiation exposure: Vitamin E,[34,39] vitamin A and β-carotene,[38] selenium,[35] and superoxide dismutase.[37] Protection of immune responses in irradiated mice may contribute to enhanced survival.[38,39] Inhibition of radiation-induced apoptosis of lymphoid cells, which is mediated by membrane lipid peroxidation, may play a role in postirradiation protection by some antioxidants.[46]

TABLE 2 provides evidence that postirradiation administration of vitamin E and selenomethionine can significantly increase the survival of irradiated mice. Results indicate a long window of protection. Selenomethionine was equally protective when administered at −24 h, −1 h, and +15 min with respect to radiation exposure. Sodium selenite provided a similar degree of protection,[33,35] although selenomethionine is a less toxic alternative. It should be emphasized that these studies were conducted at a low dose rate. Vitamin E, for example, has a lower protective effect at a higher dose rate (1 Gy/min) than at a lower dose rate (0.2 Gy/min).[34] There is evidence that radiation-induced lipid peroxidation is greater after lower dose rates compared to higher dose rates,[47] and it is possible that comparatively greater protec-

TABLE 2. Protection of male CD2F1 mice against low dose-rate irradiation by selenomethionine and vitamin E

		30-day survivors	
Treatment	Treatment schedule	9 Gy	10 Gy
Saline		23%	0%
Selenomethionine (4 mg/kg Se, IP)			
	−24 h	93%	36%
	−1 h	71%	21%
	+15 min	79%	35%
Vitamin E (100 IU/kg, SC)			
	−1 h	83%	—
	+15 min	100%	42%

$N = 14$–28/group, unilateral cobalt-60, 0.2 Gy/min.

tion would be observed at lower dose rates with natural antioxidants, such as vitamin E. Greater protection by vitamin E is obtained when it is administered parenterally compared to when it is provided in the diet, especially if the diet already contains adequate[48] or minimal levels.[49,50]

It is not clear how the radioprotective effect of selenium compounds relates to induction of glutathione peroxidase and other selenium-containing proteins, or whether there is also a direct enzyme-mimetic activity of compounds, such as selenomethionine. Cellular studies on mechanisms of protection by selenium against radiation-induced oxidative damage indicate that selenium is not likely to have a direct effect on preventing DNA damage through induction of glutathione peroxidase.[51] A decrease in glutathione peroxidase activity in mouse tissues was observed in the postirradiation period (TABLE 3).[50] There was less depression when the diet was supplemented with either selenium or vitamin E. The greatest effect and complete protection of glutathione peroxidase activity was in mice that were fed both increased vitamin E and selenium. An increase in postirradiation survival was also greatest in mice that were fed increased vitamin E and selenium in comparison to mice fed either nutrient alone. This study supports other evidence for the synergistic effects of vitamin E and selenium in protection from oxidative damage.[52]

Although it is probable that antioxidant enzymes, such as glutathione peroxidase, manganese superoxide dismutase (MnSOD), copper-zinc superoxide dismutase, and catalase, are important in providing protection from radiation exposure,[32,53,54] the proper balance of the enzymes in specific cells and in the whole organism required for maximum radioprotection is far from clear. For example, a large increase in MnSOD in some model systems may have a radiosensitizing effect rather than a protective effect,[55] probably related to the inability of the cell to cope with overproduction of H_2O_2 or ·OH.

Radiation exposure itself results in changes in antioxidant enzyme levels; most consistently reported are elevations in MnSOD,[53] although the pattern of enzymatic changes may be related to radiation dose.[56] Cytokines, which are increased during radiation exposure, may also modulate endogenous antioxidants. Interleukin-1 (IL-1), a radioprotective cytokine, provides varying degrees of protection, depending on the time of administration before or after radiation exposure.[57,58] The mech-

TABLE 3. Effect of increased levels of vitamin E and selenium in mouse diet on tissue glutathione peroxidase levels 24 h after radiation exposure (percent difference from non-irradiated controls)

Diet	Serum	Liver
minimal vitamin E diet	−30%	−20%
increased vitamin E diet	−5%	−5%
increased selenium diet	−18%	+40%
increased vitamin E and selenium	+12%	+55%

Mice were fed (1) control diet with minimal recommended dietary allowance of vitamin E,[49] (2) diet with 10X minimal vitamin E levels, (3) drinking water ad libitum with 4 ppm Se as sodium selenite, or (4) combined diet of increased vitamin E and selenium for 2 months before cobalt-60 irradiation (9 Gy, 0.4 Gy/min).

anisms of protection may include induction of a variety of hematopoietic growth factors as well as induction of endogenous antioxidants such as metallothionein, ceruloplasmin, and MnSOD.[57,59,60] MnSOD induction also plays a role in the activities of tumor necrosis factor and lymphotoxin, other cytokines important in radiation responses.[61] Some synthetic radioprotectors, e.g., nitroxides [6] and metal complexes,[36] can also mimic SOD activity.

The role of antioxidant enzymes in carcinogenesis, tumor metabolism, and resistance to cancer therapy has been studied for a number of years.[62,63] Although these studies have yet to be translated to clinical utility, one of the most promising areas appears to be manipulation of MnSOD levels. One example involves overexpression of a transgene for human MnSOD delivered by plasmid-liposome or adenovirus to mouse lungs prior to irradiation, which decreases late effects of whole lung irradiation such as fibrosis.[64] Such animal models are needed to provide a rational basis for the development of antioxidant therapy for protection against acute and chronic damage from radiotherapy.

Natural antioxidants (vitamins C and E) show protective effects against radiation-induced chromosomal damage in mice even when administered after radiation exposure,[65,66] and selenium has antimutagenic properties *in vitro*.[67] Vitamins A and E, selenium and MnSOD also inhibit radiation-induced transformation *in vitro*.[68-70] Some epidemiological studies, based on animal studies that indicate an anticarcinogenic potential of antioxidant nutrients, suggest an association of certain vitamins and decreased cancer incidence.[71] Carefully structured chemoprevention trials[72-75] can be viewed as models for the study of radioprotective agents (administered at various points in relation to radiation exposure) since they provide the required information on toxicity and possible unexpected results. For example, contrary to expectations, a trial of ß-carotene supplementation indicated an increase in lung, prostate, and stomach cancer in smokers[74] There appear to be beneficial effects due to supplementation with vitamin E[74] or selenium,[75] associated with decreased incidence of cancer at several sites, including the prostate. Ongoing and projected chemoprevention trials include a number of other antioxidants with varying degrees of radioprotective efficacy. If these trials identify specific antioxidants that protect against cancer development, it would be reasonable to assess their protective effect against radiation-induced cancer. Future chemoprevention trials will be strengthened by information from the Human Genome Project, which should identify cohorts at risk and provide specific gene targets for developing intervention strategies.[72] Molecular epidemiological studies in human populations exposed to acute or chronic radiation exposures will also provide information on treatment strategies.

There is evidence that "natural" antioxidants (superoxide dismutase, *Gingko biloba* extract, and mixtures of antioxidant plant phenols, vitamins and minerals) may protect against long-term effects of radiation exposure occurring in human populations exposed to radiation.[76-78] Emerit and coworkers described the appearance of clastogenic factors in the plasma of Chernobyl emergency workers (liquidators) and exposed children many years after radiation exposure and the suppression of these factors by antioxidant supplements.[76-78] The clastogenic factors that cause chromosomal breakage are thought to be lipid peroxidation products, cytokines, especially tumor necrosis factor, and altered nucleotides. It is believed that the clastogenic factors, which also occur in chronic inflammatory disease and other conditions, are produced via superoxide and themselves generate superoxide and are therefore self-sustaining. In another study, lipid peroxidation products were detected in children exposed to radiation from the Chernobyl nuclear reactor many years after the accident, and supplementation with natural β-carotene resulted in decreased levels of these products.[79]

It is apparent that a large number of natural antioxidants can protect against radiation-induced DNA damage in some way, but it is unlikely that this is due to direct interaction with DNA. A possible explanation for the postirradiation protective effects of antioxidants is a model based on electron spin resonance studies suggesting that radiation exposure results in short-lived radicals, such as ·OH, responsible for cell death, and long-lived radicals, which can cause mutations and transformations.[80] This hypothesis is also supported by experiments using cells irradiated with a microbeam.[81] Besides radiation-induced nuclear effects, the experiments indicate that reactive oxygen species are formed in the cytoplasm, resulting in minimal cell killing but significant mutagenic potential.

OTHER DRUGS AND AGENTS WITH POTENTIAL ANTIOXIDANT ACTIVITY

Examples of drugs or agents that are not considered to be primarily antioxidants are discussed in this section. All of these are approved drugs and along with similar derivatives are currently used for various indications.[82] In addition, they all have antioxidant properties, although it is not clear how this contributes to their primary pharmacological effects. TABLE 4 provides data on several of these compounds alone or in combination with WR-151327. Radioprotection is compared to effect on locomotor activity, an index of behavioral toxicity. Data on vitamin E are given for comparison. The main purpose for this series of preclinical studies was to determine whether the protective effect of phosphorothioates could be enhanced by combining them with other agents without exacerbating the behavioral toxicity.[33] A principal problem associated with the use of phosphorothioate protectors in humans is adverse side effects, including a dose-limiting hypotensive effect.[31] The measure of decreased locomotor activity in rodents[83] reflects this hypotensive effect.[10,11]

The data in TABLE 4 summarize the protective and behavioral effects of two methylxanthines caffeine and theophylline, which are antagonists of A_1 and A_2 adenosine receptors, alone and in combination with WR-151327.[84] Caffeine reversed the depression in locomotor activity due to WR-151327 administration but did not significantly modulate the radioprotective effect of the phosphorothioate. Locomotor activity also returned to control levels when theophylline was administered before WR-151327. Theophylline improved the DMF of WR-151327 to the greatest degree,[84] increasing it from 1.51 to 1.64.

Caffeine is generally considered to result in radiosensitization in *in vitro* studies by affecting the cell cycle. However, we never observed a statistically significant radiosensitizing effect of caffeine on 30-day mouse survival when administered IP or orally. Caffeine protects mouse intestinal cells from radiation injury.[85] Furthermore, caffeine ameliorated the detrimental effects of combined radiation and indomethacin treatment on GI injury measured by 7-day survival (GI death).[86] Kesavan and

TABLE 4. Radioprotection and behavioral toxicity of agents with probable antioxidant activity. Administration alone and in combination with WR-151327 to male CD2F1 mice

Agent or combination[a]	DMF[b]	Locomotor activity[c]
WR-151327 alone	1.49	⇓⇓
caffeine (20 mg/kg)	1.00	⇑
+ WR-151327	1.54	⇔
theophylline	—	⇑
+ WR-151327	1.64	⇔
propranolol (20 mg/kg)	1.00	0
+ WR-151327	2.01	⇓
nimodipine (10 mg/kg)	1.19	0
+ WR-151327	1.67	⇓⇓⇓
vitamin E (100 IU/kg)	1.04	0
+ WR-151327	1.60	⇓⇓

[a]Agents administered IP in appropriate vehicles, except for vitamin E, which was administered SC. WR-151327 (200 mg/kg) administered 30 min. before irradiation; other agents administered before WR-151327.

[b]DMF (dose modifying factor) determined from 30-day survivals of groups of mice (8–16) exposed to various doses of cobalt-60 (1 Gy/min) = radiation dose at 50% mortality of group receiving drug(s) divided by radiation dose at 50% mortality of group receiving vehicle(s).

[c]Behavioral toxicity evaluated by automated measurement of horizontal locomotor activity for 4 h after drug injection to groups of nonirradiated mice.[8] Overall behavioral effects relative to vehicle are: depression (⇓), stimulation (⇑), counteraction (⇔), or no behavioral effect (0) over this time period. The number of arrows reflects the magnitude of the effect.

coworkers suggested that antioxidant properties of caffeine contribute to its radioprotective effects, such as decreased chromosomal aberrations in mice when caffeine was administered either before or after radiation exposure.[87] They also observed protection against radiation-induced lethality using a very high dose of caffeine,[88] although a DMF was not determined. Radiotherapy studies also suggest that caffeine consumption may decrease severe late toxicity after radiation to the pelvis.[89] Pentoxyfylline, another antioxidant methylxanthine used to treat peripheral vascular disease, has been shown to protect against late radiation-induced injury to the mouse extremity.[90]

Protection by methylxanthines may be due to increased cellular cyclic AMP levels.[85, 86] Elevation of extracellular adenosine by combined administration of dipyridamole, a drug inhibiting the cellular uptake of adenosine, and adenosine monophosphate has radioprotective effectiveness.[91] Dipyridamole, used clinically as a vasodilator, is also an effective antioxidant. It decreases lipid peroxides in mouse tissues along with increasing 30-day survival (DMF = 1.15).[92]

Propranolol is a nonselective β-adrenergic blocker used in the treatment of hypertension, angina, and certain types of arrhythmias. Whereas L-propranolol is the active blocking form of propranolol, both D and L enantiomers inhibit lipid peroxidation in sarcolemmal membranes.[93,94] Propranolol alone had no effect on locomotor activity in mice and significantly reduced the behavioral toxicity of WR-151327 (TABLE 4).[95] Although propranolol was not radioprotective at the dose tested, it significantly enhanced the radioprotective effect of WR-151327.

Nimodipine, a 1,4-dihydropyridine-derivative calcium-channel blocking agent, inhibits the influx of extracellular calcium ions through voltage-dependent and receptor-mediated slow calcium channels in the membranes of myocardial, vascular smooth muscle, and neuronal cells. Nimodipine treatment alone (10 mg/kg) yielded a statistically significant DMF of 1.19 (TABLE 4).[96] The DMF was significantly enhanced when nimodipine was combined with WR-151327 (200 mg/kg); DMF increased from 1.46 to 1.67. In contrast to the results with propranolol, this appeared to be an additive effect. Nimodipine alone did not affect locomotor activity. Although there was an additive radioprotective effect when nimodipine was combined with WR-151327, the behavioral toxicity of WR-151327 was also significantly enhanced.[96]

Studies by Floersheim[97,98] indicated protection of normal tissue by calcium antagonists, including nimodipine, whereas the radiotherapeutic efficacy in three human tumor xenografts was not reduced by drug treatment. In addition to modulation of calcium homeostasis, calcium channel blockers protect against free-radical–mediated injury of cardiovascular cells, suggesting another possible mode of radioprotection.[94] Studies on the interrelationship of calcium, oxidative damage and radiation-induced apoptosis may also provide insights into the radioprotective mechanisms of calcium channel blockers.

The data in TABLE 4 and other studies have indicated that the radioprotective effects of phosphorothioates are increased by treatment with vitamin E.[34] Vitamin E has very low toxicity, including no effect on locomotor activity at the dose tested. It did not ameliorate the behavioral effect of WR-151327, although the lethal toxicity of high doses of WR-151327 was decreased by prior administration of vitamin E. This effect is similar to the decreased lethal toxicity and improved radioprotective efficacy of phosphorothioates obtained when mice are pretreated with selenium and other metals such as zinc.[33,35,36] The behavioral toxicity of the phosphorothioates is not diminished by combined treatment with metals. Another approach for decreasing the toxicity and improving the protective effect of the major aminothiol radioprotectors is combined treatment with dithiol chelating agents, such as dimercaptopropanesulfonic acid (unithiol).[99]

There appears to be a large number of approved drugs with antioxidant properties that may be secondary to other pharmacological effects. The results discussed above suggest that a number of these agents could be considered for use in radiotherapy patients to modulate tissue and tumor responses to radiation.

REFERENCES

1. PATT, H.M., E.B. TYREE, R.L. STRAUBE & D.E. SMITH. 1949. Cysteine protection against x-irradiation. Science **110:** 213–214.
2. GERSCHMAN, R., D.L. GILBERT, S.W. NYE, P. DWYER & W.O. FENN. 1954. Oxygen poisoning and x-irradiation: A mechanism in common. Science **119:** 623–626.

3. SWEENEY, T.R. 1979. A Survey of Compounds from the Antiradiation Drug Development Program of the U.S. Army Medical Research Development Command, Walter Reed Army Institute of Research, Washington, DC.
4. WEISS, J.F. 1997. Pharmacologic approaches to protection against radiation-induced lethality and other damage. Environ. Health Perspect. **105** (Suppl. 6): 1473–1478.
5. MURRAY, D. & W.H. MCBRIDE. 1996. Radioprotective agents. In Kirk-Othmer Encyclopedia of Chemical Technology, Fourth edition, Vol. **20:** 963–1006. John Wiley & Sons, New York, NY.
6. MITCHELL, J.M., A. RUSSO, P. KUPPUSAMY & M.C. KRISHNA. 2000. Radiation, radicals, and images. Ann. N.Y. Acad. Sci. **899:** this volume.
7. BROWN, D.Q., W.J. GRAHAM, L.J. MACKENSIE, J.W. PITTOCK & L. SHAW. 1988. Can WR-2721 be improved upon? Pharmac. Ther. **39:** 157–168.
8. WEISS, J.F., M.R. LANDAUER, P.J. GUNTER-SMITH & W.R. HANSON. 1995. Effect of radioprotective agents on survival after acute intestinal radiation injury. In Radiation and the Gastrointestinal Tract. A. Dubois, G.L. King & D.R. Livengood, Eds. :183–199. CRC Press. Boca Raton, FL.
9. LANDAUER, M.R., H.D. DAVIS, JA. DOMINITZ & J.F. WEISS. 1988. Long-term effects of the radioprotector WR-2721 on locomotor activity and body weight of mice following exposure to ionizing radiation. Toxicology **49:** 315–323.
10. LANDAUER, M.R., H.D. DAVIS, J.A. DOMINITZ & J.F. WEISS. 1987. Dose and time relationships of the radioprotector WR-2721 on locomotor activity in mice. Pharmacol. Biochem. Behavior **27:** 573–576.
11. LANDAUER, M.R., H.D. DAVIS, K.S. KUMAR & J.F. WEISS. 1992. Behavioral toxicity of selected radioprotectors. Adv. Space Res. **12**(2): 273–283.
12. CAPIZZI, R.L. & W. OSTER. 1995. Protection of normal tissue from the cytotoxic effects of chemotherapy and radiation by amifostine: clinical experience. Eur. J. Cancer **31A**, (Suppl 1): S8– S13.
13. LANDAUER, M.R., H.D. DAVIS, J.A. DOMINITZ & J.F. WEISS. 1988. Comparative behavioral toxicity of four sulfhydryl radioprotective compounds in mice: WR-2721, cysteamine, diethyldithiocarbamate and N-acetylcysteine. Pharmac. Ther. **39:** 97–100.
14. WEISS, J.F., V. SRINIVASAN, A.J. JACOBS & W. A. RANKIN. 1984. Effect of sulfur compounds on radiation-induced suppression of delayed-type hypersensitivity to oxazolone in mice. Proc. Am. Assoc. Cancer Res. **25:** 235.
15. SRINIVASAN, V. & J.F. WEISS. 1984. Suppression of delayed-type hypersensitivity to oxazolone in whole-body-irradiated mice and protection by WR-2721. Radiat. Res. **98:** 438–444.
16. RENOUX, G. & M. RENOUX. 1977. Thymus-like activities of sulphur derivatives on T-cell differentiation. J. Exp. Med. **145:** 466–471.
17. HEIKKILA, R.E., F.S. CABBAT & G. COHEN. 1976. In vivo inhibition of superoxide dismutase in mice by diethyldithiocarbamate. J. Biol. Chem. **251:** 2182–2185.
18. KUMAR, K.S., A.M. SANCHO & J.F. WEISS. 1986. A novel interaction of diethyldithiocarbamate with the glutathione/glutathione peroxidase system. Int. J. Radiat. Oncol. Biol. Phys. **12:** 1463–1467.
19. EVANS, R.G. 1985. Tumor radiosensitization with concomitant bone marrow radioprotection: a study in mice using diethyldithiocarbamate (DDC) under oxygenated and hypoxic conditions. Int. J. Radiat. Oncol. Biol. Phys. **11:** 1163–1169.
20. UMA DEVI, P. & B. THOMAS. 1988. Bone marrow cell protection and modification of drug toxicity by combination of protectors. Pharmac. Ther. **39:** 213–214.
21. MAISIN, J. R. 1998. Chemical radioprotection: past, present and future prospects. Int. J. Radiat. Biol. **73:** 443–450.
22. MURLEY, J.S. & D.J. GRDINA. 1998. Chemoprevention with WR-2721 and its metabolites. In Radioprotectors: Chemical, Biological, and Clinical Perspectives. E.A. Bump & K. Malaker, Eds. :299–313. CRC Press. Boca Raton, FL.
23. MILAS, L., N. HUNTER, L.C. STEPHENS & L.J. PETERS. 1984. Inhibition of radiation carcinogenesis in mice by S-2-(3-aminopropylamino)-ethylphosphorothioic acid. Cancer. Res. **44:** 5567–5569.

24. GRDINA, D.J, B.J. WRIGHT & B.A. CARNES. 1991. Protection by WR-151327 against late-effect damage from fission-spectrum neutrons. Radiat. Res. **128:** S124–S127.
25. HOLWITT, E.A., E. KODA & C.E. SWENBERG. 1990. Enhancement of topoisomerase I-mediated unwinding of supercoiled DNA by the radioprotector WR-33278. Radiat. Res. **124:** 107–109.
26. SAVOYE, C., C. SWENBERG, S. HUGOT, D. SY, R. SABATTIER, M. CHARLIER & M. SPOTHEIM-MAURIZOT. 1997. Thiol WR-1065 and disulphide WR-33278, two metabolites of the drug Ethyol (WR-2721), protect DNA against fast neutron-induced strand breakage. Int. J. Radiat. Biol. **71:** 193–202.
27. GRDINA, D.J., N. SHIGEMATSU, P. DALE, G.L. NEWTON, J.A. AGUILERA & R.C. FAHEY. 1995. Thiol and disulfide metabolites of the radiation protector and potential chemopreventive agent WR-2721 are linked to both its anti-cytotoxic and anti-mutagenic mechanisms of action. Carcinogenesis **16:** 767–774.
28. MITCHELL, J. L., J. RUPERT. A. LEYSER & G. G. JUDD. 1998. Mammalian cell polyamine homeostasis is altered by the radioprotector WR1065. Biochem. J. **335:** 329–334.
29. DRAB-WEISS, E.A., I.K. HANSRA, E.R. BLAZEK & D.B. RUBIN. 1998. Aminothiols protect endothelial cell proliferation against inhibition by lipopolysaccharide. Shock **10:** 423–429.
30. KUNA, P., K. VOLENEC, I. VODICKA & M. DOSTAL. 1983. Radioprotective and hemodynamic effects of WR-2721 and cystamine in rats: time course studies. Neoplasma **30:** 349–357.
31. RYAN, S.V., S.L. CARRITHERS, S.J. PARKINSON, C. SKURK, C. NUSS, P.M. POOLER, C.S. OWEN, A.M. LEFER & S.A. WALDMAN. 1996. Hypotensive mechanisms of amifostine. J. Clin. Pharmacol. **36:** 365–373.
32. WEISS, J.F. & K.S. KUMAR. 1988. Antioxidant mechanisms in radiation injury and radioprotection. *In* Cellular Antioxidant Defense Mechanisms, Vol. II. C.K. Chow, Ed. :163–189. CRC Press. Boca Raton, FL.
33. WEISS, J.F., K.S. KUMAR, T.L. WALDEN, R. NETA, M.R. LANDAUER & E.P. CLARK. 1990. Advances in radioprotection through the use of combined agent regimens. Int. J. Radiat. Biol. **57:** 709–722.
34. SORENSON, J.R. 1992. Essential metalloelement metabolism and radiation protection and recovery. Radiat. Res. **132:** 19–29.
35. WEISS, J.F., V. SRINIVASAN, K.S. KUMAR, M.R. LANDAUER & M.L. PATCHEN. 1994. Radioprotection by selenium compounds. *In* Trace Elements and Free Radicals in Oxidative Diseases. A.E. Favier, J. Neve J & P. Fauve, Eds. :211–222. AOCS Press, Champaign, IL.
36. SRINIVASAN, V., K.S. KUMAR & J.F. WEISS. 1991. Radioprotection by metals: Zinc. Radiation Research: A Twentieth-Century Perspective, Vol. I: Congress Abstracts. J.D. Chapman, W.C. Dewey & G.F. Whitmore, Eds. :369. Academic Press, San Diego, CA.
37. PETKAU, A. 1987. Role of superoxide dismutase in modification of radiation injury. Brit. J. Cancer **55** (Suppl. VIII): 87–95.
38. SEIFTER, E., G. RETTURA, J. PADAWER, F. STRATFORD, J. WEINZWEIG, A.A. DEMETRIOU & S.M. LEVENSON. 1984. Morbidity and mortality reduction by supplemental vitamin A or β-carotene in CBA mice given total body γ-radiation. J. Natl. Can. Inst. **73:** 1167–1177.
39. ROY, R.M., M. PETRELLA & H. SHATERI. 1988. Effects of administering tocopherol after irradiation on survival and proliferation of murine lymphocytes. Pharmac. Ther. **39:** 393–395.
40. RAMAKRISHNAN, N.N., W.W. WOLFE & G.N. CATRAVAS. 1992. Radioprotection of hematopoietic tissues in mice by lipoic acid. Radiat. Res. **130:** 360–365.
41. VIJAYALAXMI, R. J. REITER, M. L. MELTZ & T. S. HERMAN. 1998. Melatonin: possible mechanisms involved in its 'radioprotective' effect. Mutat. Res. **404:** 187–189.
42. WHITNALL, M.H., T.B. ELLIOTT, M.R. LANDAUER, C.L. WILHELMSEN, K.S. KUMAR, V. SRINIVASAN, W.E. JACKSON, G.D. LEDNEY & T.M. SEED. 1999. 5-Androstenediol as a radioprotectant. FASEB J. **13:** A636.

43. BLUMENTHAL, M. 1998. The Complete German Commission E Monographs. Therapeutic Guide to Herbal Medicines. American Botanical Council, Integrative Medicine Communications. Austin, TX.
44. UMA DEVI, P., K.S. BISHT & M. VINITHA. 1998. A comparative study of radioprotection by Ocimum flavinoids and synthetic aminothiol protectors in the mouse. Br. J. Radiol. **71:** 782–784.
45. UMA DEVI, P., A. GANASOUNDARI, B.S.S. RAO & K.K. SRINIVASAN. 1999. In vivo radioprotection by Ocimum flavinoids: survival of mice. Radiat. Res. **151:** 74–78.
46. RAMAKRISHNAN, N., J.F. KALINICH & D.E. MCCLAIN. 1998. Radiation-induced apoptosis in lymphoid cells: Induction, prevention, and molecular mechanisms. In Radioprotectors: Chemical, Biological, and Clinical Perspectives. E.A. Bump & K. Malaker, Eds. :253–273. CRC Press. Boca Raton, FL.
47. KONINGS, A.W., J. DAMEN & W.B. TRIELING. 1979. Protection of liposomal lipids against radiation induced oxidative damage. Int. J. Radiat. Biol. Stud. Phys. Chem. Med. **35:** 343–350.
48. KROLAK, J.M., H. MAY, R.L. HOOVER & J.F. WEISS. 1986. Effect of nutritional supplements on postirradiation survival of mice. Presented at the Twenty-fifth Hanford Life Science Symposium, Radiation Protection—A Look to the Future, Richland, WA.
49. SRINIVASAN, V., A.J. JACOBS, S.A. SIMPSON & J.F. WEISS. 1983. Radioprotection by vitamin E: effects on hepatic enzymes, delayed type hypersensitivity and postirradiation survival of mice. In Modulation and Mediation of Cancer by Vitamins. F.L. Meyskens & K.N. Prasad, Eds. :119–131. Karger. Basel.
50. JACOBS, A. J., W. A. RANKIN, V. SRINIVASAN & J. F. WEISS. 1983. Effects of vitamin E and selenium on glutathione peroxidase activity and survival of irradiated mice. Proceedings of the 7th International Congress of Radiation Research. J.J. Broerse, G.W. Barendsen, H.B. Kal & A.J. van der Kogel, Eds. :D5-15. Martinus Nijhoff Publishers, Amsterdam.
51. SANDSTROM, B. E., J. CARLSSON & S. L. MARKLUND. 1989. Selenite-induced variation in glutathione peroxidase activity of three mammalian cell lines: no effect on radiation-induced cell killing or DNA strand breakage. Radiat. Res. **117:** 318–325.
52. CHOW, C.K. 1988. Interrelationships of cellular antioxidant defense systems. In Cellular Antioxidant Defense Mechanisms, Vol. II. C.K. Chow, Ed. :217–237. CRC Press. Boca Raton, FL.
53. KUMAR, K. S., Y. N. VAISHNAV & J. F. WEISS. 1988. Radioprotection by antioxidant enzymes and enzyme mimetics. Pharmac. Ther. **39:** 301–309.
54. SUN, J., Y. CHEN, M. LI & Z. GE. 1998. Role of antioxidant enzymes on ionizing radiation resistance. Free Radic. Biol. Med. **24:** 586–593.
55. SCOTT, M.D., S.R. MESHNICK & J.W. EATON. 1989. Superoxide dismutase amplifies organismal sensitivity to ionizing radiation. J. Biol. Chem. **264:** 2498–2501.
56. YAMAOKA, K., S. KOJIMA, M. TAKAHASHI, T. NOMURA & K. IRIYAMA. 1998. Change of glutathione peroxidase synthesis along with that of superoxide dismutase synthesis in mice spleens after low-dose X-ray irradiation. Biochim. Biophys. Acta **1381:** 265–270.
57. NETA, R., J.J. OPPENHEIM & S.D. DOUCHES. 1988. Interdependence of the radioprotective effects of human recombinant interleukin 1, tumor necrosis factor, granulocyte colony-stimulating factor, and murine recombinant granulocyte-macrophage colony stimulating factor. J. Immunol. **140:** 108–111.
58. NETA, R. 1997. Insights into the mechanisms of radiation damage and radioprotection through their modulation by cytokines. Environ. Health Perspect. **105**(Suppl. 6): 1463–1465.
59. VAISHNAV, Y.N., K.S. KUMAR, J.F. WEISS & R. NETA. 1989. Induction of superoxide dismutase: a mechanism for radioprotection by interleukin-1 (IL-1)? Abstracts of Papers for the 37th Annual Meeting of the Radiation Research Society. :185. Nashville, TN.
60. MASUDA, A., D.L. LONGO, Y. KOBAYASHI, E. APPELLA, J.J. OPPENHEIM & K. MATSUSHIMA. 1988. Induction of mitochondrial manganese superoxide dismutase by interleukin 1. FASEB J. **2:** 3087–3091.

61. WONG, G.H. 1995. Protective roles of cytokines against radiation: induction of mitochondrial MnSOD. Biochim. Biophys. Acta **1271:** 205–209.
62. WESTMAN, N.G. & S.L. MARKLUND. 1981. Copper- and zinc-containing superoxide dismutase and manganese-containing superoxide dismutase in human tissues and human malignant tumors. Cancer Res. **41:** 2962–2966.
63. OBERLEY, T. D. & L. W. OBERLEY. 1997. Antioxidant enzyme levels in cancer. Histol. Histopathol. **12:** 525–535.
64. EPPERLY, M., J. BRAY, S. KRAEGER, R. ZWACKA, J. ENGELHARDT, E. TRAVIS & J. GREENBERGER. 1998. Prevention of late effects of irradiation lung damage by manganese superoxide dismutase gene therapy. Gene Ther. **5:** 196–208.
65. KONOPACKA, M., M. WIDEL & J. RZESZOWSKA-WOLNY. 1998. Modifying effect of vitamin C, E and beta-carotene against gamma-ray-induced DNA damage in mouse cells. Mutat. Res. **417:** 85–94.
66. SARMA, L. & P.C. KESAVAN. 1993. Protective effects of vitamins C and E against gamma-ray-induced chromosomal damage in mouse. Int. J. Radiat. Biol. **63:** 759–764.
67. DIAMOND, A.M., P. DALE, J.L. MURRAY & D.J. GRDINA. 1996. The inhibition of radiation-induced mutagenesis by the combined effects of selenium and the aminothiol WR-1065. Mutat. Res. **23:** 147–154.
68. BOREK, C., A. ONG, H. MASON, L. DONAHUE & J.E. BIAGLOW. 1986. Selenium and vitamin E inhibit radiogenic and chemically induced transformation *in vitro* via different mechanisms. Proc. Natl. Acad. Sci. USA **83:** 1490–1494.
69. KENNEDY, A.R. 1984. Prevention of radiation transformation *in vitro*. *In* Vitamins, Nutrition and Cancer. K.N. Prasad, Ed. :166–179. Karger. Basel.
70. ST. CLAIR, D.K., X.S. WAN, T.D. OBERLEY, K.E. MUSE & W.H. ST. CLAIR. 1992. Suppression of radiation-induced neoplastic transformation by overexpression of mitochondrial superoxide dismutase. Mol. Carcinogen. **6:** 238–242.
71. BERTRAM, J.S., L.N. KOLONEL & F.L. MYSKENS. 1987. Rationale and strategies for chemoprevention of cancer in humans. Cancer Res. **47:** 3012–3031.
72. KELLOFF, G.J., C.W. BOONE, J.A. CROWELL, V.E. STEELE, R.A. LUBET, L.A. DOODY, W.F. MALONE, E.T. HAWK & C.C. SIGMAN. 1996. New agents for cancer chemoprevention. J. Cell Biochem. Suppl. **26:** 1–28.
73. KELLOFF, G.J., E.T. HAWK, J.E. KARP, J.A. CROWELL, C.W. BOONE, V.E. STEELE, R.A. LUBET & C.C. SIGMAN. 1997. Progress in clinical chemoprevention. Semin. Oncol. **24:** 241–252.
74. ALBANES, D., O.P. HEINONEN, J.K. HUTTUNEN, P.R. TAYLOR, J. VIRTAMO, B.K. EDWARDS, J. HAAPAKOSKI, M. RAUTALAHTI, A.M. HARTMAN, J. PALMGREN *et al.* 1995. Effects of alpha-tocopherol and beta-carotene supplements on cancer incidence in the Alpha-Tocopherol Beta-Carotene Cancer Prevention Study. Am. J. Clin. Nutr. **62** (Suppl. 6): 1427S–1430S.
75. COMBS, G.F., L.C. CLARK & B.W. TURNBULL. 1997. Reduction of cancer risk with an oral supplement of selenium. Biomed. Environ. Sci. **10:** 227–234.
76. EMERIT, I., N. OGANESIAN, T. SARKISIAN, R. ARUTYUNYAN, A. POGOSIAN, K. ASRIAN, A. LEVY & L. CERNJAVSKI. 1995. Clastogenic factors in the plasma of Chernobyl accident recovery workers: anticlastogenic effect of Ginkgo biloba extract. Radiat. Res. **144:** 198–205.
77. EMERIT, I., M. QUASTEL, J. GOLDSMITH, L. MERKIN, A. LEVY, L. CERNJAVSKI, A. ALAOUI-YOUSSEFI, A. POGOSIAN & E. RIKLIS. 1997. Clastogenic factors in the plasma of children exposed at Chernobyl. Mutat. Res. **373:** 47–54.
78. EMERIT, I., N. OGANESIAN, R. ARUTYUNIAN, A. POGOSSIAN, T. SARKISIAN, L. CERNJAVSKI, A. LEVY & J. FEINGOLD. 1997. Oxidative stress-related clastogenic factors in plasma from Chernobyl liquidators: protective effects of antioxidant plant phenols, vitamins and oligoelements. Mutat. Res. **377:** 239–246.
79. BEN-AMOTZ, A., S. YATZIV, M. SELA, S. GREENBERG, B. RACHMILEVICH, M. SHWARZMAN & Z. WESHLER. 1998. Effect of natural beta-carotene supplementation in children exposed to radiation from the Chernobyl accident. Radiat. Environ. Biophys. **37:** 187–193.

80. KOYAMA, S., S. KODAMA, K. SUZUKI, T. MATSUMOTO, T. MIYAZAKI & M. WATANABE. 1998. Radiation-induced long-lived radicals which cause mutation and transformation. Mutat. Res. **421:** 45–54.
81. WU, L. J., G. RANDERS-PEHRSON, A. XU, C. A. WALDREN, C. R. GEARD, Z. YU & T. K. HEI. 1999. Targeted cytoplasmic irradiation with alpha particles induces mutations in mammalian cells. Proc. Natl. Acad. Sci. USA **96:** 4959–4964.
82. DRUG INFORMATION. 1999. American Hospital Formulary Service, American Society of Health-System Pharmacists, Bethesda, MD.
83. MACPHAIL, R.C., D.B. PEELE & K.M. CROFTON. 1989. Motor activity and screening for neurotoxicity. J. Amer. Coll. Toxicol. **8:** 117–125.
84. LANDAUER, M.R., J.B. HOGAN, C.A. CASTRO, K.A. BENSON, C.W. SHEHATA & J.F. WEISS. 1996. Behavioral toxicity and radioprotective efficacy of WR-151327 in combination with adenosine receptor antagonists. In Proceedings of the 1995 ERDEC Scientific Conference on Chemical and Biological Defense Research. D.A. Berg, Ed. :665–669. Research and Technology Directorate. Aberdeen Proving Ground, MD.
85. LEHNERT, S. 1979. Radioprotection of mouse intestine by inhibitors of cyclic AMP phosphodiesterase. Int. J. Radiat. Oncol. Biol. Phys. **5:** 825–833.
86. WEISS, J.F., M.R. LANDAUER, J.B. HOGAN, P.J. GUNTER-SMITH, K.A. BENSON, R. NETA & W.R. HANSON. 1997. Modification of radiation-induced gastrointestinal and hematopoietic injury in mice by combinations of agents: effects of indomethacin and caffeine. Adv. Exp. Med. Biol. **400B:** 865–872.
87. FAROOQUI, Z. & P. C. KESAVAN. 1992. Radioprotection by caffeine pre- and post treatment in the bone marrow chromosomes of mice given whole-body γ-irradiation. Mutat. Res. **269:** 225–230.
88. GEORGE, K.C., S.A. HEBBAR, S.P. KALE & P.C. KESAVAN. 1999. Caffeine protects mice against whole-body lethal dose of gamma-irradiation. J. Radiol. Prot. **19:** 171–176.
89. STELZER K. J., W-J. KOH, H. KURTZ, B. E. GREER & T. W. GRIFFIN. 1994. Caffeine consumption is associated with decreased severe late toxicity after radiation to the pelvis. Int. J. Radiat. Oncol. Biol. Phys. **30:** 411–417.
90. DION, M.W., D.H. HUSSEY & J.W. OSBORNE. 1989. The effect of pentoxifylline on early and late radiation injury following fractionated irradiation in C3H mice. Int. J. Radiat. Oncol. Biol. Phys. **17:** 101–107.
91. POSPISIL M., M. HOFER, J. NETIKOVA, I. PIPALOVA, A. VACEK, A. BARTONICKOVA & K. VOLENEC. 1993. Elevation of extracellular adenosine induces radioprotective effects in mice. Radiat. Res. **134:** 323–330.
92. UEDA, T., Y. TOYOSHIMA, T. MORITANI, K. RI, N. OTSUKI, T. KUSHIHASHI, H. YASUHARA & T. HISHIDA. 1996. Protective effect of dipyridamole against lethality and lipid peroxidation in liver and spleen of the ddY mouse after whole-body irradiation. Int. J. Radiat. Biol. **69:** 199–204.
93. MAK, I.T. & W.B. WEGLICKI. 1992. Membrane antiperoxidative activities of D- propranolol, L-propranolol and dimethyl quaternary propranolol (UM-272). Pharmacol. Res. **25:** 25–30.
94. MAK, I.T. & W.B. WEGLICKI. 1990. Comparative antioxidant activities of propranolol, nifedipine, verapamil, and diltiazem against sarcolemmal membrane lipid peroxidation. Circ. Res. **66:** 1449–1452.
95. BENSON, K.A , M.R. LANDAUER, J.B. HOGAN, C.W. SHEHATA, C.A. CASTRO & J.F. WEISS. 1994. Propranolol enhances the radioprotective efficacy and reduces the behavioral toxicity of WR-151327. Abstracts of Papers for the 42nd Annual Meeting of the Radiation Research Society. :211. Nashville, TN.
96. LANDAUER, M.R., C.A. CASTRO, K.A. BENSON, J.B. HOGAN & J.F. WEISS. 1997. Radiation protection and locomotor performance with a combination of calcium antagonist and a phosphorothioate. In Proceedings of the 1996 ERDEC Scientific Conference on Chemical and Biological Defense Research. D.A. Berg, Ed. :535–541. Research and Technology Directorate, Aberdeen Proving Ground, MD.
97. FLOERSHEIM, G.L. 1992. Calcium antagonists protect mice against the lethal effects of ionizing radiation. Brit. J. Radiol. **65:** 1025–1029.

98. FLOERSHEIM, G.L. & C. RACINE. 1995. Calcium antagonist radioprotectors do not reduce radiotherapeutic efficacy in three human tumor xenografts. Strahlenther. Onkol. **171:** 403–407.
99. GRACHEV, S.A., A.G. SVERDLOV, N.G. NIKANOROVA, O.I. BOL'SHAKOVA & I.K. KOROLVA. 1994. Decrease of toxic effects of aminothiol radiation-protective agents and increase of chemical protection action against ionizing radiation by the use of unithiol (Russian). Radiats. Biol. Radioecol. **34:** 424–429.

Ionizing Radiation Potentiates the Induction of Nitric Oxide Synthase by Interferon-γ and/or Lipopolysaccharide in Murine Macrophage Cell Lines

Role of Tumor Necrosis Factor-α

LESLIE C. MCKINNEY,[a,c] ELIZABETH M. AQUILLA,[a] DEBORAH COFFIN,[b] DAVID A. WINK,[b] AND YORAM VODOVOTZ[b,d]

[a]*Department of Radiation Pathophysiology and Toxicology, Armed Forces Radiobiology Research Institute, Bethesda, Maryland 20889-5145, USA*

[b]*Radiation Biology Branch, National Cancer Institute, Bethesda, Maryland 20892, USA*

ABSTRACT: Macrophages respond to infection or injury by changing from a "resting" cellular phenotype to an "activated" state defined by the expression of various cytotoxic effector functions. Regulation of the transition from a resting to an activated state is effected by cytokine and/or pathogenic signals. Some signals do not directly induce activation, but instead "prime" the macrophage to respond more vigorously to a second signal. One example of this priming phenomenon involves induction of nitric oxide (NO) synthesis by the enzyme nitric oxide synthase (NOS2). Our experiments indicate that low doses (1–5 Gy) of ionizing radiation can enhance the induction of enzymatically active NOS2 by IFN-γ or LPS in J774.1 and RAW264.7 murine macrophage cell lines. Radiation alone did not produce this induction, rather, it was effective as a priming signal; cells exposed to radiation produced more NO when a second signal, either IFN-γ or LPS, was applied 24 h later.

INTRODUCTION

Macrophages respond to infection or injury by changing from a "resting" cellular phenotype to an "activated" state that is defined by the expression of various cytotoxic effector functions. The transition from a resting to an activated state is regulated by cytokine and/or pathogenic signals. The macrophage response varies depending on both the nature of the incoming signals and the order in which they are presented. Some signals do not directly induce activation, but instead "prime" the macrophage to respond more vigorously to a second signal. One example of this priming phenomenon involves induction of nitric oxide (NO) synthesis by the enzyme nitric oxide synthase (NOS2).[1] For example, unstimulated thioglycollate-elicited macrophages do not produce NO constitutively.[2,3] Moderate levels of NO

---

[c]Address for correspondence: Voice: 301-295-2902; fax: 301-295-0313.
e-mail: mckinney@mx.afrri.usuhs.mil

[d]Present address: Cardiology Research Foundation and Medlantic Research Institute, Washington Hospital Center, Washington, DC 20010.

can be induced directly by exposure to LPS in most types of macrophages. However, if macrophages are pre-exposed to IFN-γ, which itself only induces low levels of NO, there is a synergistic effect of the two signals that leads to a greater output than is observed with either alone.[4-6]

Of interest to us were several reports that ionizing γ- or X-radiation could serve as either a priming or an activating stimulus for macrophages *in vitro*. As early as 1985, Gallin *et al.*[7] showed that irradiation of J774.1 macrophages with 20 Gy γ-radiation produced morphological and enzymatic changes consistent with increased cell activation. In these cells and in human monocytes, radiation also produced an increase in PMA-induced H_2O_2 production that was evident 1 day post-irradiation.[7,8] In RAW264.7 macrophages, 50 Gy of γ-radiation induced MHC class I antigen expression and sensitivity to LPS, and irreversibly primed the cells to become cytotoxic, as assessed by tumor cell killing.[9] Similarly, Duerst and Werberig[10] showed that 10 Gy of γ-radiation primed J774.1 cells for antibody-dependent cell-mediated cytotoxicity. O'Brien-Ladner *et al.*[11] showed increased secretion of IL-1α and IL-1β from human alveolar macrophages following treatment with 2 Gy of γ-radiation *in vitro*.

Because cytokine-induced priming leads to increased NO production in the macrophage, it is not surprising that radiation-induced priming has the same effect. Recently, Ibuki and Goto[12,13] demonstrated enhanced NO production from mouse peritoneal macrophages by doses of 6 or 12 Gy applied *in vitro*. The goal of our project was to systematically investigate the effect of lower doses of radiation (0.5–10 Gy), alone and in combination with other signals that activate macrophages (IFN-γ and LPS), on induction of NOS2 *in vitro*. Furthermore, since radiation is known to induce the expression of TNF-α,[14,15] we examined the role of this cytokine in the co-stimulatory effect of radiation.

MATERIALS AND METHODS

Cell Culture

J774.1 (J774A.1) and RAW264.7 cells were obtained from American Type Culture Collection (Rockville, MD) and maintained in culture as described elsewhere.[16]

Experimental Design

Cells were plated overnight in 24-well (for nitrite [NO_2^-] measurements) or 6-well plates (for immunoblots) at either $2-4 \times 10^5$ or 1×10^6 cells/well, respectively. Cells were then either irradiated or exposed to IFN-γ (recombinant rat or mouse, used interchangeably; Genzyme, Cambridge, MA) and/or LPS (*Escherichia coli* lipopolysaccharide 0127:B8; Sigma, St. Louis, MO). The culture medium was replaced just prior to irradiation or upon treatment with IFN-γ/LPS. One, 4, or 24 h later IFN-γ/LPS-treated cells were irradiated or, alternatively, the irradiated cells were treated with IFN-γ/LPS; the culture medium was replaced for the latter case only. In some cases, other reagents (TNF-α, anti-TNF-α antibody [R&D Systems, Minneapolis, MN]) or anti-CNTF (ciliary neurotrophic factor) antibody [Sigma Immunochemicals, St. Louis, MO]) were added in place of or just prior to irradiation.

Aliquots of cell supernatant were taken 24 h after the last treatment (radiation or IFN-γ/LPS, depending on the protocol; see below) for assay of NO_2^-, the stable reaction product of NO in aqueous solution, using the Griess reaction.[17] Nitrite values were normalized to cellular protein, which was assayed from cells in 24-well plates *in situ* using a fluorescent reagent[18] or by colorimetric assay[19] of aliquots of lysed cells from 6-well plates that had been harvested for NOS2 immunoblots.[16]

Radiation

Cells were irradiated in 6- or 24-well tissue culture plates in a bilateral γ-radiation field of the Armed Forces Radiobiology Research Institute (AFRRI) 60Co facility.

Data Analysis

For each experiment (designated as N = 1), NO_2^- measurements (means of 8–11 values per condition and given as nmol NO_2^-/well ± SEM) were normalized to protein (means of 3–6 values per condition and given as mg protein/well ± SEM) to yield nmol NO_2^-/mg protein. The SEM for the normalized value was determined from the error propagation formula $SEM(z) = [1/|y_{mean}|]*[SEM^2(x_{mean}) + z^2 SEM^2(y_{mean})]1/2$ where $z = x_{mean}/y_{mean}$; x_{mean} = nmol NO_2^-/well; and y_{mean} = μg protein/well. In some figures, mean nmol NO_2^-/mg protein values from multiple experiments were pooled (N > 1). In those cases, the error bars represent the SEM of the combined means.

RESULTS

Radiation Primes Macrophage Cell Lines to Produce More NO in Response to IFN-γ or LPS

To determine whether radiation can "sensitize" or "prime" cells to other NO-inducing signals, cells were first irradiated and then left without further treatment or exposed to maximal levels of IFN-γ (100 U/mL) and/or LPS (1 μg/mL) 24 h post-irradiation. Radiation alone (1–20 Gy; 1 Gy/min) did not induce measurable production of NO (N = 6; data not shown). Irradiated cells treated with IFN-γ or LPS showed increased production of NO over that of controls for all treatment conditions, at doses as low as 1 Gy (FIG. 1A). In FIGURE 1B, values at each dose of radiation were normalized to 0 Gy for each treatment condition, yielding a measure of fold increase due to radiation for either IFN-γ, LPS, or the combination. The effect of radiation was maximal at 5 Gy. The priming effect of radiation on NO production was only manifest if the radiation was applied 24 h prior to the addition of IFN-γ or LPS; addition of IFN-γ or LPS 1 h or 4 h post-irradiation or 1, 4, or 24 h pre-irradiation resulted in no radiation-dependent increase in NO. Experiments on RAW264.7 cells yielded results similar to those shown in FIGURE 1 (data not shown).

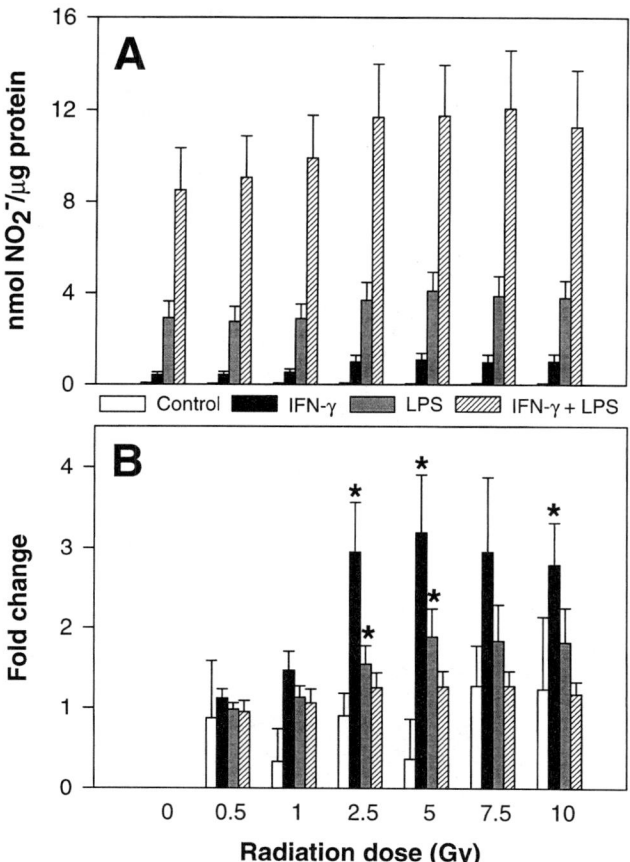

FIGURE 1. Priming of IFN-γ- and/or LPS-stimulated NO production by radiation in J774.1 cells. **A.** J774.1 cells were plated for 24 h and exposed to radiation. 24 h post-irradiation, cells were exposed to either nothing (*open bars*), 100 U/mL IFN-γ (*black bars*), 1 μg/mL LPS (*gray bars*), or both IFN-γ and LPS (*hatched bars*). NO2- was assayed 24 h later (48 h post-irradiation). N = 12, 8, 10, and 3 for control, IFN-γ-, LPS-, or IFN-γ + LPS-treatments, respectively. **B.** Data from A expressed as fold increase over 0 Gy for each treatment condition. *$p < 0.05$ vs. 0 Gy.

Radiation Increases the Amount of NO Induced by IFN-γ or LPS but Does Not Shift Dose-Sensitivity

To determine whether radiation acts to shift the dose-sensitivity of J774.1 cells to IFN-γ or LPS, dose-response curves were constructed at 4 radiation doses: 0.5, 1, 2.5, and 5 Gy. FIGURES 2A and B show that for both IFN-γ and LPS, radiation did not shift the dose sensitivity, but rather increased the absolute magnitude of the response.

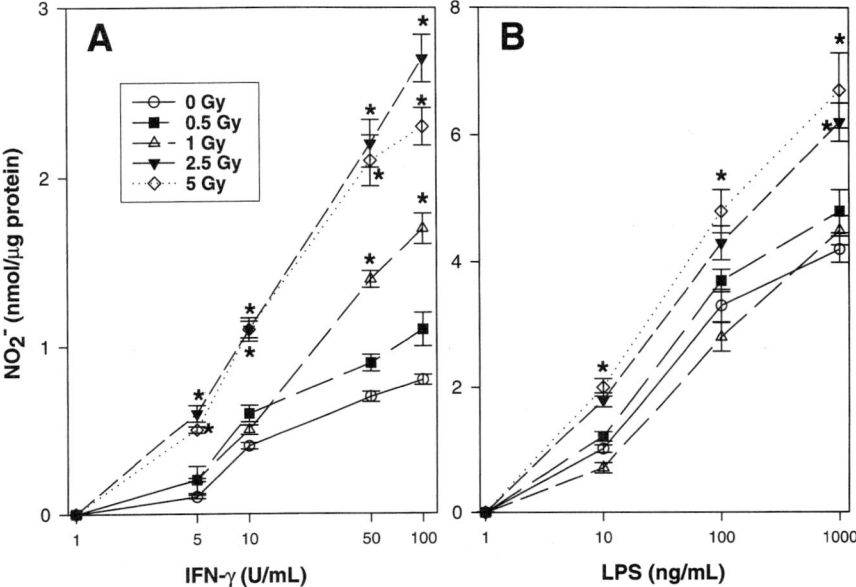

FIGURE 2. Augmentation of IFN-γ- or LPS-stimulated NO production by radiation. J774.1 cells were plated for 24 h, irradiated at 0.5, 1, 2.5, or 5 Gy, then exposed to various doses of (**A**) IFN-γ or (**B**) LPS 24 h post-irradiation (medium replaced). Nitrite was assayed 24 h following stimulation. $N = 1$. $*p < 0.01$ vs. 0 Gy.

Irradiated Cells Express Higher Levels of NOS2 Protein in Response to IFN-γ and/or LPS

FIGURES 3A and B show immunoblots from J774.1 and RAW264.7 cells, respectively, that were irradiated and then exposed to IFN-γ and/or LPS 24 h later. Induction of NOS2 was clearly dependent on radiation in both cell lines, but to varying extent depending on treatment condition.

Anti-TNF-α Antibody Blocks Radiation-Enhanced NO Production

One possible mechanism by which radiation may enhance NO production is via the induction of autocrine factors that can synergize with IFN-γ and/or LPS. Since radiation is known to induce TNF-α in some macrophages,[14,15] and since TNF-α is known to induce NO production synergistically with IFN-γ,[2,5,6] we tested the effect of anti-TNF-α antibody in our system. FIGURE 4A (*solid bars*) shows that anti-TNF-α antibody completely blocked the radiation-enhanced response to IFN-γ but only slightly decreased the radiation-enhanced response to LPS (FIG. 4B, *solid bars*). The anti-TNF-α antibody had no effect on NO production induced by IFN-γ and/or LPS at 0 Gy and, as expected, completely inhibited the TNF-α–induced increase in NO production. Control experiments using an isotype-matched goat IgG (anti-CNTF antibody) had no effect (FIG. 4, *hatched bars*).

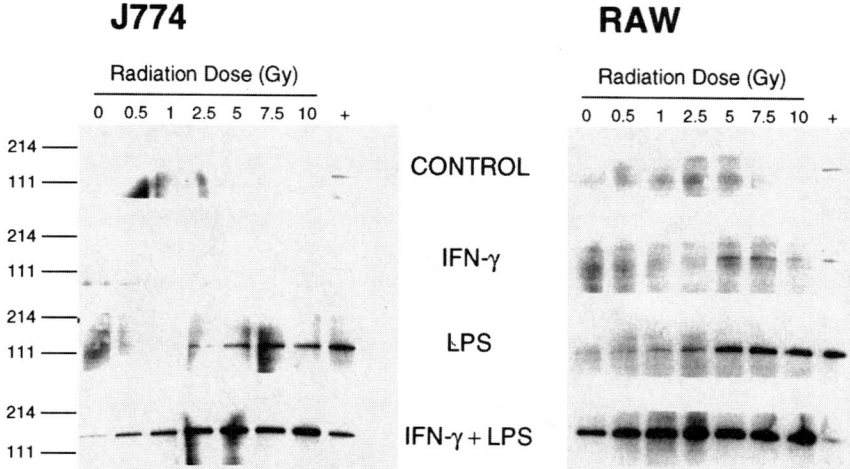

FIGURE 3. Expression of NOS2 protein following irradiation. J774.1 cells (*left panels*) or RAW264.7 cells (*right panels*) were plated for 24 h, irradiated with various doses of radiation (indicated above the lanes), then exposed to either nothing, 100 U/mL IFN-γ, 1 μg/mL LPS, or both IFN-γ and LPS 24 h post-irradiation. Cells were harvested and prepared for immunoblots 12 h following exposure to cytokine. Each well contained 60 mg of total cellular protein. The positive control ("+") lane contained 20 mg protein from unirradiated J774.1 cells exposed to IFN-γ + LPS. Numbers on the left indicate apparent molecular mass in kD. Representative of 3 experiments for each cell type.

DISCUSSION

Our experiments indicate that low doses (1–5 Gy) of ionizing radiation can enhance the induction of enzymatically active NOS2 by IFN-γ or LPS in J774.1 and RAW264.7 murine macrophage cell lines. Radiation alone was not sufficient for this induction, rather, it was effective as a priming signal; cells exposed to radiation produced more NO when a second signal, either IFN-γ or LPS, was applied 24 h later. Perhaps because IFN-γ, LPS, and IFN-γ + LPS are weak, moderate and strong inducers of NO respectively,[2] the fold-increase in NO production induced by radiation was largest (~3) with IFN-γ as the co-stimulus. Radiation was a comparatively weak priming signal, since saturating doses of radiation (5 Gy) + LPS (1 μg/mL) induced only about half the maximum inducible NO observed with IFN-γ + LPS. Radiation did not change the dose sensitivity of the cells to IFN-γ or LPS, since cytokine dose response curves were shifted vertically, not right or left, as radiation dose increased. The dose-dependent increase in NO following irradiation was paralleled by increased expression of NOS2.

The finding that anti-TNF-α antibody completely blocked the increase in NO following treatment with radiation + IFN-γ is indirect evidence that the effect of radiation is mediated by TNF-α. TNF-α can synergize with IFN-γ (but not LPS) to induce NO,[5,6] probably since LPS itself strongly induces the expression of TNF-α. This result is consistent with data indicating that radiation induces production of TNF-α in

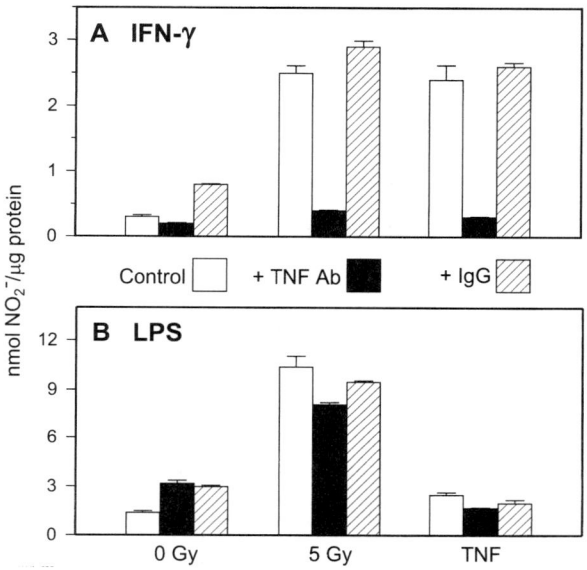

FIGURE 4. Effect of anti-TNF-α antibody on radiation-induced NO production. J774.1 cells were plated for 24 h and either left untreated (*open bars*), exposed to anti-TNF-α antibody (10 µg/mL; *black bars*), or exposed to anti-CNTF antibody (isotype matched goat IgG; 10 µg/mL; *hatched bars*) just prior to 5 Gy irradiation. Cells were then exposed to either nothing (data not shown), (**A**) 100 U/mL IFN-γ or (**B**) 1 µg/mL LPS 24 h post-irradiation. (Culture medium replaced and antibodies re-added). The protocol for the positive control was the same except that cells were exposed to 5 ng/mL TNF-α (with or without antibodies) instead of radiation. Nitrite was assayed 24 h following stimulation. Anti-TNF-antibody tested 5 times, anti-CNTF antibody tested twice with similar results.

monocytes/macrophages, which synergizes with IFN-γ applied at a later time to induce NO.[14,15] However, questions about the time course and magnitude of secretion of TNF-α in J774.1 and RAW264.7 cells remain. In other macrophage cell types in which induction of TNF-α by ionizing radiation has been characterized, TNF-α is produced transiently between 2 and 6 h post-irradiation at doses ranging from 2–20 Gy with or without co-stimuli.[14,15] In J774.1 and RAW264.7 cells, if secretion of TNF-α is induced by radiation, it occurs at low doses and is persistent. Alternatively, radiation treatment may increase either the surface expression of TNF-α or the number of TNF-α receptors possibly increasing sensitivity to this cytokine.

ACKNOWLEDGMENTS

We thank Joan Chen and Catriona Miller for technical assistance, Dr. Robert Webber (Research and Diagnostics Antibodies, Richmond, CA) for anti-NOS2 antibody, and Mr. William Jackson for advice on data analysis.

REFERENCES

1. MACMICKING, J., Q.W. XIE & C. NATHAN. 1997. Nitric oxide and macrophage function. Annu. Rev. Immunol. **15:** 323–350.
2. NATHAN, C.F. & Q.-W. XIE. 1994. Nitric oxide synthases: Roles, tolls, and controls. Cell **78:** 915–918.
3. VODOVOTZ, Y., C. BOGDAN, J. PAIK, Q.-W. XIE & C. NATHAN. 1993. Mechanisms of suppression of macrophage nitric oxide release by transforming growth factor-β. J. Exp. Med. **178:** 605–613.
4. STUEHR, D.J. & M.A. MARLETTA. 1987. Synthesis of nitrite and nitrate in murine macrophage cell lines. Cancer Res. **47:** 5590–5594.
5. DING, A.H., C.F. NATHAN & D.J. STUEHR. 1988. Release of reactive nitrogen intermediates and reactive oxygen intermediates from mouse peritoneal macrophages: comparison of activating cytokines and evidence for independent production. J. Immunol. **141:** 2407–2412.
6. DRAPIER, J., J. WIETZERBIN & J.B. HIBBS. 1988. Interferon-gamma and tumor necrosis factor induce the L-arginine-dependent cytotoxic effector mechanism in murine macrophages. Eur. J. Immunol. **18:** 1587–1592.
7. GALLIN, E.K., S.W. GREEN & P.A. SHEEHY. 1985. Enhanced activity of the macrophage-like cell line J774.1 following exposure to gamma radiation. J. Leukoc. Biol. **38:** 369–381.
8. GALLIN, E.K. & S.W. GREEN. 1987. Exposure to gamma-irradiation increases phorbol myristate acetate-induced H_2O_2 production in human macrophages. Blood **70:** 694–701.
9. LAMBERT, L.E. & D.M. PAULNOCK. 1987. Modulation of macrophage function by gamma-irradiation. Acquisition of the primed cell intermediate stage of the macrophage tumoricidal activation pathway. J. Immunology **139:** 2834–2841.
10. DUERST, R. & K. WERGERIG. 1991. Cells of the J774 macrophage cell line are primed for antibody-dependent cell-mediated cytotoxicity following exposure to gamma-irradiation. Cellular Immunology **136:** 361–372.
11. O'BRIEN-LADNER, A., M.E. NELSON, B.F. KIMLER & L.J. WESSELIUS. 1993. Release of interleukin-1 by human alveolar macrophages after in vitro irradiation. Radiat. Res. **136:** 37–41.
12. IBUKI, Y. & R. GOTO. 1995. Augmentation of NO production and cytolytic activity of macrophages obtained from mice irradiated with a low dose of gamma-rays. J. Radiat. Res. **36:** 209–220.
13. IBUKI, Y. & R. GOTO. 1997. Enhancement of NO production from resident peritoneal macrophages by in vitro gamma-irradiation and its relationship to reactive oxygen intermediates. Free Radical. Biol. Med. **22:** 1029–1035.
14. CHIANG, C.S. & W.H. MCBRIDE. 1991. Radiation enhances tumor necrosis factor a production by murine brain cells. Brain Res. **566:** 265–269.
15. SHERMAN, M.L., R. DATTA, D.E. HALLAHAN, R.R. WEICHSELBAUM & D.W. KUFE. 1991. Regulation of tumor necrosis factor gene expression by ionizing radiation in human myeloid leukemia cells and peripheral blood monocytes. J. Clin. Invest. **87:** 1794–1797.
16. MCKINNEY, L.C., E.M. AQUILLA, D. COFFIN, D.A. WINK & Y. VODOVOTZ. 1998. Ionizing radiation potentiates induction of nitric oxide synthase by interferon-gamma and/or lipopolysaccharide in murine macrophage cell lines: Role of tumor necrosis factor-alpha. J. Leukoc. Biol. **64:** 459–466.
17. GREEN, L.C., D.A. WAGNER, J. GLOGGLOSKI, P.L. SKIPPER, J.S. WISHNOK & S.R. TANNENBAUM. 1982. Analysis of nitrate, nitrite and [^{15}N]nitrate in biological fluids. Anal. Biochem. **126:** 131–138.
18. LORENZEN, A. & S.W. KENNEDY. 1993. A fluorescence-based protein assay for use with a microplate reader. Anal. Biochem. **214:** 346–348.
19. BRADFORD, M.M. 1976. A rapid and sensitive method for the quantitation of microgram quantities of protein utilizing the principle of protein-dye binding. Anal. Biochem. **72:** 248–254.

Genetic Responses to Free Radicals

Homeostasis and Gene Control

BEATRIZ GONZÁLEZ-FLECHA[a] AND BRUCE DEMPLE[b,c]

[a]*Physiology Program, Department of Environmental Health, and*
[b]*Division of Toxicology, Department of Cancer Cell Biology,*
Harvard School of Public Health, Boston, Massachusetts 02115, USA

ABSTRACT: Gene regulation mechanisms have evolved allowing cells to fine-tune the level of "endogenous" oxidative stress and to cope with increased free radicals from external sources. Levels of H_2O_2 are tightly controlled in *E. coli* by OxyR, which is activated by H_2O_2 to increase scavenging activities and limit H_2O_2 generation by the respiratory chain. Sub-micromolar levels of H_2O_2 are maintained in mammalian tissues, though the regulatory systems that govern this control are unknown. Excess superoxide triggers the *soxRS* system in *E. coli*, which is controlled by the oxidant-sensitive iron-sulfur centers of the SoxR protein. Nitric oxide activates SoxR by a different modification of the iron-sulfur centers. The *soxRS* regulon mobilizes diverse functions to scavenge free radicals and repair oxidative damage in macromolecules, and other mechanisms that exclude many environmental agents from the cell. Mammalian cells also sense and respond to sub-toxic levels of nitric oxide, activating expression of heme oxygenase 1 through stabilization of its mRNA. These inductions give rise to adaptive resistance to nitric oxide in neuronal and other cell types.

REGULATION OF INTRACELLULAR H_2O_2

Although catalase was one of the first enzymes purified and characterized, its physiological role remained obscure for many years. This situation was due to a lack of experimental evidence showing the presence of adequate concentrations of both substrate and hydrogen donors in cells. In 1952 B. Chance reported the first determinations of catalase-hydrogen peroxide complex I in the aerobic *Micrococcus lysodeikticus* cells.[1] Chance's paper pioneered the idea of a regulated steady-state concentration of hydrogen peroxide, which has since been extended to a variety of biological systems (TABLE 1).

The regulatory role of H_2O_2, and therefore the significance of controlling the H_2O_2 concentration, was advanced in 1960 by R. Clayton, who reported the induction of catalase synthesis in *Rhodopseudomonas spheroides* by single doses of H_2O_2 (initial concentrations 4–40 µM).[2] In this early report Clayton put forth the concept of an "internal substance or condition that promotes catalase synthesis." We know now that, in many bacteria, this "condition" is the oxidation of the transcription factor OxyR, which orchestrates a well-studied cellular response to oxidative stress.

[c]Address for correspondence: Voice: 617-432-3462; Fax: 617-432-2590/432-0377.
e-mail: bdemple@hsph.harvard.edu

Regulation of Intracellular H_2O_2 in Escherichia coli

The enterobacterium *Escherichia coli* constitutes the best understood model system in studies of the regulation of the metabolic sources of H_2O_2. Aerobic growth of bacterial cells in rich medium follows a characteristic profile. A first phase of adaptation to the availability of nutrients (lag phase), is characterized by an increase in protein synthesis with no change in the number of cells. When cells have acquired the necessary biochemical machinery, they enter a phase of active growth where the cell number increases exponentially over time. Finally, when nutrients are limiting, cell growth is arrested and the cells enter stationary phase.

TABLE 1. Hydrogen peroxide concentrations in biological systems

System		[H_2O_2] (µM)	Reference
Bacteria	*Micrococcus lysodeikticus*	0.01	1
	Escherichia coli		
	(logarithmic phase)	0.25 ± 0.03	4
	(stationary phase)	0.15 ± 0.01	4
Parasites	*Tripanosoma cruzi*		
	(epimastigotes)	1.5 ± 0.5	84
Plants	Soybean embryonic axes		
	(germination phase I)	0.3	85
	(germination phase II)	0.9	85
Rat	Isolated perfused rat liver		
	(cytosol)	0.1	86
	(peroxisomes)	0.001	86
	Liver slices	0.09 ± 0.01	16
	Hepatocytes		
	(primary cultures)	0.08 ± 0.01	84
	Kidney cortex slices	0.09 ± 0.01	18
	medulla slices	0.07 ± 0.01	18
	papilla slices	0.08 ± 0.01	18
	Alveolar epithelial cells	0.60 ± 0.05	87
	(primary cultures)		
Human			
	Melanoma cells		
	(logarithmic phase)	3.3 ± 0.2	88
	(stationary phase)	2.1 ± 0.2	88

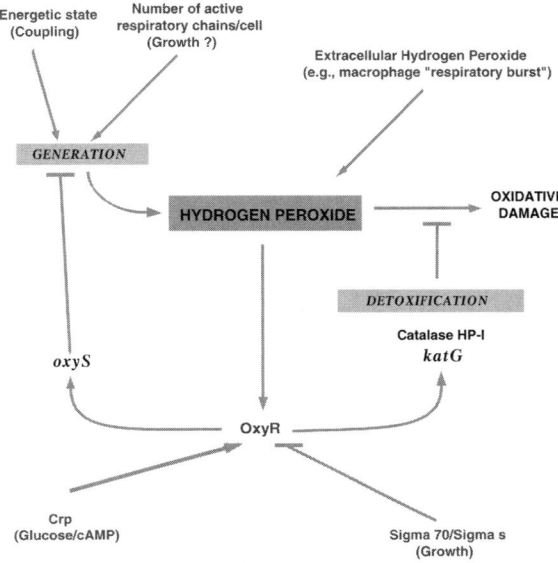

FIGURE 1. Levels of regulation of the intracellular concentration of hydrogen peroxide in *Escherichia coli*. (→): activation, (⊥): inhibition. For explanations see text.

The metabolic changes occurring in bacterial cells during aerobic growth prompted us to evaluate the changes in the production and elimination of H_2O_2, as well as the putative occurrence of oxidative stress during exponential growth. We have found that both the generation and detoxification of H_2O_2 are tightly regulated as cells progress through the growth phases and in response to environmental changes. Changes in growth conditions (e.g., availability or changes of carbon source), as well as the intracellular concentration of H_2O_2 itself, affect both generation and detoxification processes. Many of these metabolic regulations are mediated by OxyR (FIGURE 1).

The Aerobic Respiratory Chain As a Source of H_2O_2.

Measurements of the rates of H_2O_2 generation in the presence of electron transport inhibitors such as rotenone (inhibitor of NADH dehydrogenase) or antymicin (inhibitor of electron transfer between ubiquinone and cytochrome *b*) allowed us to show that univalent reduction of O_2 can occur at complexes I and III *in vivo* to form superoxide anion ($\cdot O_2^-$) (FIG. 2).[3] These effects parallel those seen with mammalian mitochondria treated with these inhibitors. In addition, we found that the rate of H_2O_2 production in wild-type *E. coli* increased markedly as the cells entered exponential growth. This regulation was effected at two levels: changes in the number of active respiratory chains per cell, which was maximal during exponential growth; and changes in the energy status of the cell and in the arrangement of respiratory chain components corresponding to particular growth conditions.

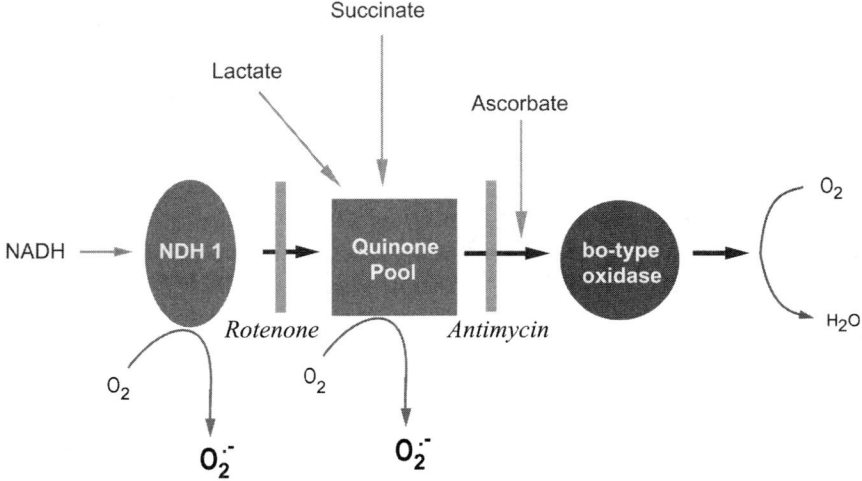

FIGURE 2. Schematic representation of the bacterial respiratory chain indicating the point of entry of reducing equivalents from the different substrates (NADH, lactate, succinate, and ascorbate) and the sites of inhibition of the electron flux by rotenone and antimycin A. We have chosen for this cartoon the most typical composition for exponentially growing cells. NDH 1: NADH dehydrogenase 1; Quinone pool: mostly ubiquinone; *bo*-type oxidase: cytochrome oxidase *o*.

Feedback to OxyR

We then looked at changes in the rate of H_2O_2 generation, catalase activity, and the H_2O_2 steady-state concentration in strains defective in respiratory chain components. We found that *E. coli* continuosly regulate the activity of catalase to compensate for changes in the H_2O_2 production rates during the transition from lag phase to exponential growth.[3,4] Such changes in H_2O_2 production with compensatory catalase expression were observed in control cells during the transition from the lag to the exponential phases of growth, and also in strains with defects in the respiratory chain. In this way cells maintained a nearly constant intracellular H_2O_2 concentration of 0.1–0.2 µM in the face of 10-fold changes in the rate of H_2O_2 production.[3,4]

This regulation was achieved in part by a transient increase in OxyR-dependent transcription of the catalase gene *katG* (monitored using a *katG::lacZ* operon fusion that reports transcription of the catalase-hydroperoxidase I gene), which directly correlated with the increased rate of H_2O_2 generation.

The OxyR protein is the sensor and direct transcriptional activator of the OxyR regulon. The co-regulated group of ~10 genes encodes catalase-hydroperoxidase I (*katG*), NADPH-dependent alkyl hydroperoxidase (*ahpFC*), glutathione reductase (*gorA*), glutaredoxin (*grxA*), a protective DNA binding protein (Dps), and several other genes.[5] The expression of these genes occurs rapidly in response to micromolar levels of H_2O_2 added instantaneously to the growth medium.[5-8] As discussed below, the continuous regulation of OxyR activity during growth is governed by much smaller changes in the H_2O_2 concentrations, of the order of 0.1 µM.

In our experimental model, induction of OxyR-dependent genes proved to be essential to preserve cellular integrity. During normal aerobic growth, strains lacking *oxyR* showed throughout growth a frequency of spontaneous mutation 8- to 10-fold higher than that seen for wild-type bacteria, indicative of an extra load of oxidative damage in OxyR-deficient bacteria. These defective strains also had elevated levels of oxidative damage to proteins and membranes in exponential phase.[4]

Regulation of the Regulator

While studying the metabolic production of hydrogen peroxide in aerobically growing *E. coli*, we found that the response to exogenous H_2O_2 is quantitatively different in lag-, exponential- and stationary-phase cultures.[3] This observation prompted us to look for possible changes in the level of OxyR protein as a function of growth. Immunoblotting revealed the amount of OxyR protein in cells varied 6-fold, with the highest value at 4–6 h of growth.[9] Parallel changes in the expression of the *oxyR* transcript were observed.

We showed that the upregulation of *oxyR* in exponential phase depends on the transcription factor Crp, presumably in response to changes in glucose and cAMP levels. In fact, the *oxyR* promoter region contains a highly probable Crp binding site, and we showed genetically that Crp-regulated OxyR levels affect expression of the *oxyR* regulon. Mutations that inactivated the *crp* system abolished the exponential phase induction of *oxyR* and lowered the expression of both catalase HP-I and glutathione reductase (two members of the OxyR regulon). These changes in turn increased the cellular sensitivity to exogenously added hydrogen peroxide.[9] On the other hand, the decrease in *oxyR* expression after 6 h of growth was due to changes in the sigma subunit of RNA polymerase, which accompany the transition from exponential growth to stationary phase.[10] These multiple transcriptional systems operate continuously to fine-tune OxyR activity,[9] and in turn the disposal of H_2O_2 in *E. coli*.

oxyS-Mediated Regulation of the Metabolic Production of H_2O_2

The intracellular concentration of H_2O_2, and the susceptibility to oxidative stress imposed by exogenous H_2O_2, are much higher in Δ*oxyR* strains than in strains mutated only in *katG*.[4] This observation suggested that other OxyR-regulated activities, in addition to catalase HP-I, are involved in controlling H_2O_2 levels in *E. coli*. Mutation of the *ahpFC* genes encoding alkyl hydroperoxidase showed that this enzyme does not contribute detectably to the H_2O_2 regulation. Another candidate for such a role was the *oxyS* gene, a member of the *oxyR* regulon that specifies a small RNA with post-transcriptional regulatory activity.[11] This analysis showed that *oxyS* is essential for the homeostatic regulation of intracellular H_2O_2.[12] The activity of *oxyS* affects the metabolic output of H_2O_2,[12] rather than H_2O_2 removal by catalase, although the specific target(s) have not yet been identified. One possibility is that an *oxyS*-regulated gene product directs the electron flux through the respiratory chain to components with higher energetic efficiency and less "leakage" of $\cdot O_2^-$.[12] This hypothesis is supported by the finding that cellular respiration is significantly increased in an *oxyS* deficient strain.[12]

In summary, we have found to date six different mechanisms for regulation of the intracellular H_2O_2 concentration in *E. coli* (FIG. 1):

1. the regulation of the energetic state of the cell by switching from coupled to uncoupled isoforms of the components of the respiratory chain;
2. the change in the number of active respiratory chain units per cell, which occurs in association with changes from low to high metabolic states;
3. the *oxyS*-dependent inhibition of the rate of $\cdot O_2^-$ production in response to increases in OxyR activation;
4. the OxyR-dependent regulation of the *katG* gene and concomitant change in catalase activity in response to increases in H_2O_2;
5. the Crp-dependent up-regulation of the transcription of the *oxyR* gene in response to changes in glucose and cAMP levels; and
6. the σ^s-dependent inhibition of *oxyR* transcription upon transition from exponential growth to stationary phase.

The activity of OxyR is the central component of this complex system that allows both immediate, short-term responses to oxidative stress through its redox-regulation, and overall integration with energy metabolism and cell growth through controlled expression by Crp and σ^s.

Regulation of Intracellular H_2O_2 in Mammalian Tissues

Mitochondrial Production of H_2O_2

The production of H_2O_2 in mammalian mitocondria is regulated by some of the same mechanisms that are relevant to the control of free radicals in bacteria. These mechanisms include:

1. the transition from the resting state 4 to the active state 3 (see below);
2. the supply of oxygen or mitochondrial substrates; and
3. the inhibition of the electron flow at different points with increased leakage of single electrons to form $\cdot O_2^-$.[13,14]

State 4 is the "energized" condition with a highly reduced steady state for the electron carriers, and a relatively high rate of H_2O_2 generation. State 3 is the "de-energized" condition with a more oxidized steady state for the respiratory chain components and a relatively low H_2O_2 production.[14]

Of special interest is the regulated production of free radicals due to inhibition of the respiratory chain. This effect was initially described in isolated mitochondria treated with specific inhibitors.[15] Later studies found similar regulation to underlie the increased production of free radicals in the rat liver and kidney after ischemia-reperfusion *in vivo*. González-Flecha *et al.*[16,17] showed that mitochondrial production of free radicals accounted for ~95% of the increase in the total rate of H_2O_2 production after ischemia-reperfusion cycles in these tissues. The increased rate of intra-mitochondrial production of $\cdot O_2^-$ upon reperfusion was due to the inhibition of electron transfer at mitochondrial complex I during ischemia. The superoxide production was aggravated by blockage of electron transport through the ubiquinone-cytochrome *c* segment of the respiratory chain when oxygen was reintroduced in the tissue.[17] Although the mechanism(s) by which such an inhibition occurs is still unknown, we can speculate on three possibilities: a) structural dearrangement of the mitochondrial membrane with changes in the distance between carries, and a concomitant decrease in the efficiency of electron transfer; b) production of an inhibitor which would intercalate between carriers; and c) damage or alteration of the components with changes in their affinity or redox potential.

Tissue-wide Homeostasis

A remarkable example of the concept of homeostatic regulation of H_2O_2 is the rat kidney. In this system, metabolic differences in the three anatomic/functional regions (cortex, medulla and papilla) determine differences in the profile of H_2O_2 sources and antioxidant activities yielding about the same final value for the steady-state concentration of H_2O_2 (FIG. 3).[18] Tissue oxygen concentration, mitochondrial mass, and the activity of enzymes related to the TCA cycle are higher in cortex and medulla than in papilla.[19-21] As expected from the higher respiration rates in cortex and medulla, both the total rate of H_2O_2 production, and the generation of H_2O_2 at the mitochondria were higher in those regions (FIG. 3 and ref. 17). On the other hand, papillary H_2O_2 was mostly generated in association with prostaglandin metabolism. Zonal distribution of SOD and catalase directly correlated with the metabolic rates of H_2O_2 production and compensated for the differences in the rates of production. the final result was the same steady-state concentration of hydrogen peroxide in all three zones. In contrast, the level of glutathione peroxidase and α-tocopherol were negatively correlated with the rate of H_2O_2 generation in each zone. The levels of these antioxidants correlated, instead, with the susceptibility of the tissue to oxidative and other stresses.[22,23] For example, the renal papilla had a higher content of α-tocopherol and a lower H_2O_2 concentration than cortex and medulla, and was more resistant to ischemia/reperfusion damage.[18]

Although the biochemical mechanisms underlying these regulations are still unknown, and the involvement of specific genetic responses remains to be determined, several lines of evidence suggest that there exists genetic control on the level of eukaryotic antioxidant enzymes in response to changes in the level of reactive oxygen species. For example, expression of antioxidant enzymes in the lung changes during development, following changes in oxygen tension. Messenger RNA levels for Mn-SOD, CuZnSOD, and catalase increase during fetal development and after birth, toward adulthood.[24] Regulation of antioxidant enzymes by oxidants has been also reported in epithelial cells *in vitro*[25] and *in vivo*,[26] Hela cells exposed to UV-B light,[27] and hepatocytes treated with tumor necrosis factor.[28] However, in many experimental models, the increases in the levels of mRNAs for antioxidant enzymes are not reflected in increased enzymatic activities,[24,26,27,29] indicating that post-transciptional and post-translational regulation play an important role. Additionally, under high rates of free radical output, protein inactivation prevails and the enzymatic activities of SODs, catalase and GPx are reduced (discussed in ref. 30).

In summary, mammalian cells regulate intracellular hydrogen peroxide. As in the case of bacterial cells, the regulation occurs at the levels of production and detoxification. Studies are under way to define the mechanisms involved in such regulation and the biological relevance of having a narrowly set concentration of H_2O_2.

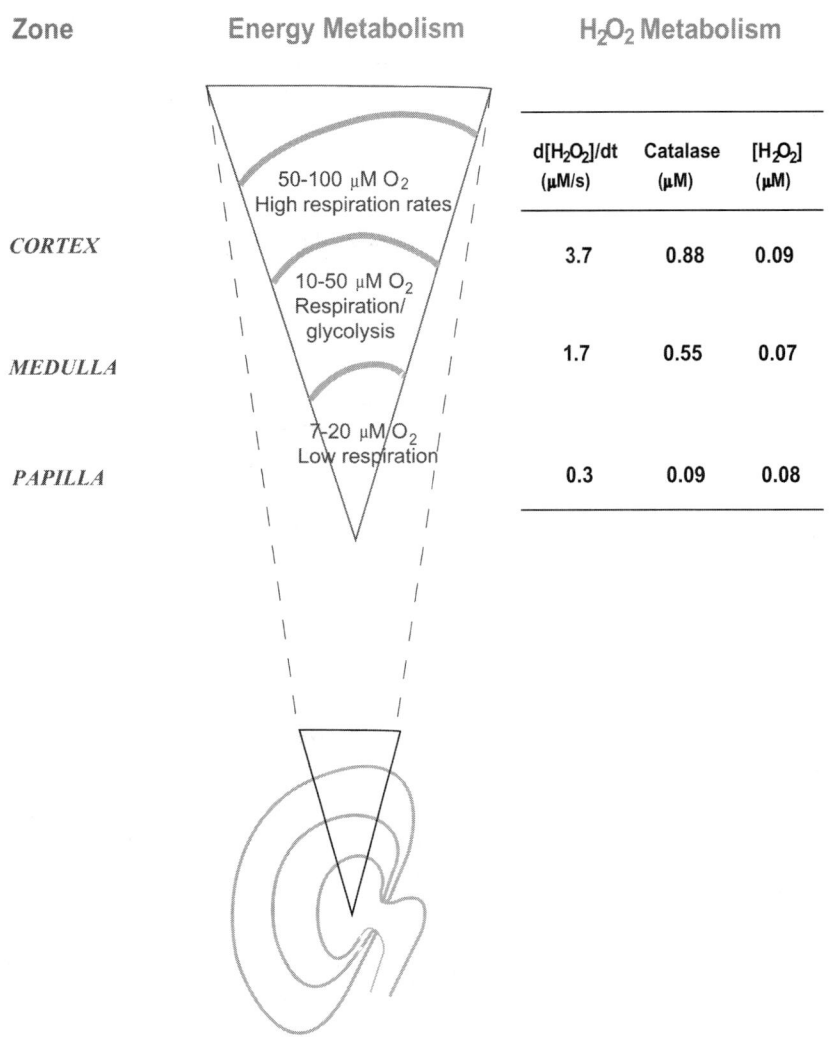

FIGURE 3. Energetic and metabolic profiles in the three anatomic zones of the rat kidney. Note that the steady-state level of H_2O_2 is the same in all three zones.

CELLULAR RESPONSES TO SUPEROXIDE

As noted above (*Regulation of Intracellular H_2O_2 in E. coli*), $\cdot O_2^-$ is the primary product of oxidation of respiratory chain components. Still other autoxidation reactions unavoidably generate $\cdot O_2^-$ in cells,[31] and this production is strongly increased by various environmental compounds, such as paraquat or quinones, which undergo "redox cycling" to generate $\cdot O_2^-$ catalytically.[32] Paraquat and other redox-cycling agents (e.g., naphthoquinones and aromatic nitro compounds) have long been used experimentally to generate oxidative stress in the form of excess superoxide.[32] Their use has revealed novel regulatory systems that respond to oxidative stress.

The soxRS *system of* E. coli *and* Salmonella typhimurium

In addition to the *oxyR* system described above, *E. coli* harbors a second genetic response to oxidative stress that is activated by redox-cycling agents, but not by H_2O_2: the *soxRS* regulon.[5,33] Activation of the *soxRS* regulon is governed by two proteins that trigger consecutive stages of transcription (FIG. 4). SoxR protein is a unique transcription factor that contains a pair of redox-active [2Fe-2S] centers, one in each 17-kD subunit of a homodimer.[33] These [2Fe-2S] centers have a midpoint redox potential of -285 mV,[34,35] which is poised near the redox state thought to prevail in cells. When the metal centers are in the reduced state, SoxR is largely inactive in transcription, but their one-electron oxidation unleashes a powerful and very specific regulatory activity.[36,37] As will be described below (*NO and the* soxRS *System*), SoxR is also activated by nitric oxide in a fundamentally different reaction. Activated SoxR protein targets just one known gene, *soxS*, which itself encodes a second transcriptional activator (see below). This induction of a second transcription-activating protein amplifies the signal to induce ≥ 15 genes involved in resistance to free radical damage and other functions.

Functions of SoxR: soxS *Activation and Redox Signal Transduction*

The *soxR* and *soxS* genes are arranged head-to-head in both the *E. coli* and the *S. typhimurium* genomes.[5,38] SoxR binds a region between the genes, in the center of the conserved RNA polymerase binding site of the soxS promoter. Unusually, from this position more typical of repressor than activator proteins, activated SoxR stimulates soxS transcription up to 100-fold. It is also very unusual that non-activated SoxR binds the same site in the *soxS* promoter, and just as tightly as does the activated form.[39] Thus, the activation of *soxR* transcription by SoxR proceeds through an allosteric transition in the SoxR-promoter DNA complex (FIG. 4), which is driven by the oxidation of the [2Fe-2S] centers. A structural explanation of this transition in SoxR is not yet available; one possibility is that it is driven by the development of net +1 charges on the metal centers during oxidation.[40] The activation overcomes a sub-optimal spacing of sequences within the *soxS* promoter, suggesting that SoxR reorients the promoter elements in a way that eases formation of the "open" complex by RNA polymerase.[41]

At least *in vitro*, SoxR activity can also be controlled by the assembly and disassembly of the [2Fe-2S] centers. Apo-SoxR is generated upon exposure of Fe-SoxR to monothiols, including the biological compound glutathione.[42] The reactions that destroy the SoxR [2Fe-2S] centers are complex ones that may involve thiyl radicals.[42] Unexpectedly, stripping away the [2Fe-2S] centers from SoxR does not affect its oligomeric state (still a dimer even at low concentration [43]) or its affinity for the *soxS* promoter.[39] Apo-SoxR can be reactivated enzymatically by incubation with iron and the sulfide-generating NifS protein,[44] while non-enzymatic reactivation by iron and sulfide is dramatically stimulated by dithiols, including the protein thioredoxin.[45] Thus, SoxR activity in cells may also be regulated to some degree, or under specific conditions, by thiol-mediated disassembly and reassembly of the critical

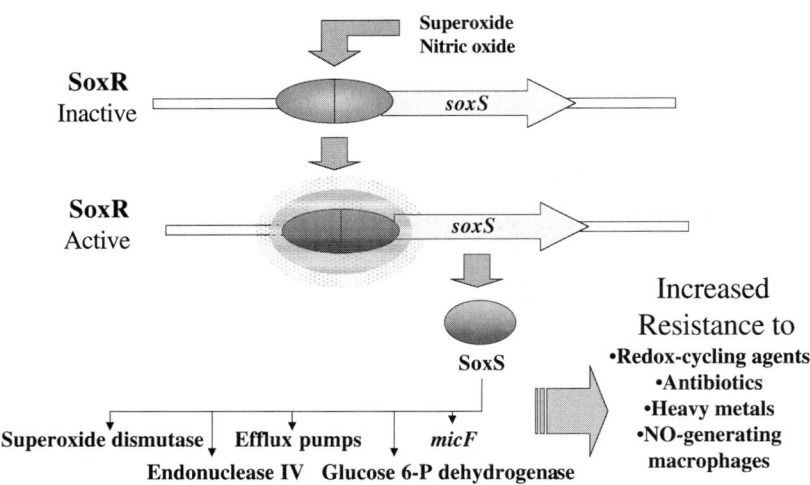

FIGURE 4. Regulation and functions of the *soxRS* regulon. SoxR in the inactive (reduced) state remains bound to the soxS promoter (*upper complex*). Oxidation or reaction with NO modifies the [2Fe-2S] centers of SoxR to activate transcription of *soxS* (*lower complex*). The resulting SoxS protein binds the promoters of regulon genes and activates their expression to give the multiple phenotypes associated with the *soxRS* regulon. We are grateful to Dr. Pablo Pomposiello for developing this figure.

iron-sulfur centers (FIG. 5). Some genetic experiments are consistent with this possibility.[42,45]

Although SoxR seems to be oxidized very readily upon extraction from the cell,[35,39] in cells without oxidative stress the metal centers are virtually entirely in the reduced state.[46–48] Upon exposure of the cells to paraquat, oxidation occurs rapidly (< 2 min) and is sustained as long as the oxidative stress is imposed.[48] However, upon removal of the oxidative stress, re-reduction also occurs within a few minutes.[48] The rapid re-reduction of oxidized SoxR in intact cells, as well as general principles,[49] suggest that SoxR is linked to cellular reducing donors such as NADPH, most likely by reductase enzymes (FIG. 5). A candidate SoxR reductase was recently identified.[50] Genetic studies will have to verify the significance of this or other SoxR reductase activities.

SoxS Protein and Defense Gene Induction

The 13-kD SoxS protein is related to the large AraC/XylS family of ~30-kD transcription regulators,[51] and a newer family typified by *E. coli* Rob protein and several other small, SoxS-like proteins.[52,53] Induction of the *soxRS* regulon depends on the *soxS* gene, and expression of SoxS from a plasmid independently of *soxR* activates all the genes and phenotypes associated with the *soxRS* system, even in the absence of oxidative stress.[54] The *soxR* gene is also required, but is not sufficient.[54,55] The

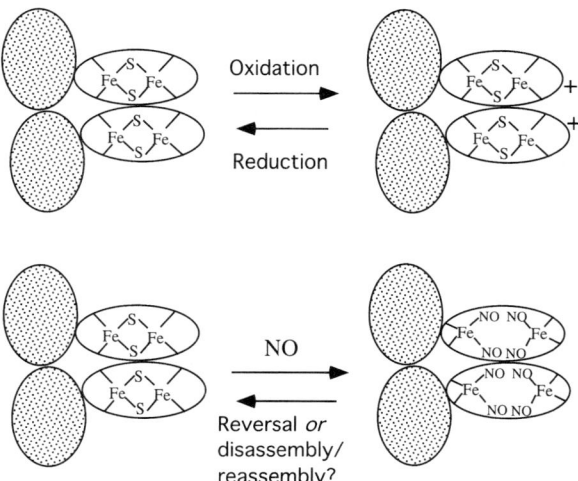

FIGURE 5. Regulation of SoxR activity. The two domains of SoxR are indicated as stippled ovals (DNA binding) and the [2Fe-2S] clusters; the protein is shown as a dimer. Oxidation (*upper scheme*) of the iron-sulfur centers activates SoxR, perhaps through the development of net positive charges, and reduction likely counteracts this process. Nitric oxide (*lower scheme*) reacts to form dinitrosyl-iron-dithiol complexes, and may be reversed either directly, or by disassembly and the reassembly of fresh clusters.

action of SoxS is direct: the purified protein binds the promoter regions of the target genes, enhancing the binding of RNA polymerase[56] and thus stimulating transcription.[57] Some information has accrued on the sequence specificity for SoxS binding,[53,58] but degeneracy in the profile has so far made the prediction of SoxS binding sites difficult.[59] A further complication is the overlap in specificity among SoxS and the homologous MarA and Rob proteins.[52,53,60,61] This overlap implies that the different regulatory systems may form a complicated network of regulation.

The *soxRS* regulon orchestrates the induction of diverse classes of genes.[5] Many of these genes are related to antioxidant defense: *sodA* (Mn-containing superoxide dismutase); *nfo* (DNA repair endonuclease IV); *zwf* (glucose-6-phosphate dehydrogenase); *fpr* (ferredoxin:NADPH oxidoreductase); *ribA* (GTP cyclohydrolase II[62]); and *nfsA* (nitroreductase A[63]). Other genes appear geared to maintaining central metabolism in the face of oxidative damage, such as *fumC* (oxidation-resistant fumarase) and *acnA* (aconitase).[5]

We were surprised to find that *soxR*-constitutive strains display multiple antibiotic resistance.[64] In fact, paraquat induces anitbiotic resistance via *soxRS* in *E. coli*[65] and *S. typhimurium*.[38,66] We now understand that this phenotype depends on the same resistance genes as the *marRAB*-regulated antibiotic resistance[65,67]: *micF* (antisense RNA that down-regulates expression of the OmpF porin); *acrAB* (efflux pump that eliminates many compounds from the cell). Undoubtedly other *soxRS*-regulated genes remain to be found.

Eukaryotic Cells

Genetic systems that govern gene expression in response to superoxide-generating agents are only beginning to be identified. Several groups have shown that yeast homologs of the mammalian AP-1 transcription factor (Yap1 and Yap2 proteins) contribute to defense against oxidative stress in *S. cerevisiae* and *Schizosacchar-mocyes pombe* through the induction of the glutathione reductase and other defense functions.[68,69] Physiological experiments demonstrate a differential response of *S. cerevisiae* to H_2O_2 and to menadione, a redox-cycling agent, which suggests the existence of more than one pathway for signaling and counteracting oxidative stress in these model eukaryotes.[68,69] Still missing are the biochemical mechanisms to explain the sensing pathways that govern oxidative stress gene induction in yeast; however, increased interest is being focused on this problem, and it can be hoped that this level of understanding will soon be reached.

CELLULAR RESPONSES TO NITRIC OXIDE

Nitric oxide (NO) is produced biologically in a variety of contexts. Several NO synthase enzymes are responsible for this NO production, and they are regulated in distinct fashions[70] for intercellular signaling molecule at low (nanomolar) levels[71] or at higher levels as a cytotoxic weapon of macrophages and other cells activated by inflammatory pathways.[72] In response to the toxicity of NO,[72,73] cells have evolved defenses to counteract it.

NO and the soxRS System

Although *soxRS* is not activated by extracellular $\cdot O_2^-$, exposure to pure NO induces the system.[74] This activation is actually enhanced under anaerobic conditions,[74,75] which rules out a role for adventitiously generated $\cdot O_2^-$ or any O_2-dependent products of NO (nitrite, peroxynitrite, etc.). The *soxRS* response to NO has a physiological role: *soxRS* enhances the survival of *E. coli* within NO-generating murine macrophages,[74] with kinetics that parallel perfectly the activation of *soxRS* in the phagocytosed cells.[75]

The mechanism by which NO activates SoxR has recently become clear. It was formally possible that NO somehow oxidizes the [2Fe-2S] centers of SoxR, but the chemistry of such a process is unknown, and an effort to detect the products of such an oxidation by mass spectrometry yielded negative results (E. Hidalgo, H. Ding, P. Wishnok, S.R. Tannenbaum, and B.Demple, unpublished data). We applied electron paramagnetic (epr) spectroscopy of intact *E. coli* to investigate the state of SoxR in NO-treated cells; such an approach successfully demonstrated the *in vivo* oxidation and re-reduction of SoxR in paraquat-treated cells.[48] With NO-treatment of *E. coli*, the epr signal for reduced SoxR [2Fe-2S] centers ($g = 1.92$) disappeared and was replaced by a new signal at $g = 2.03$.[76] The $g = 2.03$ signal was nearly identical to the epr signature of iron-sulfur centers modified by NO to form dinitrosyl-iron-dithiol adducts (FIG. 5). Moreover, the dinitrosyl-iron-dithiol signal disappeared rapidly after the NO-treatment and was replaced by the $g = 1.93$ signal for unmodified, reduced SoxR. Transcription of *soxS* paralleled these changes exactly, with the highest rate of synthesis when the highest amount of NO-SoxR was present.[76]

Dinitrosyl-iron-dithiol centers were formed in purified SoxR treated *in vitro* with pure NO gas (dissolved anaerobically in water). Analysis by epr spectroscopy showed that no unmodified [2Fe-2S] centers remained in these samples.[76] These modified centers *in vitro* were very stable, allowing purification of NO-modified SoxR, and withstanding repeated rounds of partial reduction and reoxidation. Most surprisingly, the *in vitro* activity of NO-modified SoxR in stimulating transcription of the *soxS* gene was, on a molar basis, equivalent to SoxR with oxidized [2Fe-2S] centers,[76] the "gold standard" for SoxR activity.[37,40]

As can be seen from the foregoing, a modification that disrupts the essential iron-sulfur centers of SoxR (nitrosylation) gives evidently full biological activity. Thus, at least two different forms of SoxR (oxidized or nitrosylated) exert the same transcription-activating effect. Although the mechanism by which SoxR stimulates RNA polymerase at the *soxS* gene is not known in detail, the strongest current hypothesis is that the activated protein overcomes overlong spacing of the promoter elements by deforming the DNA structure.[34,41] This allosteric transition seems to be supported as well by nitrosylation as it is by oxidation.

The stability of the dinitrosyl-iron-dithiol clusters in purified, NO-treated SoxR *in vitro* contrasts with their rapid disappearance in *E. coli* after exposure to NO.[76] Along with reports of turnover of dinitrosyl-iron-dithiol centers in NO-treated mammalian cells,[77] this observation suggests that active pathways eliminate these modifications *in vivo*. The ability to follow this process in SoxR specifically, and its regulatory effects, may provide a window to dissecting the biochemistry of this process.

Gene Induction and Adaptive Resistance to NO

Although mammalian cells actively produce NO for various functions and at varying levels, mechanisms that help them avoid damage from this free radical are largely unknown. We decided to address the nature of possible defenses against endogenous NO damage by examining whether specific proteins and genes are induced by sub-toxic levels of NO, and whether such induction affects cellular NO resistance.

We initially examined protein induction in IMR90 fetal lung fibroblasts. These are non-transformed cells, and fibroblasts are likely to be faced with changing levels of NO at sites of inflammation. We used pure NO, delivered through a gas-permeable tubing,[73] in order avoid possible confounding by-products associated with NO-releasing compounds. Analysis by two-dimensional gel electrophoresis showed that a 90-min treatment with NO delivered to the growth medium at ~185 nM/sec consistently induced at least 12 proteins and repressed eight others.[78] Among the inducible proteins was a prominent 32-kD species, which we confirmed as the well-known oxidative stress protein heme oxygenase 1 (HO1)[79] by immunoblotting and northern blot analysis (see below). Another NO-inducible protein was the MAP kinase phosphatase CL-100 (a.k.a., MKP-1).[80] Other NO-inducible proteins remain to be indentified; however, the mitochondrial Mn-containinng SOD and the nuclear DNA repair protein Ape1 were not induced in NO-treated cells.[78]

Analysis of the HO1 mRNA in northern blots showed that this transcript was induced 30- to 70-fold in cells exposed to NO at 150–300 nM/sec.[78] The kinetics of this increase were such that the level of HO1 mRNA was maximal at 3–5 h after an NO exposure as described above. However, by decreasing the NO flux to ~35 nM/sec sustained expression of the HO1 message could be obtained for up to 8 h.[78] Unexpectedly, the increase in HO1 mRNA in IMR90 cells is controlled mainly at the post-transcriptional level, by a dramatic increase in the stability of the RNA following NO treatment.[78] Consistent with this observation, the induction of HO1 appeared to be independent of the NO-activated cyclic GMP pathway,[78] which controls some transcriptional regulators.[71] It will be of great interest to find the proteins that mediate this change in RNA stability and the mechanism by which NO regulates them.

Exposure of rat hepatocytes[81] and some other cell types to NO-releasing compounds was reported to increase their resistance to these same compounds. However, the role of NO in studies of this type was not always clear. We addressed this question using pure NO, and testing terminally differentiated cells that would be at risk for damage from NO generated for both signaling and inflammatory functions. The mouse cell line NSC34 differentiates in culture to form motor neuron cells, and we found that these neurons were about 10-fold more sensitive to NO than were the IMR90 fibroblasts.[82] However, the resistance of these cells could be dramatically

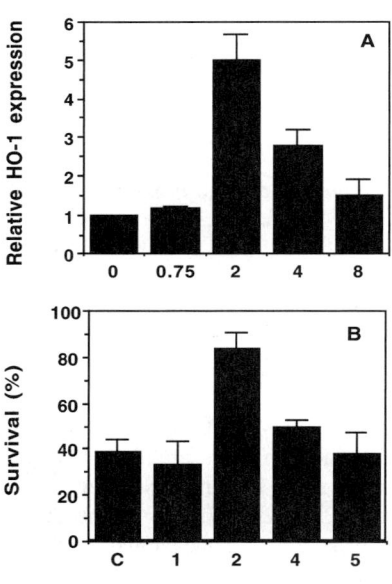

FIGURE 6. Kinetics of HO1 induction and adaptive NO resistance in NSC34 cells. NSC34 motor neurons were exposed for 60 min to NO at 25 nM/sec, then incubated for the times indicated before challenging with 277 nM/sec NO for 60 min.[82] Control cells (indicated by "C") were pretreated with argon gas instead of NO. **A:** Induction of HO1 mRNA determined by northern blotting. **B:** Cellular survival assessed by trypan blue exclusion.

increased by exposure to NO at 25 nM/sec for 60 min, followed by a 2-h incubation before a challenge with 250 nM/sec NO. This inducible "adaptive NO resistance" was accompanied by increased HO1 mRNA and protein expression, and the kinetics of HO1 induction paralleled the kinetics of the induced resistance (FIG. 6).[82] A pivotal role for HO1 in the inducible resistance was demonstrated more directly by using the enzyme inhibitor tin-protoporphyrin IX, which eliminated the adaptive resistance and sensitized the cells greatly to killing by NO.[82] There was no evidence that cyclic GMP is involved in adaptive NO resistance.[82]

The biochemical role of HO1 in NO resistance remains to be established. Our recent studies support the importance of this enzyme: spinal cord motor neurons from HO1-deficient "knockout" mice are extraordinarily sensitive to NO (A. Bishop, A.M. Lee, and B.Demple, in preparation). The enzyme catabolizes heme to release iron, CO, and biliverdin; reduction of biliverdin to bilirubin generates an antioxidant.[83] It has been argued that the Fe release may have a secondary effect elevating ferritin levels to sequester iron and prevent oxidative damage.[79] CO could also have secondary regulatory effects. Finally, the possibility has certainly not been eliminated that the *removal* of heme is somehow protective for the cell. Establishing the function of HO1 that contributes to NO resistance is an important goal.

It is likely that the constellation of genes and proteins induced by subtoxic NO treatment is much greater than the few identified by relatively crude methods so far. The adaptive NO resistance undoubtedly involves multiple biochemical and cellular functions, and almost certainly multiple regulatory systems. Defining the nature of this array of responsive genes will help us understand which NO-mediated injuries have the most biological significance, and what mechanisms have evolved to counteract them.

REFERENCES

1. CHANCE, B. 1952. The state of catalase in the respiring bacterial cells. Science **116:** 202-3
2. CLAYTON, R. K. 1960. An intermediate stage in the H2O2-induced synthesis of catalase in Rhodopseudomonas spheroides. J. Cell. Comp. Physiol. **55:** 1–8.
3. GONZÁLEZ-FLECHA, B. & B. DEMPLE. 1995. Metabolic sources of hydrogen peroxide in aerobically growing *Escherichia coli*. J. Biol. Chem. **270:** 13681–13687.
4. GONZÁLEZ- FLECHA, B. & B. DEMPLE. 1997. Homeostatic regulation of intracellular hydrogen peroxide concentration in aerobically growing *Escherichia coli*. J. Bacteriol. **179:** 382–388.
5. HIDALGO, E. & B. DEMPLE. 1996. *In* Regulation of Gene Expression in *Escherichia coli*. E.C.C. Lin & A.S. Lynch, Eds. :435–452. R.G. Landes. Austin, TX
6. DEMPLE, B. & J. HALBROOK. 1983. Inducible repair of oxidative DNA damage in *Escherichia coli*. Nature **304:** 466–468.
7. STORZ, G., L.A. TARTAGLIA & B.N. AMES. 1990. Transcriptional regulator of oxidative stress-inducible genes: direct activation by oxidation. Science **248:** 189–194.
8. DEMPLE, B. & C.F. AMÁBILE-CUEVAS. 1991. Redox redux: the control of oxidative stress responses. Cell **67:** 837–839.
9. GONZÁLEZ-FLECHA, B. & B. DEMPLE. 1997. Transcriptional regulation of the Escherichia coli oxyR gene as a function of cell growth. J. Bacteriol. **179:** 6181–6186.
10. JISHAGE, D.E. & A. ISHIHAMA. 1995. Regulation of RNA polymerase sigma subunit synthesis in Escherichia coli: intracellular levels of sigma 70 and sigma 38. J. Bacteriol. **177:** 6832–6835.

11. ALTUVIA, S., D. WEINSTEIN-FISCHER, A. ZHANG, L. POSTOW & G. STORZ. 1997. A small, stable RNA induced by oxidative stress: role as a pleiotropic regulator and antimutator. Cell **90:**
12. GONZÁLEZ-FLECHA, B. & B. DEMPLE. 1999. Role of the oxyS gene in the regulation of intracellular hydrogen peroxide in *Escherichia coli*. J. Bacteriol. **181:** 3833–3836.
13. BOVERIS, A. & B. CHANCE. 1973. The mitochondrial generation of hydrogen peroxide. General properties and effect of hyperbaric oxygen. Biochem. J. **134:** 707–716.
14. CHANCE, B., H. SIES & A. BOVERIS. 1979. Hydroperoxide metabolism in mammalian organs. Physiol. Rev. **59:** 527–605.
15. BOVERIS, A., & E. CADENAS. 1982. *Production of superoxide anion and hydrogen peroxide in mitochondria*. Superoxide Dismutase (Oberley, L. W., Ed.), II, CRC Press, Boca Raton, FL.
16. GONZÁLEZ-FLECHA, B., J.C. CUTRIN & A. BOVERIS. 1993. Time course and mechanism of oxidative stress and tissue damage in rat liver subjected to in vivo ischemia-reperfusion. J. Clin. Invest. **91:** 456–464.
17. GONZÁLEZ- FLECHA, B. & A. BOVERIS. 1995. Mitochondrial sites of hydrogen peroxide production in reperfused rat kidney cortex. Biochim. Biophys. Acta **1243:** 361–366.
18. GONZÁLEZ-FLECHA, B., P. EVELSON, N. STERIN-SPEZIALE & A. BOVERIS. 1993. Hydrogen peroxide metabolism and oxidative stress in cortical, medullary and papillary zones of rat kidney. Biochim. Biophys. Acta **1157:** 155–161.
19. APERIA, A. C., & A. A. LIEBOW. 1964. Am. J. Physiol. **206:** 499–504.
20. ROSS, B.D., J. ESPINAL & P. SILVA. 1986. Glucose metabolism in renal tubular function. Kidney International **29:** 54–67.
21. HIGGINS, E.S., H. SEIBEL, W. FRIEND & K.S. ROGERS. 1978. Heterogeneity of renal mitochondria of the rat. Proc. Soc. Exp. Biol. Med. **158:** 595–598.
22. CANAVESE, C., P. STRATTA & A. VERCELLONE. 1988. The case for oxygen free radicals in the pathogenesis of ischemic acute renal failure. Nephron **49:** 9–15.
23. BRATELL, S., P. FOLMERZ, R. HANSSON, O. JONSSON, S. LUNDSTAM, S. PETTERSSON, B. RIPPE & T. SCHERSTEN. 1988. Effects of oxygen free radical scavengers, xanthine oxidase inhibition and calcium entry-blockers on leakage of albumin after ischaemia. An experimental study in rabbit kidneys. Acta Physiol. Scand. **134:** 35–41.
24. ASIKAINEN, T.M., K.O. RAIVIO, M. SAKSELA & V.L. KINNULA. 1998. Expression and developmental profile of antioxidant enzymes in human lung and liver. Am. J. Respir. Cell & Mol. Biol. **19:** 942–949.
25. SHULL, S., N. H. HEINTZ, M. PERIASAMY, M. MANOHAR, Y.N. JANSSEN, J.P. MARSH & B.T. MOSSMAN. 1991. Differential regualtion of antioxidant enzymes in response to oxidants. Journal of Biologicla Chemistry **266:** 24398–24403.
26. HO, Y.S., M.S. DEY & J.D. CRAPO. 1996. Antioxidant enzyme expression in rat lungs during hyperoxia. Am. J. Physiol. **270:** L810–L818.
27. ISOHERRANEN, K., V. PELTOLA, L. LAURIKAINEN, J. PUNNONEN, J. LAIHIA, M. AHOTUPA & K. PUNNONEN. 1997. Regulation of copper/zinc and manganese superoxide dismutase by UVB irradiation, oxidative stress and cytokines. J. Photochem. Photobiol. **40:** 288–293.
28. ANTRAS-FERRY, J., K. MAHEO, F. MOREL, A. GUILLOUZO, P. CILLARD & J. CILLARD. 1997. Dexamethasone differently modulates TNF-alpha– and IL-1beta–induced transcription of the hepatic Mn-superoxide dismutase gene. FEBS Lett. **403:** 100–104.
29. GUPTA, M., K. DOBASHI, E.L. GREENE, J.K. ORAK & I. SINGH. 1997. Studies on hepatic injury and antioxidant enzyme activities in rat subcellular organelles following in vivo ischemia and reperfusion. Mol. Cell. Biochem. **176:** 337–347.
30. LISSI, E., L. VIDELA, B. GONZALEZ-FLECHA, C. GIULIVI & A. BOVERIS. 1991. *In* Oxidative Damage and Repair. K.J.A. Davies, Ed. :444–448. Pergamon Press. Oxford.
31. FRIDOVICH, I. 1983. Superoxide radical: an endogenous toxicant. Annu. Rev. Pharmacol. Toxico.1 **23:** 239–257.
32. KAPPUS, H. & H. SIES. 1981. Toxic drug effects associated with oxygen metabolism: redox cycling and lipid peroxidation. Experientia **37:** 1233–1241.

33. HIDALGO, E., H. DING & B. DEMPLE. 1997. Redox signal transduction: mutations shifting [2Fe-2S] centers of the SoxR sensor-regulator to the oxidized form. Cell **88**: 121–129.
34. HIDALGO, E., J.M. BOLLINGER, JR., T.M. BRADLEY, C.T. WALSH & B. DEMPLE. 1995. Binuclear [2Fe-2S] clusters in the *Escherichia coli* SoxR protein and role of the metal centers in transcription. J. Biol. Chem. **270**: 20908–20914.
35. WU, J., W.R. DUNHAM & B. WEISS. 1995. Overproduction and physical characterization of SoxR, a [2Fe-2S] protein that governs an oxidative response regulon in *Escherichia coli*. J. Biol. Chem. **270**: 10323–10327.
36. DING, H., E. HIDALGO & B. DEMPLE. 1996. The redox state of the [2Fe-2S] clusters in SoxR protein regulates its activity as a transcription factor. J. Biol. Chem. **271**: 33173–33175.
37. GAUDU, P. & B. WEISS. 1996. SoxR, a [2Fe-2S] transcription factor, is active only in its oxidized form. Proc. Natl. Acad. Sci. USA **93**: 10094–10098.
38. MARTINS, E.A., A. KOUTSOLIOUTSOU & B. DEMPLE. The *soxRS* system of enteric pathogens: multiple antibiotic resistance in a clinical *Salmonella typhimurium* infection via a *soxR*-constitutive mutation. Submitted.
39. HIDALGO, E. & B. DEMPLE. 1994. An iron-sulfur center essential for transcriptional activation by the redox-sensing SoxR protein. EMBO Journal **13**: 138–146.
40. DING, H., E. HIDALGO & B. DEMPLE. 1996. The redox state of the [2Fe-2S] clusters in SoxR protein regulates its activity as a transcription factor. J. Biol. Chem. **271**: 33173–33175.
41. HIDALGO, E. & B. DEMPLE. 1997. Spacing of promoter elements regulates the basal expression of the soxS gene and converts SoxR from a transcriptional activator into a repressor. EMBO J. **16**: 1056–1065.
42. DING, H. & B. DEMPLE. 1996. Glutathione-mediated destabilization in vitro of [2Fe-2S] centers in the SoxR regulatory protein. Proc. Natl. Acad. Sci. USA **93**: 9449–9453.
43. BRADLEY, T.M., E. HIDALGO, V.S. LEAUTAUD, H. DING & B. DEMPLE. 1997. Cysteine-to-alanine replacements in the *Escherichia coli* SoxR protein and the role of the [2Fe-2S] centers in transcriptional activation. Nucleic Acids Res. **25**: 1469–1475.
44. HIDALGO, E. & B. DEMPLE. 1996. Activation of SoxR-dependent transcription in vitro by noncatalytic or NifS-mediated assembly of [2Fe-2S] clusters into apo-SoxR. J. Biol. Chem. **271**: 7269–7272.
45. DING, H. & B. DEMPLE. 1998. Thiol-mediated disassembly and reassembly of [2Fe-2S] clusters in the redox-regulated transcription factor SoxR. Biochemistry **37**: 17280–17286.
46. HIDALGO, E., H. DING & B. DEMPLE. 1997. Redox signal transduction via iron-sulfur clusters in the SoxR transcription activator. Trends Biochem Sci **22**: 207–210.
47. GAUDU, P., N. MOON & B. WEISS. 1997. Regulation of the soxRS oxidative stress regulon. Reversible oxidation of the Fe-S centers of SoxR in vivo. J. Biol. Chem. **272**: 5082–5086.
48. DING, H. & B. DEMPLE. 1997. In vivo kinetics of a redox-regulated transcriptional switch. Proc. Natl. Acad. Sci. **94**: 8445–8449.
49. LIOCHEV, S.I. & I. FRIDOVICH. 1992. Fumarase C, the stable fumarase of *Escherichia coli*, is controlled by the soxRS regulon. Proc. Natl. Acad. Sci. USA **89**: 5892–5896.
50. KOBAYASHI, K. & S. TAGAWA. 1999. Isolation of reductase for SoxR that governs an oxidative response regulon from *Escherichia coli*. FEBS Lett. **451**: 227–230.
51. GALLEGOS, M.T., R. SCHLEIF, A. BAIROCH, K. HOFMANN & J.L. RAMOS. 1997. Arac/XylS family of transcriptional regulators. Microbiol. Mol. Biol. Rev. **61**: 393–410.
52. ARIZA, R.R., Z. LI, N. RINGSTAD & B. DEMPLE. 1995. Activation of multiple antibiotic resistance and binding of stress-inducible promoters by *Escherichia coli* Rob protein. J. Bacteriol. **177**: 1655–1661.
53. LI, Z.Y. & B. DEMPLE. 1996. Sequence specificity for DNA binding by *Escherichia coli* SoxS and Rob proteins. Mol. Microbiol. **20**: 937–945.
54. AMÁBILE-CUEVAS, C.F. & B. DEMPLE. 1991. Molecular characterization of the soxRS genes of *Escherichia coli*: two genes control a superoxide stress regulon. Nucleic Acids Res. **19**: 4479–4484.

55. WU, J. & B. WEISS. 1991. Two divergently transcribed genes, soxR and soxS, control a superoxide response regulon of *Escherichia coli*. J. Bacteriol. **173:** 2864–2871.
56. LI, Z. & B. DEMPLE. 1994. SoxS, an activator of superoxide stress genes in *Escherichia coli*. Purification and interaction with DNA. J. Biol. Chem. **269:** 18371–18377.
57. JAIR, K.W., W.P. FAWCETT, N. FUJITA, A. ISHIHAMA & R.E.J. WOLF. 1996. Ambidextrous transcriptional activation by SoxS: requirement for the C-terminal domain of the RNA polymerase alpha subunit in a subset of the *Escherichia coli* superoxide-inducible genes. Mol. Microbiol. **19:** 306–317.
58. FAWCETT, W.P. & R.E. WOLF, JR. 1995. Genetic definition of the *Escherichia coli* zwf "soxbox," the DNA binding site for SoxS-mediated induction of glucose 6-phosphate dehydrogenase in response to superoxide. J. Bacteriol. **177:** 1742–1750.
59. ROTH, F.P., J.D. HUGHES, P.W. ESTEP & G.M. CHURCH. 1998. Finding DNA regulatory motifs within unaligned noncoding sequences clustered by whole-genome mRNA quantitation [see comments]. Natl. Biotechnol. **16:** 939–945.
60. GREENBERG, J.T., J.H. CHOU, P.A. MONACH & B. DEMPLE. 1991. Activation of oxidative stress genes by mutations at the soxQ/cfxB/marA locus of *Escherichia coli*. J. Bacteriol. **173:** 4433–4439.
61. MILLER, E.K. & I. FRIDOVICH. 1986. A demonstration that O_2^- is a crucial intermediate in the high quantum yield luminescence of luminol. J. Free Rad. Biol. Med. **2:** 107–110.
62. KOH, Y.S., J. CHOIH, J.H. LEE & J.H. ROE. 1996. Regulation of the *ribA* gene encoding GTP cyclohydrolase II by the *soxRS* locus in *Escherichia coli*. Mol. Gen. Genet. **251:** 591–598.
63. LIOCHEV, S.I., A. HAUSLADEN & I. FRIDOVICH. 1999. Nitroreductase A is regulated as a member of the *soxRS* regulon of *Escherichia coli*. Proc. Natl. Acad. Sci. USA **96:** 3537–3539.
64. GREENBERG, J.T., P. MONACH, J.H. CHOU, P.D. JOSEPHY & B. DEMPLE. 1990. Positive control of a global antioxidant defense regulon activated by superoxide-generating agents in Escherichia coli. Proc. Natl. Acad. Sci. USA **87:** 6181–6185.
65. CHOU, J.H., J.T. GREENBERG & B. DEMPLE. 1993. Posttranscriptional repression of Escherichia coli OmpF protein in response to redox stress: positive control of the micF antisense RNA by the soxRS locus. J. Bacteriol. **175:** 1026–1031.
66. MARTINS, E.A., P.J. POMPOSIELLO & B. DEMPLE. Molecular characterization of the *soxRS* regulon of *Salmonella typhimurium*. Submitted.
67. MILLER, P.F. & M.C. SULAVIK. 1996. Overlaps and parallels in the regulation of intrinsic multiple-antibiotic resistance in *Escherichia coli*. Mol. Microbiol. **21:** 441–448.
68. AMIESON, D.J. 1998. Oxidative stress responses of the yeast *Saccharomyces cerevisiae*. Yeast **14:** 1511–1527.
69. TOONE, W.M. & N. JONES. 1998. Stress-activated signalling pathways in yeast. Genes Cells **3:** 485–498.
70. MARLETTA, M.A. 1993. Nitric oxide synthase: function and mechanism. Advances in Exp. Med. Biol. **338:** 281–284.
71. HOBBS, A.J. & L.J. IGNARRO. 1996. Nitric oxide-cyclic GMP signal transduction system. Methods in Enzymol. **269:** 134–148.
72. MACMICKING, J.,Q. W. XIE & C. NATHAN. 1997. Nitric oxide and macrophage function. Annu. Rev. Immunol. **15:** 323–350.
73. TAMIR, S., T. DEROJAS-WALKER, J.S. WISHNOK & S.R. TANNENBAUM. 1996. DNA damage and genotoxicity by nitric oxide. Methods in Enzymol. **269:** 230–243.
74. NUNOSHIBA, T. & B. DEMPLE. 1993. Potent intracellular oxidative stress exerted by the carcinogen 4-nitroquinoline-N-oxide. Cancer Res. **53:** 3250–3252.
75. NUNOSHIBA, T., T. DEROJAS-WALKER, S.R. TANNENBAUM & B. DEMPLE. 1995. Roles of nitric oxide in inducible resistance of *Escherichia coli* to activated murine macrophages. Infection & Immunity **63:** 794–798.
76. DING, H. & B. DEMPLE. Direct nitric oxide signal transduction via nitrosylation of iron-sulfur centers in the SoxR transcription activator. Submitted.

77. MULSCH, A., P.I. MORDVINTCEV, A.F. VANIN & R. BUSSE. 1993. Formation and release of dinitrosyl iron complexes by endothelial cells. Biochem. Biophys. Res. Commun. **196:** 1303–1308.
78. MARQUIS, J.C. & B. DEMPLE. 1998. Complex genetic response of human cells to sublethal levels of pure nitric oxide. Cancer Res **58:** 3435–3440.
79. TYRRELL, R.M. 1997. Approaches to define pathways of redox regulation of a eukaryotic gene: the heme oxygenase 1 example. Methods **11:** 313–318.
80. ALESSI, D.R., C. SMYTHE & S.M. KEYSE. 1993. The human CL100 gene encodes a Tyr/Thr-protein phosphatase which potently and specifically inactivates MAP kinase and suppresses its activation by oncogenic ras in Xenopus oocyte extracts. Oncogene **8:** 2015–2020.
81. KIM, H.Y., R.D. KLAUSNER & T.A. ROUAULT. 1995. Translational repressor activity is equivalent and is quantitatively predicted by in vitro RNA binding for two iron-responsive element-binding proteins, IRP1 and IRP2. J. Biol. Chem. **270:** 4983–4986.
82. BISHOP, A., J.C. MARQUIS, N.R. CASHMAN & B. DEMPLE. 1999. Adaptive resistance to nitric oxide in motor neurons. Free Rad. Biol. Med. **26:** 978–986.
83. STOCKER, R., Y. YAMAMOTO, A.F. MCDONAGH, A.N. GLAZER & B. N. AMES. 1987. Bilirubin is an antioxidant of possible physiological importance. Science **235:** 1043–1046.
84. GIULIVI, C., J.F. TURRENS & A. BOVERIS. 1988. Chemiluminescence enhancement by trypanocidal drugs and by inhibitors of antioxidant enzymes in *Trypanosoma cruzi*. Mol. Biochem. Parasitol. **30:** 243–251.
85. PUNTARULO, S., M. GALLEANO, R.A. SANCHEZ & A. BOVERIS. 1991. Superoxide anion and hydrogen peroxide metabolism in soybean embryonic axes during germination. Biochim. Biophys. Acta **1074:** 277–283.
86. OSHINO, N., B. CHANCE, H. SIES & T. BUCHER. 1973. The role of H_2O_2 generation in perfused rat liver and the reaction of catalase compound I and hydrogen donors. Arch. Biochem. Biophys. **154:** 117–131.
87. GONZÁLEZ-FLECHA, B., P. EVELSON, K. RIDGE & J.I. SZNAJDER. 1996. Hydrogen peroxide increases $Na^+/K^{(+)}$-ATPase function in alveolar type II cells. Biochim. Biophysi. Acta **1290:** 46–52.
88. BUSTAMANTE, J., L. GUERRA, L. BREDESTON, J. MORDOH & A. BOVERIS. 1991. Melanin content and hydroperoxide metabolism in human melanoma cells. Exp. Cell Res. **196:** 172–176.

Mapping Oxidative DNA Damage and Mechanisms of Repair

STEVEN A. AKMAN,[a] TIMOTHY R. O'CONNOR,[b] AND HENRY RODRIGUEZ[c,d]

[a]*Department of Cancer Biology, Comprehensive Cancer Center of Wake Forest University, Winston-Salem, North Carolina 27157, USA*

[b]*Department of Biology, Beckman Research Institute of the City of Hope, Duarte, California 91010, USA*

[c]*Biotechnology Division, National Institute of Standards and Technology, Gaithersburg, Maryland 20899-8311, USA*

ABSTRACT: We developed a method to map oxidative-induced DNA damage at the nucleotide level using ligation-mediated polymerase chain reaction (LMPCR) technology. *In vivo* and *in vitro* DNA base modification patterns inflicted by reactive oxygen species (ROS) in the human *P53* and *PGK1* gene were nearly identical *in vitro* and *in vivo*. In human male fibroblasts, these patterns are independent of the transition metal used (Cu (II), Fe(II), or Cr(VI)). Therefore, local probability of H_2O_2-mediated DNA base damage is determined primarily by DNA sequence. Moreover, in cells undergoing severe oxidative stress, extranuclear sites contribute metals that enhance nuclear DNA damage. The role of the base excision repair pathway in human cells responsible for the repair of the majority of ROS base damage is also discussed.

INTRODUCTION

DNA damage induced by ROS[e] is an important intermediate in the pathogenesis of human conditions such as cancer and aging.[1,2] ROS-induced DNA damage products are both mutagenic and cytotoxic.[3] One commonly studied ROS, the hydroxyl radical (·OH), is produced by H_2O_2 in the presence of transition metal ions. The mutational spectra of H_2O_2[4,5] and the transition metal ions Fe, Cu,[6,7] and Cr[8] have been studied in model systems, but the relationship of induced DNA damage to these spectra remained unknown.

Until recently, progress in this area has been hampered by the lack of damage measurement techniques with nucleotide resolution for studies in mammalian cells.

---

[d]Address for correspondence: Dr. Henry Rodriguez, Biotechnology Division, National Institute of Standards and Technology, 100 Bureau Drive, Stop 8311, Gaithersburg, MD 20899-8311. Voice: 301-975-2578;fax: 301-330-3447.

e-mail: henry.rodriguez@nist.gov; URL: http://www.nist.gov

[e]ABBREVIATIONS: µg, microgram; µl, microliter; *PGK1*, PhosphoGlycerate Kinase; kb, kilobase; CEE, continuous elution electrophoresis; LMPCR, ligation-mediated polymerase chain reaction; ROS, reactive oxygen species; H_2O_2, hydrogen peroxide; Fpg, *E. coli* FormamidoPyrimidine DNA glycosylase; Nth, *E. coli* Endonuclease III; BER, base excision repair; PCNA, Proliferating cell nuclear antigen; FEN1, Flap endonuclease I, DNase IV; FapyGua, formamidopyrimidoguanine; 8-oxoGua, 8-oxodeoxyguanine; APE protein, major AP endonuclease, HAP protein; POLβ, DNA polymerase β; dRPase, 5′deoxyribosephosphodiesterase.

LMPCR is a genomic sequencing method for mapping of rare DNA single-stranded breaks.[9,10] LMPCR is a six-step process[11] as shown in FIGURE 1A. The steps are:

1. conversion of a modified base (base damage) into a strand break, either chemically or enzymatically, followed by primer extension of an annealed gene-specific primer (upstream primer 1) to generate blunt ends;
2. ligation of a universal asymmetric double-strand linker onto the blunt ends;
3. PCR amplification using a second gene-specific primer (upstream primer 2) along with a linker primer (downstream linker primer);
4. separation of the DNA fragments on a sequencing polyacrylamide gel;
5. transfer of the DNA to a nylon membrane by electroblotting; and
6. hybridization of a radiolabeled probe.

By developing an oxidative DNA damage mapping version of LMPCR, we were able to map ROS-induced DNA base modifications sensitive to the DNA glycosylase and abasic lyase activities of the Nth and Fpg proteins.[11] Using this modified LMPCR technique, we investigated the *in vivo* and *in vitro* frequencies of DNA base modifications caused by ROS in the human *P53* and *PGK1* genes.[11–13]

Most oxidized base damage is repaired rapidly, within several hours of the initial damage via the BER system (FIG. 2).[14–18] The pathway followed depends on the DNA glycosylase excising an adduct.[19] These enzymes cleave the glycosylic bond tethering the base to the deoxyribose moiety. In the center of FIGURE 2, damages excised by a DNA glycosylase with only DNA glycosylase activity leave an abasic site in DNA that is incised by the APE protein.[20–21] POLβ is responsible for the removal of the deoxyribose 5′ phosphate and resynthesis prior to ligation.[22–23] Most repair of ROS base damage, however is initiated by DNA glycosylases with both DNA glycosylase and AP lyase activity as indicated on the left of FIGURE 2. This pathway requires the removal of a trans α,β-unsaturated aldose product following AP lyase action prior to resynthesis by POLβ.[24] In the two major BER pathways, a complex including XRCC1/DNA Ligase III/PARP completes repair.[25–27] A third pathway of BER (right hand side of FIG. 2) is PCNA dependent, and uses different enzymes from the other two pathways. This PCNA dependent pathway, however, is generally considered a minor pathway.

Using methods presented in this report, we probed the molecular mechanisms of sequence context dependent damage and repair in normal human cells.

MATERIALS AND METHODS

In Vivo H_2O_2 Treatment of Human Skin Fibroblasts

Human male foreskin fibroblasts were grown in 150 mm dishes to confluent monolayers in Dulbecco's Modified Eagle Medium containing 10% (v:v) fetal bovine serum. Fibroblasts were treated with serum-free Minimum Essential Medium with 1 mM sodium phosphate containing 50 mM H_2O_2 at 37°C for 30 minutes. After washing, cells were harvested and DNA was isolated as previously described.[28]

DNA was prepared for *in vitro* assay as previously described.[11] After phenol/chloroform extraction, DNA was precipitated in ethanol, redissolved in 10 mM

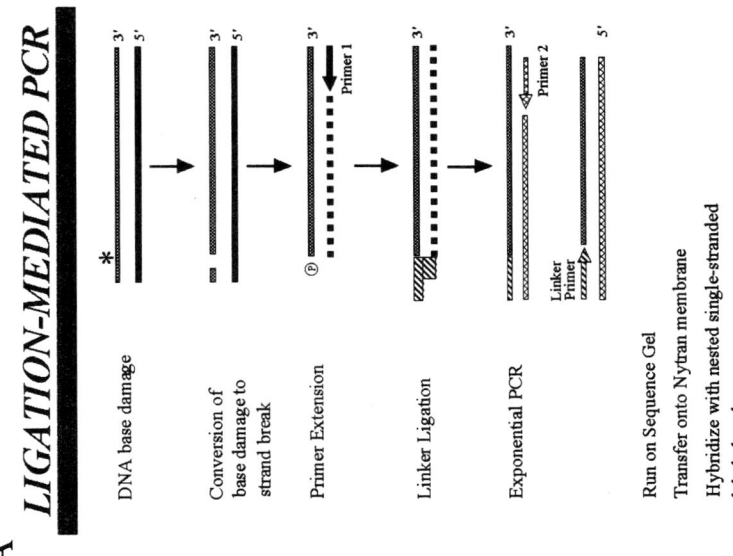

A LIGATION-MEDIATED PCR

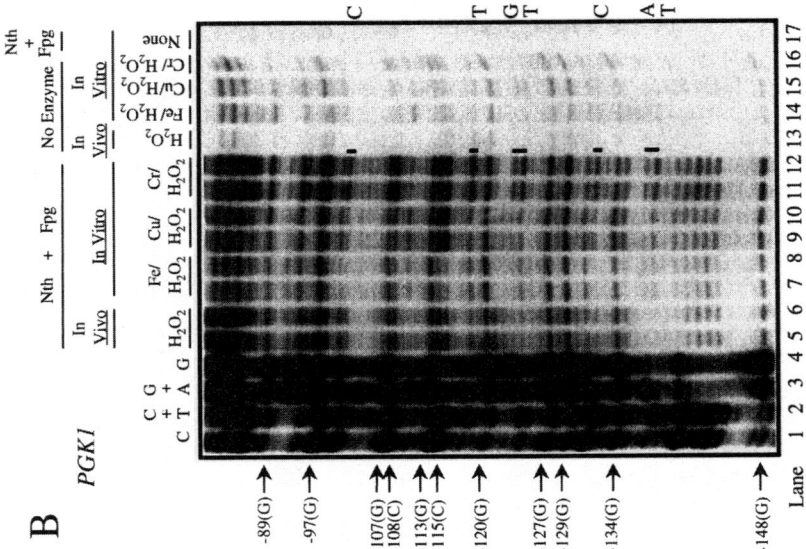

B *PGK1*

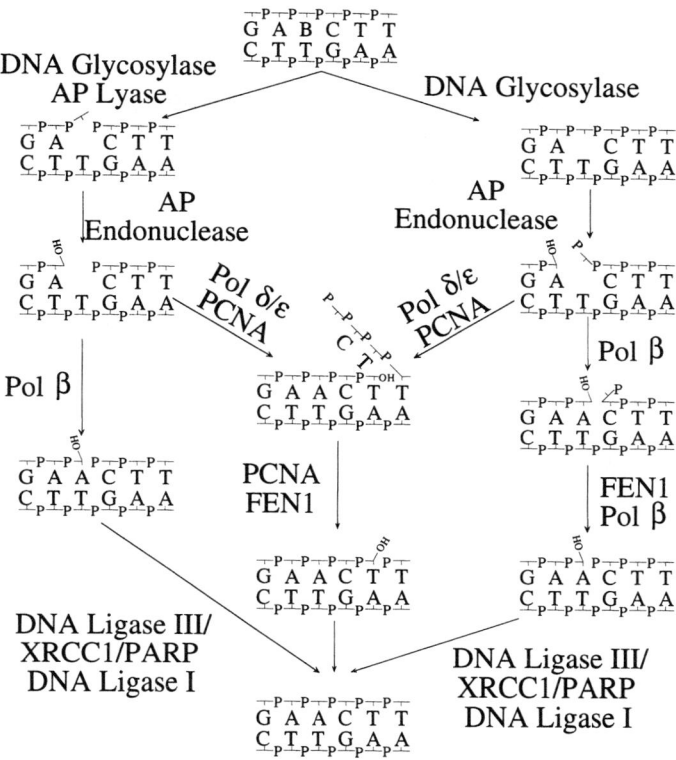

FIGURE 2. Base excision repair in mammalian cells. The left BER pathway repairs bases that are removed by DNA glycosylases with associated AP lyase activities. The center BER pathway repairs bases that are removed by DNA glycosylases alone. The right BER pathway is considered a minor pathway in mammalian cells, and is most probably associated with the repair of modified abasic sites not recognized by the APE protein.[44-46] Reproduced with permission.

FIGURE 1. A: Schematic representation of the steps in DNA base damage mapping by LMPCR. **B:** LMPCR analysis of damage induced in the promoter region of human *PGK1* using primer set A (transcribed strand). *Lanes 1–4*: DNA treated with standard Maxam-Gilbert cleavage reactions. *Lanes 5–6, 13*: DNA recovered from intact human foreskin fibroblasts exposed to 50 mM H_2O_2. *Lanes 7–8, 14*: dialyzed genomic DNA treated with 100 μM Fe(III)/100 μM ascorbate/5 mM H_2O_2 in the presence of 0.3 M sucrose. *Lanes 9–10, 15*: DNA treated with 50 μM Cu(II)/100 μM ascorbate/5 mM H_2O_2 in the presence of 1 mM potassium phosphate buffer. *Lanes 11–12, 16*: DNA treated with 100 μM Cr(VI)/100 μM ascorbate/5 mM H_2O_2. *Lane 17*: DNA incubated in potassium phosphate buffer and digested with Nth and Fpg proteins. The DNA in lanes 5–12 was digested with Nth and Fpg proteins after treatment; the DNA in lanes 13–16 was incubated in digestion buffer alone after treatment. Positions of high damage frequency bases are marked with arrows to the left of lane 1. The sequence of positions heavily damaged in the presence of chromium, but not copper or iron, is denoted by rectangles to the right of lane 12. Reproduced with permission.

HEPES, 1 mM EDTA, pH 7.4 at 70 µg/ml, then dialyzed against distilled water overnight at 4°C.

In Vitro *Metal Ion-Ascorbate-H_2O_2 Treatment*

Ten µg of dialyzed DNA dissolved in 161 µl of H_2O was incubated at room temperature for 30 min with 50 µM $CuCl_2$, 100 µM $FeCl_3$, or 100 µM $K_2Cr_2O_7$. Chelex® treated potassium phosphate, pH 7.5 (± 0.3 M sucrose), ascorbate, and H_2O_2 were added to final concentrations of 1 mM, 100 mM, and 5 mM, respectively. After 30 min at room temperature with gentle rocking, the reaction was quenched by the addition of EDTA to 2 mM, followed by precipitation of DNA in 0.3 M sodium acetate, pH 7.0 and 2 volumes of cold ethanol.

Isolation and Treatment of Fibroblast Nuclei

Human male foreskin fibroblasts were grown in 150 mm dishes to confluent monolayers in Dulbecco's Modified Eagle Medium containing 10% (v:v) fetal bovine serum. After removal of medium and washing with 25 ml of a 154 mM NaCl solution, cells were lysed by the addition of 10 ml buffer A (0.3 M sucrose, 60 mM KCl, 15 mM NaCl, 60 mM Tris-HCl, pH 7.4, 2 mM EDTA) containing 0.5% Nonidet-P40 (has now been replaced with IGEPAL-CA-620, ICN Biochemicals, Aurora, Ohio USA). Nuclei were collected by centrifugation at $1000 \times g$ for 10 min at room temperature, then gently resuspended in sucrose/phosphate buffer (1 mM potassium phosphate, pH 7.5, 60 mM KCl, 15 mM NaCl, 0.3 M sucrose). After 2 additional washes in sucrose/phosphate buffer, nuclei were exposed to 50 mM H_2O_2 ± 50 µM Cu(II) or 50 µM Fe(III) ± 100 µM ascorbate in sucrose/phosphate buffer for 30 min at 37°C. Reactions were quenched by the addition of EDTA or desferrioximine (for iron-containing samples) to 2 mM. Treated nuclei were collected by centrifugation, resuspended in 4 ml buffer A and DNA was isolated.[28]

Ligation-Mediated Polymerase Chain Reaction

Digestion of treated DNA with Nth and Fpg proteins,[11] fragment size analysis by glyoxal gel electrophoresis,[28] and the LMPCR technique[11] have been described in detail elsewhere. Key steps with recent enzyme modifications such as the incorporation of a hot-start by using AmpliTaq® Gold polymerase to provide better signal-to-noise ratio and cycling modifications are as follows:

Primer Extension

Primer 1 is extended in a siliconized 0.65 ml tube: a thermocycler (MJ Research Inc., Watertown, MA, USA) is used for all incubations. DNA (0.5–1.3 µg) is diluted in a volume of 15 µl of a solution containing 40 mM Tris-HCl, pH 7.7, 50 mM NaCl and 0.75 pmoles of Primer 1. DNA is denatured at 98°C for 3 min and the primer (Primer 1) annealed at 45°C for 30 min (primers for the *PGK1* housekeeping gene are in,[11] and *P53* gene in.[29] After cooling the sample to 4°C, 9 µl of the following mix is added: 7.5 µl of $MgCl_2$-dNTP mix (20 mM $MgCl_2$, 20 mM dithiothreitol and 0.25 mM of each dNTP), 1.1 µl dH_2O and 0.4 µl of Sequenase® 2.0 (13 U/µl, U.S. Biochemicals, Cleveland, Ohio USA). Samples are then incubated at 48°C for 15 min. The samples are placed on ice and 6 µl ice-cold 310 mM Tris-HCl, pH 7.7, is added. To inactivate Sequenase® samples are incubated at 67°C for 15 min and the samples are placed on ice.

Ligation

The primer-extended molecules that have a 5' phosphate are ligated to an unphosphorylated asymmetric double-stranded linker.[30] To each sample (consisting of 30 µl), 45 µl of the following ligation mix is added: 13.33 mM $MgCl_2$, 30 mM dithiothreitol, 1.7 mM ATP, 83.3 µg/ml BSA, 100 pmoles of linker and 5 Units of T_4 DNA ligase (5 Units/µl, Boehringer Mannheim, Gaithersburg, MD, USA). Samples are incubated overnight at 18°C. Ligase is inactivated by incubation at 70°C for 10 min. Samples are placed on ice. Next, 25 µl of 10 M ammonium acetate, 1 µl of 0.5 M EDTA, pH 8.0, 1 µl of 20 µg/µl glycogen, followed by 250 µl of ice-cold ethanol to precipitate the DNA. DNA pellets are redissolved in 50 µl of dH_2O.

PCR Amplification

50 µl of an AmpliTaq® Gold polymerase mix (2x AmpliTaq® Gold reaction buffer (Perkin Elmer Inc., Foster City, CA, USA), 1 mM $MgCl_2$, 400 µM of each dNTP, 10 pmoles of primer 2, 10 pmoles of linker primer[30] and 3.0 Units of AmpliTaq Gold polymerase (5 Units/µl, Perkin Elmer Inc., Foster City, CA, USA) are added to each sample, and reactions are overlaid with mineral oil. (Note: 2× AmpliTaq® Gold buffer contains 3 mM $MgCl_2$. Therefore, the final $MgCl_2$ concentration in the 50 µl 2× AmpliTaq® Gold polymerase mix is 4 mM, due to the added 1 mM $MgCl_2$. This translates into a 2 mM final $MgCl_2$ concentration in the 100 µl PCR reaction). Reactions undergo 1 PCR cycle of 95°C for 16 min (activating AmpliTaq® Gold polymerase), T_m of Primer 2 for 2 min, and 72°C for 3 min, and 19 PCR cycles of 95°C for 1 min, 1°C below T_m of Primer 2 for 2 min, and 72°C for 3 min. Lastly, an extension is performed at 72°C for 10 min. Following the PCR reaction, a stop mix (13 µl of 3 M sodium acetate, pH 5.2, 3 µl of 0.5 M EDTA, pH 8.0, and 9 µl of dH_2O) is added under the mineral oil layer. Samples are extracted with 170 µl of premixed phenol:chloroform (50 µl:120 µl), then ethanol precipitated by adding 370 µl ice-cold ethanol. Air-dried DNA pellets are dissolved in 7.0 µl of premixed formamide-dye [2.3 µl dH_2O, 4.7 µl formamide loading dye: 95% (v/v) formamide, 10 mM EDTA, pH 8, 0.05% (w/v) xylene cyanol, 0.05% (w/v) bromophenol blue] in preparation for sequencing gel electrophoresis.

Continuous Elution Electrophoresis

The CEE procedure has been described in detail.[31] Several key steps are:

Loading the DNA Sample and Running the Continuous Elution Electrophoresis Apparatus

30 µl of 10× agarose loading dye (Sigma, St. Louis, MO, USA) were added to 300 µg of *Bam*H I-digested (Gibco BRL, Gaithersburg, MD, USA) genomic DNA in a volume of 300 µl TE (final concentration of 1 µg/ml). The DNA sample was loaded onto a Model 491 Prep Cell (Bio-Rad Laboratories, Hercules, CA, USA), containing a 0.5% preparative agarose gel (SeaKem® Gold; FMC BioProducts, Rockland, ME, USA) and a 0.25% agarose stacking gel (SeaKem® Gold; FMC BioProducts, Rockland, ME, USA). The sample was run in 50 mM TBE (Tris-borate/

EDTA), at 55 constant volts (PowerPac 300 Power Supply; Bio-Rad Laboratories, Hercules, CA, USA) at 4°C. An elution flow rate of 50 µl per minute was maintained by a peristaltic pump (Model EP-1 Econo Pump; Bio-Rad Laboratories, Hercules, CA, USA), with fraction collection times of 20 minutes (1 ml final volume).

Fraction Screening for Gene of Interest by Dot-Blot Analysis

After continuous elution electrophoresis, 10 µl from each fraction (approximately 30 ng of DNA in 50 mM TBE buffer) were added to 190 µl of 0.4 M NaOH, 10 mM EDTA solution. Samples were heated to 95°C for 5 minutes in a thermocycler (PTC-100, MJ Research, Watertown, MA, USA), placed on ice, and loaded onto a Dot-Blot apparatus (Bio-Dot; Bio-Rad Laboratories, Hercules, CA, USA) containing a positively charged nylon plus membrane (Qiabrane; Qiagen, Santa Clarita, CA, USA). Wells were rinsed with 200 µl of 0.4 M NaOH, 10 mM EDTA solution. Membrane was then soaked in 2× SSC (NaCl/sodium citrate) for 5 minutes, UV-cross linked (1200 joules/m^2) (Stratalinker; Stratagene, La Jolla, CA, USA), and placed into a hybridization tube containing hybridization solution (0.25 M NaPO$_4$, 1 mM EDTA, 7% SDS (sodium dodecyl sulfate), 1% BSA (Bovine Serum Albumin) (fraction V; Sigma, St. Louis, MO, USA)), and radiolabeled probe.[31] This was placed into a hybridization oven (HB 1100D Red Roller II, Pharmacia Biotech, Piscataway, NJ, USA) overnight.[11] After overnight hybridization at 66°C, membranes were washed 5 minutes in Buffer A (20 mM NaPO$_4$, 1 mM EDTA, 2.5% SDS, 0.25% BSA (fraction V; Sigma, St. Louis, MO, USA)) at 66°C, followed by 5 minutes in Buffer B (20 mM NaPO$_4$, 1 mM EDTA, 1% SDS) at 66°C. Buffer B wash was repeated two times. Air-dried membranes were exposed to Kodak XAR-5 x-ray films (Eastman Kodak, Rochester, NY, USA) with intensifying screens (Optex, Cedar Knolls, NJ, USA) at −70°C. The intensity of each dot was quantitated by PhosphorImager analysis (Model 425S; Molecular Dynamics, Sunnyvale, CA, USA).

Cleavage of Enriched Samples at ROS-induced Modified Bases

Following ROS treatment, continuous elution electrophoresis and fraction screening by dot-blot analysis, fractions containing the highest percentage of the gene of interest were precipitated by making two tubes, each consisting of 500 µl eluted fraction, 50 µl 3M NaOAc, pH 7, 1 µl glycogen (20 µg/µl), and 1 ml ice-cold ethanol. DNA was ethanol precipitated by a 10 min incubation on dry ice. Air-dried pellets were resuspended in 25 µl 1× TE, pH 8. Each respective pair was pooled to yield a 50 µl total volume consisting of 3 µg DNA. Respective fractions (containing 3 µg DNA) were then mixed with 50 µl of a unique 2× Nth/Fpg reaction buffer (91.4 mM Tris-HCl, pH 7.7, 200 mM KCl, 1.1 mM EDTA, 0.2 mM DTT, 200 µg/µl BSA-fraction V), yielding a final volume of 100 µl. To this, 400 ng Fpg protein and 100 ng Nth protein were added and samples were digested at 37°C for 60 min as previously described.[11] Fpg and Nth proteins were purified as described.[19] Control samples (No enzyme) were incubated in buffer alone (data not shown). Digestions were terminated[11] and, the DNA pellets were dissolved in Sequenase buffer (40 mM Tris-HCl, pH 7.7 and 50 mM NaCl) for LMPCR.[11]

RESULTS

Distribution of DNA Damage and Repair Induced in Vitro by Cu(II), Fe(III), or Cr(VI) plus H_2O_2/Ascorbate and in Vivo by H_2O_2

The distribution of oxidative damage induced in exons 5 and 9 of human *p53* and the promoter region of human *PGK1* was assessed by LMPCR.[13] A representative autoradiogram indicating the damage distributions induced in the region of the human *PGK1* gene covered by primer set A^{11} is shown in FIGURE 1B (for human *p53* primer sets, see ref. 29). In these regions of the genome, the base damage frequency distributions induced in dialyzed DNA *in vitro* by Cu(II) and Fe(III) plus ascorbate/ H_2O_2 were nearly identical. Sucrose was included in the Fe(III) reaction to suppress the direct strand break signal. Sucrose has no effect on the LMPCR-derived damage distribution signals.[13] The base damage distribution associated with these two transition metal ions was non-uniform, confirming our previous observations[11] with Cu(II). Prominent base damage hotspots were observed in both the *PGK1* and *p53* gene.[13] The distribution of damage caused by Cr(VI)/ascorbate/H_2O_2 was similar, but not identical, to that mediated by copper or iron ions in these regions. The unique chromate sensitive positions were often thymines (FIG. 1B).

The distribution of DNA base damage occurring *in vivo* induced by exposure of cultured human male fibroblasts to 50 mM H_2O_2 was also determined in *PGK1* and *p53*.[13] Here, 50 mM H_2O_2 induced a global damage frequency in human male fibroblast DNA *in vivo* equivalent to the damage frequency induced *in vitro* by 50 µM Cu(II)/100 µM ascorbate/5 mM H_2O_2 (data not shown). FIGURE 1B demonstrates that base damage distribution induced *in vivo* in the assessed regions of *PGK1* and *P53* were identical to the damage distributions induced *in vitro* by Cu(II) or Fe(III) plus H_2O_2/ascorbate and similar to the damage distribution induced *in vitro* by Cr(VI)/H_2O_2/ascorbate. Similar results were observed for the *P53* gene.[13]

Comparison of damage intensity among sequence contexts was made by applying sequential Wilcoxon rank-sum tests.[13] Guanine was the most heavily modified base associated with H_2O_2-mediated DNA damaging reactions both *in vivo* and *in vitro*. The triplet d(pCGC) was the principal hotspot sequence, with guanine stretches also hit.

The *in vivo* base damage and strand breaks of the HIF1 (hypoxia-inducible factor 1 binding site) footprint of *PGK1* were examined at 5 mM H_2O_2. Base damage and strand breaks are repaired but at different rates.[12] The base damage at positions G_{-214} and C_{-213} are 52% and 91% repaired in 24 h, respectively. In contrast, strand breaks at the same positions are repaired more slowly, 39% repair for position G_{-214} and 55% repair for position C_{-213} in 24 h (FIG. 3).

Damage Induced by H_2O_2 in Isolated Fibroblast Nuclei

Exposure of human male fibroblasts to a concentration of H_2O_2 several orders of magnitude higher than those generated under basal metabolic conditions was required in order to generate sufficient DNA base damage for the purpose of damage frequency mapping. Therefore, we assessed to what extent artifacts caused by

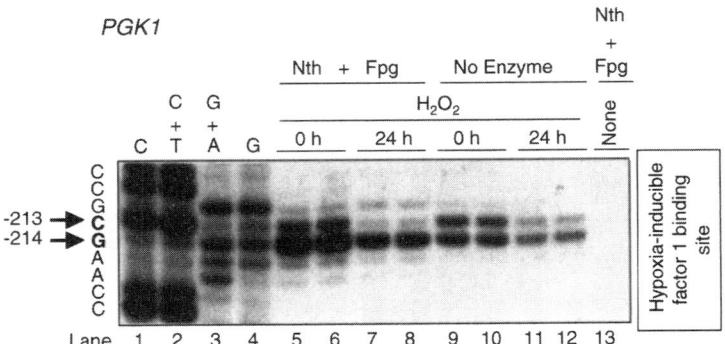

FIGURE 3. Repair of HIF-1 (hypoxia-inducible factor 1 binding site) footprint. Cells were treated as described in MATERIALS AND METHODS, except 5 mM H_2O_2 was used and half of the treated cells were replaced in fresh complete culture medium for 24 h of repair before DNA extraction.[12] *Lanes 1–4*: DNA treated with standard Maxam-Gilbert cleavage reactions. The DNA in lanes 5–8, 13 was digested with Nth and Fpg proteins after treatment; the DNA in lanes 9–12 was incubated in digestion buffer alone after treatment. Reproduced with permission.

exposure to high concentration H_2O_2 contributed non-physiologic distortion of the observed damage frequency patterns. The principal effect of high concentration H_2O_2, causing severe cellular oxidative stress, turned out to be release of cellular transition metal ions from normally sequestered extranuclear sites. Evidence for this was obtained by assessing DNA damage in isolated human male fibroblast nuclei. DNA damage was assessed globally by neutral denaturing agarose gel electrophoresis.[28] FIGURE 4A shows that isolated nuclei behave similar to naked DNA in that neither H_2O_2 alone nor Cu(II)/Fe(III) plus ascorbate cause detectable DNA damage in isolated human male fibroblast nuclei. However, nuclei are different from naked DNA or cells in that significant DNA damage was observed only if H_2O_2 and Cu(II)/Fe(III) were added. Thus, isolated nuclei have the following two notable properties: (1) They contain endogenous reducing agents capable of reducing transition metals such that the metals redox cycle in the presence of H_2O_2, and (2) They do not contain

FIGURE 4. A: Global frequency of direct strand breaks ("No Enzyme" lanes) and direct strand breaks plus modified bases ("Nth + Fpg" lanes) observed after exposure of isolated human male fibroblast nuclei to 50 mM H_2O_2 (lanes 3–4), 50 μM Cu(II) + 100 μM ascorbate (lanes 5–6), 50 μM Fe(III) + 100 μM ascorbate (lanes 7–8), or combinations of these reagents (lanes 9–16). *Lane 17*: 500 ng lambda DNA digested with *Hin*d III and 500 ng PhiX174 digested with *Hae* III. *Lanes 1–2*: nuclei not treated (Controls). Nuclear isolation was performed as described in ref.13. Exposures were for 30 min at 37°C, after which DNA was isolated. Electrophoresis was carried out as described in ref. 28. (*) 0.8 is a lower limit of estimated lesion frequency in these lanes; a more precise value could not be determined. **B:** Schematic representation of the steps in target gene enrichment by CEE. Reproduced with permission.

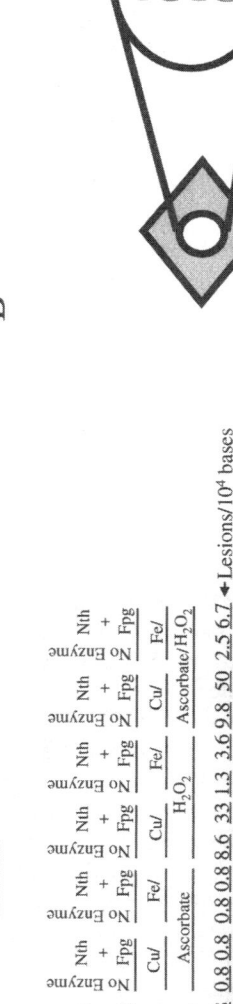
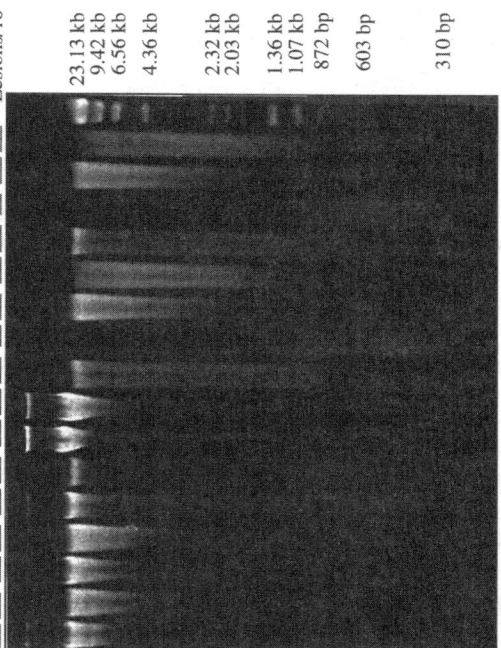

sufficient bound metals (or metal-like ligands) to cause significant base damage in the presence of H_2O_2.

Enhancement of LMPCR Damage Detection Sensitivity by Genomic Gene Enrichment

The requirement for exposure of fibroblasts to cytotoxic concentrations of H_2O_2 in order to map ROS-induced DNA base damage *in vivo* by LMPCR spurred us to develop methods to enhance the sensitivity of LMPCR in order to map non-cytotoxic concentrations of H_2O_2. One approach to enhancing LMPCR-generated base damage signal intensity is to increase the relative copy number of the target gene in the substrate genomic DNA. This was accomplished by size-fractionating restriction endonuclease-digested, ROS-exposed genomic DNA by continuous elution electrophoresis (CEE) through a preparative agarose gel (FIG. 4B[31]). Use of this genomic gene-enriched DNA containing the target as an LMPCR substrate enhanced by 24-fold the LMPCR-derived base damage signal intensity compared to non-enriched total genomic DNA for the *PGK1* and *p53* gene.

DISCUSSION

The frequency patterns of DNA base damage in two genes on separate chromosomes induced *in vivo* by H_2O_2 and *in vitro* by H_2O_2 in the presence of copper, iron, or chromate ions plus ascorbate were determined. The nucleotide-resolution maps of DNA base damage induced *in vitro* in the presence of Cu(II), Fe(III), or Cr(VI) transition metal ions in two genes (*PGK1* and *p53*), were similar, as well to the *in vivo* base damage induced by H_2O_2. The *in vitro* similarity suggests a model in which the local binding site occupancy rate and the local geometry of the metal ion-DNA-peroxo coordination complex determine the probability of a damage event at each position. Additionally, the damage frequency patterns indicates that the principal determinant of the probability of a H_2O_2-associated damaging event occurring at any position is the primary DNA sequence. Chromatin structure, with the exception of transcription factor footprints,[12] is only a minor determinant of base damage probability.

Experiments with isolated fibroblast nuclei indicated that nuclei have insufficient bound transition metal to induce base damage detectable by neutral denaturing gel electrophoresis; exogenous transition metal ions must be added to the nuclei to produce measurable base damage. The extreme oxidative stress caused by exposure of cells to 50 mM of H_2O_2 may have liberated normally sequestered extranuclear transition metal ions into the nucleus, allowing chromatin to show a transition metal ion-saturated pattern. Oxidant-mediated liberation of cytoplasmic sequestered transition metal ions has been observed in other model systems.[32]

Furthermore, damage caused by H_2O_2 *in vivo* is mediated by DNA-bound transition metal ions (or other redox cycling ligands with similar characteristics), that in addition to residing within a cell's nucleus, also reside at extranuclear sites. The damage frequency patterns do not permit definitive determination of which transition metal ions are involved in *in vivo* DNA damage production because the damage

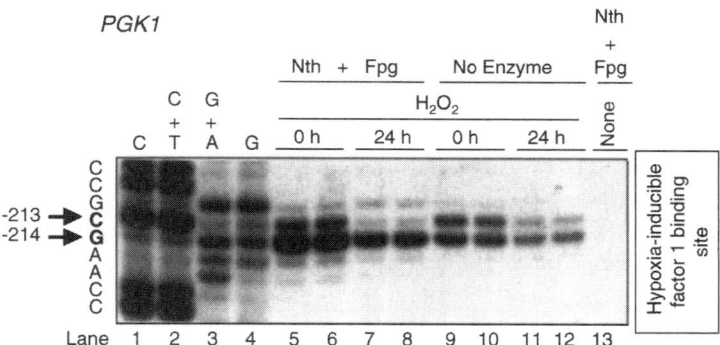

FIGURE 5. Substrates of the Fpg and Nth proteins. **A:** Adducts excised by the Fpg protein. **B:** Adducts excised by the Nth protein. The relative excision efficiencies observed for *in vitro* damage are indicated.[34] Reproduced with permission.

frequency pattern is independent of the type of transition metal ion, at least with respect to copper, iron, and chromium ions.

Both the base damage and strand break components of the HIF1 footprint were repaired but at different rates. The base damage at this footprint was 52–91% repaired in 24 h, which was similar to the global base damage repair rate. However, strand breaks at this footprint were only 39–55% repaired in 24 h or approximately 100-fold slower than the global strand break repair rate, likely due to the presence of the HIF1 protein.[12]

Recent improvements, such as the enhancement of LMPCR-derived damage signal intensity by genomic gene enrichment, permits the mapping of base damage induced by non-cytotoxic exposures of fibroblasts to H_2O_2 *in vivo*; such studies are currently ongoing in our laboratory. In addition, other DNA isolation techniques such as eliminating phenol/chloroform extractions are used to reduce background, providing DNA yielding better signal-to-noise ratios for LMPCR.

The majority of base damage is repaired by the BER system. Most adducts are repaired between 1–4 h post-exposure to H_2O_2 at low concentrations. At the higher concentrations used for our LMPCR repair studies, however, the mechanism of BER may be significantly altered as a consequence of increased metal transport to the nucleus as described above. The HIF1 factor region repair showed that most of the damage was repaired in under 24 h for the base damage, but repair of strand breaks (either single- or double-strand breaks), is significantly slower. Reduced break repair could also result from the high damage levels needed for LMPCR detection without use of genomic gene enrichment. Therefore, to understand how both BER and strand repair occur in mammalian cells at nucleotide resolution, enrichment of damage containing fragments is necessary to permit detection of damage at low H_2O_2 concentrations.

For H_2O_2-induced oxidative damage, numerous base damages have been identified.[33,34] LMPCR monitors a number of DNA base damages dependent on the DNA glycosylase used to incise the DNA leaving a 5′ phosphoryl group subject to ligation. FIGURE 5 shows some substrates recognized by the Fpg (FIG. 5A) and Nth (FIG. 5B) proteins. Both FapyGua and 8-oxoGua are recognized by the Fpg protein. Nth substrates in FIGURE 5B are arranged according to the enzyme's specificity for those substrates according to gas chromatography-mass spectrometry data. Some damaged bases are efficiently removed by the Nth protein, whereas other damaged bases such as 5,6-dihydrothymine are not efficiently excised. LMPCR is unable to distinguish which of the damaged bases is excised by these DNA glycosylases. Although we cannot determine which of the damaged bases is removed at a particular position, we are able to determine the pattern of enzyme sensitive damage along a gene. Moreover, other techniques are unable to yield sequence context damage patterns in mammalian DNA. Mammalian homologues of the Fpg and Nth proteins are responsible for the excision of similar damages.[35–40]

The capacity of LMPCR to yield sequence context damage and repair patterns renders the technique essential to understanding sequence context damage and repair patterns. Damage and repair patterns at nucleotide resolution are more heterogeneous than patterns of gene-specific and strand-specific repair. Thus, use of LMPCR to study damage and repair provides a key to understanding mutations at specific sites observed in tumor suppressor genes such as P53 that are frequently observed in human cancers.[41–43]

REFERENCES

1. AMES, B.N. 1987. Oxidative DNA damage, cancer, and aging. Ann. Intern. Med. **107:** 526–545.
2. GUYTON, K.Z. & T.W. KENSLER. 1993. Oxidative mechanisms in carcinogenesis. Br. Med. Bull. **49:** 523–-544.
3. WALLACE, S.S. 1994. DNA damage processed by base excision repair: biological consequences. Int. J. Radiat. Biol. **66:** 579–584.
4. MORAES, E.C., S.M. KEYSE, M. PIDOUX & R.M. TYRRELL. 1989. The spectrum of mutations generated by passage of a hydrogen peroxide damaged shuttle vector plasmid through a mammalian host. Nucleic Acids Res. **17:** 8301–8312.
5. AKMAN, S.A., G.P. FORREST, J.H. DOROSHOW & M. DIZDAROGLU. 1991. Mutation of potassium permanganate- and hydrogen peroxide-treated plasmid pZ189 replicating in CV-1 monkey kidney cells. Mutat. Res. **261:** 123–130.
6. LOEB, L.A., E.A. JAMES, A.M. WALTERSDORPH & S.J. KLEBANOFF. 1988. Mutagenesis by the auto-oxidation of iron with isolated DNA. Proc. Natl. Acad. Sci. USA. **85:** 3918–3922.
7. TKESHELASNVILI, L.K., T. MCBRIDE, K. SPENCE & L.A. LOEB. 1991. Mutation spectrum of copper-induced DNA damage. J. Biol. Chem. **266:** 6401–6406.
8. KAWANISHI, S., S. INOUE & K. YAMAMOTO. 1994. Active oxygen species in DNA damage induced by carcinogenic metal compounds. Environ. Health Perspect. **102** (Suppl. 3): 17–20.
9. MUELLER, P.R. & B. WOLD. 1989. *In vivo* footprinting of a muscle specific enhancer by ligation mediated PCR. Science **246:** 780–786.
10. PFEIFER, G.P., R. DROUIN, A.D. RIGGS & G.P. HOLQUIST. *In vivo* mapping of a DNA adduct at nucleotide resolution: Detection of pyrimidine (6-4) pyrimidonephotoproducts by ligation-mediated polymerase chain reaction. Proc. Natl. Acad. Sci.USA. **88:** 1374–1378.

11. RODRIGUEZ, H., R. DROUIN, G.P. HOLMQUIST, T.R. O'CONNOR, S. BOITEUX, J. LAVAL, J.H. DOROSHOW & S.A. AKMAN. 1995. Mapping of copper/hydrogen peroxide-induced DNA damage at nucleotide resolution in human genomic DNA by ligation-mediated polymerase chain reaction. J. Biol. Chem. **270**: 17633–17640.
12. RODRIGUEZ, H., R. DROUIN, G.P. HOLMQUIST & S.A. AKMAN. 1997. A hot spot for hydrogen peroxide-induced damage in the human hypoxia-inducible factor 1 binding site of the PGK1 gene. Arch. Biochem. Biophys. **338**: 207–212.
13. RODRIGUEZ, H., G.P. HOLMQUIST, R. D'AGOSTINO, JR., J. KELLER & S.A. AKMAN. 1997. Metal ion-dependent hydrogen peroxide-induced DNA damage is more sequence specific than metal specific. Cancer Res. **57**: 2394–2401.
14. COOPER, P.K., T. NOUSPIKEL, S.G. CLARKSON & S.A. LEADON. 1997. Defective transcription-coupled repair of oxidative base damage in Cockayne syndrome patients from XP group G. Science **275**: 990–993.
15. GRISHKO, V.I., W.J. DRIGGERS, S.P. LEDOUX & G.L. WILSON. 1997. Repair of oxidative damage in nuclear DNA sequences with different transcriptional activities [published erratum appears in Mutat. Res. 1998, Feb; **407**(1): 85]. Mutat. Res. **384**: 73–80.
16. LEADON, S.A. & P.K. COOPER. 1993. Preferential repair of ionizing radiation-induced damage in the transcribed strand of an active human gene is defective in Cockayne syndrome. Proc. Natl. Acad. Sci. USA **90**: 10499–10533.
17. TAFFE, B.G., F. LARMINAT, J. LAVAL, D.L. CROTEAU, R.M. ANSON & V.A. BOHR. 1996. Gene-specific nuclear and mitochondrial repair of formamidopyrimidine DNA glycosylase-sensitive sites in Chinese hamster ovary cells. Mutat. Res. **364**: 183–192.
18. YAKES, F.M. & B. VAN HOUTEN. 1997. Mitochondrial DNA damage is more extensive and persists longer than nuclear DNA damage in human cells following oxidative stress. Proc. Natl. Acad. Sci. USA **94**: 514–519.
19. O'CONNOR, T.R. 1996. The use of DNA glycosylases to detect DNA damage. In Technologies for Detection of DNA Damage and Mutation. G. Pfeifer, Ed. :155-167. Plenum Press, New York.
20. DEMPLE, B., T. HERMAN & D.S. CHEN. 1991. Cloning and expression of APE, the cDNA encoding the major human apurinic endonuclease: definition of a family of DNA repair enzymes. Proc. Natl. Acad. Sci. USA **88**: 11450–11454.
21. ROBSON, C.N. & I.D. HICKSON. 1991. Isolation of cDNA clones encoding a human apurinic/apyrimidinic endonuclease that corrects DNA repair and mutagenesis defects in *E. coli* xth (exonuclease III) mutants. Nucleic Acids Res. **19**: 5519–5523.
22. MATSUMOTO, Y. & K. KIM. 1995. Excision of deoxyribose phosphate residues by DNA polymerase b during DNA repair. Science **269**: 699–702.
23. WIEBAUER, K. & J. JIRICNY. 1990. Mismatch-specific thymine DNA glycosylase and DNA polymerase beta mediate the correction of G.T mispairs in nuclear extracts fromhuman cells. Proc. Natl. Acad. Sci. USA **87**: 5842–5845.
24. BHAGWAT, M. & J.A. GERLT. 1995. 3' and 5'-Strand cleavage reactions catalyzed by the Fpg protein from *Escherichia coli* occur via successive b- and d- elimination-mechanisms, respectively. Biochemistry **35**: 659–665.
25. CALDECOTT, K.W., S. AOUFOUCHI, P. JOHNSON & S. SHALL. 1996. XRCC1 polypeptide interacts with DNA polymerase beta and possibly poly (ADP-ribose) polymerase, and DNA ligase III is a novel molecular 'nick-sensor' in vitro. Nucleic Acids Res. **224**: 4387–4394.
26. CAPPELLI, E., R. TAYLOR, M. CEVASCO, A. ABBONDANDOLO, K. CALDECOTT & G. FROSINA. 1997. Involvement of XRCC1 and DNA ligase III gene products in DNA base excision repair. J. Biol. Chem. **272**: 23970–23975.
27. NASH, R.A., K.W. CALDECOTT, D.E. BARNES & T. LINDAHL. 1997. XRCC1 protein interacts with one of two distinct forms of DNA ligase III. Biochemistry **36**: 5207–5211.
28. DROUIN, R., H. RODRIGUEZ, S. GAO, Z. GEBREYES, T.R. O'CONNOR, G.P. HOLMQUIST & S.A. AKMAN. 1996. Cupric ion/ascorbate/hydrogen peroxide-induced DNA damage: DNA-bound copper ion primarily induces base modifications. Free Radic. Biol. Med. **21**: 261–273.

29. TORNALETTI, S., D. ROZEK & G.P. PFEIFER. 1993. The distribution of UV photoproducts along the human p53 gene and its relation to mutations in skin cancer. Oncogene **8:** 2051–2057.
30. GAO, S., R. DROUIN & G.P. HOLMQUIST. 1994. DNA repair rates mapped along the human PGK1 gene at nucleotide resolution. Science **263**: 1438-1440.
31. RODRIGUEZ, H. & S.A. AKMAN. 1998. Large scale isolation of genes as DNA fragment lengths by continuous elution electrophoresis through an agarose matrix. Electrophoresis **19:** 646–652.
32. CALDERARO, M., E.A. MARTINS & R. MENEGHINI. 1993. Oxidative stress by menadione affects cellular copper and iron homeostasis. Mol. Cell Biochem. **126:** 17–23.
33. BOITEUX, S., E. GAJEWSKI, J. LAVAL & M. DIZDAROGLU. 1992. Substrate specificity of the *Escherichia coli* Fpg protein: excision of purine lesions in DNA produced by ionizing radiation or photosensitization. Biochemistry **31:** 106–110.
34. DIZDAROGLU, M., J. LAVAL & S. BOITEUX. 1993. Substrate specificity of the *Escherichia coli* Endonuclease III: Excision of thymine- and cytosine-derived lesions in DNA produced by radiation-generated free radicals. Biochemistry **32:** 12105–12111.
35. ARAI, K., K. MORISHITA, K. SHINMURA, T. KOHNO, S.R. KIM, T. NOHMI, M. TANIWAKI, S. OHWADA & J. YOKOTA. 1997. Cloning of a human homolog of the yeast OGG1 gene that is involved in the repair of oxidative DNA damage. Oncogene **14:** 2857–2861.
36. ASPINWALL, R., D.G. ROTHWELL, T. ROLDAN-ARJONA, C. ANSELMINO, C. J. WARD, J. P. CHEADLE, J. SAMPSON, R., T. LINDAHL, P.C. HARRIS & I.D. HICKSON. 1997. Cloning and characterization of a functional human homolog of *Escherichia coli* endonuclease III. Proc. Natl. Acad. Sci. USA **94:** 109–114.
37. HILBERT, T.P., W. CHAUNG, R.J. BOORSTEIN, R.P. CUNNINGHAM & G.W. TEEBOR. 1997. Cloning and Expression of the cDNA encoding the human homologue of the DNA repair enzyme, *Escherichia coli* Endonuclease III. J. Biol. Chem. **272:** 6733–6740.
38. LU, R., H.M. NASH & G.L. VERDINE. 1997. A mammalian DNA repair enzyme that excises oxidatively damaged guanines maps to a locus frequently lost in lung cancer. Curr. Biol. **7:** 397–407.
39. RADICELLA, J.P., C. DHERIN, C. DESMAZE, M.S. FOX & S. BOITEUX. 1997. Cloning and characterization of hOGG1, a human homolog of the OGG1 gene of *Saccharomyces cerevisiae*. Proc. Natl. Acad. Sci. USA **94:** 8010–8015.
40. ROLDAN-ARJONA, T., Y.F. WEI, K.C. CARTER, A. KLUNGLAND, C. ANSELMINO, R.P. WANG, M. AUGUSTUS & T. LINDAHL. 1997. Molecular cloning and functional expression of a human cDNA encoding the antimutator enzyme 8-hydroxyguanine-DNA glycosylase. Proc. Natl. Acad. Sci. USA **94:** 8016–8020.
41. DENISSENKO, M.F., A. PAO, M. TANG & G.P. PFEIFER. 1996. Preferential formation of benzo[a]pyrene adducts at lung cancer mutational hotspots in p53. Science **274:** 430–432.
42. TORNALETTI, S. & G.P. PFEIFER. 1994. Slow repair of pyrimidine dimers at p53 mutation hotspots in skin cancer. Science **263:** 1436–1438.
43. YE, N., G.P. HOLMQUIST & T.R. O'CONNOR. 1998. Heterogeneous repair of N-methylpurines in normal human cells at the nucleotide level. J. Mol. Biol. In press.
44. FROSINA, G., P. FORTINI, O. ROSSI, F. CARROZZINO, G. RASPAGLIO, I.S. COX, D.P. LANE, A. ABBONDANDOLO & E. DOGLIOTTI. 1996. Two pathways for base excision repair in mammalian cells. J. Biol. Chem. **271:** 9573–9578.
45. KLUNGLAND, A. & T. LINDAHL. 1997. Second pathway for completion of human DNA base excision-repair: reconstitution with purified proteins and requirement for DNase IV (FEN1). EMBO J. **16:** 3341–3348.
46. MATSUMOTO, Y., K. KIM & D.F. BOGENHAGEN. 1994. Proliferating cell nuclear antigen-dependent abasic site repair in *Xenopus laevis* oocytes: an alternative pathway of base excision repair. Mol. Cell Biol. **14:** 6187–6197.

Post-Transcriptional Regulation of Lung Antioxidant Enzyme Gene Expression

LINDA BIADASZ CLERCH[a]

Georgetown University School of Medicine, Lung Biology Laboratory, Department of Pediatrics, Washington, DC 20007, USA

> ABSTRACT: It is an honor, and indeed fitting, to have a chapter on pulmonary oxygen toxicity included in a Festschrift for Dan Gilbert, whose contributions to the free radical theory of oxygen toxicity have been a catalyst to the last half-century of investigation in this field. There is cellular damage that results in pulmonary edema and even death if the increase in reactive oxygen species produced in the lung during exposure to hyperoxia is not counterbalanced by an increase in the cell's antioxidant defense systems. In this chapter experimental evidence will substantiate the importance of post-transcriptional regulation of antioxidant enzyme gene expression in animal models of pulmonary oxygen toxicity and tolerance to hyperoxia with special emphasis given to the role of manganese superoxide dismutase (MnSOD) synthesis, specific activity, and RNA half-life and to a proposed function of a MnSOD RNA-binding protein as a positive regulator in the control of translational efficiency.

PULMONARY OXYGEN TOXICITY

The lung, as the organ exposed to the highest oxygen concentration, is particularly at risk to the toxic effects of oxygen.[1] This is a clinically important problem because lung cells experience a rapid step up of pO_2 during birth where the lung cells go from a pO_2 of 20 torr in the fluid filled lung to a pO_2 of 100 torr upon air breathing and a premature lung, in particular, may not be biochemically ready to handle this rapid increase in oxygen tension. Also, there are medical conditions that require oxygen therapy including: 1) adult respiratory distress syndrome, and 2) and premature babies with respiratory distress syndrome who are ventilated with increased oxygen concentration at a time when their lung's antioxidant enzyme defenses are not adequately developed.

During exposure to hyperoxia the production of oxygen radicals and hydrogen peroxide is increased in lung tissue.[2,3] Tolerance is the ability to resist the potential damaging effects of hyperoxia and requires that the increased production of reactive oxygen species be counterbalanced by increases in the cell's antioxidant defense system including the antioxidant enzymes (AOEs). The antioxidant enzymes that are the subject of this chapter are copper,zinc superoxide dismutase (Cu,ZnSOD), manganese superoxide dismutase (MnSOD), catalase, and glutathione peroxidase (GPx).

---

[a]Address for Correspondence: Linda Biadasz Clerch, Ph.D., Georgetown University Medical Center, Lung Biology Laboratory, Preclinical Science Bldg., GM12, 3900 Reservoir Rd., NW, Washington, DC 20007. Voice: 202-687-4984;fax: 202-687-8538.
e-mail: clerchlb@gusun.georgetown.edu

The SODs catalyze the conversion of superoxide to hydrogen peroxide that can be eliminated either by catalase or GPx. One experimental approach to the problem of oxygen toxicity has been to determine the mechanisms responsible for the different levels of lung antioxidant enzyme gene expression in order to understand the molecular basis of lung tolerance to oxidant stress and gain insight into possible means of treatment to prevent pulmonary oxygen toxicity. Without question, there is regulation of AOE expression at the level of transcription that is responsible for the synthesis of AOE mRNAs and, for that reason, many scientists are studying essential cis promoter elements as well as pertinent transcription factors. However, our laboratory and others have found that AOEs are also significantly regulated at a post-transcriptional level by alterations in

(a) the degradation rate of AOE mRNA, i.e. mRNA stability,
(b) the rate of AOE protein synthesis, i.e. translation, and
(c) post-translational modifications, i.e. specific activity.

The remainder of this chapter will review some of the evidence supporting the hypothesis that AOE post-transcriptional regulation and post-transcriptional MnSOD regulation, in particular, play a critical role in rat models of both pulmonary oxygen toxicity and tolerance to hyperoxia.

ANIMAL MODELS OF PULMONARY OXYGEN TOXICITY AND TOLERANCE TO HYPEROXIA

Several animal models are useful in examining the relationship between AOE regulation and oxygen toxicity. Examples of oxygen toxicity models include a) hyperoxia-exposed adult rats,[4] b) premature baboons exposed to hyperoxia that results in a homolog for human premature infants with bronchopulmonary dysplasia,[5,6] and c) pertussis toxin-treated rats breathing room air that develop morphological and biochemical characteristics of pulmonary oxygen toxicity.[7] One model of endogenous tolerance to hyperoxia is that experienced by neonatal animals of some species. Newborn rats, mice and rabbits are able to survive periods of normobaric hyperoxia that are lethal to adults of the same species.[4] There are also experimental and pharmacological ways of inducing tolerance to hyperoxia in adult rats including

(a) pre-exposure to 85% O_2[8] or 10% O_2,[9]
(b) the administration of low doses of lipopolysaccharide (endotoxin) to adult rats,[10]
(c) the introduction of a "rest period" in air between hyperoxic exposures of adult rats,[11] and
(d) over expression of MnSOD in lung epithelial cells of transgenic mice.[12]

POST-TRANSCRIPTIONAL REGULATION OF ANTIOXIDANT ENZYME GENE EXPRESSION

To illustrate the role of AOE post-transcriptional regulation in oxygen toxicity and tolerance to hyperoxia, I will summarize AOE gene expression data obtained from two models of toxicity [hyperoxia-exposed adult rats and air-breathing pertussis toxin-treated rats] and two models of tolerance [hyperoxia-exposed neonatal rats and endotoxin-treated hyperoxia-exposed adult rats].

In the model of endogenous tolerance wherein neonatal rats are able to survive periods of normobaric hyperoxia that are lethal to adult rats, it is worth noting that survival of neonatal rats requires an increase above the basal levels of AOE activity in the lung. Put differently, the difference between the neonates and the adults is not in the basal level of AOE activity, which is higher in adult rats, rather it is the ability of the neonates to increase AOE activity during hyperoxic exposure that contributes to their tolerance to hyperoxia. The results of studies in neonatal rats are shown in TABLE 1.[13,14] Four day-old rats were exposed to air or 72 h of > 95% O_2. The arrows indicate significant differences between the air and hyperoxia-exposed lung. The enzymatic activity of catalase, GPx, Cu,ZnSOD, and MnSOD was significantly elevated and the increase in activity was associated with an increased concentration of each AOE mRNA. Measurement of the degradation rates of AOE mRNAs showed hyperoxia caused increased stability of each AOE mRNA in the lungs from the neonatal hyperoxia-exposed rats. The magnitude of the change of RNA concentration and half-life was approximately 2-fold. For catalase, the rate of RNA synthesis by nuclear run-on was also measured and there was no change in transcription.[13] To determine the specificity of the AOE response, we examined the expression of an endogenous lung 14 kD beta-galactoside binding protein, referred to here as lectin; lung lectin is not known to have any antioxidant property and neither the RNA concentration nor RNA half-life of this protein were changed during hyperoxia suggesting the up-regulation of AOE RNAs is at least partly specific.[13] Thus, in the endogenously tolerant neonatal rat, AOE activity is increased during hyperoxic exposure and the AOEs appear to be coordinately regulated at the post-transcriptional level of mRNA stability.

The results obtained in the tolerant neonates (TABLE 1) can be compared with studies of non-tolerant adult rats (TABLE 2). The arrows (TABLE 2) indicate significant differences in lung AOE gene expression between adults rats treated for 48 h with > 95% O2 compared with air breathing controls.[15] There was no change in the activity of Cu,ZnSOD; the activities of catalase and GPx increased during hyperoxic exposure but the activity of MnSOD fell. Thus, a major difference in the AOE response between the tolerant neonate and the non-tolerant adult rat is the ability of

TABLE 1. Effect of hyperoxia on neonatal rat lung antioxidant enzyme expression

Enzyme	Activity	mRNA Concentration	mRNA half-life
MnSOD	↑	↑	↑
Cu,ZnSOD	↑	↑	↑
GPx	↑	↑	↑
Catalase	↑	↑	↑
Lectin	NA	NC	NC

Arrows indicate significantly higher values in lungs from 4 day old rats exposed to 72 h of >95% O_2 compared with air-breathing rats.
NA = not applicable; NC = no change.

the neonate to increase the level of MnSOD activity while MnSOD activity was decreased in the lungs of the hyperoxia-exposed adult rats. The reason for this decline in activity is two-fold: first, the specific activity of the enzyme fell indicating a post-translational decrease in activity due, at least in part, to protein oxidation,[15] and second the there was fall in the absolute rate of MnSOD synthesis measured by the incorporation of phenylalanine into MnSOD protein.[16] MnSOD RNA increased in hyperoxia-exposed adult rats and this increase was associated with an increased half-life of the mRNA (TABLE 2), but this increased RNA was not efficiently translated into active protein.[15,16] These results indicate that in adult hyperoxia-exposed rats the AOEs are not coordinately regulated and that among the AOEs tested, an increase in MnSOD activity is critical for tolerance to hyperoxia The failure to increase MnSOD activity in the adult hyperoxia-exposed rat was caused mainly by decreased translational efficiency of MnSOD RNA, i.e. amino acids polymerized per unit time per MnSOD RNA molecule. The rate of MnSOD synthesis was 2.1 pmol/mg DNA/MnSOD RNA/h in air-breathing rats and 0.4 pmol/mg DNA/MnSOD RNA/h in rats after breathing > 95% O_2 for 48 h.[16] Thus, there was a 5-fold decrease in translational efficiency during exposure to hyperoxia. From these data, it is also important to point out a) the danger of only measuring RNA levels when looking for a particular response when post-transcriptional mechanisms are exerting a controlling effect, and b) that when considering gene transfer or gene therapy approaches one must keep in mind that increasing the mRNA level may not be enough to effect a cure if the translational machinery is compromised.

As noted above, a low dose of endotoxin (~ 1/50 LD_{50}) confers remarkable tolerance to hyperoxia on adult rats. After 72 h of > 95% O_2 there is nearly 100% survival of adult rats treated with endotoxin compared with ~ 32% survival of non-treated rats.[10] TABLE 3 shows the results of studies examining AOE gene expression in lungs from saline-treated air-breathing adult rats compared with rats treated with endotoxin (500 µg/kg) and exposed to 48 h of > 95% O_2.[15] MnSOD, catalase and GPx enzyme activities were significantly elevated in hyperoxia-exposed endotoxin-treated rats compared with air-saline control rats; there was a correlative increase in their mRNA concentrations. Cu,ZnSOD activity was unchanged. The increase of MnSOD, catalase and GPx mRNAs was due, at least in part, to increased mRNA stability indicated in TABLE 3 by elevated mRNA half-life. Thus, in this pharmacologically-induced tolerant model, endotoxin was able to overcome the hyperoxia-

TABLE 2. Effect of hyperoxia on adult rat lung antioxidant enzyme expression

Activity	mRNA Concentration	mRNA Half-life	Enzyme Activity	Enzyme Synthesis
MnSOD	↑	↑	↓	↓
GP	NC	NC	↑	ND
Catalase	NC	NC	↑	ND
Cu,ZnSOD	NC	ND	NC	NC

Arrows indicate the change in values in lungs from adult rats exposed to 48 h > 95% O_2 compared with air-breathing control rats.
NC = no change; ND = not determined.

TABLE 3. Effect of endotoxin and hyperoxia on adult rat lung antioxidant enzyme expression

Enzyme	Activity	mRNA Concentration	mRNA half-life
MnSOD	↑	↑	↑
GP	↑	↑	↑
Catalase	↑	↑	↑
Cu,ZnSOD	NC	ND	ND

Arrows indicate a significant increase in values in lungs from adult rats treated with endotoxin (500 µg/kg) and exposed to 48 h of > 95% O_2 compared with air-breathing control rats. NC = no change; ND = not determined.

induced block in MnSOD translation and the increased RNA was effectively translated into protein.

Although it might appear as though the endotoxin treatment converted the biochemical response of the adult rats into that of neonates with respect to tolerance to hyperoxia, data (below) indicate that lung MnSOD gene expression in these two rat models of tolerance to hyperoxia [neonates and endotoxin-treated adults] is regulated differently.

To examine the molecular mechanism responsible for the increase in MnSOD we asked how environmental signals, in this case high oxygen, are transduced into altered MnSOD expression. Our hypothesis was that in the neonate the hyperoxia-induced increases in MnSOD mRNA and activity were mediated by G-proteins. To test this hypothesis, rats were injected with pertussis toxin (PTX) which acts by catalyzing the transfer of ADP-ribose from cytoplasmic NAD to the α subunit of heterotrimeric G-proteins of the Gi and Go subclass. The ADP ribosylation prevents the interaction of the G-proteins with receptors on plasma membranes thereby blocking the G-protein mediated transduction of environmental signals to cellular effector molecules. TABLE 4 provides a summary of the morphological and biochemical effects of PTX treatment in air-breathing animals.[7] These findings are consistent with the notion that PTX treatment caused oxygen toxicity in air-breathing rats and, once again, among the AOEs studied, MnSOD is the major participant. The data in TABLE 5 summarize the effect of PTX treatment on lung MnSOD gene expression in both air-breathing and hyperoxia-exposed young rats (< 30 d old) that, if otherwise-untreated, are still tolerant to hyperoxia.[7] At 72 h after the initial injection of PTX

TABLE 4. Effect of pertussis toxin treatment on the lungs of air-breathing rats

Lung edema : Increased wet weight:dry weight ratio, and a widening of interstitial space with floccular material

Pleural effusion and intra-alveolar floccular material

Decrease in lung reduced glutathione

Increase in the carbonyl content of lung proteins

Decrease in MnSOD activity with no change in the activity of Cu,ZnSOD, catalase, or GPx

TABLE 5. Effect of pertussis toxin and endotoxin treatment on lung MnSOD expression in air- and hyperoxia-exposed young rats

	Air-PTX	O$_2$	O$_2$-PTX	Air-Endo	Air-Endo + PTX
Activity	↓	↑	↓	↑	↑
mRNA	↑	↑	↑	↑	↑

Arrows indicate a significant change in MnSOD activity and mRNA concentration in lungs from 20–30 day-old air-breathing rats treated with saline compared to those treated with PTX (50 µg/kg) (column 1); > 95% O$_2$ (column 2); PTX + hyperoxia (column 3); endotoxin (500 µg/kg) (column 4); and endotoxin + PTX (column 5).

there was a 50% fall in MnSOD activity in the air-PTX group of rats in spite of the fact that there was a 2-fold increase in MnSOD RNA concentration. This decrease in activity was due to a decrease in the absolute rate of MnSOD synthesis; there was no change in specific activity.[17] Thus, under these conditions, lung MnSOD was down-regulated because of a decrease in relative translational efficiency and the physiological consequence was pulmonary oxygen toxicity.

Hyperoxia-exposed diluent-treated neonatal rats had elevated lung MnSOD activity and mRNA concentration as expected in the endogenously tolerant neonate (TABLE 5). However, the hyperoxia-exposed PTX-treated rats did not have an increase in lung MnSOD activity; rather the activity in the O$_2$-PTX rats was depressed to the same extent as in air-PTX. These data indicate that both the steady state level of MnSOD in air-breathing rats and the hyperoxia-induced increase in MnSOD in neonates is mediated, at least in part, through a PTX-sensitive heterotrimeric G-protein. These observations raise the important possibility of the presence of an O$_2$-responsive membrane protein or perhaps ion channel that serves as an oxygen sensor and interacts with a G-protein that in turn mediates changes in MnSOD gene expression in response to changes in environmental O$_2$ concentration.

An examination of the effect of PTX in endotoxin-treated rats showed that with endotoxin treatment alone, the activity and RNA of MnSOD were significantly elevated and, importantly, this induction was not blocked by PTX treatment (TABLE 5).[7] From these data, it appears that endotoxin is not working through a G-protein mediated mechanism. Just as endotoxin treatment is capable of protecting the adult rat against damage during hyperoxia, endotoxin was also able to protect against the oxygen toxicity caused by PTX in air breathing rats. Thus, while the biological effect of endotoxin treatment is to make the adult rats respond to hyperoxia like neonates, the signal transduction mechanism in the two models of tolerance is different. Hyperoxia-induced tolerance in the neonate operates, at least in part, through a G-protein coupled mechanism whereas the mechanism of action of endotoxin-induced tolerance is not regulated by a PTX-sensitive G protein.

MNSOD RNA-BINDING PROTEIN (MNSOD-BP) ACTIVITY

In the data presented thus far, a recurrent theme has been that, in rat models of oxygen toxicity and tolerance to hyperoxia, MnSOD gene expression is regulated at different post-transcriptional levels, namely mRNA stability and translational efficiency. In addition, toxicity appears to occur in those models where MnSOD protein synthesis and translational efficiency are down-regulated, for example, the hyperoxia-exposed otherwise-untreated adult rat and the air-breathing PTX-treated rat. The next problem then was to determine what mechanism might be responsible for altering mRNA stability or translation. We decided to test the hypothesis that binding of trans-acting factors to MnSOD mRNA alters RNA stability and/or translation. In the case of RNA stability, the trans-acting factors might include proteins that bind to MnSOD RNA rendering the RNA either more susceptible to or more protected against the action of a ribonuclease. In the case of translation, an RNA-protein complex might inhibit or enhance protein synthesis. We have begun to test our hypothesis by:

(a) testing for the presence of rat lung proteins that bind MnSOD mRNA,
(b) identifying specific *cis* elements in the 3′ untranslated region (3′ UTR) of MnSOD mRNA that are involved in protein binding, and
(c) examining the effect of MnSOD RNA-binding protein (MnSOD-BP) activity on *in vitro* translation.

Gel retention and competition assays were used to show the presence of specific binding between rat lung protein and the 3′ UTR of MnSOD RNA. The initial studies on the 3′ UTR of MnSOD were done using a probe that was 450 bases long, located 111 bases downstream of the stop codon.[18] The binding was redox-sensitive and required the presence of free sulfhydryl groups on the protein; if the lung extract was treated with N-ethylmaleimide or diamide, MnSOD-BP activity did not occur. In addition, MnSOD-BP activity was greater in $12,000 \times g$ supernatant fractions (S12) from neonatal lung extracts than in S12 from adult rat lungs. This developmental regulation was, at least in part, lung-specific because there was no difference in MnSOD-BP activity between neonatal and adult brain extracts.[18] To determine the mechanism underlying this developmental difference, lung subcellular fractions were tested for MnSOD-BP activity; protein in the $130,000 \times g$ supernatant (S130) of lung extracts bound the 3′ UTR, but, the developmental difference in binding was not present in S130.[19] The $130,000 \times g$ pellet (P130) did not bind the 3′ UTR; rather, it contained an inhibitor of MnSOD-BP activity. Furthermore, adult rat lung P130 was a more potent inhibitor of RNA-binding activity than neonatal P130. These data indicate the developmental difference in MnSOD-BP activity is due, in part, to the presence of an inhibitor resulting in less MnSOD-BP activity in adult than in neonatal rat lung. Biochemical characterization revealed the inhibitor is an RNA molecule that may participate in the post-transcriptional control of MnSOD gene expression.[19]

The region of protein binding in the MnSOD 3′ UTR has been further delimited to a 41 base *cis*-element located at the NdeI site, 111 bases downstream of the stop codon.[20] The conclusion that this region is the *cis* element involved in protein binding was verified by gel retention, cross-competition, and RNaseH studies.[20] This region is present in the 3′ UTR of all rat MnSOD mRNAs and contains the first polyadenylation signal "ATTAAA". A comparison of this sequence with the information available in GenBank revealed ~75% conserved sequence homology between the rat *cis*-element and regions in the 3′ UTR of mouse, cow, and human MnSOD RNAs at approximately the same distance downstream of the stop codon.[20]

TABLE 6. Effect of MnSOD RNA-binding protein activity on MnSOD RNA translation and translation initiation in rabbit reticulocyte lysate

MnSOD-BP Activity	Translation Relative Densitometry Units	Translation Initiation Relative Densitometry Units
Present	1.00	1.00
Absent	0.35 ± 0.10*	0.49 ± 0.02**

*$p < 0.05$; **$p < 0.01$.

In addition, MnSOD-BP activity is present in cells of all species tested including rat, baboon, human, mouse, monkey and hamster (unpublished data) suggesting MnSOD-BP may be ubiquitous and perform a similar function across many species.

The role of the 3′ UTR protein-binding region in RNA translation was assessed in an *in vitro* cell-free rabbit reticulocyte lysate (RRL) system. Translation of MnSOD RNA from which the 3′ UTR element was deleted decreased 60% compared with translation of MnSOD RNA containing the 3′ UTR *cis* element.[20] In the presence of a specific competitor oligoribonucleotide that inhibits MnSOD RNA-protein binding activity, translation of MnSOD RNA containing the 3′ UTR was decreased 65% (TABLE 6). Thus, both the *cis* element and RNA-protein binding activity were required for efficient translation of the MnSOD. An analysis of ribosomal profiles suggests the MnSOD RNA-binding protein participates in the formation of the translation initiation complex; when MnSOD-RNA binding activity was inhibited, initiation complex formation was decreased 51% (TABLE 6). From the data obtained in this study, it appears that the 3′ UTR cis element of MnSOD through its interaction with MnSOD RNA-binding protein may function as a translational enhancer.

In summary, an analysis of lung AOE gene expression in experimental models of pulmonary oxygen toxicity and tolerance to hyperoxia indicates that the ability to withstand the damaging effects of oxygen toxicity appears to be regulated, at least in part, by the translational efficiency of MnSOD RNA. Thus, understanding the mechanism whereby MnSOD translational yield can be increased through the action of an RNA-binding protein is important because an inability to defend against free radical damage to the mitochondria [the intracellular location of MnSOD] is responsible, in large part, for the pulmonary oxygen toxicity that may result in mortality. Future studies will be directed at isolating and purifying MnSOD-BP in order to test the function of the RNA-binding protein *in vivo*.

ACKNOWLEDGMENTS

I thank Don Massaro both for his long-time collaboration in the study of oxygen toxicity and for his review of this paper. This work was supported by NIH HL47413, HL20366, and an American Lung Association Career Investigator Award.

REFERENCES

1. CLARK, J.M. & C.J. LAMBERTSON. 1971. Pulmonary oxygen toxicity: a review. Pharmacol. Rev. **23:** 37–133.

2. FREEMAN, B.A., M.K. TOPOLOSKY & J.D. CRAPO. 1982. Hyperoxia increases oxygen radical production in rat lung homogenates. Arch. Biochem. Biophys. **216:** 477–484.
3. FREEMAN, B.A. & J.D. CRAPO. 1981. Hyperoxia increases oxygen radical production in rat lungs and lung mitochondria. J. Biol. Chem. **256:** 10986–10992.
4. FRANK, L., J.R. BUCHER & R.J. ROBERTS. 1979. Oxygen toxicity in neonatal and adult animals of various species. J. Appl. Physiol. **45:** 699–704.
5. ESCOBEDO, M.B., J.L. HILLIARD, F. SMITH, K. MEREDITH, W. WALSH, D. JOHNSON, J.J. COALSON, T.J. KUEHL, D.M. NULL JR. & J.C. ROBOTHAM. 1982. A baboon model of bronchopulmonary dysplasia. I. Clinical features. Exp. Molec. Pathol. **37:** 323–334.
6. COALSON, J.J., T.J. KUEHL, M.B. EXCOBEDO, J.L. HILLIARD, F. SMITH, K. MEREDITH, D.M. NULL JR., W. WALSH, D. JOHNSON & J.L. ROBOTHAM. 1982. A baboon model of bronchopulmonary dysplasia. II. Pathologic features. Exp. Molec. Pathol. **37:** 335–350.
7. CLERCH, L.B., G. NEITHARDT, U. SPENCER, J.A MELENDEZ, G.D. MASSARO & D. MASSARO. 1994. Pertussis toxin treatment alters manganese superoxide dismutase activity in lung: evidence for lung oxygen toxicity in air-breathing rats. J. Clin. Invest. **93:** 2482–2489.
8. CRAPO, J.D., B.E. BARRY, H.A. FOSCUE & J. SHELBURNE. 1980. Structural and biochemical changes in rat lungs occurring during exposure to lethal and adaptive doses of oxygen. Am. Rev. Respir. Dis. **122:** 123–143.
9. SJOSTROM, K., & J.D. CRAPO. 1983. Structural and biochemical adaptive changes in rat lungs after exposure to hypoxia. Lab. Invest. **48:** 68–79.
10. FRANK, L., J. YAM & R.J. ROBERTS. 1978. The role of endotoxin in protection of adult rats form high oxygen lung toxicity. J. Clin. Invest. **61:** 269–275.
11. FRANK, L., J. IQBAL, M. HASS, & D. MASSARO. 1989. New "rest period" protocol for inducing tolerance to high O_2 exposure in adult rats. Am. J. Physiol **257:** L226–L231.
12. WISPE, J.R., B.B. WARNER, J.C. CLARK, C.R. DEY, J. NEUMAN, S.W. GLASSER, J.D. CRAPO, L.Y. CHANG & J. A. WHITSETT. 1992. Human Mn-superoxide dismutase in pulmonary epithelial cells of transgenic mice confer protection from oxygen injury. J. Biol. Chem. **267:** 23927–23941.
13. CLERCH, L.B., J. IQBAL & D. MASSARO. 1991. Perinatal rat lung catalase gene expression: influence of a corticosteroid and hyperoxia. Am. J. Physiol. **260:** L428–L433.
14. CLERCH, L.B. & D. MASSARO. 1992. Rat lung antioxidant enzymes: differences in perinatal gene expression and regulation. Am. J. Physiol. **263:** L466–L470.
15. CLERCH, L.B. & D. MASSARO. 1993. Tolerance of rats to hyperoxia: lung antioxidant enzyme gene expression. J. Clin. Invest. **91:** 499–508.
16. CLERCH, L.B., D. MASSARO & A. BERKOVICH. 1998. Molecular mechanisms of antioxidant enzyme expression in lung during exposure to and recovery from hyperoxia. Am. J. Physiol. **274:** L313–L319.
17. BERKOVICH, A., MASSARO & L. B. CLERCH. 1996. Pertussis toxin alters the concentration and turnover of manganese superoxide dismutase in rat lung. Am. J. Physiol. **271:** L875–L879.
18. FAZZONE, H., A. WANGNER & L.B. CLERCH. 1993. Rat lung contains a developmentally-regulated manganese superoxide dismutase mRNA-binding protein. J. Clin. Invest. **92:** 1278–1281.
19. CHUNG, D.J. & L.B. CLERCH. 1997. RNA in polysomes is an inhibitor of manganese superoxide dismutase RNA-binding protein activity. Am. J. Physiol. **272:** L714–L719.
20. CHUNG, D.J., A.E. WRIGHT & L.B. CLERCH. 1998. The 3' untranslated region of manganese superoxide dismutase RNA contains a translational enhancer element. Biochemistry **37:** 16298–16306.

Oxidative Stress, Mitochondrial Respiration, and Parkinson's Disease

GERALD COHEN[a]

Department of Neurology and Center for Neurobiology, Mount Sinai School Medicine, New York, New York 10029, USA

ABSTRACT: When either oxidizing species, such as H_2O_2 or oxy-radicals, are present in excess or cellular anti-oxidant defenses are lowered, a state of oxidative stress exists. Parkinson's disease is characterized by the loss of dopamine (DA) neurons, which leads to overactivity of the surviving DA neurons and an increase in neurotramsmitter release and turnover. The increased metabolism of DA neurotransmitter by monoamine oxidase (MAO) can be looked upon as an endogenous oxidative stress, leading to damage to Complex I-linked mitochondrial respiration. It remains an open question to what extent the mitochondrial damage seen in Parkinson's disease is of genetic origin and how much is caused by H_2O_2 generated during enhanced turnover of DA, especially during treatment with L-dopa.

PROLOGUE

It is a pleasure to contribute to this Festchrift volume for Dr. Daniel Gilbert. I first met Dan at a specialty meeting on "Implications of Organic Peroxides in Radiobiology" held at the Argonne National Laboratories, outside of Chicago, in May of 1962. A role for free radicals in biological processes was not in vogue at the time. That awaited the description of superoxide dismutase by Joe McCord and Irwin Fridovich in 1969, and the research and events spawned by that discovery. The prevailing assumption in 1962 was that oxy-radicals were of theoretical interest, but without important roles in biology. The exception was radiobiology because it was clear that the scission of water by high-energy radiation gave rise to superoxide and hydroxyl radicals, associated with tissue damage. But, this was clearly a specialized area of research. The major emphasis at the Argonne conference was on more stable products, such as hydrogen peroxide and organic hydroperoxides.

Such a highly specialized meeting, lying off more-traveled tracks, drew only a slender group of specialists, and a few outsiders, like myself, who had come to listen and learn. Among the key participants were Cheves Walling, well-known for his contributions to the chemistry of free radicals, Dean Burk, who left a lasting mark in enzymatic biochemistry with his Lineweaver-Burk plot, and Frederick Bernheim, who organized and oversaw the "Peroxide Club" that met yearly at the FASEB meeting in Atlantic City. The crucial opening session of the conference laid out the basic

---

[a]Address for correspondence: Dr. Gerald Cohen, Department of Neurology (Box 1137), Mount Sinai School of Medicine, One Gustave L. Levy Place, New York, NY 10029. Voice: 212-241-7312.

e-mail: gerald.cohen@mssm.edu

facts, nuts, bolts, and pieces of chemistry, biology, and physics that go together in approaches to more complex biologic phenomena. With many of the world's experts in attendance, the opening session provided overviews by Walling & Bernheim, and a masterly review by Gilbert on the role of free radicals and peroxides in oxygen toxicity. In the group photograph of the speakers at the meeting (Radiation Research, Supplement 3, 1963), Dan is the "kid" in the first row (FIG. 1). With his grasp of both chemical-physics and biology, and his interest in the important events and personalities in the field, Dan demonstrated a keen knack for providing the audience with the essential facts and keeping them in historical perspective. He is still doing that today with his overview of 50 years of free radical research in this, his own Festschrift volume, sponsored by the New York Academy of Sciences.

As a footnote, it was Frederick Bernheim's wife, Mary, who discovered and characterized tyramine oxidase (Mary L.C. Hare, Tyramine oxidase. I. A new enzyme in liver. Biochem. J. **22:** 968–979, 1928) as part of her doctoral research at Newnahm College in Cambridge (England) at about the time that the visiting Bernheim swept her off her feet and they were married. "Tyramine oxidase" is now better known as monoamine oxidase. It is the subject of the material that follows.

****

Front row: Titus C. Evans, B. A. Kihlman, Dean Burk, Hugo Aebi, Paul Kotin, Daniel L. Gilbert, E. L. Powers, Jr., F. H. Sobels, Nicholas A. Milas.
Back row: Elwood V. Jensen, Frederick Bernheim, J. St. L. Philpot, Howard I. Adler, John F. Thomson, Cheves Walling, Orville Wyss, Walter R. Guild.
Kneeling: Robert N. Feinstein.
Missing from photograph: Raymond Latarjet, John H. Pomeroy, Bernard Smaller.

FIGURE 1. Reproduced with permission of Academic Press.

INTRODUCTION

A state of oxidative stress exists when either oxidizing species, such as H_2O_2 or oxy-radicals, are present in excess, or cellular antioxidant defenses are lowered. Oxidative stress drives cellular systems to an oxidized state. One cellular target is the redox state of protein sulfhydryls (Pr-SH). Many enzymes contain essential thiol groups, which are derived from cysteine residues and are essential for biologic function. When these thiols are blocked by disulfide formation, such as protein-glutathione mixed disulfides (PrSSG), enzymatic function is suppressed. This change represents an oxidative step that is reversible catalyzed by enzymes such as thioredoxins and protein disulfide isomerase.[1,2]

One factor that markedly affects the redox state of protein thiols is the removal of H_2O_2 via the enzyme glutathione (GSH) peroxidase (Eqn 1). Detoxification of H_2O_2 results in the formation of glutathione disulfide (GSSG). In turn, GSSG reacts with protein thiols to form protein mixed disulfides (Eqn. 2).

$$H_2O_2 + 2GSH \rightarrow GSSG + 2H_2O \qquad (1)$$

$$GSSG + PrSH \rightarrow PrSSG + GSH \qquad (2)$$

Parkinson's disease is characterized by the loss of dopamine (DA) neurons from the substantia nigra, a heavily pigmented region of the human brain. The pigment is derived from DA itself and therefore loss of DA neurons is accompanied by a loss of pigmentation in this region of the brain. The surviving DA neurons are overactive. They increase their release of neurotransmitter and turnover of DA in a compensatory manner.[3] Indeed, this compensation effectively delays overt motor abnormalities until the loss of DA neurons is 80% or greater.

Most of the released DA is normally recaptured by the presynaptic nerve terminal and stored in vesicles, permitting reutilization of the neurotransmitter. However, a portion of the DA does not reach the safety of the storage vesicles, but instead encounters mitochondria, which are the site of metabolism of monoamine neurotransmitters by the enzyme monoamine oxidase (MAO). The metabolic turnover of DA closely parallels neuronal activity and increases as neuronal activity increases. This is evident from the rise in the acid metabolites of DA, namely DOPAC and homovanillic acid (HVA), in brain, which can be seen at autopsy in Parkinson's disease[3] and in animal models. MAO itself is an H_2O_2-generating enzyme (Eqn. 3). Therefore, the increased turnover of DA neurotransmitter and its O-methylated metabolite (3-O-methyl-DA) can be looked upon as an endogenous oxidative stress, increasing the steady state level of H_2O_2 and evoking oxidation of GSH to GSSG both pre- and post-synaptically in the region of DA neurons and nerve terminals.

$$\text{tyramine} + O_2 + H_2O \rightarrow H_2O_2 + NH_3 + p\text{-hydroxyphenylacetaldehyde} \qquad (3)$$

It was previously demonstrated that oxidation of DA and other monoamines (viz., tyramine, benzylamine) by isolated brain mitochondria can detrimentally affect mitochondrial electron transport.[4] Damage to electron transport, measured by a dye-reduction method, was evident when either pyruvate or succinate was used as the mitochondrial substrate. However, damage was more severe when pyruvate was used, compared to succinate. Pyruvate initiates electron flow at Complex I, while succinate initiates electron flow at Complex II. In more recent experiments, we measured directly the oxygen consumption (respiration) associated with electron flow. These experiments[5] showed that respiration is impaired by MAO activity and that this change is accompanied by a marked accumulation of mitochondrial PrSSG.

Parkinson's disease is also characterized by a significant loss (~35%) of Complex I activity from mitochondria.[6] Complex I takes reducing equivalents from pyruvate dehydrogenase and transfers them to coenzyme Q of the respiratory chain. Complex I activity is dependent upon protein thiol groups[7] and therefore, it can be affected by changes in the thiol redox state of associated proteins. Similarly, pyruvate dehydrogenase is also SH-dependent.[8] In the study described here, MAO activity evoked both a rise in PrSSG and inhibition of pyruvate-linked mitochondrial respiration.[5]

METHODS

Mitochondria were isolated from the pooled whole brain (less the cerebellum) of groups of three Sprague-Dawley rats (250–275 g) by a minor modification of the method of Clark and Nicklas.[9] The isolation medium consisted of 5 mM Mops, containing 0.225 M mannitol, 0.075 M sucrose, and 1.0 mM EGTA, adjusted to pH 7.4 with KOH. Isolation was carried out in the cold with a refrigerated Sorvall RC24 centrifuge, equipped with an SS 34 rotor for 10 min at $15,000 \times g$. The isolated mitochondria were suspended in cold Mops buffer at a concentration of 15–20 mg mitochondria protein/ml and maintained in an ice bath until used. The yield was 3–4 mg mitochondrial protein per rat brain.

Incubations were conducted by dilution of an aliquot of the mitochondrial preparation to 0.5 mg or 1.0 mg mitochondrial protein/ml in the respiration buffer (pH 7.2), which consisted of 5 mM Hepes, 125 mM sucrose, 50 mM KCl, 2 mM KH_2PO_4, and 1 mM $MgCl_2$ with 0.5 mg bovine serum albumin/ml at 27°C. Incubations were carried out at 27°C in a volume of 1 ml in plastic tubes (12-ml) on a water bath with gentle shaking for 15 min. Samples were processed individually with immediate assessment of respiration and rotation among the experimental groups. Samples not incubated with MAO inhibitors (2 µM clorgyline plus 2 µM pargyline) received additions of the inhibitors after the incubation was complete, just prior to the measurement of respiration or electron flow. Each experiment consisted of 10–12 samples (3–4 samples per group).

Respiration was measured in a miniature chamber system (0.6 ml capacity; Instech Labs. Plymouth Meeting, PA.), equipped with a magnetic stirrer and maintained at 27°C. Oxygen consumption was assessed with a YSI Model 5300 Biological Oxygen Monitor (Yellow Springs Instrument Co., Yellow Springs, OH). Measurements were made sequentially after the addition of pyruvate/malate (5 mM each) (State 4 respiration), followed by 0.4 mM ADP (State 3) and, lastly, 10 µM FCCP (State 5). In some experiments, ADP was omitted and State 5 respiration was measured directly. Respiratory activity of the stock mitochondrial preparation, which was held on ice, was well maintained and did not change over the course of 3–4 h.

PrSSG was measured with a modification of the method of Akerboom and Sies.[10] The liberated GSH was measured on a plate reader with a modification of the enzymatic recycling method of Tietze.[11] Protein was measured by the method of Lowry

TABLE 1. Effect of MAO activity on mitochondrial state 3 and state 5 respiration

Respiratory state	natoms oxygen/min/mg mitochondrial protein (% untreated control)			
	Without MAO inhibitors		With MAO inhibitors	
	Control	Tyramine	Control	Tyramine
State 3	100.0 ± 3.3	67.2 ± 1.7^a	108.8 ± 3.8	95.2 ± 3.7^b
State 5	100.0 ± 2.5	59.9 ± 1.9^a	100.3 ± 2.7	99.5 ± 2.9^b

NOTE: Isolated rat brain mitochondria were incubated with 500 μM tyramine for 15 minutes at 27°C and, subsequently, State 3 and State 5 respiration were measured with an oxygen electrode and with 5 mM pyruvate plus 5 mM malate as substrate. Where indicated, the MAO inhibitors clorgyline and pargyline (2 μM each) were present. Data are the mean ± SEM for N = 5–6 per group for State 3 respiration and N = 10–11 for State 5 respiration.
$^a p < 0.001$ vs. untreated control.
$^b p < 0.001$ vs. tyramine without MAO inhibitors.

et al.[12] and was used to normalize data for both PrSSG and respiration, which were expressed per mg protein. Data are expressed as the mean ± SEM. Statistical assessment was conducted by the Tukey-Kramer multiple comparison test or, where appropriate, by the 2-tailed Student t-test.

RESULTS

Prior to conducting experiments, the preparations of rat brain mitochondria were characterized. The respiratory control ratios (State 3/State 4) of freshly isolated mitochondria were in the range 5.5–7.0 with 5 mM pyruvate plus 5 mM malate as substrate. State 3 respiration was in the range 72–103 ng-atoms oxygen/min/mg protein. These parameters are in good agreement with reports in the literature.[9]

Rat brain mitochondria were exposed to 500 μM tyramine for 15 min at 27°C in either the absence of presence of MAO inhibitors. The concentration of tyramine corresponds to the level of neurotransmitter in the cytosol of catecholamine neurons

TABLE 2. Effect of MAO activity with either tyramine or dopamine as substrate on mitochondrial levels of protein-glutathione mixed disulfides (PrSSG)

Time (min)	mitochondrial PrSSG (μmoles/mg protein)		
	Control	Tyramine	Dopamine
0	0.03 ± 0.02	0.03 ± 0.02	0.03 ± 0.02
7.5	0.05 ± 0.01	0.53 ± 0.03	0.41 ± 0.03
15	0.07 ± 0.01	0.84 ± 0.05	0.65 ± 0.01
22.5	0.05 ± 0.02	0.94 ± 0.06	0.77 ± 0.04

NOTE: Isolated rat brain mitochondria were incubated with either 500 μM tyramine or 500 μM dopamine for 22.5 minutes at 27°C. Subsequently a protein pellet was isolated with perchloric acid and the level of PrSSG was measured. Data are the mean ± SEM for $N = 4$. All elevations induced by tyramine or dopamine were significant ($p < 0.001$).

(discussed in Cohen, Farooqui & Kesler[4]). The results of these experiments (TABLE 1) show that State 3 respiration was suppressed by 32.8% and State 5 respiration by 40.1% after exposure to tyramine. Tyramine is a mixed MAO-A/MAO-B substrate. Inhibition of both isoforms of MAO with a mixture of 2 μM clorgyline (selective MAO A inhibitor) and 2 μM pargyline (selective MAO B inhibitor) prevented the damage to mitochondrial respiration (TABLE 1). These observations are consistent with damage by H_2O_2, which is a product of MAO activity. Because pyruvate was used as substrate, electron flow was initiated at pyruvate dehydrogenase/Complex I.

Mitochondria were also evaluated for the effects of the incubation conditions alone. This was done because it is well known that isolated mitochondria are susceptible to damage by agitation, particularly at 37°C. Indeed, most investigations of respiration are carried out at lower temperature, such as 30°C or room temperature. We used 27°C in order to limit damage by environmental conditions. Simply shaking the mitochondria at 27°C decreased respiration by 27.3% ± 0.7% for State 3 and 21.7% ± 1.5% for State 5 ($p < 0.01$, n = 5/group, 2 experiments). However, the data in TABLE 1 are expressed as the effects of tyramine relative to incubated controls and, therefore, changes due to experimental conditions, unrelated to tyramine, cancel out. Nonetheless, the loss of a highly vulnerable fraction of respiratory activity may cause the effect of tyramine to be underestimated in TABLE 1. The stock, concentrated suspension of mitochondria (15–20 mg protein/ml) in Mops buffer, held on ice, was stable and did not lose respiratory activity over the course of the experiments (3–4 h).

The results shown in TABLE 2 show that the levels of PrSSG rose rapidly during the first 7.5 minutes of incubation. The rise was greater than 10-fold. Further accumulation of PrSSG was seen at 15 minutes and 22.5 minutes. In separate experiments it was observed that inhibition of MAO by a mixture of clorgyline and pargyline suppressed the rise in PrSSG: while PrSSG achieved levels of 0.4 to 1.2 μmoles/ mg protein with either tyramine or dopamine, the MAO inhibitors suppressed PrSSG to control levels (less then 0.04 μmoles/mg protein). As in the study of respiration, this result was not due to the added tyramine per se, nor to possible effects of autoxidizing DA, because MAO inhibitors completely suppressed the rise in PrSSG.

DISCUSSION

The main observations are that incubation of intact rat brain mitochondria with tyramine results in suppression of both State 3 and State 5 respiration, accompanied by a rise in mitochondrial PrSSG. The result is not due to the added tyramine per se, but rather, to a product of MAO activity, because inhibition of MAO activity by a combination of clorgyline and pargyline completely suppressed both PrSSG accumulation and damage to respiratory activity. Because pyruvate was used as substrate, electron flow was initiated at pyruvate dehydrogenase/Complex I.

Mitochondrial defects associated with Complexes I–IV of the respiratory chain occur in a number of neurodegenerative diseases, including Parkinson's disease, Huntington's disease, Friedreich's ataxia, hereditary spastic paraplegia, Alzheimer's disease, and amyotrophic lateral sclerosis.[13,14] Moreover, several animal models of neurodegenerative disease are based on mitochondrial toxins, such as MPTP, which inhibits Complex I of the electron transport chain, producing an animal model for Parkinson's disease, or 3-nitroproprionic acid and malonate, which inhibit Complex II, producing models for Huntington's disease.[15] Therefore, mitochondrial defects appear to play primary roles in disease expression and progression. Defects in cellular respiration lead to diminished ATP production, increased sensitivity to oxidative stress and, eventually, to apoptotic or necrotic neuronal cell death.[16] The defect in Parkinson's disease is localized to Complex I and is seen at autopsy as a 35% decrement in the substantia nigra, which is the region of the brain containing the affected DA neuron cell bodies.

Mitochondrial respiratory defects can be directly inherited or may be acquired as the result of exposure to stressors. The experiments described here identify a mitochondrial enzyme, monoamine oxidase, and the turnover of monoamine neurotransmitters by MAO, as a source of oxidative stress that can suppress mitochondrial respiration. MAO is a flavo-enzyme, localized to the outer mitochondrial membrane.[17] It plays an essential metabolic role in the turnover of dopamine, serotonin, norepinephrine, and epinephrine in the central nervous system. As discussed earlier, oxidative deamination of monoamines by MAO is accompanied by the reduction of molecular oxygen to H_2O_2,[18,19] a potentially toxic agent that can evoke changes in the cellular thiol status, as well as direct damage to mitochondrial DNA.

H_2O_2 is also formed naturally during mitochondrial respiration. It is estimated that 1–3% of consumed oxygen is converted to H_2O_2.[20] H_2O_2 that "leaks" from the electron transport chain damages both mitochondrial proteins and mitochondrial DNA.[21,22] It is widely believed that the H_2O_2 generated in this way is responsible for the decline in mitochondrial function in aging, reperfusion injury, and certain disease states.[23,24] However, the quantity of H_2O_2 generated by mitochondrial MAO exceeds by a wide margin the amount generated during electron flow. Hauptmann et al.[19] studied the oxidation of 2 mM tyramine by rat brain mitochondria and reported that H_2O_2 production was 48-fold greater than that from succinate during electron transport in the presence of antimycin A. Hence, MAO possesses a considerable toxic potential. Moreover, the mitochondrial localization of MAO makes this enzyme uniquely situated to evoke selective mitochondrial damage.

It remains an open question how much of the mitochondrial damage seen in Parkinson's disease is of strictly genetic origin and how much is derived from damage by H_2O_2 generated during enhanced turnover of DA, particularly during treatment with L-dopa. Moreover, it remains to be seen whether or not mitochondrially generated H_2O_2 interacts with genetic factors in subjects predisposed to Parkinson's disease. For example, recent studies have described genetic defects, such as the alpha synuclein gene and the parkin gene in select families with Parkinson's disease[25,26]; other defects affecting Complex II of the respiratory chain have been described in Huntington's disease. How these genes interact with endogenous or environmental factors to produce different mitochondrial lesions and neurodegenerative states is as yet unclear. For Parkinson's disease, the production of H_2O_2 during the natural turnover of DA, or the enhanced turnover associated with overt symptomatology, may place genetically susceptible subjects at risk for damage to mitochondrial respiratory activity. If such events are mediated by PrSSG formation, they may be reversible.

Therefore, further studies of the relationships between MAO activity, PrSSG accumulation, and defects in mitochondrial respiration are clearly warranted.

ACKNOWLEDGMENT

This study was supported by a grant DAMD17-98-1-8624 from the U.S. Army Medical Research and Materiel Command (USAMRMC) and by a grant from the Parkinson's Disease Foundation. Support by USAMRMC does not constitute endorsement by the U.S. Government or the U.S. Army.

REFERENCES

1. HOLMGREN, A. 1985. Thioredoxin. Annu. Rev. Biochem. **54:** 237–271.
2. RABENSTEIN, D.L. & K.K MILLIS. 1995. Nuclear magnetic resonance study of the thioltransferase-catalyzed glutathione/glutathione disulfide interchange reaction. Biochim. Biophys. Acta **1249:** 29–36.
3. HORNYKIEWICZ, O. & S.J. KISH. 1996. Biochemical pathophysiology of Parkinson's disease. Adv. Neurol. **45:** 19–34.
4. COHEN, G., R. FAROOQUI & N. KESLER. 1997. Parkinson disease: a new link between monoamine oxidase and mitochondrial electron flow. Proc. Natl. Acad. Sci. USA **94:** 4890-4894.
5. COHEN, G. & N. KESLER. 1999. Monoamine oxidase and mitochondrial respiration. J. Neurochem. **73:** 2310–2315.
6. SCHAPIRA, A.H. 1998. Mitochondrial dysfunction in neurodegenerative disorders. Biochim. Biophys. Acta **1366:** 225–233.
7. GUTMAN, M., H. MERSMANN, J. LUTHY & T.P. SINGER. 1970. Action of sulfhydryl inhibitors on different forms of the respiratory chain-linked reduced nicotinamide-adenine dinucleotide dehydrogenase. Biochemistry. **9:** 2678–2687.
8. ALI, M.S., T.E. ROCHE & M.S. PATEL. 1993. Identification of the essential cysteine residue in the active site of bovine pyruvate dehydrogenase. J. Biol. Chem. **268:** 22353–22356.
9. CLARK, J.B. & W.J. NICKLAS. 1970. The metabolism of rat brain mitochondria. Preparation and characterization. J. Biol. Chem. **245:** 4724–4731.
10. AKERBOOM, T.P.M. & H. SIES. 1981. Assay of glutathione, glutathione disulfide, and glutathione mixed disulfides in biological samples. Meth. Enzymol. **77:** 373-382.
11. TIETZE, F. 1969. Enzymic method for the quantitative determination of nanogram amounts of total and oxidized glutathione: Applications to mammalian blood and other tissues.. Anal. Biochem. **27:** 502–522.
12. LOWRY, O., N.J. ROSEBROUGH., A.L .FARR & R.J. RANDALL. 1951. Protein measurement with the Folin phenol reagent. J. Biol. Chem. **193:** 265–-275.
13. SCHAPIRA, A.H. 1999. Mitochondrial involvement in Parkinson's disease, hereditary spastic paraplegia and Friedreich's ataxia. Biochim. Biophys. Acta **1410:** 159–170.
14. CASSARINO, D.S. & J.P. BENNETT JR. 1999. An evaluation of the role of mitochondria in neurodegenerative diseases: mitochondrial mutations and oxidative pathology, protective nuclear responses, and cell death in neurodegeneration. Brain Res. Rev. **2:** 1–25.
15. SCHULZ, J.B., R.T. MATTHEWS, T. KLOCKGETHER, J. DICHGANS & M.F. BEAL. 1997. The role of mitochondrial dysfunction and neuronal nitric oxide in animal models of neurodegenerative disease. Mol. Cell. Biochem. **174:** 193–197.
16. ZAMZAMI, N., T. HIRCH, B. DALLAPORTA, P.X. PETIT & G. KROEMER. 1997. Mitochondrial implication in accidental and programmed cell death: apoptosis and necrosis. J. Bioenerg. Biomembr. **29:** 185–193.

17. RAGAN, C.I., M.T. WILSON, V.M. DARLEY-USMAR & P.N. LOWE. 1987. Sub-fractionation of mitochondria and isolation of the proteins of oxidative phosphorylation. *In* Mitochondria. A Practical Approach. V.M. Darley-Usmar, D. Rickwood & M.T. Wilson, Eds. :79–112. IRL Press. Oxford.
18. SINET, P.M., R.E. HEIKKILA & G. COHEN. 1980. Hydrogen peroxide formation by rat brain in vivo. J. Neurochem **34:** 1420–1428.
19. HAUPTMANN, N., J. GRIMSBY, J.C. SHIH & E. CADENAS. 1996. The metabolism of tyramine by monoamine oxidase A/B causes oxidative damage to mitochondrial DNA. Arch. Biochem. Biophys. **335:** 295–304.
20. CHANCE, B., H. SIES & A. BOVERIS. 1979. Hydroperoxide metabolism in mammalian organs. Physiol Rev. **59:** 527–605.
21. SOHAL, R.S., B.H. SOHAL & W.C. ORR. 1995. Mitochondrial superoxide and hydrogen peroxide generation, protein oxidative damage, and longevity in different species of flies. Free Radic. Biol. Med. **19:** 499–504.
22. GIULIVI, C. & E. CADENAS. 1998. The role of mitochondrial glutathione in DNA base oxidation. Biochim. Biophys. Acta **1366:** 265–274.
23. KU, H.H., U.T. BRUNK & R.S. SOHAL. 1993. Relationship between mitochondrial superoxide and hydrogen peroxide production and longevity of mammalian species. Free Radic. Biol. Med. **15:** 621–627
24. RICHTER, C., V. GOGVADZE, R. LAFFRANCHI, R. SCHLAPBACH, M. SCHWEIZER, M. SUTER, P. WALTER & M. YAFFEE. 1995. Oxidants in mitochondria: From physiology to diseases. Biochim. Biophys. Acta **1271:** 67–74.
25. POLYMEROPOULOS, M.H., C. LAVEDAN, E. LEROY, S.E. IDE, A. DEHEJIA, A. DUTRA, B. PIKE, H. ROOT, J. RUBENSTEIN, R. BOYER, E.S. STENROOS, S. CHANDRASEKHARAPPA, A. ATHANASSIADOU, T. PAPAPETROPOULOS, W.G. JOHNSON, A.M. LAZZARINI, R.C. DUVOISIN, G. DI IORIO, L.I. GOLBE & R.L. NUSSBAUM. 1997. Mutation in the alpha-synuclein gene identified in families with Parkinson's disease. Science **276:** 2045–2047.
26. KITADA, T., S. ASAKAWA, N. HATTORI, H. MATSUMINE, Y. YAMAMURA, S. MINOSHIMA, M. YOKOCHI, Y. MIZUNO & N. SHIMIZU. 1998. Mutations in the parkin gene cause autosomal recessive juvenile parkinsonism. Nature **392:** 605–608.

Regulation of Mitochondrial Respiration by Oxygen and Nitric Oxide

ALBERTO BOVERIS,[a,b] LIDIA E. COSTA,[c] JUAN J. PODEROSO,[d]
MARIA C. CARRERAS,[d] AND ENRIQUE CADENAS[e]

[a]*Laboratory of Free Radical Biology, School of Pharmacy and Biochemistry*
[c]*Institute of Cardiological Research, and*
[d]*University Hospital, School of Medicine,*
University of Buenos Aires, Buenos Aires, Argentina

[e]*Department of Molecular Pharmacology and Toxicology, School of Pharmacy,*
University of Southern California, Los Angeles, California, USA

ABSTRACT: Although the regulation of mitochondrial respiration and energy production in mammalian tissues has been exhaustively studied and extensively reviewed, a clear understanding of the regulation of cellular respiration has not yet been achieved. In particular, the role of tissue pO_2 as a factor regulating cellular respiration remains controversial. The concept of a complex and multisite regulation of cellular respiration and energy production signaled by cellular and intercellular messengers has evolved in the last few years and is still being researched. A recent concept that regulation of cellular respiration is regulated by ADP, O_2 and NO preserves the notion that energy demands drive respiration but places the kinetic control of both respiration and energy supply in the availability of ADP to F_1-ATPase and of O_2 and NO to cytochrome oxidase. In addition, recent research indicates that NO participates in redox reactions in the mitochondrial matrix that regulate the intramitochondrial steady state concentration of NO itself and other reactive species such as superoxide radical (O_2^-) and peroxynitrite ($ONOO^-$). In this way, NO acquires an essential role as a mitochondrial regulatory metabolite. NO exhibits a rich biochemistry and a high reactivity and plays an important role as intercellular messenger in diverse physiological processes, such as regulation of blood flow, neurotransmission, platelet aggregation and immune cytotoxic response.

INTRODUCTION

The regulation of mitochondrial respiration and energy production in mammalian tissues has been exhaustively studied and extensively reviewed. However, a clear understanding of the regulation of cellular respiration is not yet complete. In particular, the role of tissue pO_2 as a factor regulating cellular respiration is a matter of controversy. It was considered that maximal rates of mitochondrial respiration could be maintained in a wide range of tissue pO_2, from the usual 5 to 30 μM O_2 to about 0.8 μM O_2[1–3] based on the $[O_2]_{0.5}$ values (the $[O_2]$ that sustains half maximal respiratory rate) of 0.02 to 0.3 μM O_2 that were measured with isolated mitochondria and cells.[4–5] Considering this high affinity of mitochondrial respiration for O_2, the rates

---

[b]Author to whom correspondence should be addressed.

at which mitochondria perform oxidative phosphorylation were supposed to be rather independent of tissue pO_2 and limitation of respiration by $[O_2]$ was thought to occur only under severe hypoxia. However, recent $[O_2]_{0.5}$ values carefully determined turned out to be higher than the ones previously measured, especially for active (state 3) respiration.[6–10] The $[O_2]_{0.5}$ determined by Costa et al.[10] using high-resolution respirometry in liver and heart mitochondria were 0.30–0.40 µM in state 4 and 1.57–1.69 µM in state 3,[10] which implies that the intracellular $[O_2]$ prevailing in some tissues, e.g., 3–8 µM in the heart,[2–3] would be regulatory under normoxia with respiration slowed below its maximal rate. Defining a critical rate (Vc) as 80% of V_{max}[9] and considering a classical Michaelis-Menten kinetics it follows that the critical $[O_2]$ will be reached at 6 µM O_2 ($V_c = V_{max} [O_2]/([O_2]_{0.5} + [O_2])$) and that limitation of active mitochondrial respiration may occur at physiological normoxia in the heart; higher values of pO_2, in the range of 22 to 32 µM O_2 and well above the critical $[O_2]$, have been reported for skeletal muscle and liver, however.[1,2,4,11]

Cytochrome oxidase was not usually considered a regulatory enzyme, yet evidence has been obtained that non-catalytic subunits alter enzyme kinetics.[12] Direct spectral analysis have indicated that a substantial fraction of cytochrome oxidase is reduced in intact tissues,[13] even though this reduction is not observed in isolated mitochondria. It has been suggested that this difference is due to regulatory mechanisms that are lost in vitro.[14] It is worth noting that the reduced cytochrome oxidase-NO complex is spectroscopically similar to reduced cytochrome oxidase.[15–16]

The concept of a complex and multisite regulation of cellular respiration and energy production signaled by cellular and intercellular messengers has evolved in the last few years and is still under development. The classical and elegant concept of the regulation of cellular oxygen uptake by ADP put forward by Lardy and Wellman,[17] Chance and Williams,[18] and Estabrook[19] considers that energy needs drive respiration and that availability of ADP to mitochondrial F_1-ATPase exerts the kinetic control of respiration and energy production over a wide range of O_2 concentration that certainly includes the physiological conditions (FIG. 1). The new concept of regulation of cellular respiration by ADP, O_2 and NO keeps the idea that energy

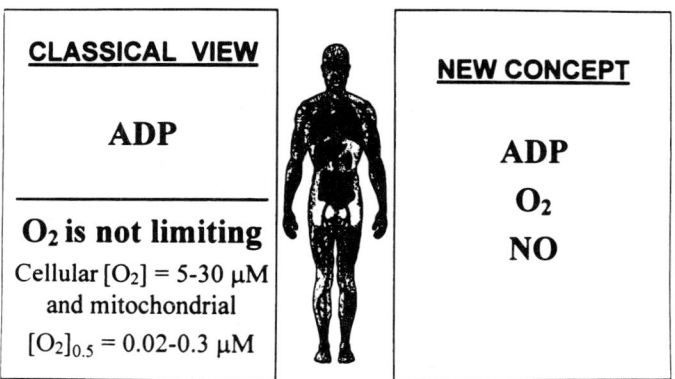

FIGURE 1. Classical view and new concept of the regulation of cellular O_2 uptake.

demands drive respiration but places the kinetic control of both respiration and energy supply in the availability of ADP to F_1-ATPase and of O_2 and NO to cytochrome oxidase (FIG. 1). In addition, recent reports by Poderoso et al.[20–22] and by Boveris et al.[23] indicate that NO participates in redox reactions in the mitochondrial matrix that regulate the intramitochondrial steady state concentration of NO itself and other reactive species such as superoxide radical (O_2^-) and peroxynitrite ($ONOO^-$).

In this way, NO acquires an essential role as a mitochondrial regulatory metabolite. Nitric oxide, a diatomic gas with an unpaired electron in its external orbitals, exhibits a rich biochemistry and a high reactivity and plays important roles as intercellular messenger in diverse physiological processes, such as regulation of blood flow, neurotransmission, platelet aggregation and immune cytotoxic response.[24] Endothelium produced NO stimulates the guanylate cyclase activity of the underlying vascular smooth muscle cells and the increased level of cGMP produces muscle relaxation and vasodilation.[25–26] This action slows down blood flow and increases the equilibration time of HbO_2 with tissue O_2 resulting in better tissue oxygenation. The coupling of the endothelial production of NO and blood flow to changes in tissue pO_2 may be thought as dependent on cellular metabolites that are produced when mitochondrial metabolism becomes O_2 limited and that reach endothelial cells.

Since Granger and Lehninger[27] early recognized macrophage cytotoxicity as exerted partly on mitochondrial respiration, the recognition of NO production in activated macrophages and neutrophils[28–29] prompted the assay of NO effects on mitochondrial function. Two British research groups, Cleeter et al.[30] and Brown and Cooper,[31] simultaneously reported the inhibitory effects of NO on cytochrome oxidase activity using skeletal muscle mitochondria[30] and brain synaptosomes.[31] The inhibition is reversible and competitive with O_2[31] suggesting that tissue NO may raise the $[O_2]_{0.5}$ and therefore become the first known physiological regulator to act directly on the mitochondrial respiratory chain.

Cytosolic Ca^{2+} has been frequently suggested as a regulator of mitochondrial function, likely as part of a role as intracellular second messenger that signals activation of diverse and specialized cell responses, many of which require increased energy production.[14] However, cytosolic Ca^{2+}, as a cation able to discharge mitochondrial membrane potential, seems to act as antagonist for the coordinated activation of mitochondrial respiration and energy supply.[32]

OXYGEN DEPENDENCE OF MITOCHONDRIAL RESPIRATION

The dependence of the respiratory rate of isolated mitochondria and cells on $[O_2]$ is hyperbolic; therefore it is conveniently described by $[O_2]_{0.5}$, the oxygen concentration that provides half-maximal respiration rate. The $[O_2]_{0.5}$ values are also sometimes referred to as $K_{0.5}$ and Km_{O2}; however, since, owing to their operational nature they are different with mitochondria in state 4 or in state 3 and depend on [NO], it is convenient to use only the $[O_2]_{0.5}$ notation and concept.

The use of high resolution respirometers is advisable to precisely measure the rates of O_2 uptake that occur in the 0.01 to 5 µM O_2 range which is essential to determine $[O_2]_{0.5}$ values. The $[O_2]_{0.5}$ of active (state 3) and resting (state 4) respiration were recently reexamined in tightly coupled mitochondria isolated from rat liver and

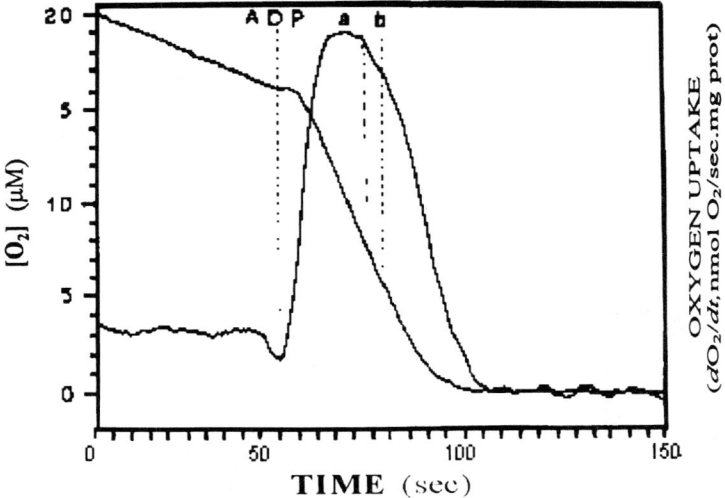

FIGURE 2. Oxygen concentration ($[O_2]$) and its first derivative (rate of O_2 uptake; $d[O_2]/dt$) during the respiratory activity of rat liver mitochondria (0.35 mg prot/ml) suspended in 0.25 M sucrose, 0.5 mM EGTA, 5 mM $MgCl_2$, 1.5% BSA, 10 mM HEPES, and 6 mM phosphate buffer (pH 7.35). ADP pulse: 0.3 mM. **A:** initial rate of O_2 uptake, $\Delta[O_2]/\Delta t$, used to produce one data point in FIGURE 3. **B:** time course of the differential rates of O_2 uptake, $d[O_2]/dt$, that generate the data points of FIGURE 4. Modified from Costa et al.[10]

TABLE 1. Oxygen dependence of active (state 3) mitochondrial respiration determined by four approaches

	Aerobic mitochondria		Anaerobic mitochondria	
	ADP pulses		O_2 pulses	
	Initial rates	Differential rates	Air-saturated medium	H_2O_2
Liver mitochondria				
Vmax (nmol O_2/sec·mg prot)	1.80 ± 0.02	1.78 ± 0.09	1.84 ± 0.07	1.82 ± 0.08
$[O_2]_{0.5}$ (µM)	2.45 ± 0.32	1.69 ± 0.09	1.04 ± 0.02	1.09 ± 0.04
Heart mitochondria				
Vmax (nmol O_2/sec·mg prot)	1.47 ± 0.10	1.55 ± 0.09	1.51 ± 0.09	1.53 ± 0.10
$[O_2]_{0.5}$ (µM)	3.02 ± 0.30	1.45 ± 0.10	1.07 ± 0.04	1.12 ± 0.10

heart by Costa et al.[10] The $[O_2]_{0.5}$ values corresponding to state 4 respiration (0.30–0.40 μM) were not different from previously reported values obtained using advanced instrumentation[6–9] and are given, as well as methodological details, elsewhere.[10,23] The new $[O_2]_{0.5}$ values corresponding to state 3 respiration were determined using four different experimental approaches and opened the way to the new concept of respiratory regulation by pO_2.[10,23]

In the first approach, termed "initial rates," ADP pulses were added to mitochondria in the resting state 4, at a different $[O_2]$ in different runs, collecting one data point in each run, *i.e.* the maximal rate, $\Delta O_2/\Delta t$, usually constant during several sec after ADP addition (FIG. 2). Hyperbolic fitting of the individual data points gave the $[O_2]_{0.5}$ (FIG. 3 and TABLE 1, initial rates), which tend to be overestimated because only a few points could be obtained in the range of limiting $[O_2]$.

In the second approach, termed "differential rates," ADP pulses were also added to mitochondria in resting state 4, with excess ADP to exhaust O_2 in the reaction medium and taking advantage of the 0.2–1 sec period in which dO_2/dt can be measured in high resolution respirometers (FIG. 2). The dO_2/dt values were fitted to a single hyperbola as a function of $[O_2]$ and gave the highest possible accuracy for the determination of $[O_2]_{0.5}$ (FIG. 4 and TABLE 1, differential rates). The third approach, termed "O_2 pulses," consisted of the injection of air-saturated reaction medium to mitochondria incubated in anaerobiosis for 3–5 min in the presence of non-limiting [ADP] (FIG. 5). In order to reach easily anaerobiosis without accumulation of metabolic by-products, mitochondria were suspended in a reaction medium in which O_2

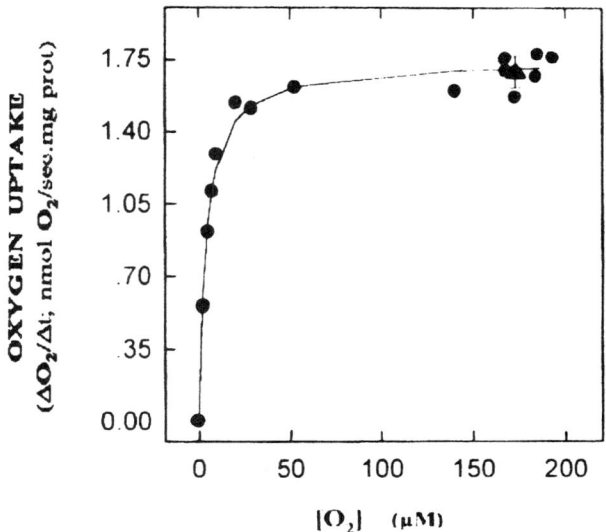

FIGURE 3. Oxygen dependence of the state 3 respiration of rat liver mitochondria showing the hyperbolic fitting of the initial rates ($\Delta[O_2]/\Delta t$). For details see text and FIGURE 2. The triangle indicates the mean value of the O_2 uptake rate measured in air-saturated buffer.

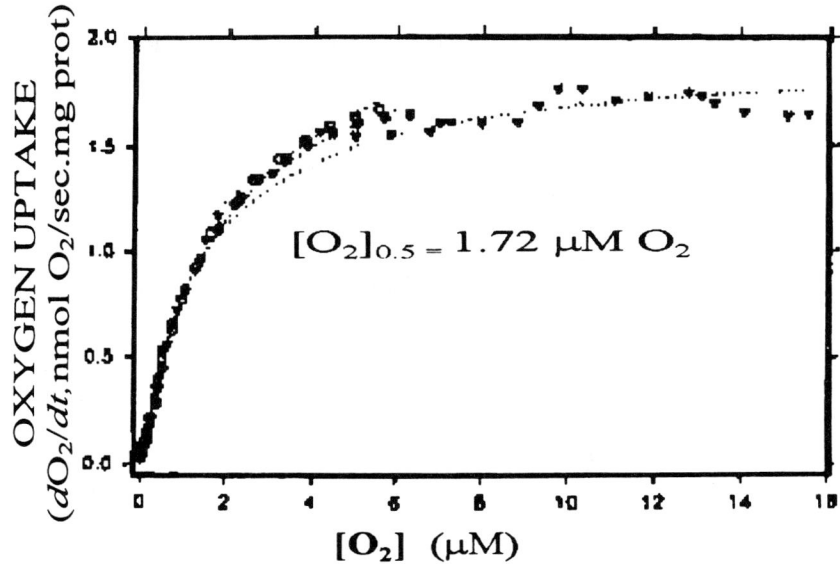

FIGURE 4. Oxygen dependence of the state 3 respiration of rat liver mitochondria. The dotted line is the hyperbolic fitting of the data points ($d[O_2]/dt$) collected at 1 sec intervals after an ADP pulse. For details see text and FIGURE 2. Different symbols indicate different runs. The calculated $[O_2]_{0.5}$ is 1.7 µM. Modified from Costa et al.[10]

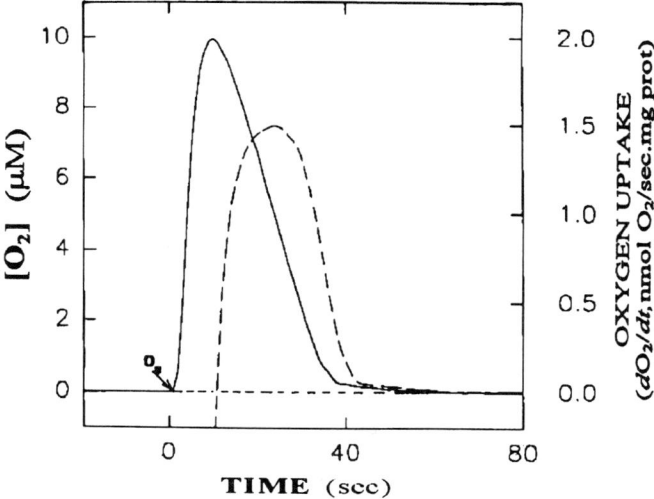

FIGURE 5. Oxygen concentration ($[O_2]$) and its first derivative ($d[O_2]/dt$) during the respiration of rat liver mitochondria (0.30 mg prot/ml) that were incubated in anaerobiosis for 3–5 min and added with an O_2 pulse. For details see text. Experimental conditions as in FIGURE 2.

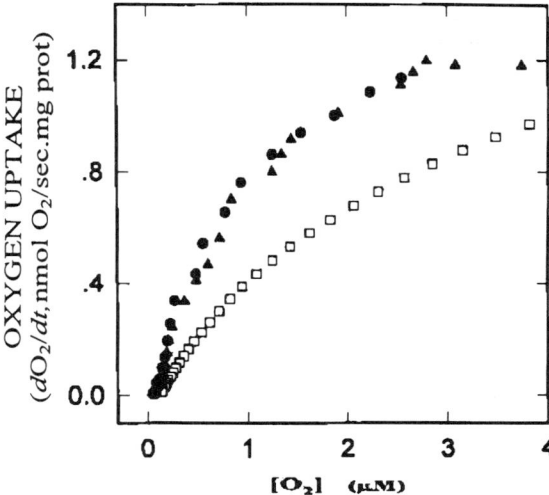

FIGURE 6. Oxygen dependence of the state 3 respiration of rat liver mitochondria that were pulsed with ADP in aerobiosis (□) or with O_2 after anaerobiosis (●, ▲). The data points ($d[O_2]/dt$) were collected at 1 sec intervals. For details see text and FIGURES 2 and 5. The calculated $[O_2]_{0.5}$ are 1.60 μM O_2 (ADP pulse) and 1.05 μM O_2 (O_2 pulse).

was diminished to 10–20 μM O_2 by flushing N_2. The non-linear regression fitting of dO_2/dt as a function of $[O_2]$ from two consecutive O_2 pulses and from an ADP pulse show that the $[O_2]_{0.5}$ was significantly lower in the case of the O_2 pulses than with the ADP pulse (FIG. 6, TABLE 1). Because of the relatively low $[O_2]$ in air-saturated reaction medium and the fast rate of state 3 respiration, a fourth approach delivering more O_2 to the reaction medium and termed "H_2O_2 pulse" was introduced. In this condition 10 μM O_2, enough to reach the V_{max}, is easily added to anaerobic catalase-supplemented mitochondria by the injection of a few μl of a H_2O_2 solution. The $[O_2]_{0.5}$ values obtained by this approach were similar to the ones obtained by the O_2 pulses (TABLE 1). The lower $[O_2]_{0.5}$ values obtained after anaerobiosis suggest the existence of an O_2-dependent system that is able to modulate mitochondrial respiration at low $[O_2]$ and that the regulator is competitive with O_2, since $[O_2]_{0.5}$ is increased in its presence while V_{max} is not changed. The production of NO by mitochondrial nitric oxide synthase (mtNOS)[33–34] during active state 3 respiration seems to afford the referred regulation.[35]

THE EFFECTS OF NO ON THE MITOCHONDRIAL RESPIRATORY CHAIN

The rich biochemical properties of NO make possible its multiple effects on the mitochondrial respiratory chain: (a) NO inhibits cytochrome oxidase activity competitively with oxygen; (b) NO inhibits electron transfer between cytochromes *b* and

c and increases the mitochondrial production of O_2^-; and (c) NO inhibits electron transfer and NADH-dehydrogenase function in Complex I.

Cytochrome Oxidase (Complex IV)

Concerning the inhibition of cytochrome oxidase activity, NO binds to the enzyme in its reduced and oxidized forms; in the reduced form NO binds to the binuclear reaction center formed by cytochrome a_3 heme and Cu_B. The cytochrome oxidase activity of the isolated enzyme[15,31] and of rat heart submitochondrial particles,[21] as well as the active respiration of mitochondria isolated from rat muscle,[30] liver,[36] heart[21] and brown adipose tissue[37] and of rat brain synaptosomes[31] are effectively inhibited by 0.05–2 µM NO (see FIG. 7). Binding and inhibition are reversible and removable by washing[21,30] or by addition of excess myoglobin or hemoglobin.[21,36] The degree of inhibition of cytochrome oxidase activity by NO depends on the O_2 concentration in the reaction medium [31,36,37]; NO and O_2 compete for the binding site at the reaction center of cytochrome oxidase. Then, the inhibition of mitochondrial respiration by NO can be expressed as a function of the ratio [O_2]/[NO]; half-maximal inhibition of state 3 respiration is reached at a ratio of 150 O_2/NO (FIG. 7, inset) which clearly indicates the very high affinity of NO for cytochrome oxidase. Ratios of 400–500 O_2/NO and 500–1000 O_2/NO have been reported to inhibit 50% the respiration of rat brain synaptosomes[31] and of rat brown adipose tissue,[37] respectively.

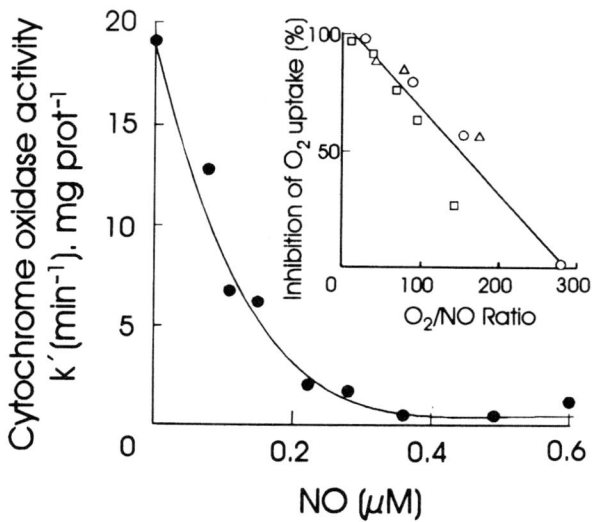

FIGURE 7. Inhibition of cytochrome oxidase activity and active respiration by NO. The inhibition of O_2^- uptake, normalized for the different experiments, is plotted against the ratio of O_2 and NO concentrations. Rat liver mitochondria in the presence of: ○, ADP (state 3), and △, 20 µM dinitrophenol (state 3u); data from Takahara et al.[36] Rat heart mitochondria, △, in the presence of ADP; data from Poderoso et al.[20]

The Ubiquinol-Cytochrome b-Cytochrome c Space (Complex III)

Rat heart submitochondrial particles added with NO show a marked inhibition of their succinate-cytochrome c activity with half maximal effect at about 0.7 mM NO (FIG. 8 and ref. 22) with a NO-induced reduction of cytochrome b.[21,22] This second effect of NO on the mitochondrial respiratory chain results in increased O_2^- production in submitochondrial particles and H_2O_2 generation in whole mitochondria, being about 0.5 mM NO the concentration required for half maximal effect (FIG. 8 and ref. 21). The interaction of NO with the NO-reactive component of the ubiquinone-cytochrome b area of the mitochondrial respiratory chain, likely an iron-sulfur center, is also reversible but is not affected by the O_2/NO ratio. The time course of the inhibition of cytochrome oxidase produced by a 1 mM NO pulse as a function of $[O_2]$ shows a marked O_2 dependence (a steeper slope) when ascorbate-TMPD, reductants of cytochrome oxidase, were added, and a lower O_2 dependence (a less marked slope) when succinate and antimycin, reductants of the ubiquinone-cytochrome b isopotential pool, were supplied (FIG. 8, *inset*). In the first case, the marked O_2 dependence indicates the competition of NO and O_2 for cytochrome oxidase. In the latter case, the lower O_2 dependence of the inhibition of the ubiquinol-cytochrome c electron transfer reflects the O_2 dependence of O_2^- production by ubisemiquinone autoxidation.[38,39]

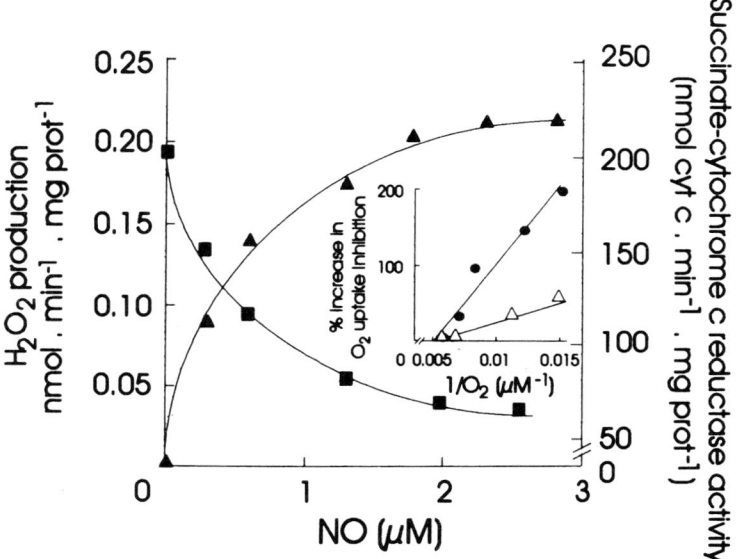

FIGURE 8. Inhibition of succinate–cytochrome c reductase activity and increase in the generation of O_2^- produced by NO in rat heart submitochondrial particles. Modified from Poderoso *et al*.[20] *Inset*: Effect of $[O_2]$ on the NO-dependent inhibition of cytochrome oxidase activity.

NADH-Dehydrogenase (Complex I)

Prolonged exposure of cells to NO results in a persistent inhibition of complex I activity [26,40] simultaneously with a decrease in the cellular content of reduced glutathione. The inhibition is reversible by exposing the cells to high intensity light and appears to result from S-nitrosylation of thiol groups in the enzyme.[40] It has been claimed that S-nitrosylation of complex I may play a role in neurodegenerative diseases.[40]

MITOCHONDRIAL PRODUCTION OF NO AND INHIBITION OF CYTOCHROME OXIDASE ACTIVITY

Nitric oxide is produced during the oxidation of L-arginine (Arg) to citrulline catalyzed by nitric oxide synthase (NOS). The recent finding of a mitochondrial enzyme (mtNOS) in the inner mitochondrial membrane by Giulivi et al.[33,34] and by Ghadoufar and Richter[41] supports the idea of a physiological role for NO in mitochondrial respiration. The production of NO has been measured in whole mitochondria and in mitochondrial membranes isolated from a few rat and mouse organs (FIG. 9 and TABLE 2). Giulivi[35] assayed the effects of the endogenous NO production on the rates of mitochondrial respiration and cytochrome oxidase activity by supplementing mitochondria with either the substrate Arg or the inhibitor of NOS N(G)-monomethyl-L-arginine (NMMA). The $[O_2]_{0.5}$ of the respiration of tightly coupled mitochondria oxidizing succinate in the presence of ADP ranged from 2 to 3 µM in agreement with those obtained by the ADP pulses (TABLE 1, initial rates). In the presence of Arg the rates of O_2 uptake decreased significantly at all measured O_2

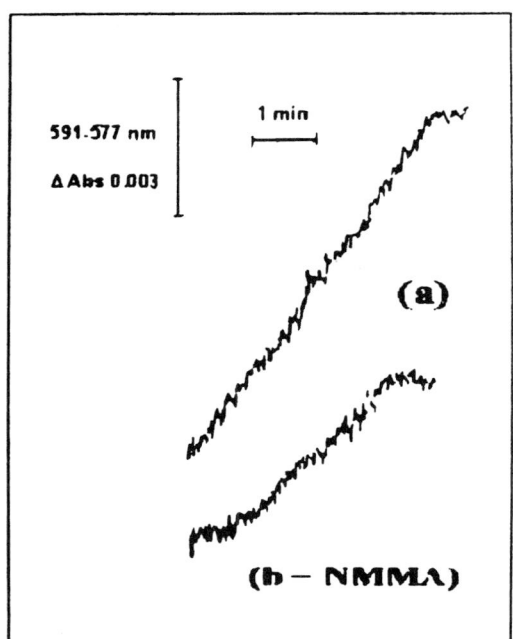

FIGURE 9. Mitochondrial production of NO. Rat thymus mitochondrial membranes (0.25 mg protein/ml) were supplemented with 0.1 mM NADPH, 1 mM arginine, 1 mM Cl_2Ca, 1 µM superoxide dismutase, 0.5 µM catalase and 10 µM oxyhemoglobin in 50 mM phosphate buffer (pH 7.4). **a:** complete reaction mixture; **b:** as in a *plus* NMMA.

concentrations. Two concentrations of Arg were used: 5 µM (about the K_m of mt-NOS for Arg[33]) and 0.1 mM. The observed values of $[O_2]_{0.5}$ were 18 and 40 µM at low and high concentrations of Arg, respectively. Conversely, addition of NMMA to intact mitochondria increased the O_2 uptake by 40–50% of the control values at low O_2 levels and decreased the $[O_2]_{0.5}$ to 1.2 µM. Concerning cytochrome oxidase, the enzyme activity as a function of $[O_2]$ was determined in mitochondria supplemented with either Arg or NMMA. The effects of NOS substrate and inhibitor were similar to those observed in mitochondrial state 3 respiration. The $[O_2]_{0.5}$ were 0.9, 5 and 45 µM with NMMA, endogenous and exogenous Arg, respectively. Giulivi plotted the data concerning the effect of Arg and NMMA on state 3 respiration and cytochrome oxidase with the Eadie-Hofstee treatment and found them to fit to a single line indicating that cytochrome oxidase inhibition by endogenous NO entirely accounts for respiratory regulation. The unchanged V_{max} and the different $[O_2]_{0.5}$ indicate a competitive inhibition kinetics in which NO inhibited the respiratory chain competing with O_2 for cytochrome oxidase.[35] Furthermore, the rate of ATP synthesis in intact mitochondria, evaluated by measuring the amount of ATP in aliquots taken at different time points during state 3 respiration, was similarly decreased in the presence of Arg, as well as the respiratory rates in state 4.[35]

Assuming the classical view of "all or nothing," where mitochondria are in state 3 or in state 4, and that tissue O_2 uptake is accounted by the sum of the O_2 uptakes of mitochondria respiring in state 4 and in state 3, the fraction of mitochondria in state 3 and in state 4 under physiological conditions were estimated as 28% and 72%, respectively, for rat heart.[42] Alternatively, considering for heart mitochondria an $[O_2]$ of 6 µM and a [NO] of 30 nM, the corresponding ratio 200 O_2/NO indicates an inhibition of 33% for both state 3 and state 4 respiration (FIG. 7, *inset*) and the actual respiratory rate can be estimated as 0.67 × maximal respiratory rate measured at saturating $[O_2]$. In such case, with kinetic NO control, the fraction of mitochondria in state 3 (X) and in state 4 (1–X) are estimated as: (X) × (NO-inhibited state 3 O_2 uptake) + (1–X) × (NO-inhibited state 4 O_2 uptake) = O_2 uptake of the perfused organ / content of mitochondria in the tissue. The data used for the calculation are (0.67 × 135 nmol O_2/min.mg prot) and (0.67 × 28 nmol O_2/min.mg prot) for NO-inhibited state 3 and state 4 respiration,[42] 3.05 µ mol O_2/min.g heart[21] and 53 mg mitochondrial prot/g heart.[43] The fraction of mitochondria in state 3 and state 4 are 54% and 46%, respectively.

TABLE 2. **Mitochondrial production of nitric oxide**

Mitochondria	NO production (nmol/min.mg prot)	Reference
Rat liver mitochondria	1.4	a
Rat liver mitochondrial membranes	0.9–4.2	a, b, c
Rat thymus mitochondrial membranes	0.12–0.35	c
Mouse brain mitochondrial membranes	1.6	d

[a]Giulivi et al.[33]
[b]Ghafourifar and Richter.[41]
[c]J. Bustamante, personal communication.
[d]S. Lores Arnaiz, personal communication.

THE INTRAMITOCHONDRIAL STEADY STATE CONCENTRATION OF NITRIC OXIDE

The understanding of the NO regulation of mitochondrial respiration requires knowledge of its mitochondrial metabolism and intramitochondrial steady state level. For the following analysis only mitochondrial NO production and utilization are considered disregarding NO production by endothelial NOS and the fast reactions of NO with muscle cytosolic myoglobin and blood hemoglobin that certainly have an effect on mitochondrial and cellular NO steady state concentrations.

Concerning NO and O_2^- metabolism it is convenient to consider the intramitochondrial space as a specialized intracellular compartment due to the selective permeability of the inner mitochondrial membrane (FIG. 10). The key features are the impermeability of the inner membrane to O_2^- and H^+, the relative impermeability of the same membrane to $ONOO^-$, and the presence of Mn-SOD at a content which is about five times lower than CuZn-SOD in the cytosol. The reactions of O_2^- with NO ($k = 2 \times 10^{10}$ M^{-1} s^{-1}) and with Mn-SOD ($k = 2.4 \times 10^9$ M^{-1} s^{-1}) are apparently the only ones that occur in the mitochondrial matrix at rates that effectively contribute to O_2^- utilization. Considering intramitocondrial MnSOD as 3 μM^{44} and intramitocondrial [NO] as 30 nM,[45] it can be calculated that the intramitocondrial production of $ONOO^-$ will account for 8% of O_2^- utilization with the remaining 92% yielding H_2O_2 as final product. Under conditions of endothelium NOS activation by bradikinin, intramitochondrial NO reaches 100 nM[21] and consequently $ONOO^-$ formation may account for as much as 27% of O_2^- utilization. Moreover, in conditions of mt-NOS induction, intramitochondrial NO may reach 0.3 μM NO and the production of $ONOO^-$ will account for more than 50% of O_2^- utilization. Besides this oxidative pathway, intramitochondrial NO is metabolized through reductive one-electron transfer reactions from cytochrome oxidase,[46] and ubiquinol.[22] The two reductive

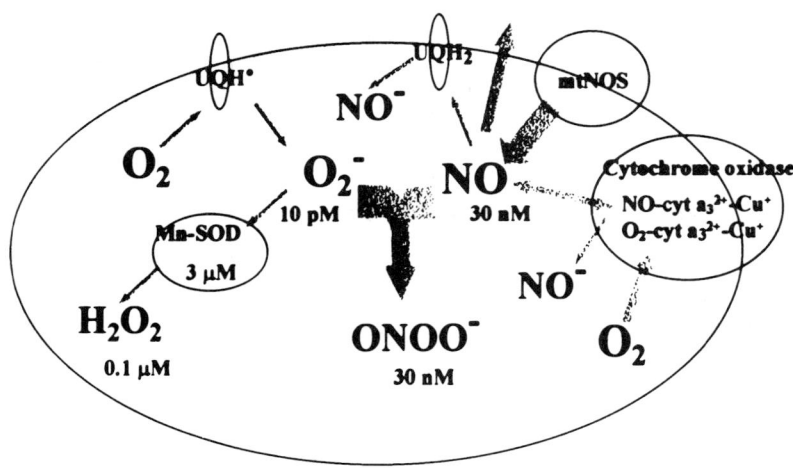

FIGURE 10. Mitochondrial metabolism of NO and its effect regulating cytochrome oxidase activity.

reactions yield nitroxyl anion (NO⁻) as intermediate and N_2O as final stable product. According to our preliminary data the oxidative reaction yielding ONOO⁻ accounts for 88% of mitochondrial NO utilization and the two reductive reactions for the remaining 12%.

The fine regulation of the steady state concentration of NO in the mitochondrial matrix, accomplished by its reactions with O_2^-, cytochrome oxidase, and ubiquinol, modulates cytochrome oxidase activity. The knowledge of the intracellular or intramitochondrial signaling that activates mtNOS is essential to fully understand the overall regulatory process.

REFERENCES

1. GAYESKI, T.E.J. & C.R. HONIG. 1988. Intracellular pO_2 in long axis of individual fibers in working dog gracilis muscle. Am. J. Physiol. **254:** H1179–H1186.
2. WITTENBERG, B.A. & J.B. WITTENBERG. 1989. Transport of oxygen in muscle. Annu. Rev. Physiol. **51:** 857–878.
3. GAYESKI, T.E.J. & C.R. HONIG. 1991. Intracellular pO_2 in individual cardiac myocytes in dogs cats rabbits, ferrets, and rats. Am. J. Physiol. **260:** H522–H531.
4. SUGANO, T., N. OSHINO & B. CHANCE. 1974. Mitochondrial functions under hypoxic conditions. The steady states of cytochrome c reduction and of energy metabolism. Biochim Biophys. Acta **347:** 340–358.
5. DE GROOT, H., T. NOLL & H. SIES. 1985. Oxygen dependence and subcellular partitioning of hepatic menadione-mediated oxygen uptake. Studies with isolated hepatocytes, mitochondria, and microsomes from rat liver in an oxystat system. Arch. Biochem. Biophys. **243:** 556–562.
6. WILSON, D.F., W.L. RUMSEY, T.J. GREEN & J.M. VANDERKOOI. 1988. The oxygen dependence of mitochondrial oxidative phosphorylation measured by a new optical method for measuring oxygen concentration. J. Biol. Chem. **263:** 2712–2718.
7. RUMSEY, W.L., C. SCHLOSSER, E.M. NUUTINEN, M. ROBIOLIO & D.F. WILSON. 1990. Cellular energetics and the oxygen dependence of respiration in cardiac myocytes isolated from adult rat. J. Biol. Chem. **265:** 15392–15399.
8. MÉNDEZ, G. & E. GNAIGER. 1994. How does oxygen pressure control oxygen flux in isolated mitochondria? A methodological approach by high-resolution respirometry and digital data analysis. *In* Modern Trends in Biothermokinetics. E. Gnaiger, F.N. Gellerich & M. Wyss, Eds. **3:** 191–194. Innsbruck University Press. Innsbruck.
9. GNAIGER, E., R. STEINLECHNER-MARAN, G. MÉNDEZ, T. EBERL & R. MARGREITER. 1995. Control of mitochondrial and cellular respiration by oxygen. J. Bioenerg. Biomembr. **27:** 583–596.
10. COSTA, L. E., G. MÉNDEZ & A. BOVERIS. 1997. Oxygen dependence of mitochondrial function measured by high-resolution respirometry in long-term hypoxic rats. Am. J. Physiol. **273:** C852–C858.
11. RICHMOND, K.N., S. BURNITE & R.M. LYNCH. 1997. Oxygen sensitivity of mitochondrial metabolic state in isolated skeletal and cardiac myocytes. Am. J. Physiol. **273:** C1613–C1622.
12. KADENBACH, B., J. BARTH, R. AKGÜN, R. FREUND, D. LINDER & S. POSSEKEL. 1995. Regulation of mitochondrial energy generation in health and disease. Biochim. Biophys. Acta **1271:** 103–109.
13. SNOW, T.R., L.H. KLEINMAN, J.C. LAMANNA, A.S. WICHSLER & F.F. JOBSIS. 1981. Response of cyt aa_3 in the situ canine heart to transient ischemic episodes. Basic Res. Cardiol. **76:** 289–304.
14. JONES D.P., X. SHAN & Y. PARK. 1992. Coordinated multisite regulation of cellular energy metabolism. Annu. Rev. Nutr. **12:** 327–343.
15. TORRES, J. & M.T. WILSON. 1997. Interaction of cytochrome c oxidase with nitric oxide. Meth. Enzymol. **269:** 3–11.

16. ROUSSEAU D.L., S. SINGH, Y.C. CHING & M. SASSAROLI. 1988. Nitrosyl cytochrome c oxidase. Formation and properties of mixed valence enzyme. J. Biol. Chem. **263:** 5681–5685.
17. LARDY, H.A. & H. WELLMAN. 1952. Oxidative phosphorylation: role of inorganic phosphate and acceptor systems in control of metabolic rates. J. Biol. Chem. **195:** 215–224.
18. CHANCE, B. & G.R. WILLIAMS. 1956. The respiratory chain and oxidative phosphorylation. Adv. Enzymol. **17:** 65–134.
19. ESTABROOK, R.W. 1967. Mitochondrial respiratory control and the polarographic measurement of ADP:O ratios. Meth. Enzymol. **10:** 41–47.
20. PODEROSO, J.J., M.C. CARRERAS, C. LISDERO, N. RIOBÓ, F. SCHOPFER & A. BOVERIS. 1996. Nitric oxide inhibits electron transfer and increases superoxide radical production in rat heart mitochondria and submitochondrial particles. Arch. Biochem. Biophys. **328:** 85–92.
21. PODEROSO, J.J., J.G. PERALTA, C.L. LISDERO, M.C. CARRERAS, M. RADISIC, F. SCHOPFER, E. CADENAS & A. BOVERIS. 1998. Nitric oxide regulates oxygen uptake and hydrogen peroxide release by the isolated beating rat heart. Am. J. Physiol. **274:** C112–C119.
22. PODEROSO, J. J., M.C. CARRERAS, F. SCHÖPFER, C. L. LISDERO, N. RIOBÓ, C. GIULIVI, A. D. BOVERIS, A. BOVERIS & E. CADENAS. 1999. The reaction of nitric oxide with ubiquinol: kinetic properties and biological significance. Free Radical Biol. Med. **26:** 925–935.
23. BOVERIS, A., L.E. COSTA, E. CADENAS & J.J. PODEROSO. 1999. Regulation of mitochondrial respiration by adenosine diphosphate, oxygen, and nitric oxide. Meth. Enzymol. **301:** 188–198.
24. BREDT, D.S. & S.H. SNYDER. 1994. Nitric oxide: a physiologic messenger molecule. Annu. Rev. Biochem. **63:** 175–195.
25. IGNARRO, L.J., G.M. BUGA, K.S. WOOD, R.E. BYRNS & G. CHAUDHURI. 1987. Endothelium-derived relaxing factor produced and released from artery and vein is nitric oxide. Proc. Natl. Acad. Sci. USA **84:** 9265–9269.
26. MONCADA, S., R.M. PALMER & E.A. HIGGS. 1988. The discovery of nitric oxide as the endogenous nitrovasodilator. Hypertension **12:** 365–372.
27. GRANGER, D.L. & A.L. LEHNINGER. 1982. Sites of inhibition of mitochondrial electron transport in macrophage-injured neoplastic cells. J. Cell. Biol. **95:** 527–535.
28. YUI, Y., R. HATTORI, K. KOSUGA, H. EIZAWA, K. KIKI & C. KAWAI. 1991. Purification of nitric oxide synthase from rat macrophages. J. Biol. Chem. **266:** 12544–12547.
29. NATHAN, C. 1992. Nitric oxide as a secretory product of mammalian cells. FASEB J. **6:** 3051–3064.
30. CLEETER, M.W.J., J.M. COOPER, V.M. DARLEY-USMAR, S. MONCADA & A.H.V. SCHAPIRA. 1994. Reversible inhibition of cytochrome c oxidase, the terminal enzyme of the mitochondrial respiratory chain, by nitric oxide. Implications for neurodegenerative diseases. FEBS Lett. **345:** 50–54.
31. BROWN, G.C. & C.E. COOPER. 1994. Nanomolar concentrations of nitric oxide reversibly inhibit synaptosomal respiration by competing with oxygen at cytochrome oxidase. FEBS Lett. **356:** 295–298.
32. DUCHEN, M.R. 1999. Contributions of mitochondria to animal physiology: from homeostatic sensor to calcium signalling and cell death. J. Physiol. (London) **516:** 1–17.
33. GIULIVI, C., J.J. PODEROSO & A. BOVERIS. 1998. Production of nitric oxide by mitochondria. J. Biol. Chem. **273:** 11038–11043.
34. TATOYAN, A. & C. GIULIVI. 1998. Purification and characterization of a nitric-oxide synthase from rat liver mitochondria. J. Biol. Chem. **273:** 11044–11048.
35. GIULIVI, C. 1998. Functional implications of nitric oxide produced by mitochondria in mitochondrial metabolism. Biochem. J. **332:** 673–679.
36. TAKEHARA, Y., T. KANNO, T. YOSHIOKA, M. INOUE & K. UTSUMI. 1995. Oxygen-dependent regulation of mitochondrial energy metabolism by nitric oxide. Arch. Biochem. Biophys. **323:** 27–32.

37. KOIVISTO, A., A. MATTHIAS, G. BRONNIKOV & J. NEDERGARD. 1997. Kinetics of the inhibition of mitochondrial respiration by NO. FEBS Lett. **417:** 75–80.
38. BOVERIS, A., E. CADENAS & A.O.M. STOPPANI. 1976. Role of ubiquinone in the mitochondrial generation of hydrogen peroxide. Biochem. J. **156:** 435–444.
39. CADENAS, E., A. BOVERIS, C.I. RAGAN & A.O.M. STOPPANI. 1977. Production of superoxide radicals and hydrogen peroxide by NADH-ubiquinone reductase and ubiquinol-cytochrome c reductase from beef-heart mitochondria. Arch. Biochem. Biophys. **180:** 248–257.
40. CLEMENTI, E., G.C. BROWN, M. FEELISCH & S. MONCADA. 1998. Persistent inhibition of cell respiration by nitric oxide: crucial role of S-nitrosylation of mitochondrial complex I and protective action of glutathione. Proc. Natl. Acad. Sci. USA **95:** 7631–7636.
41. GHAFOURIFAR, P. & C. RICHTER. 1997. Nitric oxide synthase activity in mitochondria. FEBS Lett. **418:** 291–296.
42. BOVERIS, A., L.E. COSTA & E. CADENAS. 1999. The mitochondrial production of oxygen radicals and cellular aging. *In* Understanding the Process of Aging. E. Cadenas & L. Packer, Eds. :1–16. Marcel Dekker. New York.
43. COSTA, L.E., A. BOVERIS, O.R. KOCH & A.C. TAQUINI. 1988. Liver and heart mitochondria in rats submitted to chronic hypobaric hypoxia. Am. J. Physiol. **255:** 123–129.
44. BOVERIS, A. & E. CADENAS. 1997. Cellular sources and steady-state levels of reactive oxygen species. In Oxygen, Gene Expression and Cellular Function. L. B. Clerch & D. J. Massaro, Eds. :1–25. Marcel Dekker. New York.
45. BOVERIS, A. & J.J. PODEROSO. 1999. Regulation of oxygen metabolism by nitric oxide. *In* Nitric Oxide. L. Ignarro, Ed. Academic Press, New York. In press.
46. ZHAO, X.J., V. SAMPATH & W. S. CAUGHEY. 1995. Cytochrome c oxidase catalysis of the reduction of nitric oxide to nitrous oxide. Biochem. Biophys. Res. Commun. **212:** 1054–1060.

Free Radicals and Antioxidants in the Year 2000

A Historical Look to the Future

JOHN M.C. GUTTERIDGE[a,c] AND BARRY HALLIWELL[b]

[a]*Oxygen Chemistry Laboratory, Directorate of Anaesthesia and Critical Care, Royal Brompton Hospital and Harefield NHS Trust, London SW3 6NP, UK*

[b]*Department of Biochemistry, National University of Singapore, 10 Kent Ridge Crescent, Singapore 119260*

ABSTRACT: In the late 1950's free radicals and antioxidants were almost unheard of in the clinical and biological sciences but chemists had known about them for years in the context of radiation, polymer and combustion technology. Daniel Gilbert, Rebeca Gerschman and their colleagues related the toxic effects of elevated oxygen levels on aerobes to those of ionizing radiation, and proposed that oxygen toxicity is due to free radical formation, in a pioneering paper in 1956. Biochemistry owes much of its early expansion to the development and application of chromatographic and electrophoretic techniques, especially as applied to the study of proteins. Thus, superoxide dismutase (SOD) enzymes (MnSOD, CuZnSOD, FeSOD) were quickly identified. By the 1980's Molecular Biology had evolved from within biochemistry and microbiology to become a dominant new discipline, with DNA sequencing, recombinant DNA technology, cloning, and the development of PCR representing milestones in its advance. As a biological tool to explore reaction mechanisms, SOD was a unique and valuable asset. Its ability to inhibit radical reactions leading to oxidative damage *in vitro* often turned out to be due to its ability to prevent reduction of iron ions by superoxide. Nitric oxide (NO·) provided the next clue as to how SOD might be playing a critical biological role. Although NO· is sluggish in its reactions with most biomolecules it is astoundingly reactive with free radicals, including superoxide. Overall, this high reactivity of NO· with radicals may be beneficial *in vivo*, e.g. by scavenging peroxyl radicals and inhibiting lipid peroxidation. If reactive oxygen species are intimately involved with the redox regulation of cell functions, as seems likely from current evidence, it may be easier to understand why attempts to change antioxidant balance in aging experiments have failed. The cell will adapt to maintain its redox balance. Indeed, transgenic animals over-expressing antioxidants show some abnormalities of function. There must therefore be a highly complex interrelationship between dietary, constitutive, and inducible antioxidants within the body, under genetic control. The challenge for the new century is to be able to understand these relationships, and how to manipulate them to our advantage to prevent and treat disease.

[c]Address for correspondence: Oxygen Chemistry Laboratory, Directorate of Anaesthesia and Critical Care, Royal Brompton Hospital and Harefield NHS Trust, London SW3 6NP, UK.

FROM TIES TO JEANS

The working environment in government-funded hospital laboratories during the late 1950's and early 1960's was unbelievably different from that which we know today. This was an era without computers, automated laboratory equipment, fax machines, or micro chips. Laboratory life was often formal, and a bachelors degree in science was often all that was needed to head a department. Whilst working in a hospital laboratory in the UK at this time, John Gutteridge well remembers there being different canteens for graduates and non-graduates, and an incident when a junior colleague was sent home for not wearing a neck tie. As a mild challenge to the system John came to work wearing a bow tie, although it never looked so elegant as Dan's bow ties do on him. The University environment was also much more structured than it is now and the Heads of Department reigned supreme. In his early days as a lecturer in Biochemistry, Barry Halliwell was forbidden to set foot in the Department of Chemistry for having the insolence to use a piece of equipment in that department without asking the Professor, even though the Chemistry lecturer to whom it belonged was enthusiastic about collaboration.

The formality which pervaded all aspects of society (at least in the UK) at this time was challenged and irreversibly broken by youth culture, being expressed through fashion and pop music, with the iconoclastic views of John Lennon being particularly influential. Thus, Jeans (rather wide at the ankle) entered laboratories as the preferred mode of dress, reflecting the dramatic change in attitudes to formality. Perhaps this may have gone too far—or are we being old-fashioned? We have both attended weddings and funerals where many people show up in jeans and T-shirts, and even the best restaurants in the world have minimized their dress code requirements almost to vanishing point. There are still outposts however: Barry Halliwell rejoiced when at a recent scientific meeting in a Caribbean country, over 80% of the scientists were deemed inadequately dressed to eat in the hotel restaurants (including all the American speakers). Had he been there, Dan would have surely qualified to eat with the select few in elegant splendor.

FROM JEANS TO GENES

In the late 1950's free radicals and antioxidants were almost unheard of in the clinical and biological sciences but chemists had known about them for years in the context of radiation, polymer and combustion technology. In a pioneering paper in 1954, Dan, Rebecca Gerschman and their colleagues related the toxic effects of elevated oxygen levels on aerobes to those of ionizing radiation, and proposed that oxygen toxicity is due to free radical formation.[1] This concept did not capture the imagination of most life scientists, however, until the discovery of the superoxide dismutase enzymes by McCord and Fridovich in 1968.[2] These authors had previously observed the ability of several proteins to inhibit the reduction of cytochrome c by xanthine oxidase, and in a flash of inspiration realized that the inhibitory effects were due to a novel protein contaminating the other proteins tested. They purified this contaminant and thus identified the first enzyme (copper and zinc-containing superoxide dismutase, CuZnSOD) known to act on a free radical substrate.[3,4] Barry Halliwell had at the same time observed that illuminated chloroplasts reduce cytochrome c and this was inhibited by catalase, but could not understand how removing H_2O_2 could prevent a reduction reaction. After reading the seminal papers of McCord and Fridovich, he learned the answer: the commercial catalase used was contaminated by SOD.[5] Thus Dan's pioneering "free radical theory of oxygen toxicity" became the "superoxide theory of oxygen toxicity."

Biochemistry owes much of its early expansion, after its separation from Chemistry, to the development and application of chromatographic and electrophoretic techniques, especially as applied to the study of proteins. Thus new SOD enzymes (MnSOD, FeSOD) were quickly identified. By the 1980's Molecular Biology had evolved from within biochemistry and microbiology to become a dominant new discipline, with DNA sequencing, recombinant DNA technology, cloning, and the development of PCR representing milestones in its advance.

KNOCKABOUTS AND KNOCKOUTS

The highly stable CuZnSOD enzymes are easy to purify and could be sold at high profit by suppliers. Hence they were readily available to researchers by the late 1970's, resulting in an explosion of interest in superoxide and other reduction intermediates of oxygen in biological systems. A protein antioxidant catalyst which had as its substrate a small inorganic free radical was something novel to biochemistry at that time. Now we readily accept the important biological roles of small inorganic molecules, especially the free radical nitric oxide (NO·) and increasingly, carbon monoxide (if one can call CO inorganic). One of the authors recently proposed that SO_2 might be a biological signalling system and, in investigating this, rediscovered sulfite toxicity.[6] Indeed, investigation of sulfite oxidation mechanisms was the first example of the use of SOD as a "probe" for O_2^-.[3] Things tend to come full circle in this field, as we shall see again when considering Fenton chemistry.

Not all were prepared to accept that SOD was exclusively a biological enzyme and preferred to view it more as a copper transport protein. The "copper controversy"—catalyst or non-catalyst—unnecessarily dominated several of the early conferences dedicated to this new research field. People love stand-up controversy at meetings, but it usually achieves little. Since all those involved were good scientists and generated valid data, then there must be ways of accommodating all their observations and exactly this has happened with time. CuZnSOD is an important contributor to the "pool" of total intracellular copper, and other metalloproteins can sometimes replace SOD in removing O_2^-.[7]

Nevertheless, the superoxide theory of oxygen toxicity has held its ground, although there is no general agreement on why elevated levels of O_2^- can be toxic. Is it direct selective damage by $O_2^-/HO_2\cdot$, or the O_2^--dependent formation of peroxynitrite or of the infamous hydroxyl radical? Is SOD really important *in vivo*? To answer the question directly, simply remove SOD and see what happens, as Touati *et al.* first did with *E. coli*.[8] *E. coli* mutants lacking FeSOD and MnSOD were viable, but sick (we now know they still had some SOD, as CuZnSOD in the periplasmic space). Mice lacking mitochondrial SOD (MnSOD) are also very sick: this enzyme is essential for healthy aerobic life.[9,10] But what a surprise with CuZnSOD and glutathione peroxidase! Mice can survive without them, although in the latter case they are more sensitive to certain toxins and in the former subtle defects (e.g. in reproductive potential) are gradually being discovered.[11,12]

Thus mice can adapt to live without these enzymes (but remember that mice are not men). How they do so remains to be discovered, but information is coming thick and fast. When Barry Halliwell was doing his Ph.D. in a plant biochemistry laboratory in the 1970's, everyone knew about the key role of thioredoxin in redox regulation and antioxidant defence in plants.[13] Only now is its importance in animals being fully appreciated,[14] and thioredoxin-linked peroxidases (peroxidoredoxins) are the new "flavor of the month."

"If free radicals play important roles in disease pathologies, then SOD might have remarkable properties as a drug for human medicine," was the next obvious step forward proffered by many researchers. With hindsight, we now know that cellular redox balance, under the control of a hierarchy of antioxidants and other redox-active proteins, is extremely complex and minimally influenced by extra "spoonfuls" of constitutive antioxidants. For example, oxidizing conditions can activate NFκB in some cell types, but the oxidized protein will not bind to DNA, i.e., its activity is redox-regulated. SOD, as a pharmaceutical, has made little or no impact on human medicine.

IRON TOOLS

As a biological tool to explore reaction mechanisms, SOD was a unique and valuable asset.[3] Its ability to inhibit radical reactions leading to oxidative damage *in vitro* often turned out to be due to its ability to prevent reduction of iron ions or iron chelates by superoxide. Thus SOD would inhibit superoxide-driven Fenton chemistry (reaction **A**), since removal of $O_2^{\cdot-}$ prevented the formation of ferrous ions

$$O_2^{\cdot-} + Fe(III) \rightarrow Fe^{2+} + O_2$$
$$2 O_2^{\cdot-} + 2H^+ \rightarrow H_2O_2 + O_2 \quad \textbf{(A)}$$
$$H_2O_2 + Fe^{2+} \rightarrow OH^- + \cdot OH + Fe(III)$$

However, SOD would not inhibit Fenton chemistry resulting from the mixing of Fe^{2+} with H_2O_2, or from the autoxidation of ferrous salts at physiological pH values.

$$Fe^{2+} + O_2 \rightleftharpoons Fe(III) + O_2^{\cdot-}$$
$$2 O_2^{\cdot-} + 2H^+ \rightarrow H_2O_2 + O_2 \quad \textbf{(B)}$$
$$Fe^{2+} + H_2O_2 \rightarrow OH^- + \cdot OH + Fe(III)$$

For many years the prevention of iron reduction and consequent OH· generation seemed a major explanation for the protective role of SOD. Some groups embraced it enthusiastically and took our 1984 article proposing the importance of iron *in vivo*[15] as a bible, citing it as evidence that OH· is responsible for most or all of the oxidative damage occurring *in vivo*. Indeed, we are amused to note that this 1984 paper has been often cited by groups who appear not to have read it. For example, it has been quoted as saying that OH· plays a key role as an initiator of biological iron-dependent lipid peroxidation when in fact we said the opposite.[15,16] Others vehemently opposed the concept of iron as a promoter of oxidative damage, stating that there is no iron "catalytic" for free radical reactions *in vivo* and that O_2^- couldn't reduce it even if there was, since cells are full of other reducing agents. Another controversy started, although at least the participants were polite to each other. Time has again shown us that all the data were correct: organisms carefully restrict the availability of "catalytic" iron ions *in vivo* to minimize oxidative damage.[17] Yet cells have a transit pool of low molecular mass iron (recently demonstrated directly[18]) and oxidative stress can liberate further iron catalytic for free radical reactions, as was shown over a decade ago for ferritin and heme proteins.[19,20] Superoxide can also liberate iron from certain iron-sulfur clusters[21] and the concept of oxidative stress leading to iron ion release has been repeatedly re-presented as if it were novel. Tissue injury also liberates catalytic metal ions and cells "sense" their iron levels using iron-responsive elements and regulate the turnover of mRNAs encoding iron sequestration proteins accordingly.[22] Mice lacking transferrin have a large "pool" of catalytic iron and soon die: thus iron sequestration in non-catalytic forms is almost as important as MnSOD to their survival.[23]

MALIGNANT SPIRITS

Another source of pointless controversy is the "iron wheel," the regularly resurrected circular argument about whether or not OH· *is* the key species in Fenton chemistry. In 1932 Bray and Gorin proposed that a ferrous salt plus hydrogen peroxide produced a ferryl species (FeO^{2+}), and not the hydroxyl radical (·OH), and that ferryl rather than OH· was responsible for most of the damage seen. This concept has been picked up and re-presented as novel multiple times in the past 25 years. Much of the evidence against OH· was based on the idiosyncratic behavior of a variety of "OH· scavengers" added to Fenton systems damaging biological molecules. To many of us working with iron chelators, it has been obvious from the beginning that iron-binding, both weak and strong, dictates site-specific reactions that often overrule established second-order rate constants for attack of OH· upon molecules added to Fenton reactions (for a detailed discussion see refs. 24 and 25). Other arguments have been based on the fact that the end-products expected when biomolecules were added to Fenton systems were not "typical of OH·." Walling[25] showed decades ago that iron salts influence the fate of the primary radicals generated by attack of OH· on molecules, so that the end-products will not be the same as for radiation-generated OH·. Ferryl species could conceivably be intermediates in OH· generation, e.g.

$$Fe^{2+} + H_2O_2 \rightarrow \boxed{\text{oxo-iron species}} \rightarrow Fe(III) + OH\cdot + OH^-$$

but their formation in "simple" biological Fenton chemistry has never been clearly demonstrated by classical chemical methods (although *heme* ferryl species are well-established). Walling[25] has summarized it beautifully: "The formulation of complexes to explain reactions and kinetic data is a great temptation but, particularly when they cannot be independently demonstrated, has little more intellectual justification

NITRIC OXIDE AND PEROXYNITRITE: FROM DOGMA TO DOUBT

Nitric oxide provided the next clue as to how SOD might be playing a critical biological role. Although NO· is sluggish in its reactions with most biomolecules it is astoundingly reactive with free radicals, including superoxide. Overall, this high reactivity of NO· with radicals may be beneficial *in vivo*, e.g. by scavenging peroxyl radicals and inhibiting lipid peroxidation.[26,27] However, the product of reaction of O_2^- with NO·, peroxynitrite (ONOO$^-$), is widely thought to be a "baddie," although this view is not universally shared. *In vitro* at pH 7.4, addition of ONOO$^-$ is cytotoxic and damages most biomolecules.[26] If its concentrations are great enough, nitric oxide can out-compete SOD for reaction with O_2^-. Normally, however, SOD might "control" the levels of O_2^-, balancing the action of NO· as a vasodilator.[28] Why has NO· research, after a relatively short time span, been awarded a Nobel Prize, whereas O_2^- research has not? The answer probably reflects the clinical, pharmacological and medical impact of NO· knowledge and application, which has not yet materialized for O_2^-.

Addition of peroxynitrite at physiological pH will oxidize, nitrate or nitrosylate a wide variety of biological molecules. Many researchers also observed hydroxylation of aromatic molecules to occur, which led to the suggestion that ONOOH could dissociate to yield OH· radicals. This provoked an outcry from certain chemists, who like to predict results based upon thermodynamic and other theoretical considerations. The irrelevance of these calculations to biomedical science is illustrated by the fact that merely changing one of the assumed parameters in the calculation changes the reaction

$$ONOOH \rightarrow NO_2\cdot + OH\cdot$$

from "thermodynamically impossible" to "the preferred reaction pathway."[29] One can never know the real values of thermodynamic parameters *in vivo* and thus can never make meaningful calculations. Experimental data show that ONOO$^-$ at pH 7.4 may make some OH·, but not much[30,31] and data should always be preferred over theory. One should never be dogmatic in science—look at what happened to the "central dogma"! The god-like hospital consultants of the early days of the National Health Service in the UK always "knew better" than their patients or their GPs. It has been said that "they were often in error, but never in doubt."

FROM TBA TO DIODE ARRAY

When we came to write the first edition of our textbook *Free Radicals in Biology and Medicine* in 1985,[17] the only chapter dealing with the measurement of biomolecular damage by free radicals was that on lipid peroxidation. This was mainly because there were few other simple methods available for application to clinical and biological material, and partly because lipid peroxidation, as a radical chain reaction, has intrigued scientists since its elucidation in the 1940's, incidentally, not by life scientists but by polymer scientists at the British Rubber Producers Association.[17] The key role of lipid peroxidation in causing rancidity in foods became increasingly apparent with the trend towards more pre-packaged "long life" food materials, kindling intense interest in the quantification of food oxidative deterioration and the development of protective antioxidants (of which the simplest is to pack food *in vacuo* or under nitrogen). The first toxin shown to act by a free radical mechanism, CCl_4, causes damage by stimulating lipid peroxidation, easily detected in animals, cells and microsomal fractions.[32]

Another reason perhaps why we all stayed with lipid peroxidation for so long, without looking for damage to other key biological molecules, was that almost all the methods used to detect it were so simple, such as measuring peroxides with iodide, thiobarbituric acid reactivity (TBAR) and UV spectra (the dreaded diene conjugates, which have generated almost as many artifacts as TBAR). TBAR came originally from the food industry where it was used to detect rancidity in foods containing polyunsaturated fatty acids. Just heat some food, biological tissue or fluid with 2-thiobarbituric acid under acid conditions and you will get a beautiful pink color—anybody can do it! And so they did. Free radicals could now be implicated in every disease process known to man (TBAR probably goes up in all of them). Understanding of what this method really measures took decades to penetrate the free radical community and penetration is not yet complete. To many (but not all) editors and reviewers, during the 1990's TBA has become "that bloody assay." The TBA test does have value in examining the oxidation of defined lipid systems (foods, microsomes, etc.) but cannot be used to compare peroxidation of different systems with different PUFA compositions (different PUFAs generate different levels of TBAR) and, used alone, is an unreliable index of levels of lipid peroxidation in cells, tissues or body fluids.[17]

The appeal of the microsome and its abnormal propensity to undergo lipid peroxidation *in vitro* (especially if stored frozen for long periods) eventually began to fade. During the late 1980's and early 1990's interest in free radical damage in biological systems started to move away from lipids and consider proteins and nucleic acids, especially DNA. Ironically, many basic chemical studies of DNA and protein oxidation had already been done (reviewed in refs. 33–35). The knowledge just needed application to biological material. The days of the HPLC and mass spectrometer had arrived: techniques that were also applied to get much more accurate estimates of lipid peroxidation in biological materials. This gave us for the first time an overall picture of the oxidative damage that occurs in cells and tissues.[17,34,35] We learned that (unlike CCl_4) most toxins acting by free radical mechanisms do not primarily act via lipid peroxidation, and that oxidative damage to DNA and/or proteins is often more important as a primary cytotoxic mechanism than is lipid peroxidation. This change in emphasis is reflected in the third edition of our textbook: lipid peroxidation no longer merits a chapter to itself.[17]

CAUSE OR CONSEQUENCE

TBAR-material (and, more recently, better-authenticated products of lipid peroxidation) are found in body fluids and tissues in increased amounts in almost any disease state and in animals given virtually any toxin. During the 1970's–1980's this led to the assumption that free radicals were the cause of many diseases and lipid peroxidation the primary mechanism of action of many toxins. If the former were true, the hope was, particularly for SOD, that antioxidants inhibiting peroxidation would cure many diseases. It was soon clear to many researchers that free radicals did not cause a plethora of diseases, neither were "spoonfuls" of SOD or vitamin E going to modify them, let alone cure them. In 1984, we pointed out[36] that increased free radical formation was probably an inevitable consequence of most diseases in which tissue damage occurred. Although considered by many to be heresy at the time, it is now a widely accepted (although still periodically "rediscovered") explanation for increased free radical activity. For example, reactive oxygen species are generated in increased amounts in cells undergoing apoptosis triggered by a range of mechanisms, but they are not essential constituents of the apoptotic pathway.[37] The same "consequence but not cause" concept is probably true also for many other alleged mediators of tissue injury in human disease, including NO· and cytokines. Levels of these, as well as of reactive oxygen species, all go up as part of the response to injury.[38] In some diseases, excessive production of NO· will turn out to be important (e.g. septic shock perhaps), in others, one or more of the cytokines (e.g. TNFα in rheumatoid arthritis perhaps). The same is true of the reactive oxygen species. There is good evidence for their role in atherosclerosis, suggestive (but not conclusive) evidence for their importance in the major neurodegenerative and chronic inflammatory diseases, and little evidence for an important role in the majority of human diseases.[17,38]

By the 1990's it was clear that antioxidants are not a panacea for aging and disease, and only fringe medicine still peddles this notion. What has become clearer however, is the importance of certain dietary antioxidants in *preventing* life-threatening diseases, such as heart disease and certain types of cancer (reviewed in refs. 39 and 40). Even here, however, simplistic assumptions about the biomolecules responsible (mistaking correlation for causation) may have given us a biased view of which constituents of fruits, grains, vegetables are the most protective (discussed in ref. 41). Thus diets rich in β-carotene protect against cancer development, but β-carotene itself does not (at least, not by an antioxidant mechanism). Twenty years of nutrition research have told us that for "advanced" countries the way to a healthy lifestyle is to eat more plants, a concept familiar to Hippocrates. What it has not told us is exactly why. Nevertheless, the evidence for vitamin E as an agent maintaining the health of the cardiovascular system is growing daily.[39] Research using accurate measures of oxidative damage to proteins, lipids and DNA ("biomarkers") should help us to learn more about the other protective antioxidants in a plant-rich diet.[41]

LET'S TALK

The general lack of success of antioxidant therapy (at least using SOD or chain-breaking antioxidants) may have been a major contributor to the shift away from limitation of the damage caused by reactive oxygen species to one of interest in their physiological roles, e.g. in biological signaling. Thus, sub-toxic levels of reactive oxygen species have been shown to play important roles as signal, trigger, and messenger molecules in cell culture systems.[42–45] Nitric oxide is well-known for playing such roles *in vivo*, but as yet evidence that reactive oxygen species do the same is sparse. One must be wary: cells in culture are frequently hyperoxic and usually lacking many of the antioxidants that would surround them *in vivo*. In particular, ascorbate is rarely added to culture media. Cells adapt to growth in culture. It is not impossible that, with tumor cell lines in particular, adaptations to favor growth may occur by using oxygen radicals to trigger pathways that are triggered by other means *in vivo*.[46]

An early clue that such roles for ROS and RNS might be biologically important came from comparing the different patterns of antioxidant protection used by cells, membranes, and extracellular fluids. Briefly, cells use enzymes and other antioxidants to control ROS levels so that intracellular iron can signal within cells for synthesis of iron-containing proteins to achieve iron homeostasis.[24] In extracellular fluids, however, we see the reverse situation whereby proteins remove or control iron so that reactive species persist for a short while, perhaps to act as signal molecules.[47] This is why introduction of excess iron or copper ions into the extracellular environment has the potential to cause oxidative damage.[47] *In vivo*, normal cells function in a reducing environment, and as this is changed to a more oxidizing (or less reducing) environment cell functions change. Cell proliferation followed by apoptosis, and eventually necrosis are observed *in vitro* when cells in culture are subjected to ever increasing conditions of oxidation.[44,48] Even the caspases of apoptosis are redox-regulated: oxidation by high levels of reactive oxygen species inactivates them[48] and it has been proposed that tumor cells have evolved an abnormally high "oxidation state" to do exactly this.[49]

If reactive oxygen species are intimately involved with the redox regulation of cell functions, which seems likely based on current evidence, then it is perhaps easier to understand why attempts to change antioxidant balance in aging experiments have failed. The cell will adapt to maintain its redox balance. Indeed, transgenic animals over-expressing antioxidants (e.g. CuZnSOD and glutathione peroxidase) show some abnormalities of function.[50,51] There must therefore be a highly complex interrelationship between dietary, constitutive, and inducible antioxidants within the body, under genetic control. The challenge for the next century is to be able to understand these relationships, and how to manipulate them to our advantage to prevent and treat disease.

CONCLUSION

Dan and his colleagues make a fundamental contribution to biology in their seminal paper.[1] Look how far it has taken us, and how far we have yet to go. We look forward to the future with confidence, and await the surprises that will surely come.

REFERENCES

1. GERSCHMAN, R., D.L. GILBERT, S.W. NYE, P. DWYER & W.O. FENN. 1956. Oxygen poisoning and X-irradiation: a mechanism in common. Science **119:** 623–626.
2. MCCORD, J.M. & I. FRIDOVICH. 1969. Superoxide dismutase. An enzymic function for erythrocuprein (haemocuprein). J. Biol. Chem. **244:** 6049–6055.
3. MCCORD, J.M. & I. FRIDOVICH. 1969. The utility of superoxide dismutase in studying free radical reactions. I. Radicals generated by the interaction of sulfite, dimethylsulfoxide and oxygen. J. Biol. Chem. **244:** 6056–6063.
4. MCCORD, J.M. & I. FRIDOVICH. 1968. The reduction of cytochrome c by milk xanthine oxidase. J. Biol. Chem. **243:** 5753–5760.
5. HALLIWELL, B. 1973. Superoxide dismutase: a contaminant of bovine catalase. Biochem. J. **135:** 379–381.
6. REIST, M., K.A. MARSHALL, P. JENNER & B. HALLIWELL. 1998. Toxic effects of sulphite in combination with peroxynitrite on neuronal cells. J. Neurochem. **71:** 2431–2438.
7. LEHMANN, Y., L. MILE & M. TEUBER. 1996. Rubrerythrin from Clostridium perfringens: cloning of the gene, purification of the protein and characterization of its superoxide dismutase function. J. Bact. **178:** 7152–7158.
8. TOUATI, D. 1989. The molecular genetics of superoxide dismutase in E. coli. Free Rad. Res. Commun. **8:** 1–9.
9. LEBOVITZ, R. M., H. ZHANG, H. VOGEL et al. 1996. Neurodegeneration, myocardial injury and perinatal death in mitochondrial superoxide dismutase-deficient mice. Proc. Natl. Acad. Sci. US **93:** 9782–9787.
10. LI, Y., T. T. HUANG, E. J. CARLSON et al. 1995. Dilated cardiomyopathy and neonatal lethality in mutant mice lacking manganese SOD. Nature Genet. **11:** 376–381.
11. MATZUK, M. M., L. DIONNE, Q. GUO et al. 1998. Ovarian function in superoxide dismutase 1 and 2 knockout mice. Endocrinology **139:** 4008–4011.
12. DE HAAN, J. B., C. BLADIER, P. GRIFFITHS et al. 1998. Mice with a homozygous null mutation for the most abundant glutathione peroxidase, Gpx 1, show increased susceptibility to the oxidative stress-inducing agents paraquat and H_2O_2. J. Biol. Chem. **273:** 22528–22536.
13. CHARLES, S. A. & B. HALLIWELL. 1981. Light activation of fructose bisphosphatase in isolated spinach chloroplasts and deactivation by H_2O_2. A physiological role for the thioredoxin system. Planta **151:** 242–246.
14. NAKAMURA, H., Y. NAKAMURA & J. YODOI. 1997. Redox regulation of cellular activation. Annu. Rev. Immunol. **15:** 351–369.
15. HALLIWELL B. & J. M. C. GUTTERIDGE. 1984. Oxygen toxicity, oxygen radicals, transition metals and disease. Biochem. J. **219:** 1–14.
16. GUTTERIDGE, J.M.C. 1982. The role of superoxide and hydroxyl radicals in phospholipid peroxidation catalysed by iron salts. FEBS Lett. **150:** 454–458.
17. HALLIWELL, B. & J.M.C. GUTTERIDGE. Free Radicals in Biology and Medicine. Oxford University Press. Oxford, UK. First edition 1985, second edition 1989, third edition 1999.
18. PICARD, V., S. EPSZTEJN, P. SANTAMBROGIO et al. 1998. Role of ferritin in the control of the labile iron pool in murine erythroleukemia cells. J. Biol. Chem. **273:** 15382–15386.
19. BIEMOND, P., H.G. VAN EIJK, A.J. SWAAK & J.F. KOSTER. 1984. Iron mobilization from ferritin by superoxide derived from stimulated polymorphonuclear leukocytes. J. Clin. Invest. **73:** 1576–1579.

20. GUTTERIDGE, J.M.C. 1986. Iron promoters of the Fenton reaction and lipid peroxidation can be released from haemoglobin by peroxides. FEBS Lett. **201:** 291–295.
21. KEYER, K. & J.A. IMLAY. 1996. Superoxide accelerates DNA damage by elevating free-iron levels. Proc. Natl. Acad. Sci. USA **93:** 13635–13640.
22. THEIL, E.C. 1998. The iron responsive element (IRE) family of mRNA regulators. Regulation of iron transport and uptake compared in animals, plants and microorganisms. Metal Ion Biol. Syst. **35:** 403–434.
23. SIMPSON, R.J., C.E. COOPER, K.B. RAJA *et al.* 1992. Non-transferrin-bound iron species in the serum of hypotransferrinaemic mice. Biochim. Biophys. Acta **1156:** 19–26.
24. SYMONS, M.C.R. & J.M.C. GUTTERIDGE. 1998. Free Radicals and Iron Chemistry, Biology and Medicine. Oxford University Press. Oxford, UK.
25. WALLING, C. 1975. Fenton's reagent revisited. Acc. Chem. Res. **8:** 125.
26. BECKMAN, J.S. & W.H. KOPPENOL. 1996. Nitric oxide, superoxide and peroxynitrite: the good, the bad and the ugly. Am. J. Physiol. **271:** C1424–C1437.
27. RUBBO, H., R. RADI., M. TRUJILLO *et al.* 1994. Nitric oxide regulation of superoxide and peroxynitrite-dependent lipid peroxidation. J. Biol. Chem. **269:** 26066–26075.
28. HALLIWELL, B. 1989. Superoxide, iron, vascular endothelium and reperfusion injury. Free Rad. Res. Commun. **5:** 315–318.
29. MERENYI, G., J. LIND, S. GOLDSTEIN & G. CZAPSKI. 1998. Peroxynitrous acid homolyzes into ·OH and ·NO$_2$ radicals. Chem. Res. Toxicol. **11:** 712–713.
30. KAUR, H., M. WHITEMAN & B. HALLIWELL. 1997. Peroxynitrite-dependent aromatic hydroxylation and nitration of salicylate and phenylalanine. Is hydroxyl radical involved? Free Rad. Res. **26:** 71–82.
31. RICHESON, C.E., P. MULDER, V.W. BOWRY & K.U. INGOLD. 1998. The complex chemistry of peroxynitrite decomposition: new insights. J. Am. Chem. Soc. **120:** 7211–7219.
32. SLATER, T.F. 1984. Free-radical mechanisms of tissue injury. Biochem. J. **222:** 1–15.
33. VON SONNTAG, C. 1987. The Chemical Basis of Radiation Biology. Taylor and Francis. London.
34. DAVIES, M.J. & R.T. DEAN. 1997. Radical-Mediated Protein Oxidation. From Chemistry to Medicine. Oxford University Press. Oxford, UK.
35. DIZDAROGLU, M. 1991. Chemical determination of free radical-induced damage to DNA. Free Rad. Biol. Med. **10:** 225–242.
36. HALLIWELL, B. & J.M.C. GUTTERIDGE. 1984. Lipid peroxidation, oxygen radicals, cell damage and anti-oxidant therapy. Lancet **i:** 1396–1398.
37. JACOBSON, M.D. 1996. Reactive oxygen species and programmed cell death. Trends Biochem. Sci. **21:** 83–86.
38. HALLIWELL, B., C.E. CROSS & J.M.C. GUTTERIDGE. 1992. Free radicals, antioxidants, and human disease. Where are we now? J. Lab. Clin. Med. **119:** 598–620.
39. DIPLOCK, A.T. 1997. Will the "good fairies" please prove to us that vitamin E lessens human degenerative disease? Free Rad. Res. **27:** 511–532.
40. GUTTERIDGE, J.M.C. & B. HALLIWELL. 1995. Antioxidants in Nutrition, Health, and Disease. Oxford University Press. Oxford, UK.
41. HALLIWELL, B. 1999. Establishing the significance and optimal intake of dietary antioxidants. The biomarker concept. Nutr. Rev. **57:** 104–113.
42. SEN, C.K. 1998. Redox signalling and the emerging therapeutic potential of thiol antioxidants. Biochem. Pharmacol. **55:** 1747–1758.
43. JONESON, T. & D. BAR-SAGI. 1998. A Rac 1 effector site controlling mitogenesis through superoxide production. J. Biol. Chem. **213:** 17991–17994.
44. BURDON, R.H. 1995. Superoxide and H$_2$O$_2$ in relation to mammalian cell proliferation. Free Rad. Biol. Med. **18:** 775–794.
45. SARAN, M. & W. BORS. 1989. Oxygen radicals acting as chemical messengers: a hypothesis. Free Rad. Res. Commun. **7:** 213–220.
46. HALLIWELL, B. 1996. Free radicals, proteins and DNA: oxidative damage versus redox regulation. Biochem. Soc. Trans. **24:** 1023–1027.

47. HALLIWELL, B. & J.M.C. GUTTERIDGE. 1986. Oxygen free radicals and iron in relation to biology and medicine: some problems and concepts. Arch. Biochem. Biophys. **246:** 501–514.
48. HAMPTON, M.B. & S. ORRENIUS. 1997. Dual regulation of caspase activity by H_2O_2: implications for apoptosis. FEBS Lett. **414:** 552–556.
49. CLEMENT, M.V. & S. PERVAIZ. 1999. Reactive oxygen species regulate cellular response to apoptotic stimuli: an hypothesis. Free Rad. Res. **30:** 247–252.
50. KONDO, T., F.R. SHARP, J. HONKANIEMI *et al.* 1997. DNA fragmentation and prolonged expression of c-fos, c-jun and hsp70 in kainic acid-induced neuronal cell death in transgenic mice overexpressing human Cu,Zn-superoxide dismutase. J. Cereb. Blood Flow Metab. **17:** 241–256.
51. MIROCHNITCHENKO, O., U. PALNITKAR, M. PHILBERT & M. INOUYE. 1995. Thermosensitive phenotype of transgenic mice overproducing human glutathione peroxidases. Proc. Natl. Acad. Sci. USA **92:** 8120–8124.

Reflections on the Role of the Thiol Group in Biology

NIELS HAUGAARD[a]

Department of Pharmacology and Division of Urology, University of Pennsylvania, Philadelphia, Pennsylvania 19104, USA

> ABSTRACT: This essay is concerned with the role of the thiol or sulfhydrvl group in cellular function and metabolism and with the important investigations over many years that have led us to a better understanding of the importance of this molecular moiety that plays such a vital role in biology. The tools for measuring the SH group and for inhibiting or regenerating it will be discussed as will its essential role in the actions of many enzymes. The importance of the thiol group in glycolysis and in energy production by mitochondria will be emphasized. Of special interest at present is the fact that certain low molecular weight SH-containing substances can mimic some of the actions of insulin and may become of benefit in the treatment of diabetes mellitus. Finally, the toxic effects of oxygen on metabolism and function will be discussed with particular reference to the possibility that oxidation of thiol groups may play a role in the manifestations of oxygen toxicity.

Thiol groups in the cell play extraordinarily important roles in almost all aspects of cellular function. This is true for low molecular weight soluble cell components such as glutathione as well as for numerous enzymes that are active only in the reduced state. The early work in this field was reviewed in 1951 in a fascinating and informative article by E.S. Guzman Barron who carried out extensive studies in this field and who, together with T. P. Singer, is to a great extent responsible for the concept of sulfhydryl enzymes.[1,2] Haugaard[3] pointed out that the sulfhydryl enzymes are particularly sensitive to the toxic action of oxygen and that oxygen in a sense may be considered to be an SH reagent. This article will be concerned with some aspects of the chemistry and biology of the thiol or sulfhydryl group, its involvement in glycolysis and mitochondrial function as well as in glucose metabolism. An important biological phenomenon is oxygen toxicity a subject that has fascinated biologists for a long time.[4] Here we will discuss the possible role of oxidation of thiol groups in the manifestation of oxygen toxicity. Although the essay will be to a large extent historical, some exciting recent developments demonstrating that α-lipoic acid, also known as thioctic acid, has a unique ability to stimulate glucose transport into the cell[5] will also be considered.

---

[a]Address for correspondence: Dr. Niels Haugaard, University of Pennsylvania, 3010 Ravdin Courtyard, 3400 Spruce Street, Philadelphia, PA 19104-4283. Voice: 215-662-6870; fax: 215-349-5026.

EARLY INVESTIGATIONS

The story of the thiol group in biology may well be considered to have started with the discovery of a sulfur containing substance by de Rey Pailhade in 1888 that was found to be widely distributed in nature. He named it philothion (from the Greek words love and sulfur) and this essential cellular component was rediscovered by F.G. Hopkins 36 years later and renamed glutathione. These developments in biology have been well described in a fascinating article by Meister[6] as well as in the review by Barron[1] that also considers the early findings by Rapkine in the 1930's that glutathione may play a role in the processes of cell division. The importance of thiol groups in cell division was discussed in detail later by Mazia.[7] Glutathione exists in animals almost entirely as an intracellular component and no specific coenzyme function in mammalian tissues has been ascribed to it. It is now generally accepted that the primary function of glutathione in the cell is to act as a regulator of the oxidation-reduction potential of the cell and as an antioxidant.[4,7,8] This important cell constituent may possibly be more specifically involved in maintaining sulfhydryl enzymes in their active reduced states. Two proceedings of symposia concerned specifically with glutathione have been published.[7,8] These contain a wealth of information about our early knowledge of this important cell constituent. Among other important small molecular weight sulfur containing compounds of importance in cell metabolism are coenzyme A, essential in acetyl transfer reactions, as well as thioctic acid or (α-lipoic acid), which acts as a coenzyme in the oxidation of pyruvic and α-glutaric acids. This substance, a derivative of octanoic acid, is a disulfide but is easily reduced in the body to the thiol form.

The fact that many enzymes contain free SH-groups in their active form was first observed with hydrolytic enzymes such as urease,[9] but with the work of Barron[1] and others it soon became evident that a great number of enzymes, including many kinases and oxidative enzymes, contain thiol groups and are inactivated when these groups are destroyed by the appropriate reagents. It is of considerable interest that, as far as I know, no 'sulfhydryl hormones' have been discovered. In fact insulin which contains one intrachain S-S bridge and two interchain S-S bridges is entirely inactivated when these bonds are reduced to thiol groups. In this connection it should be pointed out that the role of the S-S bridges in determining the activity of the insulin molecule remains to be elucidated.

A great number of biological processes appear to involve the participation of the thiol group. Among these is the production of insulin by the β-cells of the pancreas. The demonstration by Dunn and collaborators[10] that permanent diabetes could be produced in animals by the injection of alloxan, an agent that destroys SH groups, led to startling advances in the study of the metabolic derangements in diabetes. In the lens of the eye the content of glutathione is extremely high and the compound is considered of great importance in the prevention of cataracts.[7,8] An additional example of the participation of thiol groups in a biological process is in the function of the carotid chemoreceptors. Lahiri[11] studied carotid body function in the cat and observed that agents that combined with SH groups greatly augmented the effects of hypoxia on chemoreceptor activity.

REAGENTS THAT REACT WITH THIOL GROUPS

Progress in the field of the biology of thiol compounds, be they of small molecular weight or protein in nature, has depended to a large extent on the development of reagents that react with the thiol group. These can be divided into three classes (1) oxidizing agents such as iodine or ferricyanide (2) alkylating agents such as iodoacetate or N-methyl maleimide (NEM) and (3) mercaptide forming agents such as parachloromercuribenzoate These reagents have been used extensively in studies of enzyme action and metabolism and much progress has been made by the use of these tools. They react readily with low molecular soluble thiols but with proteins it was soon apparent that some protein SH-groups react easily with thiol reagents while others react much more sluggishly. It was also found that thiol reagents vary widely in their ability to react with protein SH-groups. There appears to be different types of SH-groups in proteins, some that are easily available, and others that are more or less buried in the protein structure. On denaturation of proteins the reactivity of the sulfhydryl group increases greatly. An interesting discussion of the reactivity of the SH group in glutathione and other peptides has been provided by Benesch et al.[7]

An important advance in this field was the introduction of a new reagent by Ellman,[12] a water soluble aromatic disulfide, 5,5'-dithiobis(2-nitrobenzoic acid). This compound, also referred to as DTNB, reacts readily with groups to form mixed disulfides and the anion p-nitrothiophenol. The latter compound is highly yellow in color and can thus be used in an excellent way to measure SH- groups in proteins or tissues. The usefulness of this reaction in biological studies should not be underestimated. One use of this agent in biochemistry has been as a component of the reaction mixture for the determination of the rate of the citrate synthase reaction. It acts by reacting with the free coenzyme A formed to produce the yellow p-nitro thiophenol anion.[13] DTNB is highly electronegative and, therefore, cannot readily penetrate cell membranes. However other disulfides are available that more electrochemically neutral and can be expected to enter the cell quite easily. Among these are disulfiram and 2,2-dithiopyridine.These agents also combine with and inactivate thiol groups and in combination with DTNB provide powerful tools in investigations concerning the role of thiol groups in biological processes.

With the development of agents that inactivate SH -groups came the introduction of substances that reversed the inhibition produced by SH-reagents or reduced S-S bonds to thiol groups. Among the first and most famous of these was British Anti-Lewisite or BAL developed by Peters and his group[14] as an antidote to the arsenical war gas Lewisite. It regenerates free thiol groups from those attacked by the arsenical and also reduces S-S bonds. A compound even more effective in regenerating SH groups from disulfides was developed by Cleland.[15] This was dithiothreitol, an agent that reacts readily with disulfides of all types. In its reaction it forms a ring compound driving the reaction to completion. It has been used widely in the purification of enzymes to maintain activity during prolonged procedures. It may be effective in this action both by reducing S-S bonds and by chelating copper ions that catalyze oxidation of SH-groups by oxygen. Another agent that is highly effective in low concentration in preventing oxidative damage to enzymes in tissue extracts is EDTA a compound that in addition to chelating calcium and magnesium also chelates a number of heavy metals including copper and iron.

ROLE OF THIOL GROUPS IN GLYCOLYSIS

A number of biochemical reaction sequences appear to depend for full activity on the presence of free thiol groups either in essential coenzymes such as coenzyme A or because a rate- limiting enzyme contains thiol groups that are highly reactive. Among these are the reactions of the tricarboxylic acid cycle, where coenzyme A plays a vital part and glycolysis in which the enzyme glyceraldehyde phosphate dehydrogenase has been found to be easily inactivated by reagents that react with the thiol group. Among the agents that combine with SH-groups of glyceraldehyde phosphate dehydrogenase and inactivate the enzyme is iodoacetate. This agent was used in the famous experiment in 1930 by Einar Lundsgaard[16] who demonstrated that when lactate formation by muscle was abolished by adding iodoacetate to the muscle bath contraction could still take place for a period of time. This led investigators to focus on ATP as the provider of energy for muscle contraction. Cain and Davies showed conclusively that ATP was the true source of energy for muscle contraction in a brilliant experiment using another enzyme inhibitor, fluoronitrobenzene. When this agent was added to muscle preparations creatine kinase, a sulfhydryl enzyme, was inhibited and replenishment of ATP from creatine phosphate was abolished. Under these conditions there was a perfect correlation, when the muscle was stimulated between the contractile change and the disappearance of high energy phosphate bonds.[17]

MITOCHONDRIAL ENERGY PRODUCTION AND PHOSPHATE UPTAKE

Several early studies provided evidence that thiol groups were in some way involved in the reactions leading to the formation of ATP by mitochondria. Fonyo and Bessman[18] for example observed that p-hydroxymercuribenzoate inhibited the respiratory response of liver mitochondria to ADP + Pi and to Ca^{2+} + Pi. Haugaard and coworkers [19] studied the action of DTNB on respiration and calcium uptake by liver mitochondria. They showed that DTNB in micromolar concentrations caused complete inhibition of the respiratory response to ADP + Pi or to Ca^{2+} + Pi but had no effect on dinitrophenol-stimulated respiration. DTNB also inhibited calcium uptake by the liver mitochondria. These effects of DTNB were completely reversed by dithiothreitol. In experiments in which both calcium and phosphate uptake by liver mitochodria were measured it was found that for each three molecules of calcium taken up by the mitochondria, two molecules of phosphate entered the mitochondria. In the presence of increasing concentrations of DTNB there was a progressive decrease in calcium uptake that was equal mole for mole to the decrease in phosphate uptake. At the highest concentration of DTNB (65 µM), phosphate entrance into the mitochondria was completely abolished while the calcium uptake was diminished by two thirds. It was apparent that in liver mitochondria the portion of calcium uptake associated with entrance of phosphate was inhibited by DTNB. It was concluded from these experiments that thiol groups were intimately involved in either the entrance of phosphate into mitochondria or in further reactions essential for oxidative phosphorylation or ion transport. In a subsequent publication[20] it was shown that the two disulfides disulfiram and 2,2′-dithiopyridine had similar effects but were less specific in that these more lipid-soluble compounds also inhibited 2,4-dinitrophenol–stimulated respiration. Later studies from different laboratories have shown conclusively that phosphate enters the mitochondrion with the facilitation of a phosphate transport protein or translocator and that this protein contains highly reactive thiol groups.[21–23] The phosphate transporter protein has been purified and its chemistry studied in considerable detail.[24] It appears to be part of a family of anion transporters essential in the proper function of mitochondria. It should be pointed out that it may be no coincidence that the glyceraldehyde phosphate dehydrogenase reaction as well as the mitochondrial reactions just discussed both involve the production of high energy phosphate bonds.

STIMULATION OF GLUCOSE METABOLISM

Investigations in the early seventies produced definite evidence that certain SH compounds acted like insulin in some *in vitro* systems. Lavis and Williams[25] studied the metabolism of isolated fat cells and observed that L-cysteine, reduced glutathione, 2-mercaptoethanol and dithiothreitol in a concentration range of 0.01 to 1 mM caused a partial suppression of lipolysis induced by epinephrine and stimulated glucose utilization in intact but not in broken cells. At concentrations of 10 mM or higher these compounds inhibited glucose uptake. The authors concluded that at the lower concentrations the thiols shared with insulin a common pathway of action. The effect of DL-α lipoic on cardiac metabolism was studied by Singh and Bowman.[26] These investigators found that among the metabolic actions of this compound in the perfused rat heart was a marked stimulation of glucose utilization both in the presence and in the absence of insulin in the perfusion fluid. The effect of thioctic or α-lipoic acid on skeletal muscle metabolism was investigated by Haugaard and Haugaard[27] who observed that this compound stimulated glucose utilization by the isolated rat diaphragm. The action appeared after the first 30 minutes of incubation and was additive to that of insulin. In a later publication,[28] Haugaard and coworkers reported that dithiothreitol (DTT) produced a large increase in glucose utilization by the rat diaphragm and decreased glycogen synthesis. When incubations were carried out in the presence and absence of insulin there was an insulin effect during the first 30 minutes of incubation, after which the highly elevated rate of glucose utilization was the same with or without insulin and the tissue appeared to be under the influence mainly of DTT. These early experiments showed conclusively that certain SH- and S-S–containing compounds stimulated glucose metabolism and that lipoic or thioctic acid in particular had a unique effect in acting like insulin in its effect on glucose metabolism in muscle. That these findings were of considerable importance was realized only many years later. Henriksen and Holloszy in 1990 published experiments demonstrating the importance of thiol groups in glucose transport by skeletal muscle incubated in vitro. They studied the action of phenylarsine oxide, a compound that combines with vicinal sulfhydryl groups, and found that this substance activated 3-O-methyl glucose uptake by epitrochlearis muscle *in vitro* but inhibited insulin stimulation of glucose transport by 80%.[29] In 1994 Henriksen and Jacob reported at a meeting of the American Diabetes Association that thioctic acid increased glucose transport activity in skeletal muscle of obese Zucker rats. These investigators working in the Unites States and Germany showed that interperitoneal injection of a racemic mixture of thioctic acid markedly increased insulin-stimulated 2-deoxyglucose uptake in epitrochlearis muscle of this obese insulin-resistant rat.[30] An extensive paper was published later elaborating on these findings.[31] The general conclusion from these experiments was that both acute (1 h, 100 mg/kg) and chronic (10 days treatment, 30 mg/kg/day) of interperitoneal injection of α-lipoic acid markedly increased skeletal muscle glucose uptake in these insulin-resistant animals. Insulin- stimulated glycogen synthesis and glucose oxidation by the muscle preparations incubated in vitro were also enhanced. *In vivo* the plasma levels of insulin and free fatty acids were decreased.[31] The action was shown to be stereo-specific in that the R(+) form of the compound was much more effective than the S(−) enantiomer.[32] Subsequently, these authors demonstrated that α-lipoic acid by itself acted like insulin in stimulating 2-deoxyglucose uptake in the epitrochlearis muscle incubated *in vitro*.[5] The mechanism of action of α-lipoic acid in this unique insulin-mimetic effect remains to be elucidated. At present our understanding of the action of α-lipoic acid is that the compound makes the glucose transporter protein (GLUT-4) more available for insertion into the cell membrane without increasing its total content in the cell. Such an action would explain both the increase in glucose metabolism by the compound itself and also the enhancement of the action of insulin. The experiments by Klip and collaborators with cultured muscle cells give strong support to this concept.[33] Many questions still remain to be answered. For example, it would be important to know more about the action of lipoic acid on glycogen synthesis. In experiments with the rat diaphragm Haugaard and Haugaard[34] demonstrated that ^{14}C-glucose in its incorporation into glycogen did not equilibrate with intracellular glucose indicating that the glucose to glycogen synthetic pathway did not include the prior entrance of glucose into the cell. Another indication that the glycogen synthetic pathway is under the influence of special regulatory factors is that lithium ions specifically directs glucose toward incorporation into glycogen rather than to simply stimulate its entrance into the muscle cell.[35,36] In addition it has been shown that in the isolated rat diaphragm the decrease in glucose uptake produced by epinephrine or phosphodiesterase inhibitors is exactly equal to the decrease in glycogen synthesis indicating that the glycogen synthetic pathway is specifically interfered with.[37] So far the actions of lipoic acid in stimulating glucose metabolism has been studied mainly in skeletal muscle. It would be important to know whether the compound also activates glucose metabolism in the bladder and other smooth muscle tissues in which insulin has minimal effects *in vitro*.[38] Lipoic acid also exerts distinct effects on metabolic reactions in the liver. These have been discussed in a review article by Bustamante *et al.*[39] It is of particular interest that lipoic acid is very effective in reducing oxidized gluthathione and that the administration of the substance to rats leads to increased levels of reduced glutathione in liver and blood.[40]

Lipoic acid is a powerful antioxidant[39,41] and the question arises whether this action is paramount in its effect on glucose metabolism or whether the action is more specific and could involve interaction with cellular SH or S-S groups Whatever the answers to these questions may be, the fact that a small molecular weight sulfur-containing compound can so clearly mimic the actions of insulin is a very exciting development in science and has led to important advances in diabetes research. Thioctic or lipoic acid has been tested in the treatment of diabetes in man and found to offer relief of diabetic neuropathy.[42,43] The establishment of its action as an insulin mimetic substance may lead to more general use of this important substance in the treatment of diabetes mellitus.

TOXIC EFFECTS OF OXYGEN ON METABOLISM

The fact that oxygen, essential as it is for life as we know it, is also toxic to living cells is among the most fascinating aspects of biology. The pioneering studies of Paul Bert are recorded in his monumental treatise La Pression Barometrique published in 1878.[44] In it he describes in detail several important attributes of oxygen poisoning: its universal nature, the variation in sensitivity among species and the involvement of the central nervous system in the manifestation of oxygen toxicity in mammals. With characteristic insight he suggested that the basis for the toxic effects of oxygen may be inhibition of "fermentative" reactions in the cell. During the Second World War the toxic effects of increased pressures of oxygen were studied systematically by Dickens in England and by Stadie, Riggs and Haugaard in the USA. The latter authors wrote a definitive review article that critically considered the evidence available at the time that showed that oxygen even at the concentration present in air can produce toxic effect on cell function and metabolism.[45] The results of the studies by Dickens and by Stadie *et al.* were in excellent agreement and showed conclusively that a great many enzymes particularly those containing essential sulfhydryl groups, are easily inactivated by increased pressures of oxygen but others are quite resistant to oxygen toxicity. These early studies are described in detail in the review by Haugaard.[4] The proceedings of a symposium held in London in 1963 has been published.[46] This deals with the larger subject of the role of oxygen in the animal organism. but also contains much material concerned with oxygen toxicity including a contribution by L.W. Bean concerned with the general manifestations of oxygen toxicity in the mammals. Two books have been published concerned with the general subject of the role oxygen in regulating function and metabolism of living cells. One, *Molecular Oxygen in Biology,* was edited by O. Hayaishi and published in 1974[47] another, *Oxygen and Living Processes,* edited by Dan Gilbert was published in 1981.[48] Both of these books summarize well the state of our knowledge in the field at the time. The latter book contains two fascinating contributions of a historical and philosophical nature written by the editor; one is concerned with the discovery of oxygen itself and the other with the consequences both of hypoxia and hyperoxia. An extremely well written and informative review concerned primarily with the effects of hyperoxia on the brain and the lung and with anti-oxidant defenses has been published by Jamieson.[49]

It is now generally accepted that free radicals or reactive intermediates formed normally during respiration and produced in excess following hyperbaric oxygenation are responsible for the damaging effects of oxygen. This was clearly recognized by Rebeca Gerschman who understood and pointed out the similarity between the cellular effects of excess oxygen and radiation. Her early work in collaboration with Daniel Gilbert was published in *Science* in a classical paper entitled "Oxygen poisoning and X-radiation: a mechanism in common."[50] Contributions of these authors are presented in references 46 and 48. Two touching tributes to the memory of Rebeca Gerschman and her pioneering work in the field of oxygen toxicity were published in 1996 in the journal *Free Radical Biology and Medicine*.[51,52] That journal has become the major vehicle for publications in the expanding field of the biology of free radicals and anti-oxidants. The discovery in 1969 of the enzyme superoxide dismutase by McCord and Fridovich[53] greatly advanced our understanding of oxygen toxicity and of the defenses developed by living cells to counteract the damaging effects of oxygen. The superoxide anion is formed normally in several metabolic reactions and in excess during hyperbaric conditions. Together with hydrogen peroxide it produces much more reactive free radicals such as OH. By destroying the superoxide anion superoxide dismutase acts as a powerful antioxidant. The toxicology of molecular oxygen and the defences developed by cells of aerobic organisms has been considered in a critical review by Di Guiseppi and Fridovich.[54]

Whatever the mechanisms involved in producing the damaging effects of excess oxygen, oxidation of essential cellular components are surely part of the phenomenon of oxygen toxicity. Among such components are thiol groups associated with enzymes or present in the molecules of coenzymes or other essential cell constituents. Tjioe and Haugaard[55] demonstrated with crystalline glyceraldehyde phosphate dehydrogenase that the enzyme was inactivated by exposure to 5 atmospheres of oxygen provided that trace amounts of ferrous or copper ions were present in the medium. The authors measured the SH groups present in the enzyme after different times of exposure to high pressure oxygen and demonstrated that the inactivation of the enzyme was proportional to the disappearance of sulfhydryl groups. In experiments with rats exposed to 5 atm of oxygen Jamieson and van den Brink observed a decrease in activity of several dehydrogenase enzymes in the lung before any signs of gross lung damage was observed.[56] Lungs damaged by hyperbaric oxygen contained fewer SH groups and higher levels of S-S groups than controls.[57] In contrast, exposure of rats to 5 atm oxygen produced no significant decrease in the SH content of the brain.

Since highly reactive thiol groups are intimately involved in glycolysis and in the essential reactions of mitochondria leading to the production of ATP, it seems reasonable to consider that these processes are very likely targets for inactivation under conditions in which the concentration of reactive oxygen species increase and the anti-oxidant defenses are overcome. In experiments concerned with the damaging effects of radiation and of hyperbaric oxygen under many different conditions much evidence has been obtained that prior administration of a variety of thiol compounds produce a considerable degree of protection4,[45,46,49] indicating that these substances may act as free radical scavengers or exert a sparing effect on glutathione or other cellular sulfhydryl compounds.

I would like to end this essay by quoting a statement by the great British biochemist F. Gowland Hopkins[58] that was cited in the introduction to the Symposium on Glutathione published in 1954.[7] He said in the Linacre Lecture in 1938 "It is true to say that in scientific borderlands not only are facts gathered that are often new in kind, but it is in these regions that wholly new concepts arise." This thought, I believe, is as true to-day as it was 60 years ago.

ACKNOWLEDGMENT

The author would like to thank Dr. Erik J. Henriksen of the University of Arizona for his very helpful comments and suggestions.

REFERENCES

1. BARRON, E.S.G. 1951. Thiol groups of biological importance. Adv. Enzymol. **11:** 201–266.
2. SINGER, T.P. 1948. On the mechanism of enzyme inhibition by sulfhydryl reagents. J. Biol. Chem. **174:** 11–21.
3. HAUGAARD, N. 1946. Oxygen poisoning XI. The relation between inactivation of enzymes by oxygen and essential sulfhydryl groups. J. Biol. Chem. **164:** 265–270.
4. HAUGAARD, N. 1968. Cellular mechanisms of oxygen toxicity. Physiol. Rev. **48:** 311–373.
5. HENRIKSEN, E.J., S. JACOB, R.S. STREEPER, D.L. FOGT, J.Y. HOKAMA & H.J. TRITSCHLER. 1997. Stimulation by α-lipoic acid of glucose transport activity in skeletal muscle of lean and obese Zucker rats. Life Sci. **61:** 805–812
6. MEISTER, A. 1988. On the discovery of glutathione. Trends Biochem. Sci. **13:** 185–188.
7. COLOWICK, S. et al., Eds. 1954. Glutathione. A Symposium. Academic Press. New York.
8. ARIAS, I.M. & W.B. JACOBY, Eds. 1976. Glutathione Metabolism and Function. Raven Press. New York.
9. HELLERMAN, L. 1937. Reversible inactivation of certain hydrolytic enzymes. Physiol. Rev. **17:** 454–484.
10. DUNN, J.S., H.L. SHEEHAN & N.G.B. MCLETCHIE. 1943. Necrosis of islets of Langerhans produced experimentally. Lancet **I:** 384–387.
11. LAHIRI, S. 1981. Chemical modification of carotid body chemoreception by sulfhydryls. Science **212:** 1065–1066.
12. ELLMAN, G.L. 1959. Tissue sulfhydryl groups Arch. Biochem. Biophys. **82:** 70–77.
13. ROBINSON, K.B., JR. R.G. BRENT, P. SUMEGI & P.A. SRERE. 1989. An enzymatic approach to the study of the Krebs tricarboxylic acid cycle. In Mitochondria: A Practical Approach. V.M. Darley Usmar, D. Rickwood & M.T. Wilson, Eds. :153–170. IRS Press. Oxford.
14. DIXON, M. & E.C. WEBB. 1964. Enzymes. Academic Press. New York, p. 344.
15. CLELAND, W.W. 1964. Dithiothreitol, a new protective reagent for SH groups. Biochemistry **3:** 480–482.
16. LUNDSGAARD, E. 1932. Weitere Untersuchungen uber die Einwirkung der Halogenessigsauren auf die Spaltung- und Oxidationsstoffwechsel. Biochem. Ztschr. **250:** 61–88.
17. CAIN, D.F. & R.E. DAVIES. 1962. Breakdown of adenosinetriphosphate during a single contraction of working muscle. Biochem. Biophys. Res. Commun. **8:** 361–366.
18. FONYO, A. & S.P. BESSMAN. 1966. The action of oligomycin and parahydroxymercuribenzoate on mitochondrial respiration stimulated by ADP, arsenate and calcium. Biochem. Biophys Res. Commun. **24:** 61–66.
19. HAUGAARD, N., N.H. LEE, R. KOSTRZEWA, R.S. HORN & E.S. HAUGAARD. 1969. The role of sulfhydryl groups in oxidative phosphorylation and ion transport by rat liver mitochondria. Biochim. Biophys. Acta **172:** 198–204.
20. HAUGAARD, N., N.H. LEE, P. CHUDAPONGSE, C.D. WILLIAMS & E.S. HAUGAARD. 1970. The actions of disulfiram and 2,2-dithiopyridine on oxidative phosphorylation and ion transport by rat liver mitochondria. Biochem. Pharmacol. **19:** 2969–2971.
21. FONYO, A. 1974. Phosphate carrier of liver mitochondria: two equivalent SH-groups in the carrier unit. Biochem Biophys. Res. Commun. **57:** 1069–1073.
22. FONYO, A. 1978. SH-group reagents as tools in the study of mitochondrial anion transport. A review. J. Bioenerg. Biomembr. **10:** 171–194

23. WEHRLE, J.P & P.L. PEDERSEN. 1981. Phosphate transport in rat liver mitochondria: location of sulfhydryl groups essential for transport activities. J. Bioenerg. Biomembr. **13:** 285–294,
24. WOHLRAB, H. & N. FLOWERS. 1982. pH-gradient-dependent phosphate transport catalyzed by the purified mitochondrial phosphate transport protein. J. Biol. Chem. **257:** 28–31.
25. LAVIS, R. & R.H. WILLIAMS. 1970. Studies of the insulin-like actions of thiols upon isolated fat cells. J. Biol. Chem. **245:** 23–31.
26. SINGH, H.P.P. & R.H. BOWMAN. 1970. Effect of DL-α-lipoic acid on the citrate concentration and fructophosphokinase activity by perfused hearts from normal and diabetic rats. Biochem. Biophys. Res. Commun. **41:** 555–561.
27. HAUGAARD, N. & E.S. HAUGAARD. 1970. Stimulation of glucose utilization by thioctic acid in rat diaphragm incubated in vitro. Biochim. Biophys. Acta **222:** 583-586.
28. HAUGAARD, E.S., M.J. SMITH & N. HAUGAARD. 1972. Effects of thiols and disulfides on glucose utilization and insulin action in the isolated rat diaphragm. Biochem. Pharmacol. **21:** 517–523.
29. HENRIKSEN, E J. & J.O. HOLOSZY. 1990. Effects of phenylarsine oxide on stimulation of glucose transport in rat skeletal muscle. Am. J. Physiol. **258:** C648–C653.
30. HENRIKSEN, E.J & S. JACOB. 1994. Chronic thioctic acid treatment increases insulin stimulated glucose transport activity in skeletal muscle of obese Zucker rats. Diabetes **43** (Suppl. 1)**:** 122A.
31. JACOB, S., R.S. STREEPER, D. FOGT, J.Y. HOKAMA, H.J. TRITCHLER, G.J. DIETZE & E.J. HENRIKSEN. 1996. The antioxidant α-lipoic acid enhances insulin-stimulated glucose metabolism in insulin-resistant rat skeletal muscle. Diabetes **45:** 1024–1029.
32. STREEPER, R.S., E.J. HENRIKSEN, S. JACOB, J.Y. HOKAMA, D.L. FOGT & H.J. TRITCHLER. 1997. Differential effects of lipoic acid stereoisomers on glucose metabolism in insulin-resistant skeletal muscle. Am. J. Physiol. **273:** E185–E191.
33. KLIP, A., A. VOLCHUK, T. RANLAL, C. ACKERLEY & Y. MITSUMOTO. 1994. *In* Molecular Biology of Diabetes, Vol. II. B. Draznin & D. Leroith, Eds. :511–528. Humana Press. Totowa, NJ.
34. HAUGAARD, E.S. & N. HAUGAARD. 1974. The action of insulin on glycogen synthesis in rat diaphragm from intracellular and extracellular glucose. Biochim. Biophys. Acta **338:** 309–316.
35. HAUGAARD, E.S., R.A. MICKEL & N. HAUGAARD. 1973. Actions of lithium ions and insulin on glucose utilization glycogen synthesis and glycogen synthase in the isolated rat diaphragm. Biochem. Pharmacol. **23:** 1675–1685.
36. FÜRNSINN, C., C. NOE, R. HERDLICKA, M. RODEN, P. NOWOTNY, B. LEIGHTON & W. WALDHXUSL. 1997. More marked stimulation by lithiuim than insulin of the glycogenis pathway in skeletal muscle. Am. J. Physiol. **273:** E514–E520.
37. DAVIDHEISER, S., E.S. HAUGAARD & N. HAUGAARD. 1979. Effects of epinephrine and the cyclic AMP phosphodiesterase inhibitor SQ 20009 on glucose and glycogen metabolism in skeletal muscle. Biochem. Pharmacol. **28:** 807–813.
38. HAUGAARD, N., R.M. LEVIN & A.J. WEIN. 1987. In vitro studies of glucose metabolism of the rabbit urinary bladder. J. Urol. **137:** 782–784.
39. BUSTAMANTE, J., J.K. LODGE, L. MARCONI, H.J. TRITCHLER, L. PACKER & B.H. RIHN. 1998. α-Lipoic acid in liver metabolism and disease. Free Rad. Biol. Med. **24:** 1023–1039.
40. BUSSE, E., G. ZIMMER, B. SCHOPOHL & B. KORNHUBER. 1992. Influence of alpha lipoic acid on intracellular glutathione in vitro and in vivo. Artzneimittelforschung **42:** 829–831.
41. PACKER, L., H.J. TRITCHLER & K. WESSEL. 1997. Neuroprotection by the metabolic antioxidant alpha-lipoic acid. Free Rad. Biol. Med. **22:** 359–378.
42. SACHSE, G. & B. WILLMS. 1980. Efficacy of thioctic acid in the therapy of peripheral diabetic neuropathy. *In* Aspects of Autonomic Neuropathy in Diabetes. F.A. Gries, H.J. Freund, F. Rabe & H. Berger, Eds. :105–108. Thieme.
43. ZIEGLER, D., M. HANEFELD, K.J. RUHNAU, H.P. MEISNER, *et al.* 1996. Treatment of symptomatic diabetic peripheral neuropathy with the antioxidant α-lipoic acid. Diabetologia **38:** 1425–1433.

44. BERT, P. 1878. Barometric Pressure: Researches in Experimental Physiology. Translated by M.A. Hitchcock & F.A. Hitchcock. College Book Co. Columbus OH.
45. STADIE, W.C., B.C. RIGGS & N. HAUGAARD. 1944. Oxygen poisoning. Am. J. Med. Sci. **207:** 84–114.
46. DICKENS, F. & E. NEIL, Eds. 1964. Oxygen in the Animal Organism. Pergamon Press. Oxford.
47. HAYAISHI, O. 1974. Molecular Oxygen in Biology. American Elsevier. New York.
48. GILBERT, D.L., Ed. 1981. Oxygen and Living Processes. An Interdisciplinary Approach. Springer Verlag. New York.
49. JAMIESON, D. 1989. Oxygen toxicity and reactive oxygen metabolites in mammals. Free Rad. Biol. Med. **7:** 87–108.
50. GERSCHMAN, R., D.L. GILBERT, S.W. NYE, P. DWYER & W.O. FENN. 1954. Oxygen poisoning and X-radiation: a mechanism in common. Science **119:** 623–625.
51. GILBERT, D.L. 1996. Rebeca Gerschman: a personal remembrance. Free Rad. Biol. Med. **21:** 1–4.
52. BOVERIS, A.A. 1996. Rebeca Gerschman: a brilliant woman scientist in the fifties. Free Rad. Biol. Med. **21:** 5–6.
53. MCCORD, J.M. & I. FRIDOVICH. 1969. Superoxide dismutase: an enzymatic function for erythrocuprein (Hemocrupein). J. Biol. Chem. **244:** 6049–6055.
54. DIGUISEPPI, J. & I. FRIDOVICH. 1984. The toxicology of molecular oxygen. Crit. Rev. Toxicol. **12:** 315–342.
55. TJIOE, G. & N. HAUGAARD. 1972. Oxygen inhibition of crystalline glyceraldehyde phosphate dehydrogenase and disappearance of enzyme sulfhydryl groups. Life Sci. **11:** 329–335.
56. JAMIESON, D. & H.A.S. VAN DEN BRINK. 1962. Pulmonary damage due to high pressure oxygen breathing in rat. 2. Changes in dehydrogenase activity of rat lung. Aust. J. Exp. Biol. Med.Sci. **20:** 51–56.
57. JAMIESON, D., K. LADNER & H.A.S. VAN DEN BRINK. 1963. Pulmonary damage due to high pressure oxygen breathing in rat. 4. Quantitative analysis of sulfhydryl and disulphide groups in rat lung. Aust. J. Exp. Biol. Med. Sci. **41:** 491–498.
58. PIRIE, N.W. 1983. Sir Frederick Gowland Hopkins (1861–1947). *In* Comprehensive Biochemistry, Vol 35. Selected Topics in the History of Biochemistry. Personal Recollections I. Elsevier Science Publishers. Amsterdam, pp. 103–127.

Differential Regulation of MAP Kinase Signaling by Pro- and Antioxidant Biothiols

YUICHIRO J. SUZUKI,[a,b] SUSAN S. SHI,[a] REGINA M. DAY,[c] AND JEFFREY B. BLUMBERG[a]

[a]*Antioxidants Research Laboratory, Jean Mayer USDA Human Nutrition Research Center on Aging at Tufts University, Boston, Massachusetts 02111, USA*

[c]*Pulmonary and Critical Care Division, New England Medical Center, Boston, Massachusetts 02111, USA*

ABSTRACT: Some biologically derived thiol-containing compounds have potential for health benefits whereas others elicit biochemical events leading to pathogenesis. Effects of two biothiols, α-lipoic acid (αLA), a therapeutic antioxidant, and homocysteine (Hcy), a risk factor for age-associated cardiovascular disease, on cell signaling events involving p44 and p42 MAP kinases (p44/42 MAPK) were evaluated in cell culture. Treatment of serum-deprived NIH/3T3 cells with Hcy (20 μM) resulted in the activation of p44/42 MAPK as determined by Western blot analysis using the phospho-specific p44/42 MAPK antibody. p44/42 MAPK phosphorylation was rapid and transient with maximal activation occurring at 10–30 min. Transient activation of p44/42 MAPK was also observed in response to treatment of serum-deprived cells with αLA. In cells grown in serum, serum-dependent p44/42 MAPK phosphorylation was transiently enhanced by Hcy or Hcy thiolactone, but inhibited by αLA. Thus, αLA and Hcy differentially influence signal transduction events depending on the state of cells. These observations may be important in understanding how some biothiols are associated with pathogenic events while others have potential as therapeutic agents.

INTRODUCTION

Thiol-containing biological molecules play various roles in physiology. Some thiols such as homocysteine (Hcy) are considered pro-oxidants and pathogenic,[1] while others such as glutathione and α-lipoic acid (αLA) serve as antioxidants important for maintaining health.[2] Redox compounds including reactive oxygen species (ROS) and antioxidants appear to regulate signal transduction,[3] but this relationship is complex and simple oxidant-antioxidant interactions do not account for many experimental observations. For example, while some studies have shown suppression of stimuli-mediated activation of NF-κB transcription factor by antioxidants,[4,5] both oxidants and antioxidants have been demonstrated to induce AP-1 activation. Activation of c-*fos* and c-*jun* proto-oncogene expression and AP-1 DNA

[b]To whom correspondence should be addressed: Dr. Yuichiro J. Suzuki, Antioxidants Research Laboratory, USDA Human Nutrition Research Center on Aging at Tufts University, 711 Washington Street, Boston, MA 02111. Voice: 617-556-3148; fax: 617-556-3344.
e-mail: YSUZUKI@HNRC.TUFTS.EDU

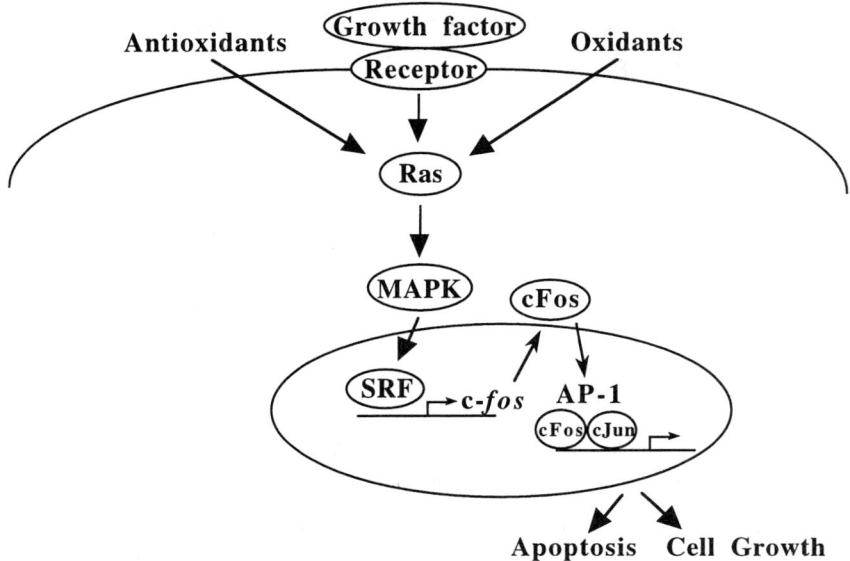

FIGURE 1. MAPK signaling pathway. Growth factor-receptor interactions elicit a cascade of signaling events involving the activation of Ras and p44/42 MAPK. Activated MAPK can elicit activation of transcription factors such as serum response factor (SRF) which regulate gene expression of c-*fos*. c-*fos* message is then translated to cFos protein which in turn dimerizes with cJun protein to become a heterodimeric AP-1 transcription factor which can regulate transcription of various genes that are involved in cell growth as well as apoptosis. Both oxidants and antioxidants have been shown to activate the MAPK signaling pathway although how markedly different outcomes are achieved is not understood.

binding occurs in response to treatment of cells with oxidants such as H_2O_2 and antioxidants such as pyrrolidine dithiocarbamate, *N*-acetyl-L-cysteine, butylated hydroxyanisole, and thioredoxin.[6,7] Recently, Müller *et al.*[8] suggested that both oxidants and antioxidants activate c-*fos*/AP-1 via Ras-dependent activation of p44 and p42 mitogen-activated protein kinases (p44/42 MAPK). These observations raise the question of how oxidants and antioxidants induce common signaling pathways yet elicit differential clinical outcomes (FIG. 1).

αLA is essential in mitochondrial dehydrogenase reactions catalyzing the oxidative decarboxylation of α-keto acids such as pyruvate, α-ketoglutarate, and branched chain α-keto acids. αLA (FIG. 2) and its reduced form dihydrolipoic acid (DHLA) appear also to serve as metabolic antioxidants[2] with DHLA preventing microsomal lipid peroxidation[9,10] and ESR spin trapping showing αLA and DHLA scavenge hydroxyl radicals, although only DHLA reacts with superoxide.[11] Therefore, αLA, either directly or via reduction to DHLA may act as an antioxidant.[12,13] Further, αLA increases intracellular GSH.[14] As αLA is efficacious as a therapeutic agent in diabetes and other diseases characterized by elevated oxidative stress, its antioxidant properties may underlie some of its efficacy.[2]

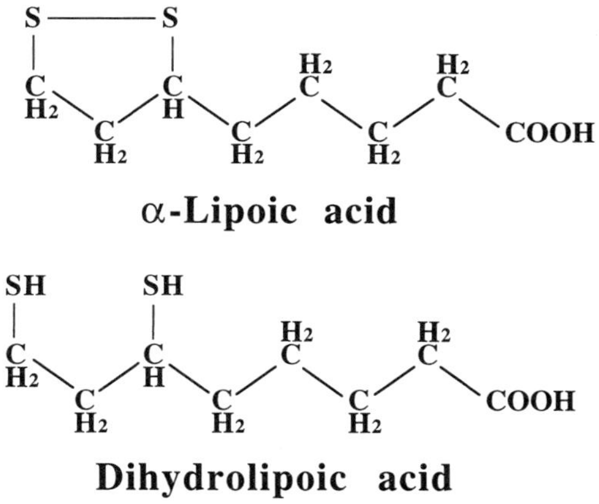

α-Lipoic acid

Dihydrolipoic acid

FIGURE 2. Structures of α-lipoic acid and dihydrolipoic acid.

Hcy (FIG. 3), another thiol-containing biological molecule, is generated during the metabolism of methionine and appears to be involved in the pathogenesis of various cardiovascular diseases.[1] Epidemiological studies indicate that moderate hyperhomocysteinemia (≥ 14 μM Hcy) is an independent risk factor for cardiovascular disease.[15–18] Hereditary hyperhomocysteinemia (> 100 μM Hcy) is associated with enzymatic defects of cystathionine synthase, Hcy methyltransferase, and methylenetetrahydrofolate reductase and results in premature arteriosclerosis, mental retardation, and other symptoms.[19] It has been postulated that the production of ROS may be involved in the pathogenesis of Hcy-mediated diseases.[19] Hcy also possesses cell growth promoting action,[20,21] possibly mediated via production of ROS which elicit proliferative signals.[22,23]

The present study examined the influences of Hcy and αLA on cell signaling for p44/42 MAPK phosphorylation in NIH/3T3 fibroblasts.

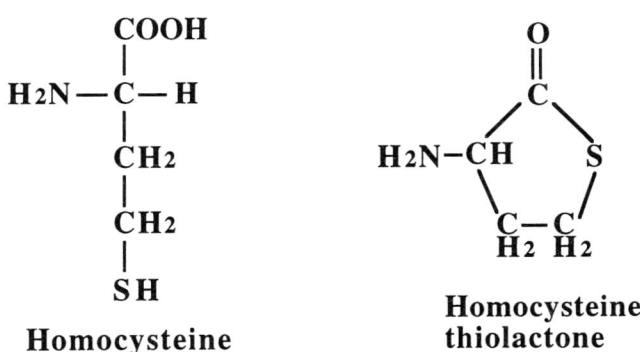

FIGURE 3. Structures of homocysteine and homocysteine thiolactone.

MATERIALS AND METHODS

Cell Culture

NIH/3T3 cells were grown in Dulbecco's modified Eagle's medium (DMEM) supplemented with 10% fetal bovine serum (FBS), 1% penicillin/streptomycin, gentamicin (0.5 mg/ml) and geneticin (0.25 mg/ml) in an atmosphere of 5% CO_2 at 37°C in humidified air. Cells grown at 80–90% confluency were used for experiments. D,L-Homocysteine (Hcy), L-homocysteine thiolactone and R,S-α-lipoic acid (αLA) (Sigma Chemical Company, St. Louis, MO) were dissolved in DMEM to make 0.2 mM stock solutions within 1 h of being used for experiments. Cell permeable farnesyltransferase inhibitor, FPT inhibitor III (Calbiochem-Novabiochem Corporation, San Diego, CA) was dissolved in H_2O.

Western Blot Analysis

Cells were solubilized with 50 mM Hepes solution (pH 7.4) containing 95 mM sodium fluoride, 2.7 mM sodium orthovanadate, 10 mM tetrasodium pyrophosphate, 4 mM EDTA, 1% Triton X-100, 0.35 mg/ml PMSF, 10 µg/ml leupeptin and 10 µg/ml aprotinin (Sigma Chemical). Following centrifugation in a microfuge, protein concentrations were determined in supernatants using the method described by Bradford.[24] Cell lysates (10 µg protein) were electrophoresed through a reducing (5% β-mercaptoethanol) 10% SDS polyacrylamide gel and electroblotted onto a nitrocellulose membrane. The membrane was incubated with the rabbit polyclonal IgG for phosphorylated p44/42 MAPK (ERK1/2) (New England Biolabs, Bevery, MA), phosphorylated Jun kinase kinase (SEK1) (New England Biolabs), or p44 MAPK protein (Santa Cruz Biotechnology, Santa Cruz, CA). The detection was made with HRP-linked secondary antibody using the ECL system (Amersham Life Science, Arlington Heights, IL).

RESULTS

Hcy and αLA Activate p44/42 MAPK in Serum-Deprived Cells

Incubation of serum-deprived NIH/3T3 cells with Hcy (20 µM) for 0–60 min resulted in phosphorylation of both p44 (ERK1) and p42 (ERK2) MAP kinases as shown in a representative Western blot (FIG. 4A), indicating the activation of these kinases.[25] The protein levels of p44/42 MAPK did not change in response to Hcy treatment as illustrated by experiments using the antibody for p44 MAPK protein which also cross-reacts with p42 MAPK. Phosphorylation of MAPK by Hcy was transient, being detected within 10 min of incubation with a peak level of activation by 30 min. Pre-incubation of cells with FPT Inhibitor III or stable transfection of dominant negative mutant p21^{ras} blocked Hcy-mediated activation of MAPK, suggesting a role for Ras in Hcy-signaling.

Similar to the actions of Hcy, αLA (20 µM) induced activation of p44/42 MAPK phosphorylation in serum-deprived NIH/3T3 cells (FIG. 4B). The kinetics of αLA-mediated p44/42 MAPK activation were also transient with the maximal activation occurring at 30 min.

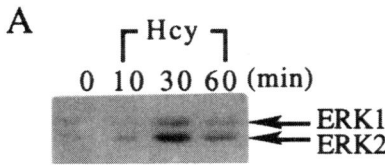

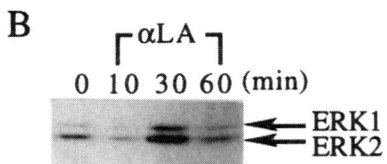

FIGURE 4. Hcy and αLA induce phosphorylation of p44/42 MAPK in serum-deprived cells. NIH/3T3 cells were serum starved for 18 h and treated for 0–60 min with (**A**) Hcy (20 μM) or (**B**) αLA (20 μM). Cell lysates were subjected to immunoblot analysis using the antibody for phospho-p44/42 MAPK (ERK1/2).

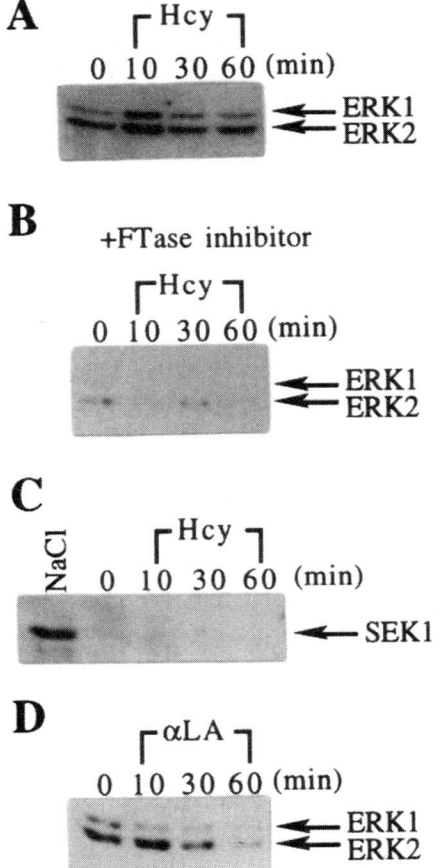

FIGURE 5. Effects of Hcy and αLA on p44/42 MAPK phosphorylation in serum-activated cells. (**A** and **C**) Cells grown in DMEM containing 10% FBS were treated for 0–60 min with Hcy (20 μM). (**B**) Cells were pretreated with FPT inhibitor III before treatment with Hcy. (**D**) Cells were treated for 0–60 min with αLA (20 μM). Cell lysates were subjected to immunoblot analysis using the antibody for (**A**, **B** and **D**) phospho-p44/42 MAPK (ERK1/2) or (**C**) phospho-Jun kinase kinase (SEK1).

Hcy and Hcy Thiolactone Activate, but αLA Inhibits, p44/42 MAPK in Serum-Stimulated Cells

Serum-deprived cells are often employed in investigations of growth factor-mediated p44/42 MAPK activation to eliminate basal background signals and to synchronize cells into a quiescent state.[26] To test whether the activation of p44/42 MAPK by Hcy and αLA could be due to a non-specific stress response, the effects of these thiols on p44/42 MAPK activation were also studied in cells grown in DMEM containing 10% FBS. Before subjecting cells to Hcy, significant p44/42 MAPK phosphorylation levels were observed due to serum-stimulation (FIG. 5A). Exposure of serum-stimulated cells to Hcy further enhanced the levels of p44/42 MAPK phosphorylation. Hcy-mediated enhancement of p44/42 MAPK phosphorylation in serum-stimulated cells was also transient. Pre-incubation of cells with FPT Inhibitor III blocked Hcy-mediated activation of MAP kinase (FIG. 5B). Similar results were obtained when stable transfectants of dominant negative p21^{ras} were treated with Hcy, suggesting a role for Ras in Hcy-signaling in serum activated cells. Hcy (20 µM) did not elicit the activation of Jun kinase kinase (SEK1), a known stress-activated kinase, whose activity was strongly stimulated by the exposure of cells with 700 mM NaCl for 30 min (FIG. 5C).

In contrast to effect of Hcy stimulating p44/42 MAPK, αLA significantly inhibited serum-dependent p44/42 MAPK phosphorylation with its levels undetectable by 60 min (FIG. 5D). Thus, the action of αLA on p44/42 MAPK phosphorylation is different in serum-stimulated cells compared to silent serum-deprived cells. Further, while αLA and Hcy both activate p44/42 MAPK in serum-deprived cells, they differentially influence p44/42 MAPK in serum-stimulated cells. The differential actions of Hcy and αLA are not due to differences in the redox states of these thiols because 20 µM homocysteine thiolactone (FIG. 3), an oxidized form of Hcy, did not inhibit, but rather stimulated, serum-dependent p44/42 MAPK phosphorylation.

DISCUSSION

p44/42 MAPK play important roles in pathogenesis of cancer and cardiovascular disease by mediating abnormal cell growth, differentiation and apoptosis.[27] We demonstrate that Hcy and αLA differentially influence cell signaling for MAPK phosphorylation depending on the state of the cells. In serum-deprived cells, Hcy and αLA elicit the activation of p44/42 MAPK. In serum-stimulated cells, Hcy induces similar signal transduction events, but αLA inhibits serum-dependent cell signaling. Importantly, the concentration of Hcy found effective in p44/42 MAPK activation are physiologically relevant and comparable to levels detected in people with moderate hyperhomocysteinemia.

The immediate-early events of Hcy-signaling to phosphorylate p44/42 MAPK are likely mediated through the activation of Ras. Inductive actions of Hcy on signal transduction in NIH/3T3 cells may be related to the actions of Hcy on vascular smooth muscle cell proliferation.[20] Dalton et al.[21] reported that Hcy induced c-*fos* gene expression through the protein kinase C mechanism. Hcy-mediated activation of the Ras-MAPK pathway may also be important in Hcy-induced signaling for c-*fos* expression and cell growth.

Inhibition of NF-κB activation by αLA[12] may be related to its potential efficacy in attenuating HIV activation[28] and cell adhesion.[29] Serum-dependent inhibition of p44/42 MAPK by αLA suggests that αLA may be useful as well in controlling the abnormal cell growth and apoptosis in cancer and cardiovascular disease.

Antioxidants appear to reduce the risk of many chronic diseases by inhibiting ROS damage to DNA, proteins, and lipids.[30] While ROS inhibit many elements of cell functions, ROS also have the capacity to stimulate signal transduction processes.[3] Irani et al.[23] reported that ROS induce oncogenic cell growth by serving as signaling molecules, and suggested that antioxidants prevent oncogenesis by eliminating oxidant signals. This hypothesis, however, does not account for the actions of antioxidants to elicit signal transduction events such as the Ras-MAPK-SRF pathway.[8] We demonstrate here that pro- and antioxidant biothiols can induce common signal transduction pathways or differentially modulate cell growth signals dependent on the state of the cell. These signaling patterns may, in part, determine the beneficial and detrimental outcomes exerted by αLA and Hcy, respectively, by affecting p44/42 MAPK-mediated cellular events such as cell growth and apoptosis.

ACKNOWLEDGMENTS

This work was supported by National Heart Foundation, a program of the American Health Assistance Foundation to Y.J. Suzuki, a gift from Yamanouchi USA Foundation and the U.S. Department of Agriculture under agreement No. 58-1950-9-001 to J.B. Blumberg. Any opinions, findings, conclusion, or recommendations expressed in this publication are those of the authors and do not necessarily reflect the view of the U. S. Department of Agriculture.

REFERENCES

1. MCCULLY, K.S. 1996. Homocysteine and vascular disease. Nature Med. **2:** 386–389.
2. PACKER, L., E.H. WITT & H.J. TRITSCHLER. 1995. Alpha-lipoic acid as a biological antioxidant. Free Rad. Biol. Med. **19:** 227–250.
3. SUZUKI, Y.J., H.J. FORMAN & A. SEVANIAN. 1997. Oxidants as stimulators of signal transduction. Free Rad. Biol. Med. **22:** 269–285.
4. SCHRECK, R., P. RIEBER & P.A. BAEUERLE. 1991. Reactive oxygen intermediates as apparently widely used messengers in the activation of the NF-κB transcription factor and HIV-1. EMBO J. **10:** 2247–2258.
5. SUZUKI, Y.J., M. MIZUNO & L. PACKER. 1994. Signal transduction for nuclear factor-κB activation. Proposed location of antioxidant-inhibitable step. J. Immunol. **153:** 5008–5015.
6. MEYER, M., R. SCHRECK & P.A. BAEUERLE. 1993. H_2O_2 and antioxidants have opposite effects on activation of NF-κB and AP-1 in intact cells: AP-1 as secondary antioxidant-responsive factor. EMBO J. **12:** 2005–2015.
7. SCHENK, H., M. KLEIN, W. ERDBRÜGGER, W. DRÖGE & K. SCHULZE-OSTHOFF. 1994. Distinct effects of thioredoxin and antioxidants on the activation of transcription factors NF-κB and AP-1. Proc. Natl. Acad. Sci. USA **91:** 1672–1676.
8. MÜLLER, J.M., M.A. CAHILL, R.A. RUPEC, P.A. BAEUERLE & A. NORDHEIM. 1997. Antioxidants as well as oxidants activate c-fos via Ras-dependent activation of extracellular-signal-regulated kinase 2 and Elk-1. Eur. J. Biochem. **244:** 45–52.

9. BAST, A. & G.R.M.M. HAENEN. 1988. Interplay between lipoic acid and glutathione in the protection against microsomal lipid peroxidation. Biochim. Biophys. Acta **963**: 558–561.
10. SCHOLICH, H., M.E. MURPHY & H. SIES. 1989. Antioxidant activity of dihydrolipoate against microsomal lipid peroxidation and its dependence on α-tocopherol. Biochim. Biophys. Acta **1001**: 256–261.
11. SUZUKI, Y.J., M. TSUCHIYA & L. PACKER. 1991. Thioctic acid and dihydrolipoic acid are novel antioxidants which interact with reactive oxygen species. Free Rad. Res. Comms. **15**: 255–263.
12. SUZUKI, Y.J., B.B. AGGARWAL & L. PACKER. 1992. α-Lipoic acid is a potent inhibitor of NF-κB activation in human T cells. Biochem. Biophys. Res. Commun. **189**: 1709–1715.
13. HANDELMAN, G.J., D. HAN, H. TRITSCHLER & L. PACKER. 1994. α-Lipoic acid reduction by mammalian cells to the dithiol form, and release into the culture medium. Biochem. Pharmacol. **47**: 1725–1730.
14. HAN, D., H.J. TRITSCHLER & L. PACKER. 1995. Alpha-lipoic acid increases intracellular glutathione in a human T-lymphocyte Jurkat cell line. Biochem. Biophys. Res. Commun. **207**: 258–264.
15. CLARKE, R., L. DALY, K. ROBINSON, E. NAUGHTEN, S. CAHALANE, B. FOWLER & I. GRAHAM. 1991. Hyperhomocysteinemia: an independent risk factor for vascular disease. N. Engl. J. Med. **324**: 1149–1155.
16. BOUSHEY, C.J., S.A.A. BERESFORD, G.S. OMENN & A.G. MOTULSKY. 1995. A quantitative assessment of plasma homocysteine as a risk factor for vascular disease. Probable benefits of increasing folic acid intakes. J. Am. Med. Assoc. **274**: 1049–1057.
17. PERRY, I.J., H. REFSUM, R.W. MORRIS, P.M. UELAND & A.G. SHAPER. 1995. Prospective study of serum total homocysteine concentration and risk of stroke in middle-aged British men. Lancet **346**: 1395–1398.
18. SELHUB, J., P.F. JACQUES, A.G. BOSTOM, R.B. D'AGOSTINO, P.W.F. WILSON, A.J. BELANGER, D.H. O'LEARY, P.A. WOLF, E.J. SCHAEFER & I.H. ROSENBERG. 1995. Association between plasma homocysteine concentrations and extracranial carotid-artery stenosis. N. Engl. J. Med. **332**: 286–291.
19. OLSZEWSKI, A.J. & K.S. MCCULLY. 1993. Homocysteine metabolism and the oxidative modification of proteins and lipids. Free Rad. Biol. Med. **14**: 683–693.
20. TSAI, J.-C., M.A. PERRELLA, M. YOSHIZUMI, C.-M. HSIEH, E. HABER, R. SCHLEGEL & M.-E. LEE. 1994. Promotion of vascular smooth muscle cell growth by homocysteine: A link to atherosclerosis. Proc. Natl. Acad. Sci. USA **91**: 6369–6373.
21. DALTON, M.L., P.F. GADSON, JR., R.W. WRENN & T.H. ROSENQUIST. 1997. Homocysteine signal cascade: production of phospholipids, activation of protein kinase C, and the induction of c-fos and c-myb in smooth muscle cells. FASEB J. **11**: 703–711.
22. SUNDARESAN, M., Z.-X. YU, V.J. FERRANS, D.J. SULCINER, J.S. GUTKIND, K. IRANI, P.J. GOLDSHMIDT-CLERMONT & T. FINKEL. 1996. Regulation of reactive-oxygen-species generation in fibroblasts by Rac1. Biochem. J. **318**: 379–382.
23. IRANI, K., Y. XIA, J.L. ZWEIER, S.J. SOLLOTT, C.J. DER, E.R. FEARON, M. SUNDARESAN, T. FINKEL & P.J. GOLDSCHMIDT-CLERMONT. 1997. Mitogenic signaling mediated by oxidants in Ras-transformed fibroblasts. Science **275**: 1649–1652.
24. BRADFORD, M.M. 1976. A rapid and sensitive method for the quantitation of microgram quantities of protein utilizing the principle of protein-dye binding. Anal. Biochem. **72**: 248–254.
25. STURGILL, T.W., L.B. RAY, E. ERIKSON & J.L. MALLER. 1988. Insulin-stimulated MAP-2 kinase phosphorylates and activates ribosomal protein S6 kinase II. Nature **334**: 715–718.
26. YORK, R.D., H. YAO, T. DILLON, C.L. ELLIG, S.P. ECKERT, E.W. MCCLESKEY & P.J.S. STORK. 1998. Rap1 mediates sustained MAP kinase activation induced by nerve growth factor. Nature **392**: 622–626.
27. DAVIS, R. J. 1993. The mitogen-activated protein kinase signal transduction pathway. J. Biol. Chem. **268**: 14553–14556.

28. MERIN, J.P., M. MATSUYAMA, T. KIRA, M. BABA & T. OKAMOTO. 1996. Alpha-lipoic acid blocks HIV-1 LTR-dependent expression of hygromycin resistance in THP-1 stable transformants. FEBS Lett. **394:** 9–13.
29. ROY, S., C.K. SEN, H. KOBUCHI & L. PACKER. 1998. Antioxidant regulation of phorbol ester-induced adhesion of human Jurkat T-cells to endothelial cells. Free Rad. Biol. Med. **25:** 229–241.
30. HALLIWELL, B. & J.M.C. GUTTERIDGE. 1985. Free Radicals in Biology and Medicine. Clarendon Press. Oxford.

Enzyme-Like Activity of Glycated Cross-Linked Proteins in Free Radical Generation

MOON B. YIM,[a,c] SA-OUK KANG,[b] AND P. BOON CHOCK[a]

[a]*Laboratory of Biochemistry, NHLBI, National Institutes of Health, Bethesda, Maryland 20892, USA*

[b]*Laboratory of Biophysics, Department of Microbiology, College of Natural Sciences, and the Research Center for Molecular Microbiology, Seoul National University, Seoul 151-742, Republic of Korea*

ABSTRACT: The structure and property of cross-linked amino acids and proteins produced by a three- carbon α-dicarbonyl methylglyoxal in glycation reaction were investigated. Our results showed that these reactions generated yellow fluorescent products and several free radical species. From the reaction with alanine, three types of free radicals were identified by EPR spectroscopy: 1) the cross-linked radical cation, methylglyoxal diaklylimine cation radical; 2) the methylglyoxal radical anion as the counterion; 3) the superoxide radical anion produced only in the presence of oxygen. Glycation of bovine serum albumin by methylglyoxal also generated the protein-bound, cross-linked free radical, probably the cation radical of the cross-linked Schiff base as observed with alanine. The glycated protein reduced ferricytochrome c to ferrocytochrome c in the absence of oxygen or added metal ions. This reduction of cytochrome c was accompanied by a large increase in the amplitude of the electron paramagnetic resonance signal originated from the protein-bound free radical. In addition, the glycated protein catalyzed the oxidation of ascorbate in the presence of oxygen while the protein-free radical signal disappeared. These results indicate that glycation of protein generates active centers for catalyzing one-electron oxidation-reduction reactions. This active center, which exhibits enzyme-like character, was suggested to be the cross-linked Schiff base/the cross-linked Schiff base radical cation of the protein. It mimics the characteristics of metal-catalyzed oxidation system. These results together indicate that glycated proteins accumulated *in vivo* provide stable active-sites for catalyzing the formation of free radicals.

INTRODUCTION

Glycation reaction (nonenzymatic glycosylation; Maillard reaction), which produces brown fluorescent compounds, can occur when a protein is in solution with a reducing sugar, such as glucose. In this reaction, free amino groups of protein react slowly with the carbonyl groups of reducing sugars to yield Schiff-base intermediates (FIG. 1), which undergo Amadori rearrangement to stable ketoamine derivatives.[1] These Schiff-bases and Amadori products subsequently degrade into α-dicarbonyl compounds such as deoxyglucosones, methylglyoxal and glyoxal.[2–5]

---

[c]Address for correspondence: Voice: 301-496-9494; fax: 301-496-0599.
e-mail: yimm@nhlbi.nih.gov

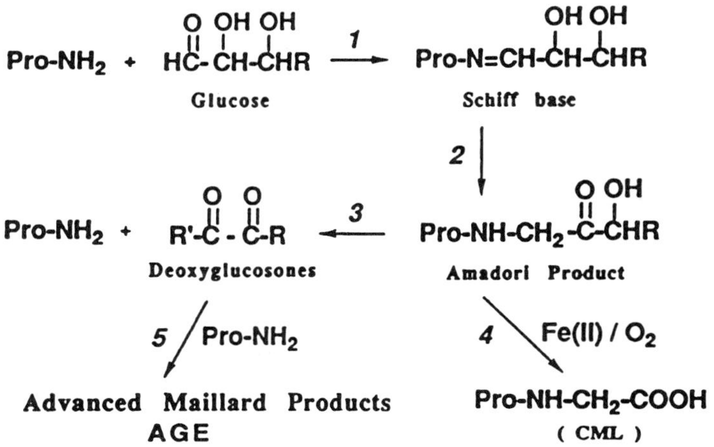

$$\text{Pro-NH}_2 + \underset{\text{Glucose}}{\overset{\overset{O}{\|}\overset{OH}{|}\overset{OH}{|}}{HC\text{-}CH\text{-}CHR}} \xrightarrow{1} \underset{\text{Schiff base}}{\overset{\overset{}{}\overset{OH}{|}\overset{OH}{|}}{\text{Pro-N=CH-CH-CHR}}}$$

$$\downarrow 2$$

$$\text{Pro-NH}_2 + \underset{\text{Deoxyglucosones}}{\overset{\overset{O}{\|}\overset{O}{\|}}{R'\text{-}C\text{-}C\text{-}R}} \xleftarrow{3} \underset{\text{Amadori Product}}{\overset{\overset{}{}\overset{O}{\|}\overset{OH}{|}}{\text{Pro-NH-CH}_2\text{-C-CHR}}}$$

$$5 \swarrow \text{Pro-NH}_2 \qquad\qquad 4 \searrow \text{Fe(II) / O}_2$$

Advanced Maillard Products   Pro-NH-CH$_2$-COOH
AGE   (CML)

FIGURE 1. A general scheme of the glycation or Maillard reaction. (From Yim et al.[37])

These compounds are more reactive than the parent sugars with respect to their ability to react with amino groups of proteins to form inter- and intra-molecular cross-links of proteins, stable end products called as advanced Maillard products or advanced glycation end products (AGEs).[1] The AGEs, which are irreversibly formed, accumulate with aging, atherosclerosis, and diabetes mellitus, especially associated with long-lived proteins such as collagens, lens crystallins, and nerve proteins.[5–10] It was suggested that the formation of AGEs not only modifies protein properties, but also induces biological damage *in vivo*.[11] For example, AGEs deposited in the arterial wall could themselves induce oxidant stress capable of oxidizing vascular wall lipids and accelerate atherogenesis in hyperglycemic diabetic patients.[12,13]

The α-dicarbonyl compounds which are essential for the cross-linking reaction are produced in a variety of ways. Fenton reaction–mediated oxidation of sugars, lipids, and proteins produces various α-dicarbonyl compounds. In accordance, the ransition metal ion-catalyzed oxidation of glucose is suggested to be a more important factor in glycation than the formation of Amadori product of glucose itself.[14–16] The α-ketoaldehydes, such as methylglyoxal, are also found as a normal metabolite in mammals and microorganisms. The methylglyoxal is formed by the non-enzymatic or enzymatic elimination of phosphate from triose phosphate, and by the oxidation of hydroxyacetone and aminoacetone.[17–19] The increased formation of methylglyoxal was observed in hyperglycemia associated with diabetes mellitus.[20,21] In addition, it was shown that methylglyoxal-modified albumin underwent receptor-mediated endocytosis by macrophage, which may suggest the involvement of methylglyoxal in pathophysiology.[22–24]

The molecular stuructures of some AGEs have been identified as pentosidines,[25–28] pyrrole-derivatives,[29] pyrazine-derivatives,[4] N$^\varepsilon$- carboxyalkyllysine (CML),[30, 31] imidazolone compounds,[32] and imidazolium cross-link species, methylglyoxal-lysine dimers.[5,33,34] In addition to these AGEs, several investigations have also shown by electron paramagnetic resonance (EPR) spectroscopy that unidentified protein-free radicals were produced during the reaction of methylglyoxal with proteins, such as bovine serum albumin (BSA) and casein.[35,36]

In this report, we describe results obtained from our studies for identifying the structure of the cross-linked radical species and their reactivity that may cause deleterious effects *in vivo*.[37,38] For this purpose, we studied a model system, the reaction between three-carbon α-dicarbonyl methylglyoxal with L-alanine to identify the radical species generated during the glycation reaction[37] and also studied oxidation-reduction property of glycated protein formed between bovine serum albumin and methylglyoxal.[38] The results showed that three types of free radicals were generated during this glycation reaction, one of which is the cross-linked radical cation.[37] This radical center or similar radical centers of the glycated proteins behave as enzyme-like active sites for catalyzing one-electron oxidation-reduction reactions.

STRUCTURAL IDENTIFICATION OF FREE RADICALS FORMED IN THE MODEL SYSTEM

Although N^ϵ of lysine residues may be the dominant sites for glycation, alanine was chosen in this study to facilitate EPR analysis because all of ^{15}N- and ^{13}C-substituted alanines are available commercially and also the electronic structure of the cross-linked radical site will probably be similar to that with lysine on the basis of the EPR, absorption, and fluorescence data.[37,38]

EPR Spectra of the Cross-Linked Radical Cation

The reaction mixture of methylglyoxal and alanine generated yellow fluorescent products which exhibited similar electronic absorption (285 nm and 334 nm) and fluorescence (385 nm when excited at 334 nm) as observed in some glycated proteins.[37,38] The EPR spectra shown in FIGURE 2 were also obtained with the anaerobic reaction mixture of methylglyoxal and various isotope-enriched L-alanines.[41] The alterations observed in the experimental EPR spectra obtained with different isotope-enriched L-alanines are entirely caused by changes of nuclear spins and their nuclear moment of the isotopes used, ^{15}N (I =1/2) in place of ^{14}N (I = 1) and their nuclear moments, $\mu(^{15}N)/\mu(^{14}N) = 1.40$, or ^{13}C (I = 1/2) in place of ^{12}C (I = 0). Using this information, the experimental spectra were simulated (lower spectra in FIG. 2A, B, and C). The simulation of spectra required hyperfine coupling constants (hfc) of two magnetically nonequivalent nitrogens (A^{14N} = 2.65 G and 8.36 G), three nonequivalent hydrogens (A^H = 7.88 G, 3.98 G, and 4.00 G), and three magnetically equivalent hydrogens (A^H = 5.87 G). In addition, extra hyperfine interactions were required for spectral simulation obtained with ^{13}C- enriched alanines: two 2-^{13}C with A^{13C} = 4.10 G and 0.20 G (spectrum 2C)); two 1-^{13}C with A^{13C} = 8.52 G and 0.30 G (spectrum, not shown);and two 3-^{13}C with A^{13C} = 3.0 G and 0.30 G (spectrum, not show). These results indicate that the amino groups of two alanines (two nitrogen hfc and two α- hydrogen hfc) are cross-linked by one methylglyoxal (three equivalent hydrogens of the methyl group and one hydrogen hfc). These results also

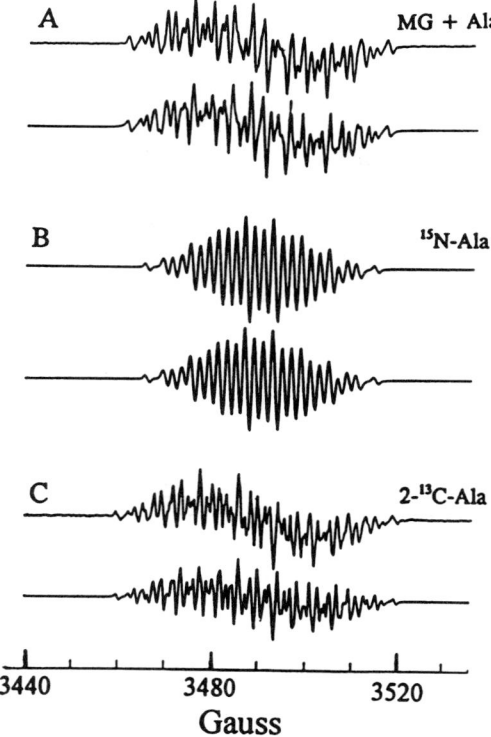

FIGURE 2. EPR spectra obtained from the reaction mixture containing methylglyoxal (0.2 M) and various isotope-enriched L-alanines (0.2 M) in carbonate buffer (0.5 M) at pH 9.5 (*upper spectra*) and simulated spectra (*lower spectra*). **A:** methylglyoxal and natural-abundance L-alanine; **B:** methylglyoxal and L-[^{15}N]alanine; **C:** methylglyoxal and L-[2-^{13}C]alanine. The hyperfine coupling constants used for the simulation are described in the text. (From Yim et al.[37])

indicate that the cross-linked radical contains all of the carbons of two alanine molecules, α-, carboxyl, and methyl carbons. In addition, the delocalized nature of the spin density suggests that this radical has an ionic character, probably a cation as shown below (structure *a*).

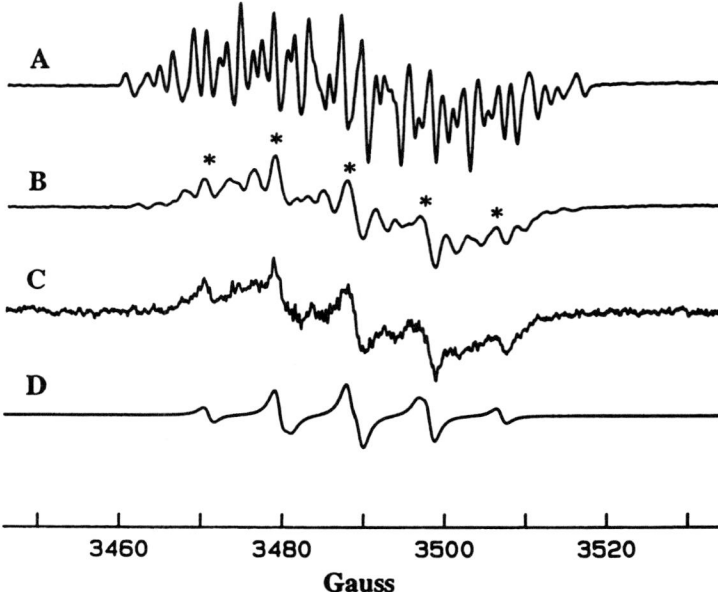

FIGURE 3. EPR spectra obtained from the reaction mixture containing methylglyoxal (0.2 M) and various isotope-enriched L-alanines (0.2 M) in carbonate buffer (0.5 M) at pH 9.5. **A:** methylglyoxal and natural-abundance L-alanine recorded at 5 min after starting reaction; **B:** methylglyoxal and 2,3,3,3-D_4 alanine in carbonate buffer solution prepared in D_2O. The spectrum was recorded at 5 min after initiating the reaction. The resonance lines marked by asterisks belong to another species as explained in the text; **C:** The spectrum was recorded at 15 min.with the same sample used for **B**; **D:** simulated spectrum obtained by using the hyperfine coupling constants described in the text. (From Yim et al.[37])

Observation of Counter Radical Anion; Methylglyoxal Radical Anion

Spectra B and C in FIGURE 3 were obtained from the reaction mixture containing natural-abundance methylglyoxal and isotope-enriched 2,3,3,3-D_4 alanine in carbonate buffer prepared in D_2O.[37] These spectra are compared with spectrum A which was obtained with the natural-abundance L-alanine in carbonate buffer prepared in H_2O.

The spectrum B, recorded at 5 min, exhibited an additional set of resonance lines (marked by asterisks), indicating the presence of another radical species. These resonance lines were not resolved in spectrum A because the signal amplitude of the cross-linked radical is much higher in spectrum A. This species is relatively more stable than the cross-linked radical as shown in the spectrum C which was obtained at 15 min. The simulation of this five-line spectrum (spectrum D) required three magnetically equivalent hydrogens with A^{H3} = 8.73 G and one hydrogen with A^H = 9.81 G, and g = 2.0043. These EPR parameters are identical to those of the *cis* form of methylglyoxal radical anion (structure *b* shown above) reported previously.[39] It suggests that the cross-linked radical observed in this reaction is a cation formed with the methylglyoxal radical anion being the radical counterion.

Direct One-Electron Transfer Reaction

The formation of these cation and anion radicals do not require molecular oxygen or transition metal ions.[37] In addition, when NaCNBH$_3$ was added to the reaction mixture, the EPR signals and yellow color typically observed with glycated products were not detected. The effect of NaCNBH$_3$, which is known to reduce Schiff base selectively and inhibits its subsquent reactions, may indicate that methylglyoxal dialkylimine, $^-O_2C(CH_3)HCN = C(CH_3)-HC = NCH(CH_3)CO_2^-$, is the intermediate for the formation of this cross-linked radical.

These results together indicate that the cross-linked radical cation and methylglyoxal radical anion are generated from direct one-electron transfer between methylglyoxal (MG) and a Schiff base, probably methylglyoxal dialkylimine (MG-DI), as shown in Reaction 1.

$$MGDI + MG \rightarrow MGDI^{\cdot +} + MG^{\cdot -} \quad (1)$$

Generation of Superoxide Radical Anions

In order to examine whether $O_2^{\cdot -}$ was generated in aerobic solutions of this model reaction, we performed NBT reduction experiment by adding varying concentration of the reaction mixture.[37] The reduction rates of NBT measured at 540 nm were increased with the addition of increasing concentration of reactants in aerobic solutions. This reduction was inhibited by Cu,Zn-SOD, but not by catalase.

These results demonstrate that although the initiation of the cross-linking reaction does not require molecular oxygen, the reaction products generate superoxide radical anions in the presence of oxygen. The methylglyoxal radical anion is most likely responsible for $O_2^{\cdot -}$ generation in this model system by its electron-transfer reaction to oxygen via Reaction 2.

$$MG^{\cdot -} + O_2 \rightarrow MG + O_2^{\cdot -} \quad (2)$$

Significance of the Results Obtained from the Model System

The formation of the cross-linked radical cation and the methylglyoxal radical anion does not require metal ions or oxygen. These results indicate that dicarbonyl compounds cross-link free amino groups by forming Schiff bases, which donate electrons directly to dicarbonyl compounds yielding the cross-linked radical cations and the methylglyoxal radical anions. Oxygen can accept an electron from anion to generate superoxide radical anion, which can initiate damaging chain reactions.

The cross-linked radical cation, which has an extensively delocalized unpaired electron, is quite stable. These radical sites in cross-linked proteins will be more persistent and could be reactive sites for putative reducing and oxidizing molecules, which produce free radicals for a long duration. We examined this hypothesis by using methylglyoxal-modified bovine serum albumin (BSA).

ENZYME-LIKE ACTIVITY OF GLYCATED CROSS-LINKED PROTEINS

The treatment of BSA with methylglyoxal produced free radical sites on the protein as shown in FIGURE 4.[38] The EPR spectra were obtained at 77 K with filtered and desalted methylglyoxal-modified BSA (MG-BSA) to remove unreacted methylglyoxal. It gives a single broad EPR line at a g value of 2.006 (FIG. 4A). However, the untreated-BSA does not give any EPR absorption (FIG. 4B).

Effect of Oxidizing Ferricytochrome c on the Protein-Free Radical

Addition of ferricytochrome c to the MG-BSA sample caused a large increase in the amplitude of the EPR signal (FIG. 4C), indicating an additional production of the protein-free radicals. This increase is derived from the interaction between ferricytochrome c and glycated BSA because with the untreated BSA one failed to observe such signal (FIG. 4D). This additional production of the protein-free radicals was accompanied by the reduction of cytochrome c, monitored at 550 nm.[38] This reduction

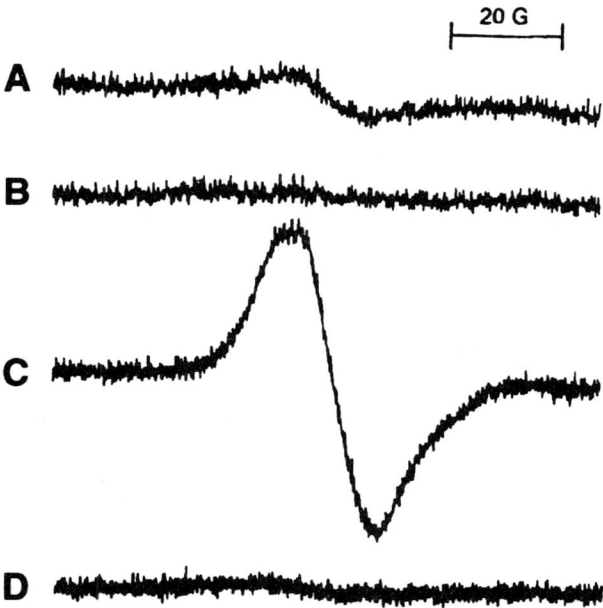

FIGURE 4. EPR spectrum of the protein-bound free radical in MG-BSA and effect by cytochrome c. The mixture containing methylglyoxal (30 mM) and BSA (20 mg/ml) in 0.1 M phosphate buffer (pH 7.4) was incubated for 5 days at 37°C. The reaction product was filtered, washed, and desalted. The resultant samples of MG-BSA did not contain unreacted free methylglyoxal. The EPR spectra were obtained with frozen samples at 77 K. **A:** 58 mg/ml MG-BSA; **B:** 58 mg/ml (0.87 mM) untreated BSA alone; **C:** mixture of 58 mg/ml MG-BSA and 0.8 mM ferricytochrome c. The EPR sample was frozen in one min. with liquid nitrogen after adding ferricytochrome c in the MG-BSA solution; **D:** 0.8 mM ferricytochrome c and untreated BSA. (From Lee et al.[38]).

did not require oxygen and it was not affected by the addition of a chelating agent, DTPA. These results together suggest that ferricytochrome c receives one electron directly from the species of glycated BSA, most likely the cross-linked Schiff bases (such as methylglyoxal dialkylimine) to produce the radical cation of the cross-linked Schiff bases (i.e., the methylglyoxal dialkylimine radical cation) on the protein.

$$\text{MGDI} + \text{Cyt c (III)} \rightarrow \text{MGDI}^{\ddagger} + \text{Cyt c (II)} \qquad (3)$$

Glycation of a protein generates heterogenous products, which may contain intra- and intermolecular cross-linked protein as well as partially glycated products with one free carbonyl group. Thus, a partially glycated proteins may also cross-link to other proteins through this free carbonyl group. Indeed, we observed that cytochrome c cross-linked to desalted MG-BSA with a very slow rate, which proceeds independent of cytochrome c reduction and the initial free radical generation (data not shown).

Effect of Reducing Ascorbate on the Protein-Free Radical

Addition of ascorbate to MG- BSA quenched the EPR signal, which indicates that ascorbate reduces the protein-radical cation of MG-BSA to the non-radical species.[38] This reduction of radical cation was accompanied by the degradation of ascorbate (FIG. 5). Under anaerobic conditions, a small portion of incubated ascorbate was consumed, which reached a plateau with time (FIG. 5A). The controlled experiments showed that unmodified BSA failed to degrade ascorbate, and a metal chelator, DTPA, had no inhibitory effect on the ascorbate degradation (data not shown). The later indicates that adventitious metal ions were not the cause of the observed ascorbate consumption. These results suggest that the ascorbate was oxidized directly by the protein-free radical cation in MG-BSA.

$$\text{MGDI}^{\ddagger} + \text{AscH}^- \rightarrow \text{MGDI} + \text{Asc}^{-} + \text{H}^+ \qquad (4)$$

In the presence of oxygen, however, the oxidation of ascorbate continued to reach far beyond one molar ratio of the degraded ascorbate to MG-BSA (FIG. 5A). The initial rates obtained from these degradation data showed that the oxidation of ascorbate proceeded linearly with respect to the MG-BSA concentration (FIG. 5B), but increased as a saturation function with respect to the ascorbate concentration (FIG. 5C). The double-reciprocal plot of the initial rates (shown in the inset) yielded a K_m of 1 mM for ascorbate.

Furthermore, superoxide dismutase, but not catalase, exerts partial inhibition on the degradation of ascorbate (approximately 20% inhibition in this experimental condition). The fact that this inhibition is only partial indicates that O_2, but not O_2^-, is directly involved in this catalytic reaction. The partial inhibition, however, suggests that superoxide radical anions are produced during this reaction and they play a role in ascorbate degradation, probably via the superoxide-scavenging reaction by ascorbate. Together, these results indicate that MG-BSA behaves as an enzyme, which has an ability to catalyze the oxidation of ascorbate in the presence of oxygen to produce superoxide radical anion and semi-dehydroascorbyl radical. This reaction is initiated by the protein-radical cation of MG-BSA.

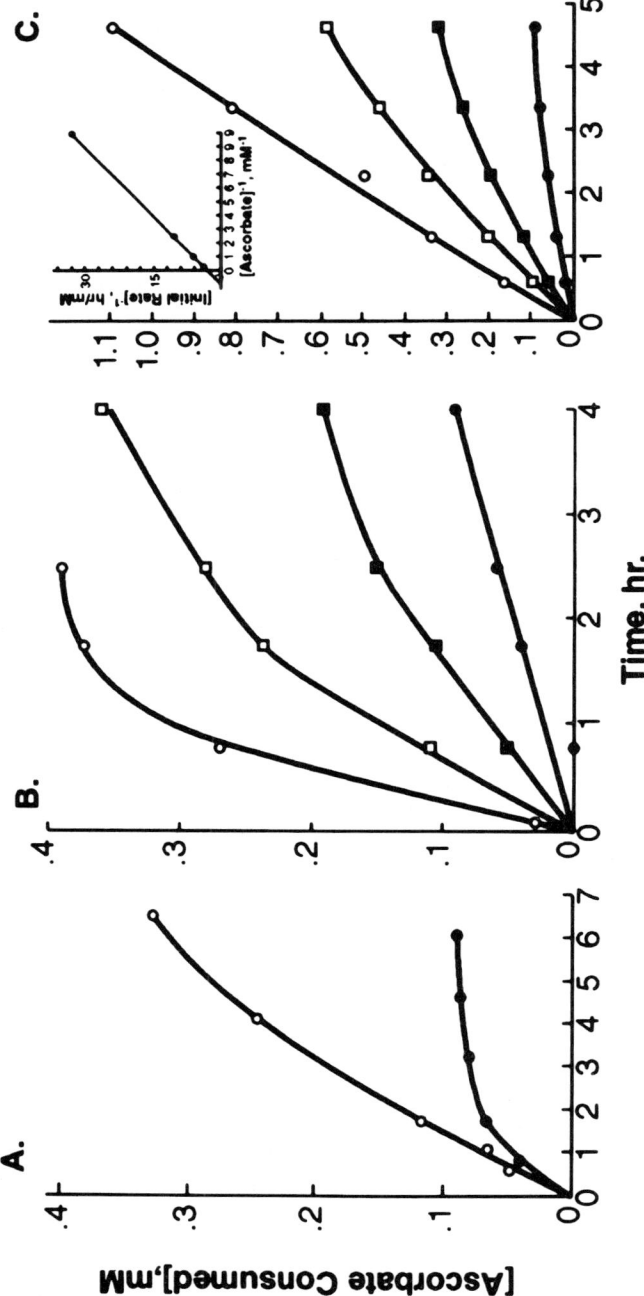

FIGURE 5. Degradation of ascorbate by MG-BSA. The MG-BSA was obtained by incubating BSA (20 mg/ml, 0.3 mM) and methylglyoxal (30 mM) in 0.1 M phosphate buffer (pH 7.4) for 5 days. The desalted MG-BSA was reacted with ascorbate in 20 mM phosphate buffer (pH 7.4) at 37°C and the remaining ascorbate was quantified at the indicated time by high performance liquid chromatography. A: MG-BSA (7 mg/ml; 0.1 mM BSA) was reacted with 0.4 mM ascorbate in nitrogen (●) or in air (○). B: 0.4 mM ascorbate was reacted with 0.3 (●), 3 (■), 10 (□), 30 (○) mg/ml of MG-BSA. C: MG-BSA (7 mg/ml) was reacted with 0.1 (●), 0.4 (■), 1 (□), and 4 mM (○) ascorbate. The *inset* in this panel depicts the double reciprocal plot. (From Lee *et al.*[38])

DISCUSSION AND CONCLUSION

FIGURE 6 summarizes our experimental results. The protein-free radical of the MG-BSA is most likely to be the radical cation of the cross-linked Schiff base (FIG. 6, species D) on the basis of our results obtained with alanine as a model system.[41] When methylglyoxal was reacted with various N^α-acetyl substituted amino acids, EPR signal was observed only with N^α-acetyl-L-lysine. In addition, previous investigations

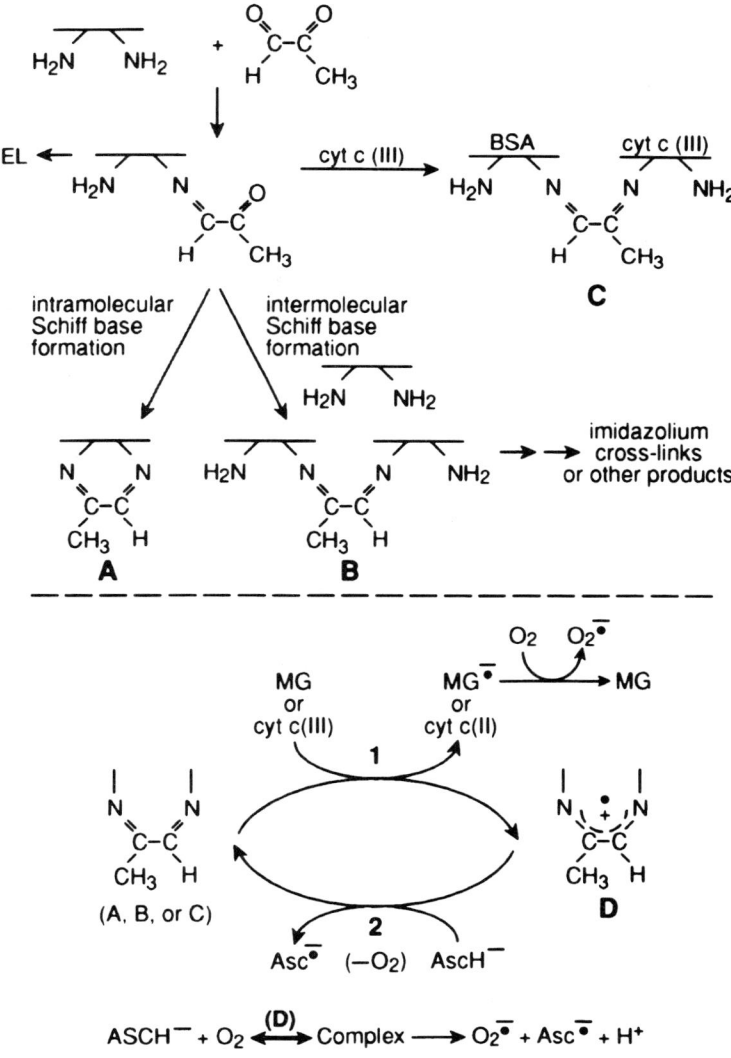

FIGURE 6. Proposed reaction scheme for glycation of protein by methylglyoxal and reaction of glycated protein. CEL; N^ε-(carboxyethyl)lysine. (From Lee *et al.*[38])

have shown that lysine residues are the main cross-linking sites in MG-treated proteins.[33,34] These results together suggest that methylglyoxal cross-links inter- or intramolecular lysine residues of the protein to form cross-linked Schiff bases (FIG. 6, species A and B). This cross-linked Schiff base of MG-BSA can donate an electron to methylglyoxal to produce the radical cation of the cross-linked Schiff base (species D). During these processes, the Schiff bases or the protein-free radicals may be oxidized to form N^ε-(carboxyethyl)lysine[31] or matured to other products such as imidazolium cross-links.[33,34]

The formation of the protein-free radical cation from the cross-linked Schiff base, or *vice versa*, depends on the electron affinity of the interacting species. Thus, in the presence of ferricytochrome *c* which has a high electron affinity, the cross-linked Schiff base of MG-BSA will lose an electron to cytochrome *c* to produce additional radical cations of the cross-linked Schiff base and ferrocytochrome *c* (reaction 3; FIG. 6, *reactions 1*). In the presence of electron-donating ascorbate, however, the radical cation of the cross-linked Schiff base accepts an electron from ascorbate to produce the cross-linked Schiff base and semidehydroascorbyl radical (reaction 4; FIG. 6, *reaction 2*). These reactions can proceed even in the absence of oxygen. Moreover, in the presence of oxygen (FIG. 5), MG-BSA behaves as an enzyme, which is capable of catalyzing the oxidation of ascorbate (K_m = 1 mM). This reaction catalyzed by MG-BSA in the presence of oxygen is similar to the transition metal ion-catalyzed oxidation of ascorbate.[40] In a transition metal ion-catalyzed oxidation system such as Fe^{3+}/ascorbate/O_2, the reaction is initiated via the one- electron reduction of Fe^{3+} by ascorbate, whereas in the MG-BSA/ascorbate/O_2 system, it is initiated via the one-electron reduction of the protein-free radical cation by ascorbate. In both reactions, superoxide radical anions are generated by the oxidation of ascorbate. A mechanism proposed for the transition metal ion-catalyzed oxidation of ascorbate in the presence of oxygen is as follows[41];

$$AscH^- + O_2 \leftrightarrow complex \xrightarrow{M^{n+}} Asc^{-\cdot} + O_2^{-\cdot} + H^+ \qquad (5)$$

$$O_2^{-\cdot} + AscH^- \rightarrow products \qquad (6)$$

In the reaction with the glycated protein, the cross-linked radical cation may function as the transition metal ion M^{n+} in reaction 5. Reactions 5 and 6 may explain the partial inhibition of ascorbate degradation observed from the reaction mixture containing superoxide dismutase.

In essence, the cross-linked Schiff base/its radical cation of glycated protein behave as an active site of an enzyme for the one-electron oxidation-reduction reactions. In addition, long term incubation of cytochrome *c* with MG-BSA produces cross-links between cytochrome *c* and MG- BSA, probably through the reaction between free carbonyl groups of MG-BSA and amino groups of cytochrome *c* (FIG. 6, species C). The cross-linked cytochrome *c* maintains its oxidation-reduction properties.

The AGEs formed *in vivo* are heterogeneous species which may have a variety of chemical structures. There is no direct evidence so far that the MG-BSA radical cation or similar radical species are formed *in vivo*. We showed, however, in this model study that one type of reactive structures (a cross-linked Schiff base/the radical cation of the Schiff base) has the enzyme-like character for catalyzing the one-electron oxidation-reduction reaction yielding free radicals including superoxide radical anions. The similar types of reactive centers, if formed *in vivo*, may exert significant effects to their biological environment by generating free radicals for a long duration.

ACKNOWLEDGMENTS

This work was supported in part by the International Collaborative Research Program of the National Heart, Lung, and Blood Institute, National Institutes of Health and of the Korea Science and Engineering Foundation.

REFERENCES

1. REYNOLDS, T.M. 1963. Chemistry of non-enzymatic browning. Adv. Food. Res. **12:** 1–52.
2. KATO, H., D.B. SHIN & F. HAYASE. 1987. 3-deoxyglucosone cross-links proteins under physiological conditions. Agric. Biol. Chem. **51:** 2009–2011.
3. WELLS-KNECHT, K.J., D.V. ZUZAK, J.E. LITCHFIELD, S.R. THORPE & J.W. BAYNES. 1995. Mechanism of autoxidative glycosylation: identification of glyoxal and arabinose as intermediates in the autoxidative modification of proteins by glucose. Biochemistry **34:** 3702–3709.
4. HAYASHI, T., S. MASE & M. NAMIKI. 1985. Formation of the N,N'-dialkylpyrazine cation radical from glyoxal dialkylimine produced on the reaction of sugar with an amine or amino acid. Agric. Biol. Chem. **49:** 3131–3135.
5. FRYE, E.B., T.P. DEGENHARDT, S.R. THORPE & J.W. BAYNES. 1998. Role of the Maillard reaction in aging of tissue proteins. Advanced glycation end product-dependent increase in imidazolium cross-links in human lens proteins. J. Biol. Chem. **273:** 18714–18719.
6. GORDILLO, E., A. AYALA, J. BAUTISTA & A. MACHADO. 1989. Implication of lysine residues in the loss of enzymatic activity in rat liver 6-phosphogluconate dehydrogenase found in aging. J. Biol. Chem. **264:** 17024–17028.
7. MONNIER, V.M., R.R. KOHN & A. CERAMI. 1984. Accelerated age-related browning of human collagen in diabetes mellitus. Proc. Natl. Acad. Sci. USA **81:** 583–587.
8. ELGAWISH, A., M. GLOMB, M. FREIDLANDER & V.M. MONNIER. 1996. Involvement of hydrogen peroxide in collagen cross-linking by high glucose in vitro and in vivo. J. Biol. Chem. **271:** 12964–12971.
9. MONNIER, V. M. & A. CERAMI. 1981. Nonenzymatic browning in vivo: possible process for aging of long-lived proteins. Science **211:** 491–493.
10. VLASSARA, H., M. BROWNLEE & A. CERAMI. 1983. Excessive nonenzymatic glycosylation of peripheral and central nervous system myelin components in diabetic rats. Diabetes **32:** 670–674.
11. BROWNLEE, M., H. VLASSARA & A. CERAMI. 1984. Nonenzymatic glycosylation and pathogenesis of diabetic complications. Ann. Int. Med. **101:** 527–537.
12. MULLARKEY, C.J., D. EDELSTEIN & M. BROWNLEE. 1990. Free radical generation by early glycation products: a mechanism for accelerated atherogenesis in diabetes. Biochem. Biophys. Res. Commun. **173:** 932–939.
13. SAKURAI, T. & S. TSUCHIYA. 1988. Superoxide production from nonenzymatically glycated protein. FEBS Lett. **236:** 406–410.
14. WOLFF, S.P. & R.T. DEAN. 1988. Aldehydes and dicarbonyls in non-enzymic glycosylation of proteins. Biochem. J. **249:** 618–619.
15. WOLFF, S.P. & R.T. DEAN. 1987. Glucose autoxidation and protein modification. The potential role of 'autoxidative glycosylation' in diabetes. Biochem. J. **245:** 243–250.

16. JIANG, Z.Y., A.C.S. WOOLLARD & S.P. WOLFF. 1990. Hydrogen peroxide production during experimental protein glycation. FEBS Lett. **268**: 69–71.
17. INOUE, Y. & A. KIMURA. 1995. Methylglyoxal and regulation of its metabolism in microorganisms. Adv. Micorbiol. Physiol. **37**: 177–227.
18. THORNALLEY, P.J., S. WOLFF, J. CRABBE & A. STERN. 1984. The autoxidation of glyceraldehyde and other simple monosaccharides under physiological conditions catalysed by buffer ions. Biochim. Biophys. Acta **797**: 276–287.
19. PHILLIPS, S.A. & P.J. THORNALLEY. 1993. The formation of methylglyoxal from triose phosphates. Investigation using a specific assay for methylglyoxal. Eur. J. Biochem. **212**: 101–105.
20. PHILLIPS, S.A., D. MIRRLEES & P.J. THORNALLEY. 1993. Modification of the glyoxalase system in streptozotocin-induced diabetic rats. Effect of the aldose reductase inhibitor Statil. Biochem. Pharmacol. **46**: 805–811.
21. MCLELLAN, A.C., P.J. THORNALLEY, J. BENN & P.H. SONKSEN. 1994. Glyoxalase system in clinical diabetes mellitus and correlation with diabetic complications. Clin. Sci. **87**: 21–29.
22. WESTWOOD, M.E., A.C. MCLELLAN & P.J. THORNALLEY. 1994. Receptor-mediated endocytic uptake of methylglyoxal-modified serum albumin. Competition with advanced glycation end product-modified serum albumin at the advanced glycation end product receptor. J. Biol. Chem. **269**: 32293–32298.
23. YAN, S.D., A.M. SCHMIDT, G.M. ANDERSON, J. ZHANG, J. BRETT, Y.S. ZOU, D. PINSKY & D. STERN. 1994. Enhanced cellular oxidant stress by the interaction of advanced glycation end products with their receptors/binding proteins. J. Biol. Chem. **269**: 9889–9897.
24. LANDER, H.M., J.M. TAURAS, J.S. OGISTE, O. HORI, R.A. MOSS & A.M. SCHMIDT. 1997. Activation of the receptor for advanced glycation end products triggers a p21(ras)-dependent mitogen-activated protein kinase pathway regulated by oxidant stress. J. Biol. Chem. **272**: 17810–17814.
25. SELL, D.R. & V.M. MONNIER. 1989. Structure elucidation of a senescence cross-link from human extracellular matrix. Implication of pentoses in the aging process. J. Biol. Chem. **264**: 21597–21602.
26. SELL, D.R. & V.M. MONNIER. 1990. End-stage renal disease and diabetes catalyze the formation of a pentose-derived crosslink from aging human collagen. J. Clin. Invest. **85**: 380–384.
27. GRANDHEE, S.K. & V.M. MONNIER. 1991. Mechanism of formation of the Maillard protein cross-link pentosidine. Glucose, fructose, and ascorbate as pentosidine precursors. J. Biol. Chem. **266**: 11649–11653.
28. DYER, D.G., J.A. BLACKLEDGE, S.R. THORPE & J.W. BAYNES. 1991. Formation of pentosidine during nonenzymatic browning of proteins by glucose. Identification of glucose and other carbohydrates as possible precursors of pentosidine in vivo. J. Biol. Chem. **266**: 11654–11660.
29. MIYATA, S. & V.M. MONNIER. 1992. Immunohistochemical detection of advanced glycosylation end products in diabetic tissues using monoclonal antibody to pyrraline. J.Clin. Invest. **89**: 1102–1112.
30. AHMED, M.U., S.R. THORPE & J.W. BAYNES. 1986. Identification of N-epsilon-carboxymethyllysine as a degradation product of fructoselysine in glycated protein. J. Biol. Chem. **261**: 4889-4894.
31. AHMED, M.U., E. BRINKMANN FRYE, T.P. DEGENHARDT, S.R. THORPE & J.W. BAYNES. (1997) N-epsilon-(carboxyethyl)lysine, a product of the chemical modification of proteins by methylglyoxal, increases with age in human lens proteins. Biochem. J. **324**: 565–570.
32. LO, T.W.C., M.E. WESTWOOD, A.C. MCLELLAN, T. SELWOOD & P.J. THORNALLEY. 1994. Binding and modification of proteins by methylglyoxal under physiological conditions. A kinetic and mechanistic study with N alpha-acetylarginine, N alpha-acetylcysteine, and N alpha-acetyllysine, and bovine serum albumin. J. Biol. Chem. **269**: 32299-32305

33. WELLS-KNECHT, K.J., E. BRINKMANN, M.C. WELLS-KNECHT, J.E. LITCHFIELD, M.U. AHMED, S. REDDY, D.V. ZYZAK, S.R. THORPE & J.W. BAYNES. 1996. New biomarkers of Maillard reaction damage to proteins. Nephrol. Dial. Transplant. **11** (Suppl. 5): 41–47.
34. NAGARAJ, R.H., I.N. SHIPANOVA & F.M. FAUST. 1996. Protein cross-linking by the Maillard reaction. Isolation, characterization, and in vivo detection of a lysine-lysinecross-link derived from methylglyoxal. J. Biol. Chem. **271**: 19338–19345.
35. GASCOYNE, P.R.C. 1981. An electron spin resonance investigation of amine models for protein-methylglyoxal interaction. Int. J. Quantum Chem: Quantum Biol. Symp. **8**: 265–270.
36. MCLAUGHLIN, J.A., R. PETHIG & A. SZENT-GYÖRGYI. 1980. Spectroscopic studies of the protein-methylglyoxal adduct. Proc. Natl. Acad. Sci. USA **77**: 499–951.
37. YIM, H.-S., S.-O. KANG, Y.-C. HAH, P.B. CHOCK & M.B. YIM. 1995. Free radicals generated during the glycation reaction of amino acids by methylglyoxal. A model study of protein-cross-linked free radicals. J. Biol. Chem. **270**: 28228–28233.
38. LEE, C., M.B. YIM, P.B. CHOCK, H.-S.YIM & S.-O. KANG. 1998. Oxidation-reduction properties of methylglyoxal-modified protein in relation to free radical generation. J. Biol. Chem. **273**: 25272–25278.
39. STEENKEN, S., W. JAENICKE-ZAUNER & D. SCHULTE-FROHLINDE. 1975. Photochem. Photobiol. **21**: 21-26
40. AMICI, A., R.L. LEVINE, L. TSAI & E.R. STADTMAN. 1989. Conversion of amino acid residues in proteins and amino acid homopolymers to carbonyl derivatives bymetal-catalyzed oxidation reactions. J. Biol. Chem. **264**: 3341–3346.
41. AFANAS'EV, I.B., V.V. GRABOVETSKII & N.S. KUPRIANOVA. 1987. Kinetics and mechanism of the reaction of superoxide ion in solution. Part 5. Kinetics and mechanism of the interaction of superoxide ion with vitamine E and ascorbic acid. J. Chem. Soc., Perkin Trans. **2**: 281–285.

Molecular Markers of Oxidative Stress Vulnerability

INGEBORG HANBAUER[a] AND MARZIA SCORTEGAGNA

Laboratory of Molecular Immunology, National Heart, Lung and Blood Institute, Bethesda, Maryland 20892, USA

ABSTRACT: In astrocyte primary cultures of trisomy 16 mice, an animal model for Down's syndrome, protein oxidation was 50% higher than in diploid littermates. Exposure to 10 μM H_2O_2 or 50 μM kainic acid incremented protein oxidation in trisomic but not in diploid cultures. Studies on stress response genes showed that metallothionein (MT) level was 2–3 times higher in trisomy 16 than in diploid cultures. Kainic acid or H_2O_2 exposure increased the MT protein level in diploid cultures but failed to increase it in trisomy 16 mouse beyond its elevated basal level. The reduced responsiveness of MT to simulated oxidative stress may result in insufficient removal of ROS, which could partially explain the further increase of protein oxidation in trisomy 16 cultures. In contrast, Pb exposure increased MT in trisomy 16 and diploid primary cultures to a similar extent. The similar metal responsiveness of MT in both phenotypes indicated that MT in trisomic glial cultures was not yet maximally stimulated. The flawed redox sensitivity in trisomy 16 mouse suggests possible alterations in the binding activity of ROS-sensitive transcription factors on the MT promoter.

INTRODUCTION

Mammalian cells continuously produce reactive oxygen species (ROS) through different metabolic pathways. Superoxide is formed during the reduction of O_2 by the mitochondria electron transport system,[1] activation of oxidases and dehydrogenases, or by redox-active endogenous compounds (catecholamines, vitamins).[2] Eukaryotic cells are equipped with an antioxidant system capable of converting ROS to H2O via different cytosolic enzymes. In brain, superoxide is converted to H_2O_2 by superoxide dismutase (SOD)-13. Whereas, H_2O_2 and fatty acid hydroxyperoxides are eliminated by Se-dependent glutathione peroxidases[4] or thioredoxine peroxidases.[1,5,6] Disturbances in O_2 or ROS metabolism may result in the accumulation of potentially "dangerous" ROS (hydroxyl radical) that could oxidize proteins, fatty acids, and DNA. A deficient antioxidant defense is believed to accelerate aging and aging-related pathologies including cancer, cardiovascular defects, different types of dementia, and early onset of Alzheimer disease (by age 35 years).[7] A combination of these pathologies occurs simultaneously in Down's syndrome. Down's syndrome is still the most frequent birth defect[8] caused by an extra copy of chromosome 21 or a part thereof. Mapping of the triplicate chromosomal element showed the presence of various genes

---

[a]Address for correspondence. Voice: 301-496-3250; fax: 301-652-8139.
e-mail: hanbaui@fido.uhlbi.nih.gov

including amyloid precursor protein (APP), superoxide dismutase-1 (SOD-1), glutamate R5 receptor, cystathione-β-synthase,[9] HMG-14 protein,[10] interferon-α, (-β, -γ) receptor,[11] and S100.[9–12] Since several of these genes had been implicated in the pathogenesis of a variety of neurodegenerative diseases,[13] it is possible that the overexpression of any of these genes might cause an early onset of dementia and perhaps also of the Alzheimer's disease-like condition in Down's syndrome patients. At the present time, little is known about the mechanisms that elicit these Down's syndrome pathologies. It is believed that in Down's syndrome, an increased SOD-1 gene dosage may alter the activity ratios between SOD-1 and other antioxidant enzymes such as glutathione peroxidase, catalase, or thioredoxin peroxidases favoring the increased production of H_2O_2. Besides SOD-1, the increased gene dosage of amyloid precursor protein, glutamate R5 receptor, and interferon (-α, -β, -γ) receptors[11] may indirectly trigger oxidative stress via increased secretion of β-amyloid, increased $[Ca^{2+}]i$ threshold,[14] or increased production of tumor necrosis factor-α and nitric oxide, respectively.

THE TRISOMY 16 MOUSE— A MODEL FOR INCREASED OXIDATIVE STRESS

Because of the ethical limitations of *in vivo* studies in Down's syndrome patients, direct evidence for increased oxidative stress in Down's syndrome brains is not available. Owing to this, the role of ROS in Down's syndrome pathology remains controversial. In order to assess the effects of an increased dosage of a combination of genes on the cellular redox balance we used the trisomy 16 mouse model. The triplicate segment of human chromosome 21 shares an extensive synteny with mouse chromosome 16.[12] The trisomy 16 mice are generated by mating female diploid mice with double heterozygous males which carry two different Robertsonian translocation chromosomes sharing one arm in common.[15] Since the mouse chromosome 16 is one third longer than human chromosome 21, extra copies exist for many other genes. Usually, about 25% to 30% of the embryos per litter are trisomic and die either *in utero* in late pregnancy or at birth. The trisomic embryos have a close to 100% incidence of atrioventricular septal defects that may result from an abnormal development of the endocardial cushions.[16] We prepared glial and neuronal primary cultures from brains of 16 d old trisomy 16 mouse embryos and their diploid littermates as described by Scortegagna *et al.*[17] In these primary cultures we studied:

1. whether measurements of protein carbonylation could reflect the increased ROS production elicited by the inherent gene dosage effect,
2. whether oxidative stress responsive genes were altered in trisomic cells, and
3. whether there was an greater vulnerability to simulated oxidative stress in trisomy 16 primary cultures.

Increased Protein Oxidation in Primary Cultures of Trisomy 16 Mouse Brain

Ample evidence in the literature showed that thiol-containing or aromatic amino acid residues in proteins are targeted by ROS and are modified into carbonyl groups.[18] This conversion can be quantitated by the oxyblot technique (Oncor) that is based on the immunodetection of carbonyl-hydrazone derivatives.[18] By comparing Western blots of different protein dilutions we found that the protein carbonyl content was 50% higher in glial primary cultures of trisomy 16 mouse brain than in cultures of diploid littermates (FIG.1, compare the lanes designated as 0). This increase of protein oxidation may serve as an indicator of increased H_2O_2 production. In primary cultures of normal littermate brains, a 1 hour incubation with 1 or 10 µM H_2O_2 failed to increase the density of protein bands above the basal level seen in nontreated cells[17] (FIG. 1). Whereas, in trisomy 16 cells, a similar incubation with H_2O_2 not only increased the density of protein bands but also oxidized additional protein bands[17] (FIG.1, *bottom*).

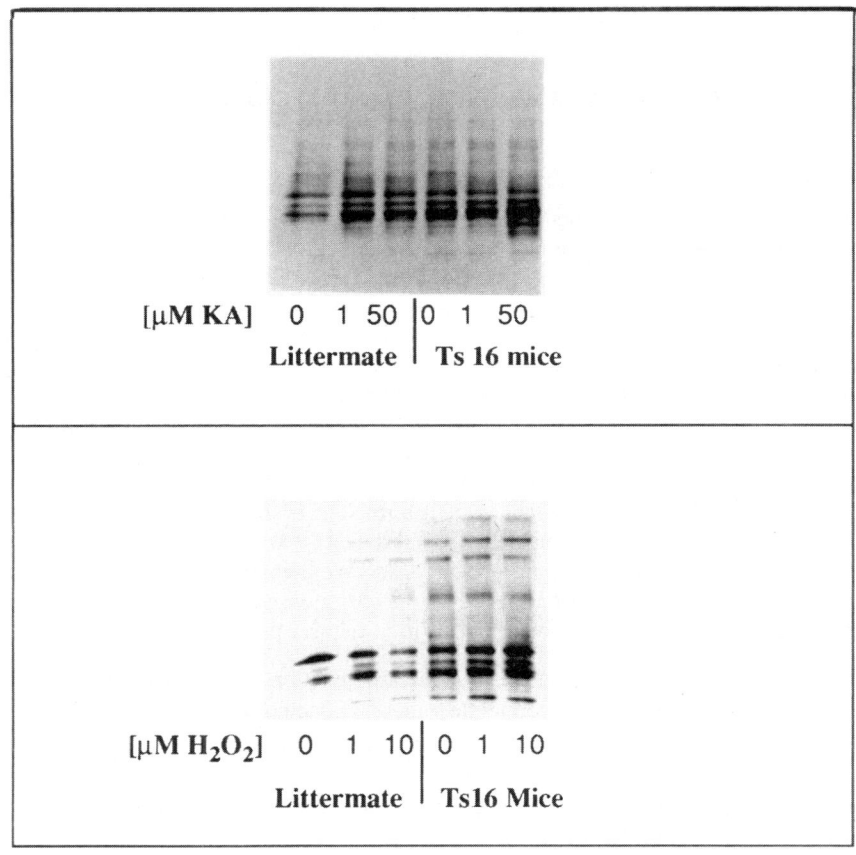

FIGURE 1. Western blot analysis (Oxyblot) of oxidized proteins in hippocampal primary cultures (14 d in culture) of trisomy 16 and diploid mouse embryos (16 d in gestation). Cells were exposed for 1 h to 1 µM and 50 µM kainic acid (*top*) or to 1 µM and 10 µM H_2O_2 (*bottom*). 30 µg protein were loaded per well.

TABLE 1. Stress response genes

Metallothioneins
Heme Oxygenase
Cytokines
Transcriptional Regulators: NF-κB, AP-1, USF

To test whether the increase of basal oxidative stress in cells could result in an increased vulnerability to excitotoxins we studied the effect of GluR5 receptor stimulation by kainic acid. The data depicted in FIGURE 1 (*top*) show that a 1 h exposure to 50 μM kainic acid increased the density and the number of protein bands in primary cultures of trisomy 16 mice when compared with cultures of diploid littermates.[19] These data are contrary to reports in the literature on transgenic mice that overexpressed SOD-1 by 10-fold. These mice were less vulnerable to kainic acid than control mice[20] and showed a similar level of lipid peroxidation products as control mice did.[21] The implications of a gene dosage effect that may result in protection or increased vulnerability emphasize the importance of the activity ratio between SOD-1 and antioxidant enzymes involved in H_2O_2 metabolism. Taken together, our findings suggest that besides the elevation of endogenously formed ROS levels, ROS scavenging may have been impaired in primary cultures of trisomy 16 mouse brain.

To consolidate this inference we studied stress response genes that could play a role in oxidative stress response. As depicted in TABLE 1, stress response genes include heat shock proteins, metallothioneins, immunomodulatory cytokines, and transcription factors, such as activator protein (AP)-1, NF-κB, upstream stimulatory factor and metal regulatory transcription factor (MTF-1). The activation of these genes may either initiate a cellular defense against increased oxidative stress or may promote further injury.

Changes of Oxidative Stress Responsive Genes in Primary Cultures of Trisomy 16 Mice

Metallothionein (MT)

MTs are small proteins consisting of 60 amino acid residues of which one third are cysteines. In the mouse, the variety for MT molecular forms are encoded by four different genes that are located on chromosome 8,[22] whereas in humans, MT is encoded by at least 17 different genes located on chromosome 16.[23] So far, only 4 MT isoforms have been found in mammalian cells. MT has 7 metal binding sites which, under physiological conditions, are occupied by Zn/Cu.[24] Although MTs were discovered 40 years ago[25] their role in heavy metal detoxification and reduction of increased oxidative stress has only recently been recognized.[22] Zn can be readily displaced from its binding sites by other metals including Cd, Pb, Bi, Ni, Ag, Hg, and Mn. The possible role of MT as a Zn donor for other metalloproteins has lately been refuted by genetic findings on MT-I/II isoform null mice which developed and reproduced normally.[26] These findings were interpreted to mean that Zn transfer from MT onto metalloproteins is essential only when cells are exposed to stressful conditions.[26] Recently, it has been shown that Zn transfer from MT-I/II onto the oxidized form of glutathione may play an important role in the glutathione redox cycle.[27] During oxidative stress triggered by environmental or pathophysiological conditions, a shift of the glutathione redox balance could increase the rate of Zn release from MTs.

The MT-I/II isoforms are rapidly induced by heavy metals, glucocorticoids, catecholamines, cytokines and ROS.[22,28,29] The transcriptional activation of MT is determined by the metal response element (MRE),[30] glucocorticoid regulatory elements,[31] interferon response elements,[32] and antioxidant response element (ARE)[30] consensus sequences on the promoter at the 5' flanking DNA. The binding affinity of MRE binding transcription factor (MTF-1) to various tandem MRE is believed to determine heavy metal toxicity[33] by regulating the basal and metal-induced transcriptional activation of MT-I/II.[34] The MTF-1 null mutant mice die *in utero* and also fail to transcribe the MT-I/II genes.[34] The ARE consensus on the MT-I/II promoter was shown to contain high affinity binding site for the AP heterodimers, AP-1, AP-2, AP-4[35] but also for an upstream stimulatory factor (USF), a member of the helix-loop-helix-Zip protein superfamily which includes Myc, Max, Mad, and TFE3 proteins.[36] Recently, it was shown that USF confers cadmium induction of MT-I via binding to the USF/ARE composite element.[36]

H_2O_2-Elicited Changes of MT-I/II Induction in Primary Cultures of Trisomy 16 and Diploid Mice Simulate Vulnerability to Increased Oxidative Stress

After observing that protein oxidation was increased in primary cultures of trisomy 16 mice, we postulated that a flawed functional antioxydant system may have shifted the redox balance and thereby may have altered stress response proteins at a transcriptional and functional level. This hypothesis was tested by studying the effect of H_2O_2 and heavy metals on MT-I/II protein expression. Immunohistochemical studies indicated that in mixed primary cultures of trisomy 16 and normal mice, MT-I/II was present in astrocytes but not in neurons.[17] Therefore, we used astrocyte-enriched primary cultures to study MT-I/II Western blots. In trisomy 16 mice, the intensity of MT-I/II protein bands was 2-times higher than in normal littermates.[17] This increase was reversible when trisomy 16 cells were cultured in presence of 0.5 mM N-acetyl-L-cysteine.[17] As a glutathione precursor, N-acetyl-L-cysteine may have helped to stabilize the redox balance that is maintained through the MT/glutathione disulfide cycle.[27] When glial primary cultures from diploid mice were exposed to 10 µM H_2O_2 for 15 min a 2-fold increase of the density of the MT-I/II protein band was measurable after 1h[17,19] (FIG. 2). In contrast, immunoblots from trisomy 16 glial cultures show that the expression of MT-I/II protein remained unchanged 1h after a 15 min incubation with 10 µM H_2O_2 (FIG. 2). This lack of response brought up the question whether MT-I/II may already have been maximally stimulated because of the inherent oxidative stress in trisomy 16 cells. This assumption was tested in dose response studies of the kainic acid and Pb effect. Similar to H_2O_2, kainic acid elicited a dose-dependent increase of MT-I/II in primary cultures of euploid littermates but failed to further increase the elevated basal MT-I/II level in trisomic cells (FIG. 2). In contrast, exposure to 25 µM Pb-acetate revealed a 5-fold increase in the density of MT-I/II protein bands in trisomy 16 and a 3-fold increase in euploid primary cultures[17] (FIG. 2). The finding of a similar MT-I/II metal responsiveness in glial cultures of trisomy 16 and euploid mice not only provided

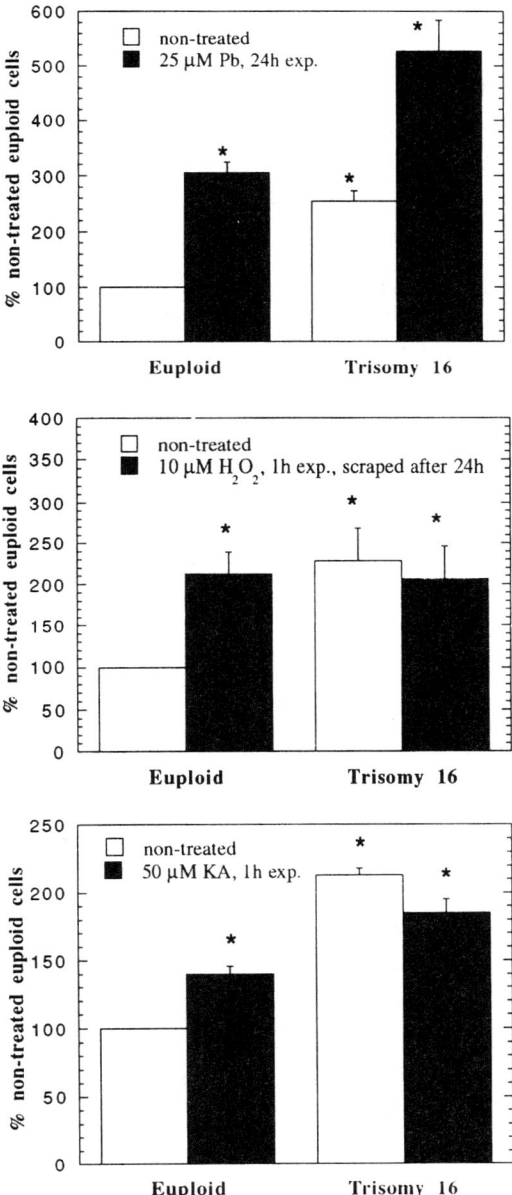

FIGURE 2. Western blot analysis of MT-I/II in astrocyte primary cultures from trisomy 16 and diploid mouse embryos (16 d in gestation). A total of 30 µg protein was loaded per well. The fluorescence intensity of the protein bands was measured by densitometry. The results are expressed in percent calculated from the ratios of trisomy 16 over diploid littermates of 4 different experiments. The statistical significance was calculated using the one-sample t-test. *$p < 0.01$ when compared with non-treated euploid cultures.

an internal control for MT-I/II oxidative stress vulnerability, but also suggested that the lack of MT induction by H_2O_2 may be due to a flawed redox-regulation in trisomy 16 mice. It is tempting to speculate that the binding activity of ROS-sensitive transcription factors on the MT promoter may have been altered and thereby prevented MT-I/II induction. A deficit of MT-I/II activity might result in elevated ROS levels which could partially explain why protein oxidation is increased in trisomy 16 cell cultures.

Activator Protein-1

To understand the lack of MT-I/II responsiveness to simulated oxidative stress in primary cultures of trisomy 16 mice, studies on the ROS-responsive transcription factors, AP-1, MTF-1, and upstream stimulatory factor may be required. AP-1 is a heterodimer complex of fos and jun protein[37] that regulates the ROS responsiveness of several genes, including MT-I, glutathione-S-transferase, interleukin-2, and quinone reductase.[35] It has not yet been established whether fos and/or jun bind directly to a site on the MT promoter (AP-1 element) or act as an enhancer for yet unknown transcriptional regulators that bind to ARE on the MT promoter. In the human MT promoter, AP-1, AP-2, and AP-4 appear to bind to two different basal level enhancers while, in the mouse MT promoter, only AP-1 DNA binding was, so far, shown to occur.[33]

THE REGULATION OF THE CELLULAR REDOX STATE

The molecular mechanisms involved in AP-1 binding are still unclear and may be mediated by yet undefined redox-sensitive activators that require the reduction of a specific Cys residue in vicinity to the DNA binding site to bind to the c-fos promoter. Both c-fos and c-jun genes are strongly induced by H_2O_2.[38] Therefore, expressed AP-1 binding activity may reflect the general redox state of the cell. The cellular redox state is modulated by ROS but also by changes of the ratio of glutathione to oxidized glutathione.[27] In fact, modification of this ratio was shown to activate several transcription factors[39] as well as specific protein kinases.[40] A redox interaction between glutathione disulfide and MT maintains protein sulfhydryl groups and zinc finger motifs in transcription factors.[24,27] Conversely, conditions that prevent the maintenance of glutathione and MT levels may cause neuronal damage by promoting oxidative stress or impairing transcriptional activity.

Taken together, the present findings support the view that in trisomy 16 glial primary cultures, the increased ROS production may have modified ROS sensitive transcriptional regulator functions. It has to be explored whether in the trisomy 16 mouse model, alterations of other target genes for AP-1, MTF-1 or USF might occur. MT and γ-glutamylcysteine synthetase are functionally linked through the glutathione redox cycling and their gene expression depends on MTF-1.[34]

REFERENCES

1. FRIDOVICH, I. 1995. Superoxide radical and superoxide dismutases. Ann. Rev. Biochem. **64:** 97–112.

2. KLEGERIS, A., L.G. KORKINA & S.A. GREENFIELD. 1995. Autoxidation of dopamine: a comparison of luminescent and spectrophotometric detection in basic solutions. Free Rad. Biol. Med. **18:** 215–222.
3. FRIDOVICH, I. 1995. Superoxide radical and superoxide dismutases. Ann. Rev. Biochem. **64:** 97–112.
4. MITCHELL, J.H., et al. 1998. Selenoprotein expression and brain development in preweanling selenium- and iodine-deficient rats. J. Mol. Endocrinol. **20:** 203-210.
5. DAVIS, J.G., et al. 1997. Murine thioredoxin peroxidase delays neuronal apoptosis and is expressed in areas of the brain most susceptible to hypoxic and ischemic injury. DNA Cell Biol. **16:** 311–321.
6. KANG S.W., et al. 1998. Mammalian peroxiredoxin isoforms can reduce hydrogen peroxide generated in response to growth factors and tumor necrosis factor-α. J. Biol. Chem. **273:** 6297–6302.
7. RAHMANI, Z., et al. 1989. Critical role of the D21S555 region on chromosome 21 in the pathogenesis of Down's syndrome. Proc. Natl. Acad. Sci. USA **86:** 5958-5962.
8. LEVY, M.J. 1993. Maps of birth defects occurrence in the US, birth defect monitoring program, 1970–1987. Teratology **48:** 551–646.
9. SKOVBY, F., N. KRASSIKOFF & U. FRANCKE. 1984. Assignment of the gene for cystathione β-synthase to human chromosome 21 in somatic cell hybrids. Hum. Genet. **65:** 291–294.
10. PASH, J., et al. 1990. Chromosomal protein HMG-14 gene maps to the Down's syndrome region of human chromosome 21 and is overexpressed in mouse trisomy 16. Proc. Natl. Acad. Sci. USA **87:** 3836–3840.
11. COX, D.R., L.B. EPSTEIN & C.J. EPSTEIN. 1980. Genes coding for sensitivity to interferon (IFREC) and soluble superoxide dismutase are linked in mouse and man and map to mouse chromosome 16. Proc. Natl. Acad. Sci. USA **77:** 2168–2172.
12. REEVES, R.H., et al. 1989. The mouse neurological mutant weaver maps within the region of chromosome 16 that is homologous to human chromosome 21. Genomics **5:** 522–526.
13. HARMAN, D. 1996. A hypothesis on the pathogenesis of Alzheimer's disease. Ann. N.Y. Acad. Sci. **786:** 152–168.
14. MÜLLER, W., U. HEINEMANN & S. SCHUCHMANN. 1997. Impaired Ca-signaling in astrocytes from the Ts16 mouse model of Down syndrome. Neurosci. Lett. **223:** 81–84.
15. GROPP, A., D. GIERS & U. KOLBUS. 1974. Trisomy in the fetal backcross progeny of male and female metracentric heterozygotes of the mouse. Cytogenet. Cell Genet. **13:** 511–535.
16. WEBB, S., R.H. ANDERSON & N.A. BROWN. 1996. Endocardialcushion development and heart loop architecture in the trisomy 16 mouse. Develop. Dynamics **206:** 301-309.
17. SCORTEGAGNA, M., et al. 1998. In cortical cultures of trisomy 16 mouse the upregulated metallothionein-I/II fails to respond to H_2O_2 exposure or glutamate receptor stimulation. Brain Res. **787:** 292–298.
18. LEVINE, R.L., et al. 1994. Methods Enzymol. **233:** 346–357.
19. HANBAUER, I., et al. 1998. Evidence of increased oxidative stress in hippocampal primary cultures of trisomy 16 mouse. Studies on metallothionein-I/II. Restorative Neurol. Neurosci. **12:** 87–93.
20. SCHWARTZ, P.J., et al. 1998. Effects of over- and under-expression of Cu,Zn-superoxide dismutase on the toxicity of glutamate analogs in transgenic mouse striatum. Brain Res. **789:** 32–39.
21. SCHWARTZ, P.J., U.V. BERGER & J.T. COYLE. 1995. Mice transgenic for copper/zinc superoxide dismutase exhibit increased markers of biogenic amine function. J. Neurochem. **65:** 660–669.
22. DALTON, T., R.D. PALMITER & G.K. ANDREWS. 1994.Transcriptional induction of the mouse metallothionein I gene in hydrogen peroxide-treated Hepa cells involves a composite major late transcription factor/antioxidant response element and metal response promoter elements. Nucl. Acids Res. **22:** 5016–5023.

23. KARIN, M., *et al.* 1984. Human metallothionein genes are clustered on chromosome 16. Proc. Natl. Acad. Sci. USA **81:** 5494–5498.
24. JACOB, C., W. MARET & B.L. VALLEE. 1998. Control of zinc transfer between thionein, metallothionein, and zinc proteins. Proc. Natl. Acad. Sci. USA **95:** 3489–3494.
25. MARGOSHES, M. & B.L. VALLEE. 1957. A cadmium protein from equine kidney cortex. J. Am. Chem. Soc. **79:** 4813–4814.
26. PALMITER, R.D. 1998. The elusive function of metallothioneins. Proc. Natl. Acad. Sci. USA **95:** 8428–8430.
27. MARET, W. 1994. Oxidative metal release from metallothionein via zinc-thiol/disulfide interchange. Proc. Natl. Acad. Sci. USA **91:** 237–241.
28. KIKUCHI, Y., T. IRIE, J. KASAHARA, *et al.* 1993. Induction of metallothionein in a human astrocytoma cell line by interleukin-1 and heavy metals. FEBS Lett. **317:** 22–26.
29. HIDALGO, J., *et al.* 1991. Metallothionein-I induction by stress in specific brain areas. Neurochem. Res. **16:** 1145–1148.
30. CULOTTA, V.C. & D.H. HAMER. 1989. Fine mapping of a mouse metallothionein gene metal response element. Mol. Cell Biol. **9:** 1376–1380.
31. KELLY, E.J., *et al.* 1997. A pair of adjacent glucocorticoid response elements regulate expression of two mouse metallothionein genes. Proc. Natl. Acad. Sci. USA **94:** 10045–10050.
32. FRIEDMAN, R.L. & G.R. STARK. 1985. Alpha-Interferon-induced transcription of HLA and metallothionein genes containing homologous upstream sequences Nature (London) **314:** 637–639.
33. SAMSON, S.L.A. & L. GEDAMU. 1998. Molecular analysis of metallothionein gene regulation. Prog. Nucleic Acid Res. **59:** 257–288.
34. GÜNES, C. *et al.* 1998. Embryonic lethality and liver degeneration in mice lacking the metal-responsive transcriptional activator MTF-1. EMBO J. **17:** 2846–2854.
35. RUSHMORE, T.H., M.R. MORTON & C.B. PICKETT. 1991. The antioxidant responsive element. Activation by oxidative stress and identification of the DNA consensus sequence required for functional activity. J. Biol. Chem. **266:** 11632–11639.
36. ABATE, C., *et al.*. 1990. Redox regulation of Fos and Jun DNA-binding activity in vitro. Science **249:** 1157–1161.
37. LI, Q., *et al.* 1998. Participation of upstream stimulator factor (USF) in cadmium-induction of the mouse metallothionein gene. Nucl. Acids Res. **26:** 5182–5189.
38. SHIBANAMU, M., T. KUROKI & K. NOSE. 1988. Induction of DNA replication and expression of protooncogenes c-myc and c-fos in quiescent Balb/3T3 cells by xanthine/xanthine oxidase. Oncogene **3:** 17–21.
39. STAAL, F.J.J., M. ROEDERER & L.A. HERZENBERG. 1990. Intracellular thiols regulate the activation of NF-kB and transcription of human immunodeficiency virus. Proc. Natl. Acad. Sci. USA **87:** 9943–9949.
40. KASS, G.E.N., S.K. DUDDY & S. ORRENIUS. 1989. Activation of protein kinase C by redox-cycling quinones. Biochem. J. **260:** 499–507.

Protein Oxidation

EARL R. STADTMAN AND RODNEY L. LEVINE

Laboratory of Biochemistry, National Heart, Lung, and Blood Institute, National Institutes of Health, Bethesda, Maryland 20892-0320 USA

ABSTRACT: The oxidative modification of proteins by reactive species, especially reactive oxygen species, is implicated in the etiology or progression of a panoply of disorders and diseases. These reactive species form through a large number of physiological and non-physiological reactions. An increase in the rate of their production or a decrease in their rate of scavenging will increase the oxidative modification of cellular molecules, including proteins. For the most part, oxidatively modified proteins are not repaired and must be removed by proteolytic degradation, and a decrease in the efficiency of proteolysis will cause an increase in the cellular content of oxidatively modified proteins. The level of these modified molecules can be quantitated by measurement of the protein carbonyl content, which has been shown to increase in a variety of diseases and processes, most notably during aging. Accumulation of modified proteins disrupts cellular function either by loss of catalytic and structural integrity or by interruption of regulatory pathways.

INTRODUCTION

In the course of their lifetime, organisms are constantly exposed to one or more systems that generate reactive oxygen species (ROS) that can damage proteins, nucleic acids and lipids. Some of the more common ROS generating systems known to modify proteins are shown in FIGURE 1. These include a number of environmental factors such as irradiation (x-rays, γ-rays, ultra violet light) and pollutants in the atmosphere (ozone, N_2O_2, NO_2, cigarette smoke); however, many are simple by-products of normal metabolic processes, such as, autoxidation of reduced forms of electron carriers (NAD(P)H, reduced flavins, cytochrome P450s), inflammatory reactions, nitric oxide synthesis, oxidase catalyzed reactions, lipid peroxidation, glycation/glycoxidation reactions, and metal catalyzed reactions. To avoid cellular damage by these processes, most biological systems have developed a battery of antioxidants that can convert ROS to unreactive derivatives. These include: a number of enzymes such as superoxide dismutase (SOD), catalase, glutathione peroxidase (GPx), glutathione-S-transferase (GST), thiol-specific peroxidase (RSH-Px), methionine sulfoxide reductase (MSR), thioredoxin reductase, and glutathione reductase; various metal-binding proteins such as ceruloplasmin, ferritin, transferrin; various metabolites and cofactors ($NADP^+$/NADPH, NAD^+/NADH, lipoic acid, uric acid, bilirubin, etc.); a number of dietary components (vitamins A, C and E,

---

Address for correspondence: Dr. Earl R. Stadtman, NIH, Building 3, Room 222, MSC 0342, Bethesda, MD 20892-0342. Voice: 301-496-4645; fax: 301-496-0599.
 e-mail: erstadtman@nih.gov

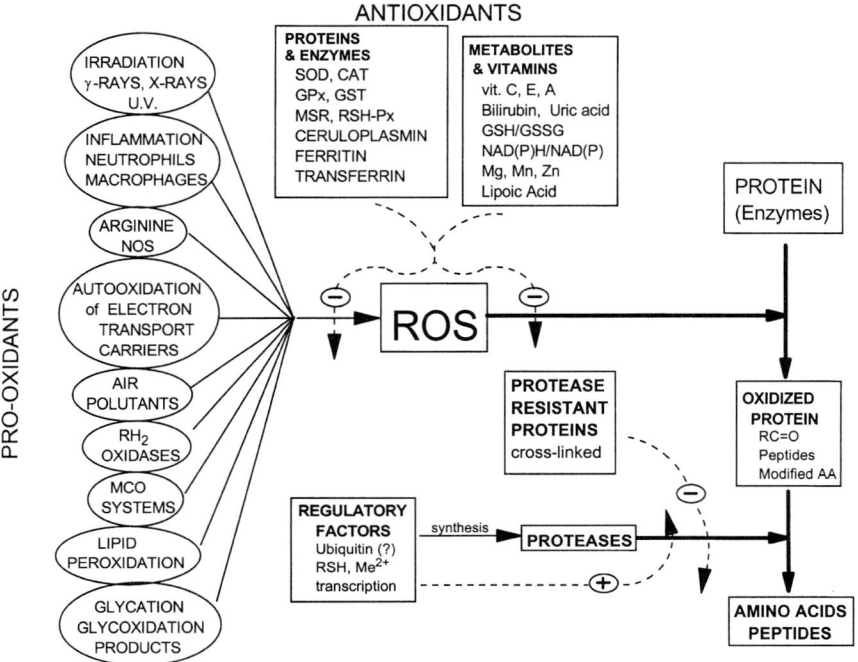

FIGURE 1. Factors which affect reactive oxygen species (ROS) and the level of oxidatively modified proteins.

quercitin), and metal ions (Mg^{2+}, Mn^{2+}, Zn^{2+}). It is therefore evident that the amount of protein oxidation that occurs under a given set of conditions will reflect the balance between the prooxidant and antioxidant activities as are dictated by prevailing environmental, genetic, and dietary factors.

Whereas the oxidative damage to nucleic acids is subject to repair by highly efficient excision/insertion mechanisms, the repair of damage to proteins appears limited to the reduction of oxidized derivatives of the sulfur-containing amino acid residues. Repair of other kinds of protein oxidation has not been demonstrated. Instead, the damaged proteins are targeted for degradation to amino acid constituents by the action of various endogenous proteases, including, cathepsin c, calpain, trypsin and especially the 20s proteosome,[1,2] whose activity is also under metabolic control by diverse regulatory factors, including the concentrations of enzyme substrates, ubiquitination, and various inhibitors (cross-linked proteins,[3–5] glycation/ glycoxidation protein conjugates,[6–9] etc.). Accordingly, as illustrated in FIGURE 1, the intracellular accumulation of oxidized forms of proteins is a complex function of prooxidant, antioxidant activities and the concentrations and activities of the proteases that degrade the oxidized forms of proteins.

METAL-CATALYZED OXIDATION OF PROTEINS

Depending upon the environmental and dietary conditions any one of the prooxidant systems illustrated in FIGURE 1 is able to damage proteins. However, there is reason to believe that under normal conditions the metal-catalyzed oxidation (MCO) systems are the major source of oxidative damage. This follows from the fact that hydrogen peroxide and alkylperoxides are the most common end products of most ROS generating systems. These peroxides by themselves are relatively unreactive compounds. However, in the presence of the transition metals, Fe(II) or Cu(I), they are converted to the highly reactive hydroxyl radical (reaction 1) or alkoxyl radical (reaction 2) which are capable of reacting with almost any organic substance.

$$H_2O_2 + Fe(II)/Cu(I) \rightarrow HO\cdot + OH^- + Fe(III)/Cu(I) \tag{1}$$

$$ROOH + Fe(II)/Cu(I) \rightarrow RO\cdot + OH^- + Fe(III)/Cu(I) \tag{2}$$

Indeed, virtually all kinds of amino acid residues of proteins are potential targets for oxidation by HO· generated by ionizing radiation or by high concentrations of H_2O_2 and Fe(II).[10,11] However, at the low concentrations of iron or copper ions and H_2O_2 present under most physiological conditions, protein damage is likely limited to the modification of amino acid residues at metal binding sites on the protein, which

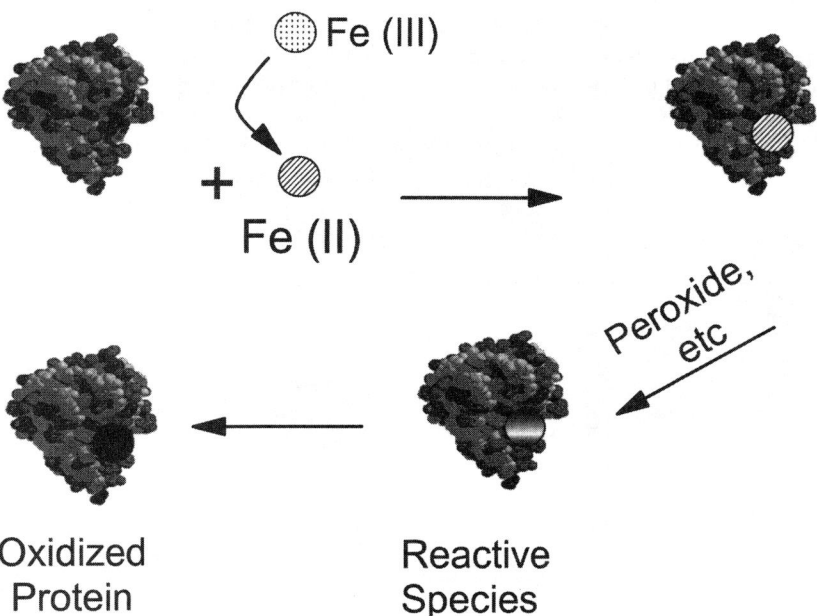

FIGURE 2. Metal catalyzed oxidation of proteins is a site specific process. Divalent cation binding sites will occasionally be occupied by redox-active cations such as Fe or Cu, rendering that site susceptible to oxidation by ROS generated *in situ*.

effectively concentrate the ions. This consideration gave rise to the proposition, illustrated in FIGURE 2, that the oxidation of proteins under physiological conditions is a site-specific process in which, the binding of Fe(II) or Cu(I) to metal binding sites on the protein is followed by reaction with peroxides to generate reactive species (OH·, RO·, perferryl radical) that will react preferentially with amino acid residues at the metal binding site.[12–15] Thus, the metal-binding site supports a biologically important "caged reaction." In the case of many enzymes, especially with those requiring a metal ion for activity, this will lead to loss of catalytic function.[16]

CARBONYL GROUPS AS MARKERS OF OXIDATIVE DAMAGE

ROS-mediated oxidation of proteins leads to the conversion of histidine residues to 2-oxohistidine,[17,18] tryptophan residues to kynurenine or N-formylkynurenine,[19–22] tyrosine residues to dihydroxy derivatives,[10,23–25] methionine residues to methionine sulfoxide or methionine sulfone derivatives,[18,26–30] leucine and valine residues to hydroxy derivatives,[27,31] and cysteine residues to disulfide derivatives.[27,32,33] Of particular significance is the fact that oxidation of some amino acid residues (lysine, arginine, and proline residues) leads to the formation of carbonyl derivatives.[34–36] In addition, carbonyl derivatives of proteins are produced as a consequence of oxidative cleavage of the peptide backbone via the α- amidation pathway,[27,33] or cleavage associated with the oxidation of glutamyl residues.[27,36] Carbonyl derivatives can also be formed as a consequence of secondary reactions of some amino acid side chains with lipid oxidation products, such as 4-hydroxy-2-nonenal (HNE),[3,11,37–39] or with reducing sugars or their oxidation products.[7,40–42] Thus, as shown in FIGURE 3, reaction of the double bond of HNE with lysyl, histidyl or sulfhydryl groups of proteins yields derivatives possessing a aldehyde function.[37,38] Likewise, reaction of one of the two aldehyde groups of the lipid peroxidation product, malondialdehyde, with lysine α-amino groups of proteins will yield a Schiff base possessing a carbonyl function.

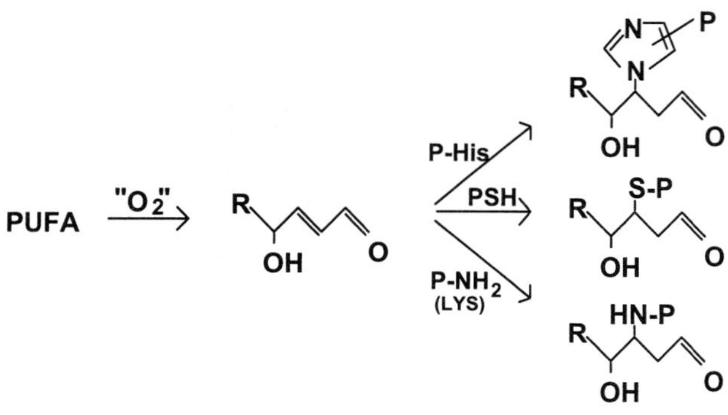

FIGURE 3. Protein carbonyl adducts produced by reaction with the lipid peroxidation product, 4-hydroxynonenal. ABBREVIATIONS: P-NH$_2$, ε-amino group of a lysine residue in protein; PUFA, polyunsaturated fatty acid; HNE, hydroxynonenal; P-SH, sulfhydryl group of a cysteine residue in protein; P-His, imidazole group of a histidine residue in protein.

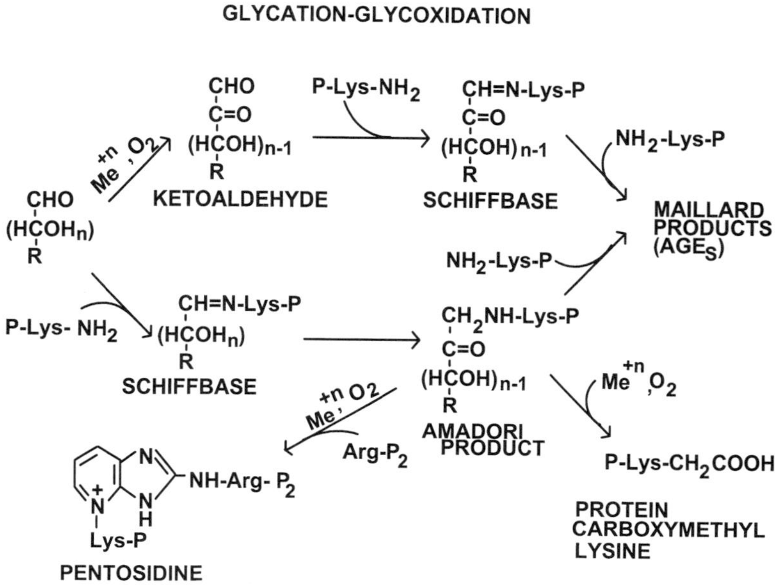

GLYCATION-GLYCOXIDATION

FIGURE 4. Generation of protein carbonyl derivatives and of protein-protein cross-linkages by glycation and glycoxidation reactions. ABBREVIATIONS: Me^{n+}, metal cation with charge n; Lys-P, a lysine residue within a protein molecule; Arg-P_2, an arginine residue within another protein molecule; AGE, advanced glycation end-product;

The interaction of reducing sugars or dicarbonyl compounds derived from the sugars can also lead to the formation of protein carbonyl derivatives, as shown in FIGURE 4. It is noteworthy the carbonyl groups of proteins generated by any one of these mechanisms may react further with the α-amino group of lysine residues in the same or another protein molecule to form intra- or inter-molecular protein cross-linked derivatives. Some such derivatives are not only resistant to proteolytic degradation by the 20s proteosome, but may also inhibit the ability of the proteosome to degrade the oxidized forms of other proteins[3,5]; they may therefore contribute to the accumulation of oxidized forms of proteins during aging and age-related diseases.

In any case, the fact that carbonyl groups are major products of ROS-mediated oxidation reactions has led to the development of several highly sensitive methods for the determination of protein carbonyl groups,[43] and the presence of carbonyl groups has become a widely accepted measure of oxidative damage under conditions of oxidative stress, aging and disease.[44] It is thus clear that protein carbonyl is a marker for oxidative modification of proteins. As outlined above, carbonyl groups arise from a variety of oxidative processes, so that the carbonyl measurement provides a generalized or integrated assessment of oxidative damage. The absolute value of the carbonyl measurement will therefore usually be much higher than the measurement of any single product. A typical value for the carbonyl content of organs from healthy young animals is 2 nmol/mg protein (~0.10 mol carbonyl/

mol protein), so that on average one of every ten protein molecules carries a carbonyl group. In contrast, a specific oxidation product may be present at very low levels. For example, dopa can be formed from the oxidation of tyrosine, and it also reacts with carbonyl reagents. In low density lipoprotein, there is approximately 1 dopa per 1,600 tyrosine residues.[45] Thus, it is not surprising that measurement of dopa in tissue proteins would give values which are orders of magnitude less than the protein-associated carbonyl in the same sample.[45] Also, as outlined in this paper, the oxidative modification giving rise to the carbonyl group is emerging as more than a marker of oxidative stress; the oxidative modification may be key to the dysfunction induced by oxidative stress. In many of the processes discussed below, the average cellular content of protein carbonyl increases markedly from basal levels, often reaching the point where at least 1 of every 3 protein molecules carries a carbonyl group. At this level, these dysfunctional molecules are likely to deleteriously affect most aspects of cellular function.

PROTEIN OXIDATION IN OXIDATIVE STRESS

Published studies demonstrate that protein oxidation, assessed by carbonyl group formation, is associated with numerous conditions characterized by oxidative stress, including; ischemia-reperfusion,[46–49] hyperoxia,[50–53] cigarette smoke,[54] estrogen administration,[55,56] artificial ventilation,[57] forced exercise,[52,58] rapid correction of hyponatremia,[59] magnesium deficiency,[60] paraquat toxicity,[53] the oxidative burst of neutrophils,[61,62] and in cell cultures exposed to stresses such as hydrogen peroxide[63] or xanthine oxidase/xanthine.[62] Protein oxidation occurs very early following exposure of cells to oxidative stress, clearly preceding the decrease in cellular ATP and subsequent cell death.[63]

PROTEIN OXIDATION AND DISEASE

Elevated levels of protein carbonyls are also associated with of a number of age-related diseases, often correlating well with the progression of the disease. An increase in the carbonyl content of protein is associated with Alzheimer's disease,[64–66] Parkinson's disease,[67,68] diabetes,[42,69,70] rheumatoid arthritis,[71] muscular dystrophy,[72] cataractogenesis,[73] induction of renal tumors,[74] bronchopulmonary dysplasia,[57] amyloidosis,[75] chronic ethanol ingestion,[76,77] acute carbon tetrachloride toxicity,[78,79] amyotrophic lateral sclerosis,[80] and the progerias.[81] In some of these diseases, more than one kind of oxidative modification has been demonstrated. For example, Alzheimer's disease has been shown to be associated with an increase in protein carbonyl content,[82,83] in advanced glycation end products,[84] in protein HNE adducts,[85–88] in nitrated tyrosine derivatives,[89,90] and in redox active iron.[91] Cataractogenesis, is associated with an increase in protein carbonyl groups[73] and methionine sulfoxide.[92,93] Parkinson's disease is associated with an increase in both protein carbonyl groups[68,75,94] and HNE adducts.[95] It is noteworthy that in the available studies of Parkinson's disease, all patients had been treated with dopa, which is known to provoke protein oxidation.[75,96] Thus, in Parkinson's disease the observed protein oxidation may be a consequence of the underlying pathophysiology or it may be an iatrogenic consequence of drug treatment.

PROTEIN OXIDATION IN AGING

Aging is accompanied by a loss of proteolytic capacity, and the accumulation of catalytically less active, more thermally sensitive forms of a number of enzymes.[97,98] In light of these observations it was proposed that the rate of protein turnover was decreased, and consequently the "dwell-time" of enzymes was lengthened, permitting them to undergo spontaneous changes in protein conformation to forms that were less active and more sensitive to heat denaturation.[99,100] More recently it was proposed that the age-related accumulation of altered enzymes is due, at least in part, to oxidative modifications of proteins. This concept is supported by the following lines of evidence:

1. Changes in catalytic activities, thermal stability and sensitivity to proteolytic degradation comparable to those associated with aging can be provoked by exposing purified enzymes to free radical generating systems *in vitro*.[15,101–108]
2. Altered forms of enzymes similar to those found in old animals can be generated *in vivo* by exposing young animals to conditions of oxidative stress.[50,51,109]
3. Factors or physiological conditions that lead to an increase in the life span of animals also cause a decrease in the intracellular levels of oxidized proteins (measured as protein carbonyl content), and vice-versa.[52,110,111]
4. Age-related loss of cognitive functions are associated with increases in the levels of oxidized protein in those areas of the brain known to be implicated in the cognitive functions.[82]
5. Proteins in tissues of old animals are more sensitive to oxidative damage than tissues from young animals.[110,112]
6. The age-related increase in surface hydrophobicity of proteins is correlated with an increase in the level of oxidized methionine residues, as discussed below.[113,114]
7. The carbonyl content of proteins in widely different animal species and tissues increases almost exponentially as a function of animal age.

Thus, as shown in FIGURE 5, intracellular level of protein carbonyl groups in cultured human fibroblasts increases as a function of the age of the fibroblast donor,[81] and similar age-related increases have been observed in human brain tissue,[82] human eye lens proteins,[115] rat hepatocytes,[51] and whole body proteins of houseflies.[52] Age-related increases in carbonyl content of proteins are not restricted to long-lived cells; one can observe an increase in human red blood cells whose life span is about 120 days.[81] A modest increase in protein carbonyl was also documented in skeletal muscle biopsies from healthy humans over age 60.[116]

Specific proteins may be particularly susceptible to oxidative damage. During aging of the housefly, the protein carbonyl content of a specific protein increased

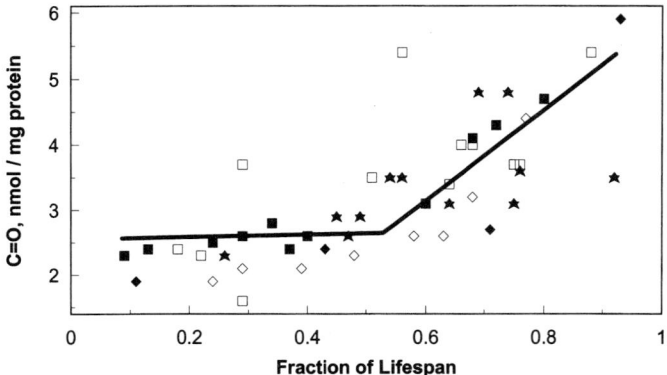

FIGURE 5. Carbonyl content of protein from different tissues. One observes a dramatic increase in oxidized protein during the last third of the lifespan. The data points were taken from published reports: ■, human dermal fibroblasts in tissue culture[81];★, human lens[115];□, human brain obtained at autopsy[82];◆, rat liver[51]; and ◇, whole fly.[52]

distinctly more than others, assessed by Western blotting.[50] The protein was shown to be aconitase, a key enzyme in the citric acid cycle. Aconitase activity decreased as its protein carbonyl increased during aging. Loss of aconitase activity may block normal electron flow to the usual final acceptor, oxygen. Consequently, reduced metabolites such as NADH will accumulate, increasing the rate of autooxidation and of metal-catalyzed oxidation. Thus, oxidative modification of key enzymes such as aconitase may initiate a self-amplifying cascade with the potential to cause a dramatic increase in the cellular burden of oxidatively modified proteins.

ANTIOXIDANT ACTIVITY OF METHIONINE RESIDUES OF PROTEINS

Methionine residues of proteins are particularly sensitive to oxidation by virtually all kinds of ROS. Ozone,[117,118] hydrogen peroxide,[119,120] alkyl peroxides,[121] peroxynitrite,[28,122,123] hypochlorous acid,[30,123] metal catalyzed reactions,[124,125] ultraviolet light,[125] and ionizing radiations[27,33] have all been shown to convert methionine residues of proteins to methionine sulfoxide (MetO) derivatives. But unlike oxidation of non–sulfur-containing amino acids, the oxidation of methionine residues by ROS (reaction 3) can be reversed. The conversion of MetO residues back to Met residues (reaction 4) is catalyzed by the thioredoxin $TR(SH)_2$-dependent peptide methionine sulfoxide reductase (msrA), which is widely distributed in most animal tissues and in bacteria. It follows that when coupled with the NADPH-dependent reduction of oxidized thioredoxin [TR(S-S)] by thioredoxin reductase [TxR] (reaction 5), the cyclic oxidation and reduction of Met residues constitutes an antioxidant system by which almost all forms of ROS can be converted to inactive derivatives (ID), as described by the over-all reaction 6.

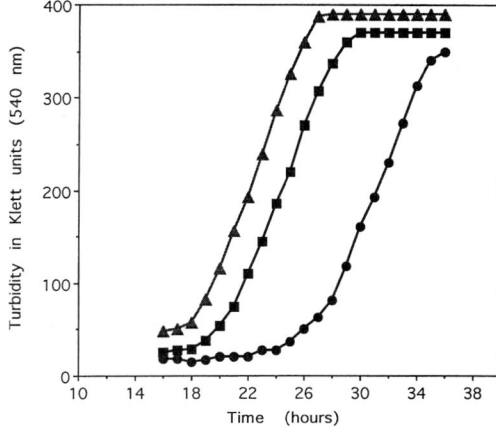

FIGURE 6. Growth of yeast strains in the presence of 2 mM hydrogen peroxide.[126] ■, wild-type parent; ●, msrA null, lacking methionine sulfoxide reductase; ▲, strain overproducing methionine sulfoxide reductase.

$$\text{ROS} + \text{Met} \rightarrow \text{MetO} + \text{ID} \quad (3)$$

$$\text{mrsA MetO} + \text{TR(SH)}_2 \xrightarrow{\text{mrsA}} \text{Met} + \text{TR(S-S)} + \text{H}_2\text{O} \quad (4)$$

$$\text{TR(S-S)} + \text{NADPH} + \text{H}^+ \xrightarrow{\text{TxR}} \text{TR(SH)}_2 + \text{NADP}^+ \quad (5)$$

$$\text{Sum:} \quad \text{ROS} + \text{NADPH} + \text{H}^+ \rightarrow \text{ID} + \text{NADP}^+ + \text{H}_2\text{O} \quad (6)$$

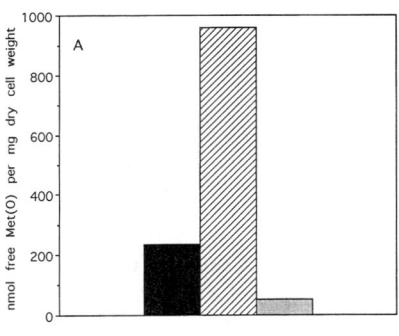

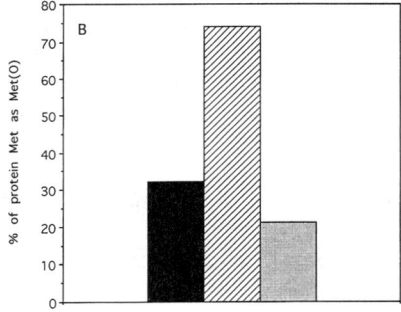

FIGURE 7. Methionine sulfoxide content in yeast strains grown in the presence of 2 mM hydrogen peroxide.[126] Free methionine sulfoxide is shown in **A** and methionine sulfoxide in protein in **B**. The *solid bars* are the wild-type parent, *hatched bars* are the msrA null strain, lacking methionine sulfoxide reductase, and *dotted bars* are the methionine sulfoxide overproducing strain.

Confirmation of the antioxidant activity of methionine residues of proteins comes from the results of studies by Moskovitz et al.[126] comparing the effects of H_2O_2 on the growth and MetO contents of normal yeast strain (WT), an msrA over-producing strain (OP), and a null mutant strain (NM) containing no peptide MetO reductase activity. As shown in FIGURE 6, the lagtime before attaining a maximal growth rate for the WT, OP, and NM strains were 19 h, 17 h, and 26 h, respectively. Moreover, the levels of MetO in both the free amino acid pool (FIG. 7A) and cellular proteins (FIG. 7B) were significantly lower in the OP strain and considerably greater in the NM strain than in the WT strain.

MODIFICATION OF REGULATORY PROTEINS BY PEROXYNITRITE

The discovery that peroxynitrite is produced endogenously by reaction of nitric oxide with superoxide anion (reaction 7), and the demonstration that peroxynitrite can modify tyrosine, tryptophan, cysteine, and methionine residues of proteins has focussed attention on the possibility that PN-mediated damage may contribute to the development of various diseases.

$$NO\cdot + O_2^{-\cdot} \rightarrow ONOO^- \tag{7}$$

Of particular concern is the fact that PN nitration of tyrosine residues of proteins might seriously compromise one of the most important mechanisms of cellular regulation, namely the cyclic interconversion of tyrosine residues of regulatory proteins. In animals, the conversion is between phosphorylated and unphosphorylated forms, especially in signal transduction pathways. In *Escherichia coli* and other gram negative bacteria, a conversion between adenylylated and unadenylylated forms occurs in the regulation of glutamine synthetase activity. Extensive studies on the regulation of glutamine synthetase in *E. coli* have shown that adenylylation of a single tyrosine residue in any one of the enzyme's 12 identical subunits converts that subunit to a form that is highly susceptible to feed-back inhibition by various end products of glutamine metabolism, and lowers the affinity of that subunit for substrates and end products. Highly sensitive assays have been developed for determination of the average number of adenylylated subunits per dodecameric enzyme, based on these characteristics and the fact that adenylylated subunits have an absolute requirement for Mn^{2+} for activity, whereas unadenylylated subunits are activated by either Mn^{2+} or Mg^{2+}.[127] Using these methods it has been established that the PN-dependent modification of glutamine synthetase converts the enzyme to a form with regulatory properties similar to those obtained by adenylylation of the enzyme.[28] Because the PN treatment led to oxidation of some methionine residues as well as to nitration of tyrosine residues, it was not possible from these results alone to know whether one or both of these modifications was responsible for the changes in regulatory characteristics. However, prompted by the discovery that PN reacts very rapidly with CO_2 to form a derivative capable of nitrating aromatic compounds 128,129, further studies revealed that the nitration of tyrosine residues in glutamine synthetase is almost completely dependent upon the presence of CO_2, whereas, the PN-dependent oxidation of methionine residues is almost completely inhibited by physiological concentrations of CO_2 (FIG. 8).[28] Accordingly, when CO_2

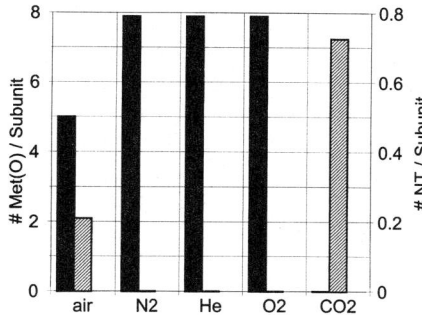

FIGURE 8. Treatment of glutamine synthetase with peroxynitrite leads to nitration of tyrosine residues in the presence of CO_2 and the oxidation of methionine residues to methionine sulfoxide in the absence of CO_2. Black and shaded bars represent methionine sulfoxide (MetSO) and nitrotyrosine (NT) residues, respectively.[28]

is completely removed from the reaction mixtures by gassing with N_2, PN oxidizes methionine residues of glutamine synthetase, but little or no nitrotyrosine is formed; conversely, when reactions are carried out in CO_2/bicarbonate buffers (pH 7.4) tyrosine residues are nitrated, but little or no MetO is formed. Surprisingly, under both sets of conditions glutamine synthetase subunits are converted to the adenylylated conformation. Nevertheless, under physiological conditions only nitration would be expected to occur, and because there are no known mechanisms for the reversal of the nitration reaction, glutamine synthetase would be locked into the adenylylated configuration, which is physiologically inactive.

In a series of parallel studies it was demonstrated by Chock and his colleagues[75] that the PN-dependent nitration of the tyrosine residue in a pentadecameric peptide substrate analog of the lymphocyte-specific tyrosine kinase (*lck*) prevents its phosphorylation by the kinase. Thus, they conclude that nitration of tyrosine residues of regulatory proteins can lead to "permanent impairment of cyclic cascades that control signal transduction processes and regulate cell cycles."

REFERENCES

1. GRUNE, T., T. REINHECKEL & K.J.A. DAVIES. 1996. Degradation of oxidized proteins in K562 human hematopoietic cells by proteasome. J. Biol. Chem. **271**: 15504–15509.
2. RIVETT, A.J. 1986. Regulation of intracellular protein turnover: Covalent modification as a mechanism of marking proteins for regulation. Current Topics Cell. Reg. **28**: 291–337.
3. FRIGUET, B., E.R. STADTMAN & L.I. SZWEDA. 1994. Modification of glucose-6-phosphate dehydrogenase by 4-hydroxy-2-nonenal: formation of cross-linked protein which inhibits the multicatalytic proteasome. J. Biol. Chem. **269**: 21639–21643.
4. GRUNE, T., T. REINHECKEL, M. JOSHI & K.J.A. DAVIES. 1995. Proteolysis in cultured liver epithelial cells during oxidative stress: Role of the multicatalytic proteinase complex, proteasome. J. Biol. Chem. **270**: 2344–2351.
5. FRIGUET, B., L. SZWEDA & E. R. STADTMAN. 1994. Susceptibility of glucose-6- phosphate dehydrogenase modified by 4-hydroxy-2-nonenal and metal-catalyzed oxidation to proteolysis by the multicatalytic protease. Arch. Biochem. Biophys. **311**: 168–173.
6. BRUENNER, B.A., A.D. JONES & J.B. GERMAN. 1994. Maximum entropy deconvolution of heterogeneity in protein modification: protein adducts of 4-hydroxy-2-nonenal. Rapid. Commun. Mass. Spectrom. **8**: 509–512.

7. MULLARKEY, C.J., D. EDELSTEIN & M. BROWNLEE. 1990. Free radical generation by early glycation products: a mechanism for accelerated atherogenesis in diabetes. Biochem. Biophys. Res. Commun. **173:** 932–939.
8. WOLFF, S.P. & R.T. DEAN. 1987. Glucose autoxidation and protein modification. The potential role of 'autoxidative glycosylation' in diabetes. Biochem. J. **245:** 243–250.
9. KRISTAL, B.S. & B.P. YU. 1992. An emerging hypothesis: synergistic induction of aging by free radicals and Maillard reactions. J. Gerontol. **47:** B107–B114.
10. HUGGINS, T.G., M.C. WELLS-KNECHT, N.A. DETORIE, J W. BAYNES & S.R. THORPE. 1993. Formation of o-tyrosine and dityrosine in proteins during radiolytic and metal-catalyzed oxidation. J. Biol. Chem. **268:** 12341–12347.
11. NEUZIL, J., J.M. GEBICKI & R. STOCKER. 1993. Radical-induced chain oxidation of proteins and its inhibition by chain-breaking antioxidants. Biochem. J. **293:** 601–606.
12. BACHUR, N.R., S.L. GORDON, M.V. GEE & H. KON. 1979. NADPH cytochrome P-450 reductase activation of quinone anticancer agents to free radicals. Proc. Natl. Acad. Sci. USA **76:** 954–957.
13. LEVINE, R. L., C. N. OLIVER, R.M. FULKS & E.R. STADTMAN. 1981. Turnover of bacterial glutamine synthetase: oxidative inactivation precedes proteolysis. Proc. Natl. Acad. Sci. USA **78:** 2120–2124.
14. CHEVION, M. 1988. A site-specific mechanism for free radical induced biological damage: the essential role of redox-active transition metals. Free Rad. Biol. Med. **5:** 27–37.
15. STADTMAN, E.R. 1990. Metal ion-catalyzed oxidation of proteins: biochemical mechanism and biological consequences. Free Rad. Biol. Med. **9:** 315–325.
16. FUCCI, L., C.N. OLIVER, M.J. COON & E.R. STADTMAN. 1983. Inactivation of key metabolic enzymes by mixed-function oxidation reactions: possible implication in protein turnover and ageing. Proc. Natl. Acad. Sci. USA **80:** 1521–1525.
17. UCHIDA, K. & S. KAWAKISHI. 1993. 2-oxohistidine as a novel biological marker for oxidatively modified proteins. FEBS Lett. **332:** 208–210.
18. BERLETT, B.S., R.L. LEVINE & E.R. STADTMAN. 1996. A comparison of the effects of ozone on the modification of amino acid residues in glutamine synthetase and bovine serum albumin. J. Biol. Chem. In press.
19. KIKUGAWA, K., T. KATO & Y. OKAMOTO. 1994. Damage of amino acids and proteins induced by nitrogen dioxide, a free radical toxin, in air. Free Rad. Biol. Med. **16:** 373–382.
20. WINCHESTER, R.V. & K.R. LYNN. 1970. X- and gamma-radiolysis of some tryptophan dipeptides. Int. J. Radiat. Biol. Relat. Stud. Phys. Chem. Med. **17:** 541–548.
21. ARMSTRONG, R.C. & A.J. SWALLOW. 1969. Pulse- and gamma-radiolysis of aqueous solutions of tryptophan. Radiat. Res. **40:** 563–579.
22. PRYOR, W.A. & R.M. UPPU. 1993. A kinetic model for the competitive reactions of ozone with amino acid residues in proteins in reverse micelles. J. Biol. Chem. **268:** 3120–3126.
23. DEAN, R.T., S. GIESEG & M.J. DAVIES. 1993. Reactive species and their accumulation on radical-damaged proteins. Trends Biochem. Sci. **18:** 437–441.
24. HEINECKE, J.W., W. LI, H.L. DAEHNKE III & J.A. GOLDSTEIN. 1993. Dityrosine, a specific marker of oxidation, is synthesized by the myeloperoxidase-hydrogen peroxide system of human neutrophils and macrophages. J. Biol. Chem. **268:** 4069–4077.
25. GIULIVI, C. & K.J.A. DAVIES. 1993. Dityrosine and tyrosine oxidation products are endogeneous markers for the selective proteolysis of oxidatively modified red blood cell hemoglobin by (the 19S) proteosome. J. Biol. Chem. **68:** 8752–8759.
26. GARRISON, W.M., M.E. JAYKO & BENNETT. 1962. Radiation-induced oxidation of proteins in aqueous solution. Radiat. Res. **16:** 487–502
27. GARRISON, W.M. 1987. Reaction mechanisms in radiolysis of peptides, polypeptides, and proteins. Chem. Rev. **87:** 381–398.
28. BERLETT, B.S., R.L. LEVINE & E.R. STADTMAN. 1998. Carbon dioxide stimulates peroxynitrite-mediated nitration of tyrosine residues and inhibits oxidation of methionine residues of glutamine synthetase: both modifications mimic effects of adenylylation. Proc. Natl. Acad Sci. USA **95:** 2784–2789.

29. PRYOR, W.A., X. JIN & G.L. SQUADRITO. 1994. One- and two-electron oxidations of methionine by peroxynitrite. Proc. Natl. Acad. Sci. USA **91:** 11173–11177.
30. VOGT, W. 1995. Oxidation of methionine residues in proteins: tools, targets, and reversal. Free Rad. Biol. Med. **18:** 93–105.
31. KOPOLDOVA, J. & J. LIEBSIER. 1963. The mechanism of radiation chemical degradation of amino acids V. Intern. J. Appl. Radiat. Isotopes **14:** 493–498.
32. BRODIE, E. & D.J. REED. 1990. Cellular recovery of glyceraldehyde-3-phosphate dehydrogenase activity and thiol status after exposure to hydroperoxide. Archives of Biochemistry and Biophysics **276:** 210–212.
33. SWALLOW, A.J. 1960. Effect of ionizing radiation on proteins, RCO groups, peptide bond cleavage, inactivation, -SH oxidation. A. J. Swallow. Ed. :211–224. Radiation Chemistry of Organic Compounds. John Wiley & Sons. New York.
34. AMICI, A., R.L. LEVINE & E.R. STADTMAN. 1989. Conversion of amino acids residues in proteins and amino acid homopolymers to carbonyl derivatives by metal-catalyzed reactions. J. Biol. Chem. **264:** 3341–3346.
35. CREETH, J.M., B. COOPER, A.S.R. DONALD & J.R. CLAMP. 1983. Studies of the limited degradation of mucous glycoproteins. Biochem. J. **211:** 323–332.
36. UCHIDA, K., Y. KATO & S. KAWAKISHI. 1990. A novel mechanism for oxidative damage of prolyl peptides induced by hydroxyl radicals. Biochem. Biophys. Res. Commun. **169:** 265–271.
37. SCHUENSTEIN, E. & H. ESTERBAUER. 1979. Formation and preparation of reactive aldehydes. Submolecular biology of cancer. Excerpta Medica/Elsevier. Amsterdam. CIBA Foundation Series **67:** 225–234.
38. UCHIDA, K. & E.R. STADTMAN. 1993. Covalent modification of 4-hydroxynonenal to glyceraldehyde-3- phosphate. J. Biol. Chem. **268:** 6388–6393.
39. NADKARNI, D.V. & L.M. SAYRE. 1995. Structural definition of early lysine and histidine adduction chemistry of 4-hydroxynonenal. Chem. Res. Toxicol. **8:** 284–291.
40. KRISTAL, B.S. & B.P. YU. 1992. An emerging hypothesis: synergistic induction of aging by free radicals and Maillard reactions. J. Gerontol. **47:** B104–B107.
41. MONNIER, V., C. GERHARDINGER, M.S. MARION & S. TANEDA. 1995. Novel approaches toward inhibition of the Maillard reaction in vivo: search, isolation, nd characterization of prokaryotic enzymes which degrade glycated substrates. *In* Oxidative Stress and Aging. R.G. Cuttler, L. Packer, J. Bertram & A. Mori, Eds. :141–149. Birkhauser Verlag. Basel.
42. BAYNES, J.W. & S.R. THORPE. 1999. Role of oxidative stress in diabetic complications: a new perspective on an old paradigm. Diabetes **48:** 1–9.
43. LEVINE, R.L., J.A. WILLIAMS, E.R. STADTMAN & E. SHACTER. 1994. Carbonyl assays for determination of oxidatively modified proteins. Methods Enzymol. **233:** 346–357.
44. BERLETT, B.S. & E.R. STADTMAN. 1997. Protein oxidation in aging, disease, and oxidative stress. J. Biol. Chem. **272:** 20313–20316.
45. DEAN, R.T., S. FU, R. STOCKER & M.J. DAVIES. 1997. Biochemistry and pathology of radical-mediated protein oxidation. Biochem. J. **324:** 1–18.
46. OLIVER, C.N., P.E. STARKE-REED, E.R. STADTMAN, G.J. LIU, J.M. CARNEY & R.A. FLOYD. 1990. Oxidative damage to brain proteins, loss of glutamine synthetase activity, and production of free radicals during ischemia/reperfusion-induced injury to gerbil brain. Proc. Natl. Acad. Sci. USA **87:** 5144–5147.
47. POSTON, J. M. & G.L. PARENTEAU. 1992. Biochemical effects of ischemia on isolated perfused rat heart tissues. Arch. Biochem. Biophys. **295:** 35–41.
48. AYENE, I.S., A.B. AL-MEDI & A.B. FISHER. 1993. Inhibition of lung tissue oxidation during ischemia/reperfusion by 2-mercaptopropionyl glycine. Arch. Biochem. Biophys. **303:** 307–312.
49. ISCHIROPOULOS, H. & A.B. AL-MEDI. 1995. Peroxynitrite-mediated oxidative protein modifications. FEBS Lett. **364:** 279–282.
50. YAN, L.J., R.L. LEVINE & R.S. SOHAL. 1997. Oxidative damage during aging targets mitochondrial aconitase. Proc. Natl. Acad. Sci. USA **94:** 11168–11172.
51. STARKE-REED, P.E. & C.N. OLIVER. 1989. Protein oxidation and proteolysis during aging and oxidative stress. Arch. Biochem. Biophys. **275:** 559–567.

52. SOHAL, R.S., S. AGARWAL, A. DUBEY & W.C. ORR. 1993. Protein oxidative damage is associated with life expectancy of houseflies. Proc. Natl. Acad. Sci. USA **90:** 7255–7259.
53. WINTER, M.L. & J.G. LIEHR. 1991. Free radical-induced carbonyl content in protein of estrogen-treated hamsters assayed by sodium boro[^{3}H]ydride reduction. J. Biol. Chem. **266:** 14446–14450.
54. REZNICK, A.Z., C.E. CROSS, M.-L. HU, Y.J. SUZUKI, S. KHWAJA, A. SAFADI, P.A. MOTCHNIK, L. PACKER & B. HALLIWELL. 1992. Modification of plasma proteins by cigarette smoke as measured by protein carbonyl formation. Biochem. J. **286:** 607–611.
55. BUTTERWORTH, M., S.S. LAU & T.J. MONKS. 1998. 2-Hydroxy-4-glutathion-S-yl-17beta-estradiol and 2-hydroxy-1-glutathion-S-yl-17beta-estradiol produce oxidative stress and renal toxicity in an animal model of 17beta-estradiol-mediated nephrocarcinogenicity. Carcinogenesis. **19:** 133–139.
56. WINTER, M.L. & J.G. LIEHR. 1991. Free radical-induced carbonyl content in protein of estrogen-treated hamsters assayed by sodium boro[^{3}H]hydride reduction. J. Biol. Chem. **266:** 14446–14450.
57. GLADSTONE, I.M., JR. & R.L. LEVINE. 1994. Oxidation of proteins in neonatal lungs. Pediatrics **93:** 764–768.
58. WITT, E.H., A.Z. REZNICK, C.A. VIGUIE, P.E. STARKE-REED & L. PACKER. 1992. Exercise, oxidative damage, and effects of antioxidant manipulation. J. Nutr. **122:** 766–773.
59. MICKEL, H.S., C.N. OLIVER & P.E. STARKE-REED. 1990. Protein oxidation and myelinolysis occur in brain following rapid correction of hyponatremia. Biochem. Biophys. Res. Commun. **172:** 92–97.
60. STAFFORD, R.E., I.T. MAK, J.H. KRAMER & W.W. WEGLICKI. 1993. Protein oxidation in magnesium deficient rat brains and kidney. Biochem. Biophys.l Res. Commun. **196:** 596–600.
61. OLIVER, C.N. 1987. Inactivation of enzymes and oxidative modification of proteins by stimulated neutrophils. Archives of Biochemistry and Biophysics **253:** 62–72.
62. KARSEK-STAPLES, J.A. & R.O. WEBSTER. 1993. Ceruloplasmin inhibits carbonyl formation in endogenous proteins. Free Rad.l Biol. Med. **14:** 115–125.
63. CIOLINO, H.P. & R.L. LEVINE. 1997. Modification of proteins in endothelial cell death during oxidative stress. Free Rad. Biol. Med. **22:** 1277–1282.
64. HENSLEY, K., J.M. CARNEY, M.P. MATTSON, M. AKSENOVA, M. HARRIS, J.F. WU, R.A. FLOYD & D.A. BUTTERFIELD. 1994. A model for beta-amyloid aggregation and neurotoxicity based on free radical generation by the peptide: relevance to Alzheimer disease. Proc. Natl. Acad. Sci. USA **91:** 3270–3274.
65. HENSLEY, K., N. HALL, R. SUBRAMANIAM, P. COLE, M. HARRIS, M. AKSENOV, M. AKSENOVA, S.P. GABBITA, J.F. WU & J.M. CARNEY. 1995. Brain regional correspondence between Alzheimer's disease histopathology and biomarkers of protein oxidation. J. Neurochem. **65:** 2146–2156.
66. SMITH, M.A., L.M. SAYRE, V.E. ANDERSON, P.L. HARRIS, M.F. BEAL, N. KOWALL & G. PERRY. 1998. Cytochemical demonstration of oxidative damage in Alzheimer disease by immunochemical enhancement of the carbonyl reaction with 2,4- dinitrophenylhydrazine. J. Histochem. Cytochem. **46:** 731–735.
67. ALAM, Z.I., S.E. DANIEL, A.J. LEES, D.C. MARSDEN, P. JENNER & B. HALLIWELL. 1997. A generalised increase in protein carbonyls in the brain in Parkinson's but not incidental Lewy body disease. J. Neurochem. **69:** 1326–1329.
68. FLOOR, E. & M.G. WETZEL. 1998. Increased protein oxidation in human substantia nigra pars compacta in comparison with basal ganglia and prefrontal cortex measured with an improved dinitrophenylhydrazine assay. J. Neurochem. **70:** 268–275.
69. JONES, R.H. & J.S. HOTHERSALL. 1993. The effect of diabetes and dietary ascorbate supplementation on the oxidative modification of rat lens beta L crystallin. Biochem. Med. Metab. Biol. **50:** 197–209.

70. UCHIDA, K., M. KANEMATSU, K. SAKAI, T. MATSUDA, N. HATTORI, Y. MIZUNO, D. SUZUKI, T. MIYATA, N. NOGUCHI, E. NIKI & T. OSAWA. 1998. Protein-bound acrolein: potential markers for oxidative stress. Proc. Natl. Acad. Sci. USA **95**: 4882–4887.
71. CHAPMAN, M.L., B.R. RUBIN & R.W. GRACY. 1989. Increased carbonyl content of proteins in synovial fluid from patients with rhematoid arthritis. J. Rheumatol. **16**: 15–18.
72. MURPHY, M.E. & J.P. KEHRER. 1989. Oxidation state of tissue thiol groups and content of protein carbonyl groups in chickens with inherited muscular dystrophy. Biochemical Journal **260**: 359–364.
73. GARLAND, D., P. RUSSELL & J.S. ZIGLER. 1988. Oxidative modification of lens proteins. M.G. Simic, K.S. Taylor, J.F. Ward & V. von Sontag, Eds. :347–353. Oxygen radicals in biology and medicine. Plenum. New York.
74. UCHIDA, K., A. FUKUDA, S. KAWAKISHI, H. HIAI & S. TOYOKUNI. 1995. A renal carcinogen ferric nitriloacetate mediates a temporary accumulation of aldehyde-modified proteins within cytosolic compartment of rat kidney. Arch. Biochem. Biophys. **317**: 405–411.
75. KONG, S.K., M.B. YIM, E.R. STADTMAN & P.B. CHOCK. 1996. Peroxynitrite disables the tyrosine phosphorylation regulatory mechanism: Lymphocyte-specific tyrosine kinase fails to phosphorylate nitrated cdc2(6–20)NH_2 peptide. Proc. Natl. Acad. Sci. USA **93**: 3377–3382.
76. GRATTAGLIANO, I., G. VENDEMIALE, C. SABBA, P. BUONAMICO & E. ALTOMARE. 1996. Oxidation of circulating proteins in alcoholics: role of acetaldehyde and xanthine oxidase. J. Hepatol. **25**: 28–36.
77. VENDEMIALE, G., I. GRATTAGLIANO, A. SIGNORILE & E. ALTOMARE. 1998. Ethanol-induced changes of intracellular thiol compartmentation and protein redox status in the rat liver: effect of tauroursodeoxycholate. J. Hepatol. **28**: 46–53.
78. SUNDARI, P.N. & B. RAMAKRISHNA. 1997. Does oxidative protein damage play a role in the pathogenesis of carbon tetrachloride-induced liver injury in the rat? Biochim. Biophys. Acta **1362**: 169–176.
79. COMPORTI, M. 1998. Lipid peroxidation and biogenic aldehydes: from the identification of 4-hydroxynonenal to further achievements in biopathology. Free Rad. Res. **28**: 623–635.
80. BOWLING, A.C., J.B. SCHULZ, R.H. BROWN, JR. & M.F. BEAL. 1993. Superoxide dismutase activity, oxidative damage, and mitochondrial energy metabolism in familial and sporadic amyotrophic lateral sclerosis. J. Neurochem. **61**: 2322–2325.
81. OLIVER, C.N., B.W. AHN, E.J. MOERMAN, S. GOLDSTEIN & E.R. STADTMAN. 1987. Age-related changes in oxidized proteins. J. Biol. Chem. **262**: 5488–5491.
82. SMITH, C.D., J.M. CARNEY, P.E. STARKE-REED, C.N. OLIVER, E.R. STADTMAN, R.A. FLOYD & W.R. MARKESBERY. 1991. Excess brain protein oxidation and enzyme dysfunction in normal aging and Alzheimer disease. Proc. Natl. Acad. Sci. USA **88**: 10540–10543.
83. BALAZS, L. & M. LEON. 1994. Evidence of an oxidative challenge in the Alzheimer's brain. Neurochem. Res. **19**: 1131–1137.
84. VITEK, M.P., K. BHATTACHARYA, J.M. GLENDENING, E. STOPA, H. VLASSARA, R. BUCALA, K. MANOGUE & A. CERAMI. 1994. Advanced glycation end products contribute to amyloidosis in Alzheimer disease. Proc. Natl. Acad. Sci. USA **91**: 4766–4770.
85. MONTINE, K.S., E. REICH, M.D. NEELY, K.R. SIDELL, S.J. OLSON, W.R. MARKESBERY & T.J. MONTINE. 1998. Distribution of reducible 4-hydroxynonenal adduct immunoreactivity in Alzheimer disease is associated with APOE genotype. J. Neuropathol. Exp. Neurol. **57**: 415–425.
86. ANDO, Y., T. BRANNSTROM, K. UCHIDA, N. NYHLIN, B. NASMAN, O. SUHR, T. YAMASHITA, T. OLSSON, M. EL SALHY, M. UCHINO & M. ANDO. 1998. Histochemical detection of 4-hydroxynonenal protein in Alzheimer amyloid. J. Neurol. Sci. **156**: 172–176.

87. MARKESBERY, W.R. & M.A. LOVELL. 1998. Four-hydroxynonenal, a product of lipid peroxidation, is increased in the brain in Alzheimer's disease. Neurobiol. Aging **19**: 33–36.
88. SAYRE, L.M., D.A. ZELASKO, P.L. HARRIS, G. PERRY, R.G. SALOMON & M.A. SMITH. 1997. 4-Hydroxynonenal-derived advanced lipid peroxidation end products are increased in Alzheimer's disease. J. Neurochem. **68**: 2092–2097.
89. GOOD, P.F., P. WERNER, A. HSU, C.W. OLANOW & D.P. PERL. 1996. Evidence of neuronal oxidative damage in Alzheimer's disease. Am. J. Pathol. **149**: 21–28.
90. SMITH, M.A., H.P. RICHEY, L.M. SAYRE, J.S. BECKMAN & G. PERRY. 1997. Widespread peroxynitrite-mediated damage in Alzheimer's disease. J. Neurosci. **17**: 2653–2657.
91. SMITH, M. A., P.L. HARRIS, L.M. SAYRE & G. PERRY. 1997. Iron accumulation in Alzheimer disease is a source of redox-generated free radicals. Proc. Natl. Acad. Sci. USA **94**: 9866–9868.
92. LUND, A.L., J.B. SMITH & D.L. SMITH. 1996. Modifications of the water-insoluble human lens alpha-crystallins. Exp. Eye Res. **63**: 661–672.
93. GARNER, M.H. & A. SPECTOR. 1980. Selective oxidation of cysteine and methionine in normal and senile cataractous lenses. Proc. Natl. Acad. Sci. USA **77**: 1274–1277.
94. ALAM, Z.I., S.E. DANIEL, A.J. LEES, D.C. MARSDEN, P. JENNER & B. HALLIWELL. 1997. A generalised increase in protein carbonyls in the brain in Parkinson's but not incidental Lewy body disease. J. Neurochem. **69**: 1326–1329.
95. YORITAKA, A., N. HATTORI, K. UCHIDA, M. TANAKA, E.R. STADTMAN & Y. MIZUNO. 1996. Immunochemical detection of 4-hydroxynonenal protein adducts in Parkinson's disease. Proc. Natl. Acad. Sci. USA **93**:.
96. LAVOIE, M.J. & T.G. HASTINGS. 1999. Dopamine quinone formation and protein modification associated with the striatal neurotoxicity of methamphetamine: evidence against a role for extracellular dopamine. J. Neurosci. **19**: 1484–1491.
97. DREYFUS, J.C., A. KAHN & F. SCHAPIRA. 1978. Post translational modifications of enzymes. Curr. Topics Cell Reg. **14**: 243–297.
98. ROTHSTEIN, M. 1977. Recent developments in the age-related alteration of enzymes: a review. Mech. Ageing Dev. **6**: 241–257.
99. GAFNI, A. 1981. Purification and comparative study of glyceraldehyde-3-phosphate dehydrogenase from the muscles of young and old rats. Biochemistry **20**: 6035–6040.
100. ROTHSTEIN, M. 1982. Biochemical Approaches to Aging. New York. Academic Press.
101. DE LA CRUZ, C.P., E. REVILLA, J.L. VENERO, A. AYALA, J. CANO & A. MACHADO. 1996. Oxidative inactivation of tyrosine hydroxylase in substantia nigra of aged rat. Free Rad. Biol. Med. **20**: 53–61.
102. GORDILLO, E., A. AYALA, M.-F. LOBATO, J. BAUTISTA & A. MACHADO. 1988. Possible involvement of histidine residues in the loss of enzymatic activity of rat liver malic enzyme during aging. J. Biol.l Chem. **263**: 8053–8057.
103. MORDENTE, A., G.E. MARTORANA, G.A. MIGGIANO, E. MEUCCI, S.A. SANTINI & A. CASTELLI. 1988. Mixed function oxidation and enzymes: kinetic and structural properties of an oxidatively modified alkaline phosphatase. Arch Biochem. Biophys. **264**: 502–509.
104. MUSCI, G., M. C.B. DI PATTI, U. FAGIOLO & L. CALABRESE. 1993. Age-related changes in human ceruloplasmin. Evidence for oxidative modifications. J. Biol. Chem. **268**: 13388–13395.
105. RIVETT, A.J. & R.L. LEVINE. 1990. Metal-catalyzed oxidation of Escherichia coli glutamine synthetase: Correlation of structural and functional changes. Arch. Biochem. Biophys. **278**: 26–34.
106. TABORSKY, G. 1973. Oxidative modification of proteins in the presence of ferrous iron and air. Effect of ionic constituents of the reaction medium on the nature of the oxidation products. Biochemistry **12**: 1341–1348.
107. TAKAHASHI, R. & S. GOTO. 1990. Alteration of aminoacyl-tRNA synthetase with age: Heat labilization of the enzyme by oxidative damage. Arch. Biochem. Biophys. **277**: 228–233.

108. ZHOU, J.Q. & A. GAFNI. 1991. Exposure of rat muscle phosphoglycerate kinase to a nonenzymatic MFO system generates the old form of enzyme. J. Gerontol. **46:** B217–B221.
109. STARKE, P.E., C.N. OLIVER & E.R. STADTMAN. 1987. Modification of hepatic proteins in rats exposed to high oxygen concentration. FASEB J. **1:** 36–39.
110. SOHAL, R.S., S. AGARWAL & B.H. SOHAL. 1995. Oxidative stress and aging in the Mongolian gerbil (Meriones unguiculatus). Mech. Ageing Dev. **81:** 15–25.
111. YOUNGMAN, L.D., J.Y. PARK & B.N. AMES. 1992. Protein oxidation associated with aging is reduced by dietary restriction of protein or calories. Proc. Natl. Acad. Sci. USA **89:** 9112–9116.
112. AGARWAL, S. & R.S. SOHAL. 1993. Relationship between aging and susceptibility to protein oxidative damage. Biochem. Biophys. Res. Commun. **194:** 1203–1206.
113. BOHLEY, P. & S. RIEMANN. 1977. Intracellular protein catabolism. IX. Hydrophobicity of substrate proteins is a molecular basis of selectivity. Acta Biol. Med. Ger. **36:** 1823–1827.
114. SEGAL, H.L., D.M. ROTHSTEIN & J.R. WINKLER. 1976. A correlation between turnover rates and lipophilic affinities of soluble rat liver proteins. Biochem. Biophys. Res. Commun. **73:** 79–84.
115. GARLAND, D. 1990. Role of site-specific, metal-catalyzed oxidation in lens-aging and cataract: a hypothesis. Exp. Eye Res. **50:** 677–682.
116. MECOCCI, P., G. FANO, S. FULLE, U. MACGARVEY, L. SHINOBU, M.C. POLIDORI, A. CHERUBINI, J. VECCHIET, U. SENIN & M.F. BEAL. 1999. Age-dependent increases in oxidative damage to DNA, lipids, and proteins in human skeletal muscle. Free Rad. Biol. Med. **26:** 303–308.
117. MUDD, J.B., R. LEAVITT, A. ONGUN & T.T. MCMANUS. 1969. Reaction of ozone with amino acids and proteins. Atmos. Environ. **3:** 669–682.
118. BERLETT, B.S., R.L. LEVINE & E.R. STADTMAN. 1996. Comparison of the effects of ozone on the modification of amino acid residues in glutamine synthetase and bovine serum albumin. J. Biol. Chem. **271:** 4177–4182.
119. KIDO, K. & B. KASSELL. 1975. Oxidation of methionine residues of porcine and ovine pepsins. Biochemistry **14:** 631–635.
120. PENNER, M.H., R.B. YAMASAKI, D.T. OSUGA, D.R. BABIN, C.F. MEARES & R.E. FEENEY. 1983. Comparative oxidations of tyrosines and methionines in transferrins: human serum transferrin, human lactotransferrin, and chicken ovotransferrin. Arch. Biochem. Biophys. **225:** 740–747.
121. CHAO, C.C., Y.S. MA & E.R. STADTMAN. 1997. Modification of protein surface hydrophobicity and methionine oxidation by oxidative systems. Proc. Natl. Acad. Sci. USA **94:** 2969–2974.
122. PRYOR, W.A., X. JIN & G.L. SQUADRITO. 1994. One- and two-electron oxidations of methionine by peroxynitrite. Proc. Natl. Acad. Sci. USA **91:** 11173–11177.
123. MORENO, J.J. & W.A. PRYOR. 1992. Inactivation of alpha 1-proteinase inhibitor by peroxynitrite. Chem. Res. Toxicol. **5:** 425–431.
124. SCHONEICH, C., F. ZHAO, G.S. WILSON & R.T. BORCHARDT. 1993. Iron-thiolate induced oxidation of methionine to methionine sulfoxide in small model peptides. Intramolecular catalysis by histidine. Biochem. Biophys. Acta **1158:** 307–322.
125. SCHONEICH, C. & J. YANG. 1996. Oxidation of methionine peptides by Fenton systems: the importance of peptide sequence, neighboring groups, and EDTA. J. Chem. Soc. Perkin Trans. **2:** 915–923.
126. MOSKOVITZ, J., E. FLESCHER, B.S. BERLETT, J. AZARE, J.M. POSTON & E.R. STADTMAN. 1998. Overexpression of peptide-methionine sulfoxide reductase in Saccharomyces cerevisiae and human T cells provides them with high resistance to oxidative stress. Proc. Natl.␣cad. Sci. USA **95:** 14071–14075.
127. STADTMAN, E.R., P.Z. SMYRNIOTIS, J.N. DAVIS & M.E. WITTENBERGER. 1979. Enzymic procedures for determining the average state of adenylylation of *Escherichia coli* glutamine synthetase. Anal. Biochem. **95:** 275–285.
128. LYMAR, S.V. & J.K. HURST. 1995. Rapid reaction between peroxynitrite ion and carbon dioxide: Implication for biological activity. J. Am. Chem. Soc. **117:** 8867–8868.

129. LYMAR, S.V. & J.K. HURST. 1996. Carbon dioxide: physiological catalyst for peroxynitrite-mediated cellular damage or cellular protectant? Chem. Res. Toxicol. **9:** 845–850.

Mechanisms of Cell Death Governed by the Balance between Nitrosative and Oxidative Stress

MICHAEL GRAHAM ESPEY,[a,b] KATRINA M. MIRANDA,[a] MARTIN FEELISCH,[c] JON FUKUTO,[d] MATHEW B. GRISHAM,[e] MICHAEL P. VITEK,[f] AND DAVID A. WINK[a]

[a]*Radiation Biology Branch, National Cancer Institute, National Institutes of Health, Bethesda, Maryland 20892, USA*

[c]*Wolfson Institute for Biomedical Research, University College London, 140 Tottenham Court Road, London, W1P 9LN, UK*

[d]*Department of Pharmacology, University of California, Los Angeles, California 90095, USA*

[e]*Department of Molecular and Cellular Physiology, Louisiana State University Medical Center, Shreveport, Louisiana 71130, USA*

[f]*Division of Neurology, Duke University Medical Center, Durham, North Carolina 27710, USA*

ABSTRACT: Many cellular functions in physiology are regulated by the direct interaction of NO with target biomolecules. In many pathophysiologic and toxicologic mechanisms, NO first reacts with oxygen, superoxide or other nitrogen oxides to subsequently elicit indirect effects. The balance between nitrosative stress and oxidative stress within a specific biological compartment can determine whether the presence of NO will be ultimately deleterious or beneficial. Nitrosative stress can be defined primarily through reactions mediated by N_2O_3, a reactive nitrogen oxide species generated by high fluxes of NO in an aerobic environment. In contrast, oxidative stress is mediated primarily by superoxide and peroxides. In addition to reactive oxygen species, several reactive nitrogen oxide species such as peroxynitrite, nitroxyl, and nitrogen dioxide can also impose oxidative stress to a cell. We here describe how the mechanisms of cell death are interwoven in the balance between the different chemical intermediates involved in nitrosative and oxidative stress.

INTRODUCTION

Nitric oxide (NO) is an endogenous mediator involved in the regulation of many physiological functions and is involved in a variety of possible pathophysiological processes.[1] In contrast to conventional signaling ligands such as cytokines, hormones, neurotransmitters which function through specific interactions with

---

[b]Address for correspondence: Michael Graham Espey, Ph.D., Radiation Biology Branch, National Institutes of Health/National Cancer Institute, Building 10, Room B3-B69, Bethesda, MD 20892. Voice: 301-496-7511; fax: 301-480-2238.
  e-mail: SP@nih.gov

receptors, the ultimate action of NO in biological systems is dictated by its chemistry. Moreover, NO and the reactive nitrogen oxide species (RNOS) derived from NO can interact with a wide variety of biochemical targets to elicit diverse biological effects.

The *chemical biology of NO* is a framework that categorizes the relevant chemical reactions of NO and predicts where they may occur *in vivo*.[2–4] In this scheme, reactions are divided into direct effects and indirect effects (FIG. 1). Direct effects are those chemical reactions in which NO reacts directly with a biological target molecule. The reaction of NO generated by vascular endothelial cells with the heme moiety of soluble guanylyl cyclase in adjacent smooth muscle cells is an example of such a direct effect.[5–8] The binding of NO to the ferrous heme moiety of guanylate cyclase activates the enzyme and leads to an increased conversion of GTP to cGMP resulting in vasorelaxation and subsequent dilation of the blood vessel. For this reaction, NO is synthesized at a low concentration (<1 µM) by the endothelial (constitutive) isoform of nitric oxide synthase (eNOS) and the reaction of NO with heme moieties of guanylate cyclase (iron-nitrosyl) is extremely fast ($k_{on} > 10^9$ M^{-1} s^{-1}).[9] This combination of a low flux of NO and fast reaction kinetics typifies the direct effects of NO. Under physiological conditions, the formation of iron-nitrosyl complexes is likely to be the predominant reaction of NO in biologic systems.

Indirect effects of NO occur during pathophysiological and toxicological events. The initial reaction of NO in an indirect effect is with molecules such as superoxide (O_2^-) or molecular oxygen (O_2) that form intermediate RNOS *prior* to the reaction with a final target.[2] These RNOS can alter a wide range of biological macromolecules such as proteins, lipids and DNA and are thought to play a pivotal role in NO-mediated cell death. In general, NO chemistry is governed by indirect effects only when the local concentration of NO is high (> 1 µM) and sustained for prolonged periods of time.[2] These attributes characterize NO production catalyzed by the inducible isoenzyme, iNOS, suggesting that indirect effects are most likely to occur in the vicinity of activated macrophages or other cells expressing iNOS after immunostimulation and in disease states.

The scheme of direct and indirect effects of NO is analogous to a river. Direct effects (e.g., the normal current enabling the transformation of physical forces into electrical energy or to transport wood downstream) take place on average days. Significant indirect effects (e.g., damage of the adjacent landscape) occur when there is a flood. Either oxidative or nitrosative stress occurs when the levy breaks. The

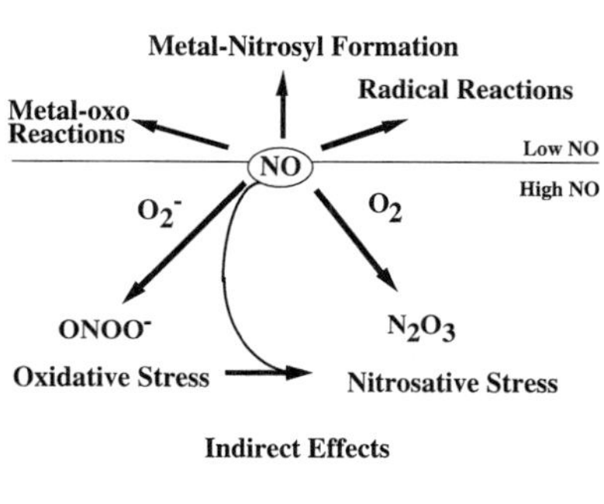

FIGURE 1. Chemical biology of NO.

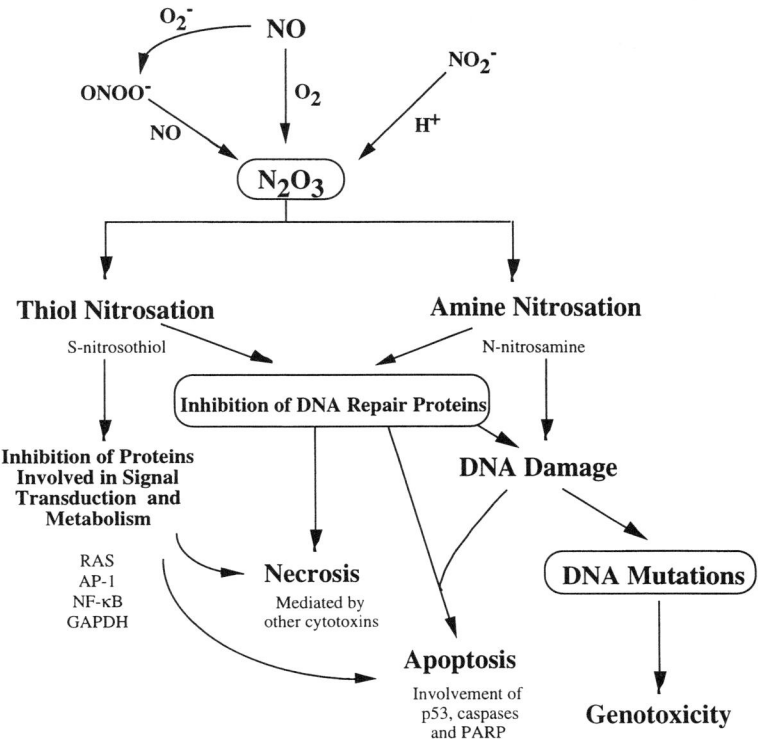

FIGURE 2. Nitrosative stress and cell death.

indirect effects of NO can be subdivided into either of two basic chemical stresses: nitrosative or oxidative (FIG. 2).[2] The chemical origins of the reactant species involved in both nitrosative and oxidative stress are paramount to understanding their potential contributions to pathophysiological and toxicological mechanisms. Using the framework of *the chemical biology of NO*, the salient inorganic reactions and biological conditions that govern whether NO elicits or protects against nitrosative and oxidative stress are discussed.

NITROSATIVE STRESS

The chemistry of nitrosation in biological systems has been studied for several decades. Consistent with the concept of nitrosation reactions as indirect effects, the NO radical (nitrogen monoxide or NO·) is usually first oxidized to generate species which can serve as a source of nitrosonium ion (NO^+), although "free" NO^+ is never specifically formed. NO^+ is exceedingly reactive in biological systems and may truly only exist in the gas phase of interstellar space. The nitrosation of nucleophiles such as thiols, alcohols or amines can occur via the generation of nitrosonium-donor species.[10–12] There are a number of NO^+ donors of the form NO-X where X can be represented by a number of species such as halides, phosphate or nitrite.[12] The most relevant NO^+ donor in biological systems is N_2O_3.[2] All other NO-X species can be derived from N_2O_3.[13] Although N_2O_3 can be formed by either acidified nitrite (HNO_2) or NO/O_2^- reactions (discussed below), the aerobic oxidation pathway may predominate *in vivo*. Nitrosation should not be confused with nitration, the oxidation of an aromatic ring by a NO_2^+ moiety (see below).

The rate equation for the NO/O_2 reaction has second order dependency on NO and a first order dependency on O_2, which indicates that the lifetime of NO is inversely proportional to its concentration.[14] Therefore, at low concentrations, the lifetime of NO is on the order of minutes, while at higher concentrations it can be as short as a few seconds indicating that the rate of this reaction depends highly on the concentration of NO. The second order dependency of the NO/O_2 reaction has several implications for biological systems. First, only cells capable of generating a high flux of NO (e.g., iNOS-catalyzed synthesis) may have the potential for causing nitrosative stress. Second, the solubility of NO and O_2 are enhanced in hydrophobic environments relative to aqueous solutions. Partitioning of these reactants into a lipid bilayer, for example, accelerates the rate of the NO/O_2 reaction 300 times.[15] These data suggest that the formation of N_2O_3 during nitrosative stress in biological systems may take place predominantly in membranes and other hydrophobic compartments.

TOXICOLOGY OF N_2O_3

Nitrosation of biological molecules can have toxic consequences (FIG. 3). Nitrosation of secondary amines by HNO_2 can occur in the acidic environment of the

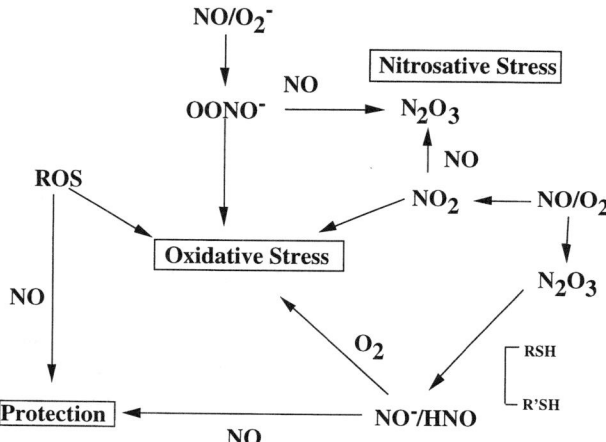

FIGURE 3. The chemistry involved in the balance between oxidative and nitrosative stress.

stomach and the formation of nitrosamines in this manner was proposed to be an etiologic component in some forms of cancer. An important initial observation in the chemical toxicology of NO was that humans produce elevated levels of nitrate upon infection.[16] Subsequent investigations showed that macrophages form nitrosamines when stimulated to express iNOS[17-19] in the presence of secondary amines.[20] The formation of nitrosamines was demonstrated in vivo using the woodchuck model of chronic hepatitis infection.[21,22] Using 2,3-diaminonaphthalene as a probe for nitrosation,[23] we have found that selective signal transduction pathways are involved in the generation of nitrosative stress by immune stimulated macrophages (personal observation).

NONOates are compounds that spontaneously release NO at physiological pH.[24] These donors permitted the evaluation of potential mechanisms leading to nitrosative stress and its associated toxicity. NONOate-mediated generation of RNOS intermediates appear to have the strongest propensity to react with thiol moieties[10,11,25] to yield S-nitrosothiols, one example for the consequences of nitrosative stress. Under conditions of intracellular glutathione depletion, the toxicity of NONOates was greatly augmented.[26] This was dependent on the presence of molecular oxygen, consistent with N_2O_3 as the effector molecule. Although reduced glutathione is often thought of as a scavenger of ROS, these data point to glutathione playing a major role in abating the toxicity of N_2O_3 during nitrosative stress.

The formation of S-nitrosothiol adducts under conditions of nitrosative stress can either promote or prevent toxicity depending on the nature of the biomolecular targets involved.[2] To illustrate this point, the thiol-rich protein metallothionein provides protection against nitrosative stress by forming iron-dinitrosyl thiolates[27] that apparently abate the cellular toxicity of N_2O_3. However, metallothionein also functions to scavenge toxic metals, such as cadmium. A price paid by S-nitrosation of metallothionein is the ejection of sequestered cadmium thereby markedly enhancing metal-mediated toxicity.[28]

The activity of a variety of enzymes are modulated by S-nitrosation. An S-nitrosothiol adduct has been suggested to inhibit glyceraldehyde phosphate dehydrogenase, a key enzyme in glycolysis, by promoting its ADP ribosylation.[29] NO inhibits DNA alkyltransferase activity both in vitro and in vivo via an S-nitrosation mechanism.[30] Formation of an S-NO complex impedes the transfer of an alkyl group from the O^6- position of guanine to the thiol residue within the protein resulting in potentiation of toxicity to alkylating agents.[30] Nitrosation of residues in zinc finger proteins that coordinate metal binding results in the disruption of the structural integrity of these motifs with subsequent inhibition of protein activity.[25,31] These findings suggested that chemotherapeutic modalities in cancer treatment may be enhanced by the induction of nitrosative stress, a concept we are currently exploring.

OXIDATIVE STRESS

The term oxidative stress has been used to describe conditions where oxygen radical species mediate deleterious effects *in vivo*. Oxidation chemistry occurs continuously as a consequence of aerobic metabolism under normal physiologic conditions. For example, oxygen metabolism (respiration), catecholamine degradation (monoamine oxidase) or monooxygenase turnover (cytochrome P450) are all involved in the basal oxidative metabolic activity of cells and are all potential sources of oxygen-derived oxidants. Thus, a variety of endogenous protective systems exist for these O_2-derived species (e.g., superoxide dismutase (SOD), catalase, glutathione peroxidase, glutathione, etc.). When oxidant species reach fluxes that overwhelm these defense systems during either a pathophysiologic or toxicologic event, chemical alterations of biological molecules may occur. This is the point at which the system is under oxidative stress. Reactive oxygen species (ROS) are the chemical intermediates that initiate oxidative stress. ROS can be derived from the reduced products of molecular oxygen (e.g., O_2^-, hydrogen peroxide (H_2O_2) and from RNOS (e.g., peroxynitrite, see below). Fenton-type reactions can occur under conditions where reduced oxygen species (i.e., O_2^-, H_2O_2) exist in the presence of redox active metals such as iron or copper. The reaction of peroxides with ferrous (Fe^{2+}) ions to form powerful oxidizing (metal-oxo) species or hydroxyl radicals typifies ROS Fenton chemistry. The metal (iron or copper) in this reaction can be catalytic if one considers that O_2^- can serve as a reducing agent for ferric ions (Fe^{3+}), converting it back to the catalytically active Fe^{2+} form (Haber-Weiss chemistry). The same toxic ROS are also generated with ionizing radiation in an aqueous environment.[32] Taken together, these reactions form a reduction-oxidation cycle that can mediate cytoxicity.

The RNOS nitroxyl anion (NO^-), nitrogen dioxide (NO_2) and peroxynitrite ($ONOO^-$) can also generate oxidative stress through substantially different chemical mechanisms. For instance, $ONOO^-$ can cause single strand breaks in supercoiled DNA[33–35] while NO^- anion generates double stranded DNA breaks.[36] The formation of NO^- may occur under a variety of biological conditions. A primary source of NO^- may be through the decomposition of S-nitrosothiols[37] where nucleophilic attack of RS-NO by reduced thiols can result in disulfide formation and NO^- release. Arnelle et al. showed that the decomposition of DTT-SNO can model this NO^- release mechanism.[38] Other reports suggest that NO^- can be formed from the decomposition of iron dinitrosyls.[39] An intriguing possibility is that NO^- is

TABLE 1. Relative concentration of redox species required to kill 99% of fibroblasts (V79)

Agent	Redox Species	Concentration of Agent (mM)
H_2O_2	metallo-oxo/OH·	0.8
t-butylhydroperoxide	lipid peroxidation	3.0
DEA/NO	NO	>>5
DEA/NO + BSO	NO/glutathione depletion	1.5
DEA/NO + XO	$NO/H_2O_2/O_2^-/ONOO^-$	>>1
SIN-1	$ONOO^-/H_2O_2$	>>1 (>5)[a]
AS	$NO^- + O_2$	2.0
AS + BSO	NO^-/glutathione depletion	0.5
AS/hypoxia	NO^-/HNO	>>4

[a]Studies have shown that SIN-1 in some other mammalian cells showed little toxicity.

produced directly from NOS[40,41] by the oxidation of the intermediate N[G]-hydroxy-L-arginine.[42]

These data raise the possibility that NO⁻ may represent a crucial portion of NO's chemical biology. Angeli's salt, sodium trioxodinitrate ($Na_2N_2O_3$), has been employed as a synthetic NO⁻ donor.[43] The toxicity of NO⁻ released by Angeli's salt was comparable to that of either H_2O_2 or alkyhydroperoxide[45,46] and was about 100-fold more cytotoxic than ONOO⁻ released from the compound SIN-1,[36,44,45] or NO released from DEA/NO[36] (TABLE 1).

The toxicity of NO⁻ may depend on the oxygenation state of the cell as cytotoxicity was greatly diminished under hypoxic conditions, raising the possibility of the existence of a NO⁻/O_2 intermediate. Moreover, the elimination of low molecular weight transition metal complexes in the presence of O_2 suggested NO⁻-mediated toxicity was not due to ROS produced by Fenton-type reactions.[36] These data are consistent with the involvement of a NO⁻/O_2 intermediate that is a two-electron oxidant and a hydroxylating agent. Unlike either ONOO⁻ or N_2O_3, this intermediate does not induce nitration, nitrosation or one-electron oxidation reactions. A possible intermediate involved in Angeli's salt–mediated cytotoxicity may be $(HO)_2N\text{-}OO^-$, which would certainly be a unique species among the chemical oxidants. Angeli's salt can form H_2O_2 dependent on the buffer system (e.g., HEPES). Therefore, it is important to be aware that Angeli's salt may produce a powerful combination of oxidants consisting of both RNOS and ROS dependent on experimental conditions (authors' personal observation).

Nitrogen dioxide is a powerful one-electron oxidant that can initiate lipid peroxidation,[47] thiol oxidation[48] and nitration of aromatic amino acids.[49] NO_2 is formed from the NO/O_2 reaction in the gas phase and in hydrophobic environments such as cellular membranes.[15] This reaction is unlikely to occur in an aqueous media such as the cytoplasm due to kinetic considerations. The reaction between ONOO⁻ and NO may also form NO_2 or N_2O_3 (discussed below,[50] as does acidified nitrite (HNO_2).[12] Gaseous NO_2 is a constituent of air pollution that may exacerbate toxicological complications of respiratory infection.[51]

Peroxynitrite, an important RNOS in the fundamental understanding of NO's inorganic chemical behavior in biology, is formed at near diffusion control in the reaction between NO and O_2^- (Eq. 1-3).[52–54]

$$O_2^- + NO \rightarrow ONOO^- \qquad (1)$$

$$ONOO^- + H^+ \rightarrow cis\text{ }ONOOH \qquad (2)$$

$$cis\text{ }ONOOH \rightarrow trans\text{ }ONOOH \rightarrow trans\text{ }ONOOH^* \qquad (3)$$

In neutral solution, ONOO⁻ is a powerful oxidant that interacts with a wide range of targets to cause tyrosine nitration, thiol oxidation, lipid peroxidation, DNA single strand breaks, and guanosine nitration/oxidation. Peroxynitrite is in equilibrium with peroxynitrous acid, which in its excited state of *trans* HOONO* may be the primary oxidant that mediates cytotoxicity.[55] In the absence of adequate substrate, peroxynitrous acid can rearrange to form nitrate (NO_3^-), an isomerization that can be considered a detoxification pathway for ONOO⁻. Nitration occurs when the concentration of ONOO⁻ (or *trans* HOONO*) and aromatic ring reactants (e.g., tyrosine, guanosine, tryptophan, catecholamines) reaches a level high enough to exceed the detoxification rate.

The rate constant for the NO/O_2^- reaction leading to $ONOO^-$ formation is extremely fast ($k > 10^9$ M^{-1} s^{-1}).[52] For this second-order reaction, the concentration of each reactant and the relative pseudo-first-order rate constants are as important as the overall rate constant (for summary of evaluating RNOS see ref. 56). The estimated intracellular concentration of NO and O_2^- under normal conditions are 1 pM[57] and 0.1–1 μM,[58] respectively. Therefore, the concentration of NO determines the pseudo-first-order rate constant for the NO/O_2- reaction. However, the limiting step in the reaction between these two radicals is the production of O_2^-. The rate constant for the consumption of O_2^- by the isoforms of SOD is comparable to that of the NO/O_2^- reaction thereby limiting formation of $ONOO^-$ in their presence. The intracellular concentration of cytoplasmic copper/zinc SOD is in the range of 4 to 10 μM[59] whereas mitochondrial manganese SOD is as high as 50 μM.[59] Other reaction partners for O_2^- such as aconitase ($k = 3 \times 10^7$ M^{-1} s^{-1})[60] and ferricytochrome c ($k = 5 \times 10^6$ M^{-1} s^{-1})[61] are found in the mitochondria, a site of putative $ONOO^-$ toxicity. These proteins could sequester O_2^- reducing its availability for subsequent reactions with NO. Despite the high rate constant of the NO/O_2^- reaction, formation of $ONOO^-$ may become significant only in cases where competing O_2^- scavenger systems are severely compromised (e.g., oxidative stress).

The toxicology of oxidative stress mediated by $ONOO^-$ has been studied using either $ONOO^-$ released from synthetic donor compounds or the concomitant flux of O_2^- and NO from hypoxanthine/xanthine oxidase (XO) and NONOates, respectively. The bolus applications of millimolar levels of synthetic peroxynitrite was shown to be cytotoxic.[62] However, the cytotoxicity resulting from O_2^- or H_2O_2 generated by XO was markedly diminished by the addition of NONOate.[63] The explanation for these data partly lies in the concept that once $ONOO^-$ is formed *in vivo*, its lifetime may be rather limited. Kinetic analysis suggests that the reaction of CO_2 with $ONOO^-$ may be the dominant pathway *in vivo*. The $CO_2/ONOO^-$ intermediate is an RNOS that can oxidize and nitrate macromolecules a bit more efficiently than the corresponding $ONOO^-$ species (authors' personal observation). Although the NO/O_2^- reaction is most often discussed in the context of oxidative stress, high fluxes of NO can convert $ONOO^-$ into a nitrosative chemical profile. The reaction of $ONOO^-$ with additional NO generates NO_2 that (as discussed above) can combine with NO to form the nitrosating species N_2O_3 (Eq. 4,5).[55,65,66]

$$ONOO^- + H^+ + NO \rightarrow NO_2 + HNO_2 \quad (4)$$

$$NO_2 + NO \rightarrow N_2O_3 \quad (5)$$

The formation of $ONOO^-$ will be limited to zones where NO and O_2^- are co-generated in a molar ratio close to one to one. While O_2^- scavenger systems compete with NO for $ONOO^-$ formation, an excess of NO relative to $ONOO^-$ can shift the oxidative chemistry associated with $ONOO^-$ to a nitrosative profile. These reactions illustrate how the balance between ROS and RNOS reactants and their respective substrates could determine the advent of either an oxidative stress or a nitrosative stress in a biologic compartment. Analysis of the temporal and spatial elaboration of the ROS and RNOS is crucial in determining the chemical toxicology of NO as being either direct or indirect, oxidative or nitrosative.

NO/RNOS AS A PROTECTOR AGAINST OXIDATIVE STRESS

The complex inter-relationship between NO and ROS *in vivo* is still being defined. Small fluxes of NO in the range of 1–2 μM[46] can significantly reduce the cellular injury caused by H_2O_2,[63] O_2^- [63] and alkyl hydroperoxides.[46] In each of these cases, the mechanism of NO's action was consistent with termination of lipid peroxidation reactions.[67–70] Independent of the exact mechanism, these data suggest that the production of NO can be a protective mechanism against oxidative stress *in vivo*.

NO is an important molecule in the use of H_2O_2 as a strategy in fighting infectious pathogens. In the study of *E. coli*, Pacelli *et al.* found that neither NO nor H_2O_2 alone was significantly bactericidal but that toxicity was markedly enhanced with the combination of these two agents.[71] The augmented toxicity of the NO and H_2O_2 combination was not observed in the presence of catalase and metal chelators, while SOD had no influence, suggesting that Fenton-type chemistry was involved in the bactericidal synergism. By disrupting heme-containing proteins in the bacterium, NO may increase the level of Fe^{2+} that can subsequently increase the number of catalytic sites for the peroxidation of proteins and DNA. These findings illustrate the concept of how NO could be protective against mammalian cell lipid peroxidation, for example in the phagolysosomal membrane, while potentiating the action of H_2O_2 on bacteria.

CONCLUSION

The role of NO in pathophysiological and toxicological mechanisms is determined by its concentration (or flux) and its temporal and spatial interaction with other ROS. NO production catalyzed by the iNOS isoform can elicit the development of indirect effects due to the potential for a high flux of NO. NO, RNOS and ROS can react with a variety of target molecules and each other through a variety of mechanisms. The relative balance between these reactions is pivotal in dictating the role NO plays in the promotion of or protection from either oxidative or nitrosative stress.

ACKNOWLEDGMENT

We thank Dr. Dan Gilbert for his influence on our understanding of redox mechanisms as a mentor, colleague and friend.

REFERENCES

1. WINK, D.A., Y. VODOVOTZ, J. LAVAL, F. LAVAL, M.W. DEWHIRST & J.B. MITCHELL. 1998. The multifaceted roles of nitric oxide in cancer. Carcinogenesis **19:** 711–721.
2. WINK, D.A. & J.B. MITCHELL. 1998. The chemical biology of nitric oxide: insights into regulatory, cytotoxic and cytoprotective mechanisms of nitric oxide. Free Rad. Biol. Med. **25:** 434–456.
3. WINK, D.A., M. GRISHAM, J.B. MITCHELL & P.C. FORD. 1996. Direct and indirect effects of nitric oxide. biologically relevant chemical reactions in biology of NO. Methods Enzymol. **268:** 12–31.

4. WINK, D.A., I. HANBAUER, M.B. GRISHAM, F. LAVAL, R.W. NIMS, J. LAVAL, J.C. COOK, R. PACELLI, J. LIEBMANN, M.C. KRISHNA, M.C. FORD & J.B. MITCHELL. 1996. The chemical biology of NO. Insights into regulation, protective and toxic mechanisms of nitric oxide. Curr. Topics Cell. Reg. **34:** 159–187
5. IGNARRO, L.J. 1989. Endothelium-derived nitric oxide: pharmacology and relationship to the actions of organic esters. Pharm. Res. **6:** 651–659.
6. FURCHGOTT, R.F. & P.M. VANHOUTE. 1989. Endothelium-derived relaxing and contracting factors. FASEB, J. **3:** 2007–2018.
7. MONCADA, S., R.M.J. PALMER & E.A. HIGGS. 1991. Nitric oxide: physiology, pathophysiology, and pharmacology. Pharmacol. Rev. **43:** 109-142.
8. MURAD, F. 1994. The nitric oxide-cyclic GMP signal transduction system for intracellular and intercellular communication. Recent Prog. Horm. Res. **49:** 239–248.
9. KHARITONOV, V.G., V.S. SHARMA, D. MAGDE & D. KOESLING. 1997. Kinetics of nitric oxide dissociation from five- and six-coordinate nitrosyl hemes and heme proteins, including soluble guanylate cyclase. Biochemistry **36:** 6814–6818.
10. SIMON, D.I., M.E. MULLINS, L. JIA, B. GASTON, D.J. SINGEL & J.S. STAMLER. 1996. Polynitrosylated proteins: characterization, bioactivity, and functional consequences. Proc. Natl. Acad. Sci. USA **93:** 4736–4741.
11. WINK, D.A., R.W. NIMS, J.F. DARBYSHIRE, D. CHRISTODOULOU, I. HANBAUER, G.W. COX, F. LAVAL, J. LAVAL, J.A. COOK, M.C. KRISHNA, W. DEGRAFF & J.B. MITCHELL. 1994. Reaction kinetics for nitrosation of cysteine and glutathione in aerobic nitric oxide solutions at neutral pH. Insights into the fate and physiological effects of intermediates generated in the NO/O_2 reaction. Chem. Res. Toxicol. **7:** 519–525.
12. WILLIAMS, D.L.H., 1988. *In* Nitrosation. Oxford, UK. Cambridge Press.
13. LEWIS, R.S., S.R.TANNENBAUM & W.M. DEEN. 1995. Kinetics of nitrosation in oxygenated nitrous solutions at physiological pH: role of nitrous anhydride and effect of phosphate and chloride. J. Am. Chem. Soc. :3933–3940.
14. FORD, P.C., D.A. WINK & D.M. STANBURY. 1993. Autooxidation kinetics of aqueous nitric oxide. FEBS Lett. **326:** 1–3.
15. LIU, X., M.J.S. MILLER, M.S. JOSHI, D.D. THOMAS & J.R.J. LANCASTER. 1998. Accelerated reaction of nitric oxide with O_2 within the hydrophobic interior of biological membranes. Proc. Natl. Acad. Sci. USA **95:** 2175–2179.
16. GREEN, L.C., K. RUIZ DE LUZURIAGA, D.A. WAGNER, W. RAND, N. ISTFAN, V.R YOUNG & S.R. TANNENBAUM. 1981. Nitrate biosynthesis in man. Proc. Natl. Acad. Sci. USA **78:** 7764–7768.
17. HIBBS, J.B., Z. VAVRIN & R.R. TAINTOR. 1987. L-arginine is required for the expression of the activated macrophage effector mechanism causing selective metabolic inhibition in target cells. J. Immunol. **138:** 550–565.
18. LANCASTER, J.R. & J.B. HIBBS. 1990. EPR demonstration of iron-nitrosyl complex formation by cytotoxic activated macrophages. Proc. Natl. Acad. Sci. USA **87:** 1223–1227.
19. STUEHR, D.J. & M.A. MARLETTA. 1985. Mammalian nitrate biosynthesis: mouse macrophages produce nitrite and nitrate in response to *Escherichia coli* lipopolysaccharide. Proc. Natl. Acad. Sci. USA **82:** 7738–7742.
20. MARLETTA, M.A. 1988. Mammalian synthesis of nitrite, nitric oxide and N-nitrosating agents. Chem. Res. Toxicol. **1:** 249–257.
21. LIU, R.H., B. BALDWIN, B.C. TENNANT & J.H. HOTCHKISS. 1991. Elevated formation of nitrate and N-nitrosodimethylamine in woodchucks (*Marmota monax*) associated with chronic woodchuck hepatitis virus infection. Cancer Res **51:** 3925–3929.
22. LIU, R H., J.R. JACOB, B.D. TENNANT & J.H. HOTCHKISS. 1992. Nitrite and nitrosamine synthesis by hepatocytes isolated from normal woodchucks (*Marmota monax*) and woodchucks chronically infected with woodchuck hepatitis virus. Cancer Res. **52:** 4139–4143.
23. MILES, A.M., D.A. WINK, J.C. COOK & M.B. GRISHAM. 1996. Determination of nitric oxide using fluorescence spectroscopy. Meth. Enzymol. **268:** 105–120.
24. KEEFER, L.K., R.W. NIMS, K.W. DAVIES & D.A. WINK. 1996. NONOates (Diazenolate-2-oxides) as nitric oxide dosage forms. Meth. Enzymol. **268:** 281–294.

25. KRONCKE, K.-D., K. FECHSEL, T. SCHMIDT, F.T. ZENKE, I. DASTING, J.R. WESENER, H. BETTERMANN, K.D. BREUNIG & V. KOLB-BACHOFEN. 1994. Nitric oxide destoys zinc-finger clusters inducing zinc release from metallothionein and inhibition of the zinc finger-type yeast transcription activator LAC9. Biochem. Biophys. Res. Commun. **200:** 1105–1110.
26. WALKER, M.W., M.T. KINTER, R.J. ROBERTS & D.R. SPITZ. 1995. Nitric oxide-induced cytotoxicity: involvement of cellular resistance to oxidative stress and the role of glutathione in protection. Pediat. Res. **37:** 41–47.
27. SCHWARZ, M.A., J.S. LAZO, J.C. YALOWICH, W.P. ALLEN, M. WHITMORE, H.A. BERGONIA, E. TZENG, T.R. BILLIAR, P.D. ROBBINS, J.R. LANCASTER & B.R. PITT. 1995. Metallothionein protects against the cytotoxic and DNA damaging effects of nitric oxide. Proc. Natl. Acad. Sci. USA **92:** 4452–4456.
28. MISRA, R.R., J.F. HOCHADEL, G.T. SMITH, M.P. WAALKES & D.A. WINK. 1996. Evidence that nitric oxide enhances cadmium toxicty by displacing the metals from metallothionein. Chem. Res. Toxicol. **10:** 326–332.
29. MOLINA Y VEDIA, L., B. MCDONALD, B. REEP, B. BRUNE, M. DISILVIO, T.R. BILLIAR & E.G. LAPETINA. 1992. Nitric oxide-induced S-nitrosylation of glyceraldehyde-3-phosphate dehydrogenase inhibits enzymatic activity and increases endogenous ADP-ribosylation. J. Biol. Chem. **267:** 24929–24932.
30. LAVAL, F. & D.A. WINK. 1994. Inhibition by nitric oxide of the repair protein O6-methylguanin-DNA-methyltransferase. Carcinogenesis **15:** 443–447.
31. WINK, D.A. & J. LAVAL. 1994. The Fpg protein, a DNA repair enzyme, is inhibited by the biomediator nitric oxide in vitro and in vivo. Carcinogenesis **15:** 2125–2129.
32. AMES, B.N., M.K. SHIGENAGA & T.M. HAGEN. 1993. Oxidants, antioxidants, and the degenerative diseases of aging. Proc. Natl. Acad. Sci. USA **90:** 7915–7922.
33. SALGO, M.G., K. STONE, G.L. SQUADRITO, J.R. BATTISTA & W.A. PRYOR, W. 1995. Peroxynitrite causes DNA nicks in plasmid pBR322. Biochem. Biophys. Res. Commun. **210:** 1025–1030.
34. YERMILOV, V., J. RUBIO & H. OHSHIMA. 1995. Formation of 8-nitroguanine in DNA treated with peroxynitrite in vitro and its rapid removal from DNA by depurination. FEBS Lett. **376:** 207–210.
35. DEROJAS-WALKER, T., S. TAMIR, H. JI, J.S. WISHNOK & S.R. TANNENBAUM. 1995. Nitric oxide induces oxidative damage in addition to deamination in macrophage DNA. Chem. Res. Toxicol. **8:** 473–477.
36. WINK, D.A., M. FEELISCH, J. FUKUTO, D. CHISTODOULOU, D. JOURD'HEUIL, M.B. GRISHAM, Y. VODOVOTZ, J.A. COOK, M. KRISHNA, W. DEGRAFF, S. KIM, J. GAMSON & J.B. MITCHELL. 1998. The cytotoxic mechanism of nitroxyl: possible implications for the pathophysiological role of NO. Arch Biochem. Biophys. **351:** 66–74.
37. WINK, D.A. & M. FEELISCH. 1996. Formation and detection of nitroxyl and nitrous oxide. *In* Methods in Nitric oxide Research. M. Feelisch & J. Stamler, Eds. :403–412. New York. John Wiley & Sons.
38. ARNELLE, D.R. & J.S. STAMLER. 1995. NO^+, NO, and NO^- donation by S-nitrosothiols: implications for regulation of physiological functions by S-nitrosylation and acceleration of disulfide formation. Arch. Biochem. Biophys. **318:** 279–285.
39. BONNER, F.T. & K.A. PEARSALL. 1982. Aqueous nitrosyliron(II) chemistry. 1. Reduction of nitrite and nitric oxide by iron(II) and (trioxodinitrato)iron(II) in acetate buffer. Intermediacy of nitrosyl hydride. Inorg. Chem. **21:** 1973–1978.
40. HOBBS, A.J., J.M. FUKUTO & L.J. IGNARRO. 1994. Formation of free nitric oxide from L-arginine by nitric oxide aynthase: direct enhancement of generation by superoxide dismutase. Proc. Natl. Acad. Sci. USA **91:** 10992–10996.
41. SCHMIDT, H.H., H. HOFMANN, U. SCHINDLER, Z.S. SHUTENKO, D. CUNNINGHAM & M. FEELISCH. 1996. No ·NO from NO synthase. Proc. Natl. Acad. Sci. USA **93:** 14492–14497.
42. PUFAHL, R.A., J.S. WISHNOK & M.A. MARLETTA. 1995. Hydrogen peroxide-supported oxidation of NG-hydroxy-L-arginine by nitric oxide synthase. Biochemistry **34:** 1930–1941.

43. WINK, D.A. & M. FEELISCH. 1996. Formation and detection of nitroxyl and nitrous oxide. *In* Methods in Nitric Oxide Research. M. Feelisch & J. Stamler, Eds. :71–118. New York. John Wiley & Sons.
44. FARIAS-EISNER, R., G. CHAUDHURI, E. AEBERHARD & J.M. FUKUTO. 1996. The chemistry and tumoricidal activity of nitric-oxide hydrogen-peroxide and the implications to cell resistance susceptibility. J. Biol. Chem. **271:** 6144–6151.
45. WINK, D.A., J.COOK, R. PACELLI, W. DEGRAFF, J. GAMSON, J. LIEBMANN, M. KRISHNA & J.B. MITCHELL. 1996. Effect of various nitric oxide-donor agents on peroxide mediated toxicity. A direct correlation between nitric oxide formation and protection. Arch. Biochem. Biophys. **331:** 241–248.
46. WINK, D.A., J.A. COOK, M.C. KRISHNA, I. HANBAUER, W. DEGRAFF, J. GAMSON & J.B. MITCHELL. 1995. Nitric oxide protects against alkyl peroxide-mediated cytotoxicty: Further insights into the role nitric oxide plays in oxidative stress. Arch. Biochem. Biophys. **319:** 402–407.
47. PRYOR, W.A. 1982. *In* Lipid Peroxides in Biology and Medicine. K. Yagi, Ed. :1–22. New York. Academic Press.
48. PRYOR, W.A., D.F. CHURCH, C.K. GOVINDAN & G. CRANK. 1982. Oxidation of thiols by nitric oxide and nitrogen dioxide: synthetic utility and toxicological implications. J. Org. Chem. **47:** 156–159.
49. PRYOR, W.A., L. CASTLE & D.F. CHURCH. 1985. Nitrosation of organic hydroperoxides by nitrogen dioxide/dinitrogen tetraoxide. J. Am. Chem. Soc. **107:** 211–217.
50. KOPPENOL, W.H. 1996. Thermodynamics of reactions involving nitrogen-oxygen compounds. Meth. Enzymol. **268:** 3–12.
51. EHRLICH, R. 1980. Interaction between environmental pollutants and respiratory infections. Environ. Health Perspect. **35:** 89–100.
52. HUIE, R.E. & S. PADMAJA. 1993. The reaction of NO with superoxide. Free Rad. Res. Commun. **18:** 195–199.
53. BECKMAN, J.S., T.W. BECKMAN, J. CHEN, P.H. MARSHALL & B.A. FREEMAN. 1990. Apparent hydroxyl radical production by peroxylnitrites: implications for endothelial injury from nitric oxide and superoxide. Proc. Natl. Acad. Sci. USA **87:** 1620–1624.
54. PRYOR, W.A. & G.L. SQUADRITO. 1996. The chemistry of peroxtynitrite and peroxynitrous acid: Products from the reaction of nitric oxide with superoxide. Am. J. Phys. **268:** L699–L721.
55. KOPPENOL, W.H., J.J. MORENO, W.A. PRYOR, H. ISCHIROPOULUS & J.S. BECKMAN. 1992. Peroxynitrite, a cloaked oxidant formed by nitric oxide and superoxide. Chem. Res. Toxicol. **5:** 834–842.
56. WINK, D.A., M.B. GRISHAM, A.M. MILES, R.W. NIMS, M.C. KRISHNA, R. PACELLI, D. TEAGUE, C.M.B. POORE & J.C. COOK. 1996. Methods for the determination of selectivity of the reactive nitrogen oxide species for various substrates. Meth. Enzymol. **268:** 120–130.
57. TYLER, D.D. 1975. Polarographic assay and intracellular distrubution of superoxide dismutase in rat liver. Biochem. J. **147:** 493.
58. LANCASTER, J. 1994. Simulation of the diffusion and reaction of endogenously produced nitric oxide. Proc. Natl. Acad. Sci. USA **91:** 8137–8141.
59. NIKANO, M., H. KIMURA, M. HARA, M. KUROIWA, M. KATO, K. TOTSUNE & T. YOSHIKAWA. 1990. A highly sensitive method for determining both Mn- and Cu-Zn superoxide dismutase activities in tissue and blood cells. Anal. Biochem. **187:** 277–280.
60. HAUSLADEN, A. & I. FRIDOVICH. 1994. Superoxide and peroxynitrite inactivate aconitases, but nitric oxide does not. J. Biol. Chem. **269:** 29405–29408.
61. FRIDOVICH, I. 1985. Cytochrome C. *In* Handbook of Methods for Oxygen Radical Research. R.A. Greenwald, Ed. :213–215. Boca Raton, FL. CRC Press, Inc.
62. ZHU, L., C. GUNN & J.S. BECKMAN. 1992. Bactericidal activity of peroxynitrite. Arch. Biochem. Biophys. **298:** 452–457.
63. WINK, D.A., I. HANBAUER, M.C. KRISHNA, W. DEGRAFF, J. GAMSON & J.B. MITCHELL. 1993. Nitric oxide protects against cellular damage and cytotoxicity from reactive oxygen species. Proc. Natl. Acad. Sci. USA **90:** 9813–9817.

64. MILES, A.M., D.S. BOHLE, P.A. GLASSBRENNER, B. HANSERT, D.A. WINK & M.B. GRISHAM. 1996. Modulation of superoxide-dependent oxidation and hydroxylation reactions by nitric oxide. J. Biol. Chem. **271:** 40–47.
65. BECKMAN, J.S., J. CHEN, H. ISCHIROPOULOS & J.P. CROW. 1994. Oxidative chemistry of peroxynitrite. Meth. Enzymol. **233:** 229–240.
66. HOGG, N., B. KALYANARAMAN, J. JOSEPH, A. STRUCK & S. PARTHASARATHY. 1993. Inhibition of low-density lipoprotein oxidation by nitric oxide. Potential role in atherogenesis. FEBS Lett. **334:** 170–174.
67. HOGG, N., A. STRUCK, S.P. GOSS, N. SANTANAM, J. JOSEPH, S. PARTHASARATHY & B. KALYANARAMAN. 1995. Inhibition of macrophage-dependent low density lipoprotein oxidation by nitric-oxide donors. J. Lipid Res. **36:** 1756–1762.
68. STRUCK, A.T., N. HOGG, J.P. THOMAS & B. KALYANARAMAN. 1995. Nitric oxide donor compounds inhibit the toxicity of oxidized low-density lipoprotein to endothelial cells. FEBS Lett. **361:** 291–294.
69. RUBBO, H., S. PARTHASARATHY, S. BARNES, M. KIRK, B. KALYANARAMAN & B.A. FREEMAN. 1995. Nitric oxide inhibition of lipoxygenase-dependent liposome and low-density lipoprotein oxidation: termination of radical chain propagation reactions and formation of nitrogen-containing oxidized lipid derivatives. Arch. Biochem. Biophys. **324:** 15–25.
70. PACELLI, R., D.A. WINK, J.A. COOK, M.C. KRISHNA, W. DEGRAFF, N. FRIEDMAN, M. TSOKOS, A. SAMUNI & J.B. MITCHELL. 1995. Nitric oxide potentiates hydrogen peroxide-induced killing of *Escherichia coli*. J. Exp. Med. **182:** 1469–1479.

Nitrone Inhibition of Age-Associated Oxidative Damage

ROBERT A. FLOYD[a–c] AND KENNETH HENSLEY[a]

[a]*Free Radical Biology and Aging Research Program, Oklahoma Medical Research Foundation and*
[b]*Department of Biochemistry and Molecular Biology, University of Oklahoma Health Sciences Center, Oklahoma City, Oklahoma 73104, USA*

ABSTRACT: The mechanistic basis of the neuroprotective activity of the nitrone-based free radical trap PBN (α-phenyl-N-*tert*-butyl nitrone) has been investigated extensively. Key observations exclude its simple mass action spin trapping of free radicals activity as the key mechanism of action. These include: A) the fact that it protects in experimental stroke even if administered several hours after the event and B) the fact that its chronic low-level administration to old experimental animals reverses their age-enhanced susceptibility to stroke even several days after the last dosage. PBN was found to inhibit gene induction in several models including stroke and an LPS-mediated septic shock model. Stoke causes inducible nitric oxide synthase (iNOS) to be expressed. High levels of nitric oxide and peroxynitrite (formed from nitric oxide), produced by iNOS, is particularly neurotoxic. PBN inhibits iNOS induction. Therefore, it seems that prevention of the formation of neurotoxic products is a rational mechanism of action of PBN in the stroke model. There is strong rationale to consider that there is an enhanced propensity for a "smoldering" neuro-inflammatory state in the old brain. Reversal of this state by PBN may explain its action in preventing age-enhanced stroke susceptibility in old experimental animals. Significant new findings underscore the importance of neuro-inflammatory processes in neuronal death or dysfunction in Alzheimer's disease. Neuro-inflammatory processes implicate enhanced signal transduction processes. Strong evidence for this is the enhanced p38 kinase activation in neurons near plaques and tangles of the Alzheimer's brain in contrast to normal aged-matched control brain which did not show p38 activation. In rat primary astrocytes p38 activation by the pro-inflammatory cytokine IL-1β, as well as by H_2O_2, was significantly suppressed by PBN. Mechanistically it was shown that PBN suppresses the amount of reactive oxygen species (ROS) produced in mitochondrial respiration. Much evidence indicates that ROS are signaling molecules and that they also are involved in maintaining brain phosphatases in an inactive state. We argue that finding a specific high affinity site mechanism for the neuroprotective action of PBN is unlikely based on the complexity of the system reflecting ROS generation and signal transduction processes that have apparently evolved to maintain adaptive responses. The promising pharmacological activity of molecules like PBN is not diminished by this however, for only excessive amounts of ROS is considered detrimental. The action of PBN in suppressing signal transduction processes, most likely by

[c]Address for correspondence: Dr. Robert A. Floyd, Free Radical Biology and Aging Research Program, Oklahoma Medical Research Foundation, University of Oklahoma Health Sciences Center, Oklahoma City, OK 73104. Voice: 405-271-7580; fax: 405-271-1795.
  e-mail: Robert-Floyd@omrf.ouhsc.edu

suppressing ROS production in mitochondrial respiration, effectively controls excessive oxidative damage and prevents induction of genes that form neurotoxic products.

INTRODUCTION AND HISTORICAL PERSPECTIVE

The mere fact that we presently conduct research with focus on free radical biology rests on the shoulders of creative pioneers like Dan Gilbert who made observations not readily explained by conventional scientific wisdom, refused to dismiss them, and had the courage to explore their veracity and implications. Michaelis had predicted biological univalent reduction of oxygen over 60 years ago (see Ref.1) and Gerschman, Gilbert and colleagues[2] made seminal observations in 1954 noting the similarities of oxygen poisoning and X-irradiation. This was the same year that free radicals in biological systems were first observed by Commoner and colleagues using electron paramagnetic resonance.[3] These findings represent the initial confluence of radiation chemistry, where the existence of oxygen free radicals had unequivocally been demonstrated, and the beginning recognition of their potential importance in biological systems. It should be noted that Thenard, the discoverer of H_2O_2, had demonstrated its metabolism in biological systems in 1818 (see Ref. 4). The discovery of superoxide dismutase by McCord and Fridovich in 1969[5] and the demonstration by Chance and colleagues that mitochondria produce H_2O_2 in 1971[6] helped lay the foundation for modern studies in free radical biology. However, the potential importance of this field was perceived by only a few people at that time because of the general skepticism of the scientific community. This skepticism was due in large part to the general lack of rigorous *in vivo* demonstrations that oxygen free radicals were formed in biological systems. Thus, the paucity of rigorous research methods held up the field for many years.

The development of spin trapping methodology, first used in analytical chemistry (see Ref. 7), showed great promise and was first applied in biochemical systems by Bolton and colleagues[8] and McCay and colleagues in our laboratory.[9] We could not demonstrate the presence of hydroxyl free radicals in biological systems using the spin traps then available. Therefore, we pioneered the use of salicylate to trap hydroxyl free radicals *in vivo*.[10,11] This approach was combined with the use of HPLC and electro-chemical detection to quantitate the small amounts of hydroxylation products of salicylate formed. The extreme sensitivity of HPLC-electrochemical detection provided us with the ability to also quantitate 8-hydroxy-2'-deoxy-guanosine,[12] an oxidation product of DNA formed by hydroxyl free radicals[13] as well as by other oxidizing species, including singlet oxygen.[14] Using the salicylate hydroxylation methodology we proved that significantly enhanced amounts of hydroxyl free radicals were formed in brain undergoing an ischemia/reperfusion event.[11] It was our attempt to trap free radicals, during brain ischemia/reperfusion using the nitrone-based free radical trap PBN (α-phenyl-N-*tert*-butyl nitrone), that led us to the serendipitous discovery that this compound protects the brain from experimental stroke,[15] even if given an hour or more after the event.[16,17] Attempts to unravel the mechanistic basis of the neuroprotective activity of PBN took us on an investigative trail where other seminal observations by Colton and Gilbert provided vital clues to the basic mechanisms involved in neurodegeneration. They demonstrated that brain microglia produce superoxide when activated.[18] This turns out to be an important observation pointing to the importance of neuro-inflammation, a process that appears to be very basic to neurodegenerative events.

In the present review we summarize some of the basic observations that have been made on the mechanisms of action of PBN and related compounds as well as the subsequent understanding that have accrued to help explain neurodegenerative processes. The neuroprotective activity of PBN and related nitrones have been summarized before,[19,20] so only the main observations will be noted here. In addition, we will briefly note the nitrone free radical trapping chemistry, even though it is probably not important to the pharmacological activity of these compounds. Most of our recent research effort indicates that the nitrones suppress signal transduction processes. Signal transduction processes are important in many normal processes but in addition significant elevation of these processes is involved in the upregulation of genes that occur in neuro-inflammation and other pro-inflammatory processes. Toxic gene products, such as nitric oxide and its oxidation products may in part account for neuron death or dysfunction, a central issue in neurodegenerative processes.

NITRONE SPIN TRAPPING CHEMISTRY

Attention to nitrones arose from their use as so-called spin trapping compounds.[7] Their value in analytical chemistry arises from the fact that they can react with and trap a free radical R·, which may have a very short half life, to form a spin adduct nitroxide that usually has a much longer half life. The electron paramagnetic resonance characteristics of the resultant nitroxide will in theory characterize and help identify the radical that has added to the nitrone. The chemical reaction involved is shown below where, in the case of PBN, X is a phenyl group and Y is a tertiary butyl group.

$$X - \underset{H}{\overset{}{C}} = \underset{}{\overset{O}{N}} - Y + R^{\bullet} \longrightarrow X - \underset{H}{\overset{}{\underset{R}{C}}} - \underset{}{\overset{O\bullet}{N}} - Y$$

PHARMACALOGICAL ACTIVITY OF NITRONES

The nitrones have been shown to have pharmacological activity in many age-associated diseases.[19,20] There are many observations illustrating the pharmacological activity of PBN in age-associated conditions, including neurodegenerative diseases[19,20] and cancer[21] as well as in aging *per se*.[22,23] The first hint that nitrones possessed pharmacological activity came to light by the observations of Novelli in 1985, who showed that PBN protected rats from shock and trauma.[24,25] These early observations have now been confirmed by many laboratories where much more defined models of septic shock have been used (see ref. 19 for a review). In general, it has been found that PBN must be administered before the initiation of shock for it to be effective.

In addition to shock, PBN has been shown to be effective in various neurodegenerative models. The neuroprotective activity of PBN was first demonstrated in brain stroke.[15] In this regard, the most surprising observation was that PBN is active if administered an hour or longer after the ischemia/reperfusion insult to the brain.[16,17] This fact clearly eliminates the notion that the pharmacological activity of PBN is due to its simple mass action spin trapping chemistry. Based on these observations we began an extensive effort to understand the mechanistic basis of the neuroprotective activity of PBN. In this process we have made surprising observations regarding the fundamental processes of neurodegeneration. The results of this research are summarized in this review.

BIOAVAILABILITY AND ANTI-OXIDANT ACTIVITY OF PBN

One primary reason that PBN has shown pharmacological activity is most likely due to its ability to readily penetrate and become available to most, if not all, tissues of the body. The rat is where most studies have been done.[19] PBN is rapidly taken up and peaks in concentration in about 20 minutes in most tissues after an intraperitoneum injection. It has a half-life in rats of about 134 minutes and is metabolized in the liver by mixed function oxidases and excreted as a metabolite in the urine. PBN appears in the brain as the authentic compound and reaches a level of about 500 µM following a 150 mg/kg dose in rats. The PBN content of brain declines after a bolus dose but its kinetics of disappearance appears to be slower than the total body loss *per se*.

The term antioxidant encompasses a multitude of diverse meanings based on the properties of the various compounds generally included in this category. PBN is not very effective as a typical free radical chain-breaking antioxidant *per se*. For instance, in 3 different models of rat liver microsomal lipid peroxidation systems, it was found that the EC_{50} of PBN was from 1.6–20 mM.[19] On the other hand, in these same systems the effectiveness of butylated hydroxy toluene (BHT) and trolox was about 1000 times more effective than PBN having an EC_{50} of between 3.4 µM to 50 µM.[19] In spite of its relative ineffectiveness in the *in vitro* lipid peroxidation systems, administration of PBN lowers the enhanced brain protein oxidation levels in old gerbils back down to levels observed in young gerbils, *i.e.* by approximately 50%.[26] These observations clearly implicate that PBN acts *in vivo* in a manner not readily explained by its actions in *in vitro* systems. Another way of viewing its action is to consider that it must interfere with *in vivo* processes that result in oxidative damage. This then was our first clue that PBN has action other than strictly as a free radical trap and/or as an antioxidant *per se*.

PBN ACTS TO INHIBIT GENE INDUCTION

The first observations regarding the action of PBN to inhibit induction of specific genes, brought on by an experimental stroke, were made by ourselves in the gerbil model.[27] It was shown that post administration of PBN inhibited the induction of heat shock protein genes and *c-fos*.[27] The importance of these early observations was not realized then. Later our attention turned to genes such as inducible nitric oxide synthase (iNOS) which produces high levels of nitric oxide and its oxidation products. It is now known that a stroke brings on at various post ischemic times suites of genes that have been termed immediate early genes, intermediate genes and late genes.[28] The first wave of genes which appear within a few minutes to an hour or so include *c-fos*, *c-jun* and Zif 268. The second wave appears within 2–6 hours and includes the heat shock protein genes 70 and 72. Finally, a battery of late genes appear within 24 hours to 3 days. These late genes include the cytokines (TNFα, IL1β, IL6), adhesion molecules (ICAM-1, ELAM-1, P-selectin) and genes involved in apoptosis as well as iNOS.[28]

Iadecola and colleagues have shown that pharmacological agents which interfere catalytically with the action of iNOS have shown neuroprotective activity in stroke models.[28] In a rat neonate model of AIDs dementia complex, we have shown that PBN inhibited the enhanced nitric oxide produced in brain and that it also inhibited the neuronal damage caused by the neurotoxic viral protein gp120.[29] Although we do not consider the rat neonate a good model of AIDs dementia complex, our observations do point to an important concept namely that the neuroprotective action of PBN may be due to its prevention of the upregulation of genes that produce neuro-

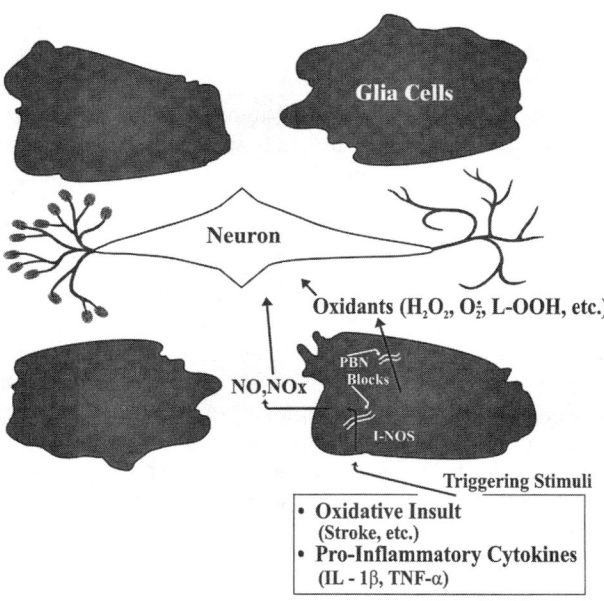

FIGURE 1. Illustration showing that glia cells which surround the neurons can be activated by oxidative stress or pro-inflammatory cytokines causing the production of reactive oxygen species and the induction of genes such as inducible nitric oxide synthase (iNOS) which then mediates the production of nitric oxide (NO) and other products such as peroxynitrite that are toxic to neurons but not glia. PBN acts to inhibit induction of iNOS as well as suppressing formation of reactive oxygen species.

toxic products such as iNOS which produces large levels of nitric oxide and subsequently peroxynitrite which is toxic to neurons. PBN was shown to prevent the induction of iNOS and the production of nitric oxide in liver of a mouse septic shock model.[30] Therefore, one important idea as to how PBN may act as a neuroprotective agent in stroke is by preventing the upregulation of iNOS. The glia cells (microglia and astrocytes) are in general terms referred to as resident macrophages and are capable of expressing iNOS. The expression of iNOS will result in high levels of nitric oxide and other oxidation products that are particularly toxic to neurons but are less toxic to the glia, the cells responsible for its production. This is an important concept that is supported by many observations and is presented in FIGURE 1. In this model the large oxidative insult created by a stroke is considered to set in motion a cascade of events which, following a lag time, will result in the induction of genes such as iNOS. It is considered that these processes can be halted or prevented if PBN is introduced within a certain time interval (approximately 1 h in the gerbil stroke model) after their initiation.

PBN REVERSAL OF AGE-ENHANCED SUSCEPTIBILITY TO STROKE

FIGURE 1 presents a conceptual working hypothesis to help explain why PBN inhibits the damage induced by a stroke. However, this model does not explain why increased age caused gerbils to be more susceptible to a stroke and why chronic low level PBN administration for 2 weeks reversed the increased age susceptibility to a stroke.[31] Data we collected illustrating this point are shown in TABLE 1. It should be noted that the PBN effect remained for 3–5 days after cessation of its administration, even though there was essentially none of the original compound present in the animals at that time. We have found this observation the most difficult to reconcile in relation to notions prevalent regarding brain aging and stroke. The data forced us to conclude that PBN administration acted upon the aged brain to mediate its reversal back to a state like that when it was young. In other words, increased age altered the brain and PBN acted to reverse the age-induced changes.

TABLE 1. Survival of old gerbils[a] following a 10 minute global stroke: effect of chronic PBN pretreatment[b]

Conditions	Survival (percentage)
No PBN pretreatment	15.0
PBN 14 days, off PBN few hours before stroke	89.3
PBN 14 days, off PBN 3 days before stroke	79.5
PBN 14 days, off PBN 5 days before stroke	30.2
PBN 14 days, off PBN 14 days before stroke	9.9

[a]18 male retired breeder (about 18 months old) animals in each group, lethality assessed at 7 days.
[b]Chronic PBN pretreatment consisted of 32 mg/kg given (I.P. twice daily for 14 days.

AGE-ENHANCED TENDENCY TOWARD NEURO-INFLAMMATION

Based on many supporting observations in the literature, we hypothesize that increasing age invokes unknown processes such that the brain has a higher risk of progressing into a smoldering pro-inflammatory state (neuro-inflammatory state) which renders it less capable of handling a large oxidative insult, such as a stroke, and that this state can be suppressed or reversed by chronic PBN administration. This is a novel idea and should be considered a working hypothesis; however, much evidence now supports this notion as detailed in our hypothesis paper[32] and in the following sections. It should be noted that many previously unexplained observations can be accounted for by these ideas. For instance, it has been noted that increased age is associated with increased oxidative damage in brain. A neuro-inflammatory state is expected to result in an enhanced level of oxidative damage since as Colton and Gilbert demonstrated the activation of microglia mediates enhanced production of reactive oxygen species.[18] We, as well as others, have found that increased age is associated with increased protein oxidation in experimental animals[26] and in humans[33] and that there is an additional enhanced amount of protein oxidation in Alzheimer's versus age-matched control brains.[33]

A central feature of neuro-inflammation is expected to be an age-associated tendency for glia activation. This research area has been studied only in a limited fashion; but one of the most thorough studies has been done by Morgan et al.[34] using the aging Fisher 344 rat model. They showed a clear increase in astrocyte activation in normal aged (24 month old) rats. Age-dependent activation was not due to an increase in astrocyte number, which remained constant with age. When they insulted the brain by 6-hydroxy-dopamine (6-OHDA) injection or by a stab wound (needle into nucleus accumbens), glia activation in the aged rats was much more robust and sustained than in the younger animals. They noted that either 6-OHDA or the stab wound caused not only the expected ipsilateral glia activation in all animals of various ages, but this caused contralateral glia activation only in the 24-month-old rats but not in the 6- or 15-month-old animals. It was considered unknown the reasons why there was: A) an age-associated low-level glia activation and B) a much longer sustained activation due to an insult by 6-OHDA or a stab wound.[34] We think these observations clearly support the hypothesis that increasing age causes an increased tendency for a "smoldering" neuroinflammatory state. It should be noted by the term "smoldering" we infer that this reflects a condition where the equilibrium of pro-inflammatory processes and anti-inflammatory processes is disturbed in favor of the former. It should also be noted that very little knowledge exists as to how these processes are controlled and how these change with age. Viewed in this manner then, the age effect can be seen as influencing events that modifies the equilibrium to favor more neuro-inflammatory events.

FIGURE 2 presents these concepts to help visualize the response of brain to an acute injury, such as stroke, and the effect of age on neuro-inflammatory events. In this model an acute injury induces a rapid increase to high levels of both reactive oxygen species (ROS) and inflammatory cytokines which is then followed by an increase in toxic gene products. In comparison to acute injury, increasing age acts to influence processes where the end result is much the same only the extent of increase in ROS and inflammatory cytokines are much less and in addition they occur over a

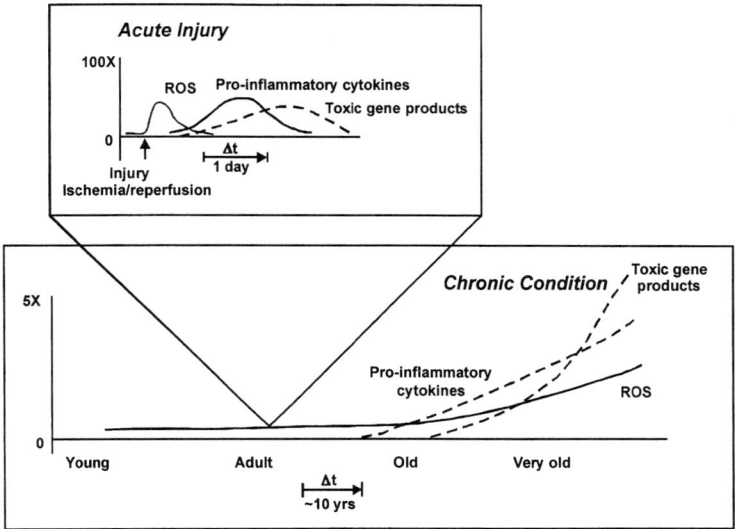

FIGURE 2. Representation of the effect of acute injury, such as a stroke, on the time course of the rapid production of reactive oxygen species (ROS), of pro-inflammatory cytokines and then toxic gene products. The acute injury represents only a minute segment of time in reference to a normal lifetime. During old to very old age there is a slow accumulation of toxic gene products. The amount of the components produced during aging is much smaller (note abscissa scale) than that produced during acute injury. The "smoldering" neuro-inflammatory state is expected to mediate the slow, steady increase in ROS, cytokines and toxic products during advanced age. (This figure is reproduced from Floyd,[32] contingent upon permission from the publishers of *Free Radical Biology & Medicine*.)

much longer time frame *i.e.* years versus hours to days. Implicated in this model is the notion that the toxic products, which increase with age, makes the brain more susceptible to acute stress. In this model it is considered that PBN if it is administered chronically to the very old, then would suppress the "smoldering" pro-inflammatory condition hence restoring the brain to a younger state and therefore capable of handling the challenge which occurs due to an acute injury.

PBN ACTS TO INHIBIT SIGNAL TRANSDUCTION PROCESSES

Baeuerle and colleagues made classical observations clearly implicating the involvement of ROS in signal transduction processes.[35] Since we observed that PBN inhibited gene induction after a stroke, we thought it a good possibility that PBN may act to suppress signal transduction processes. The results we have obtained do show that PBN suppresses signal transduction processes that are important in neurodegenerative events. The major findings of our research effort in this area will be presented here.

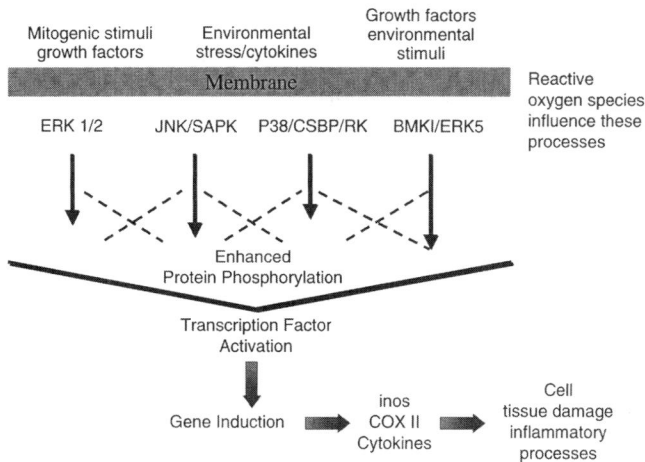

FIGURE 3. Simplified scheme showing the major MAP-kinase pathways involved in signal transduction processes.

Cells respond to various stimuli. These stimuli trigger signal transduction processes by which cells mediate a response typically by up-regulating the formation of various gene products. A central aspect of signal transduction processes involves activation of various protein kinase pathways, *i.e.* what is known as mitogen-activated protein kinase pathways or MAP-kinases. This active research area has uncovered 4 basic MAP-kinase pathways that are not totally independent but have considerable cross-talk between them (see FIG. 3). Our first experiments to test if PBN as well as NAC (N-acetyl-cysteine) influenced protein phosphorylation processes demonstrated that both were effective in this regard but that there were differences. Most of our research has focused on the p38 pathway because decisive early studies demonstrated that it was markedly activated in the gerbil stroke model[36] and also because of the availability of excellent reagents for assessment of its activation state. p38 is a cytosolic protein. It becomes a kinase once it becomes dually phosphorylated with one phosphate ester each on the threonine and tyrosine residues of the Thr 180-Gly 181-Try 182 domain. Activated p38 translocates to the nucleus where it is known to catalytically phosphorylate a transcription factor (CREB), a DNA damage-inducible gene (GADD153) and a monocyte-enhancement factor (MEF2C). The p38 kinase module has been shown to be responsible for the enhanced expression of iNOS in astrocytes.[37] Thus, study of p38 is especially important in neuro-inflammatory events involved in the production of high (neurotoxic) levels of nitric oxide.

ASTROCYTE p38 ACTIVATION AND ITS INHIBITION BY PBN

Our studies on p38 activation in rat primary astrocytes have demonstrated: A) that they are activated by various cytokines especially IL-1β and very importantly by H_2O_2 and B) that PBN as well as NAC suppresses their activation by either IL-1β or

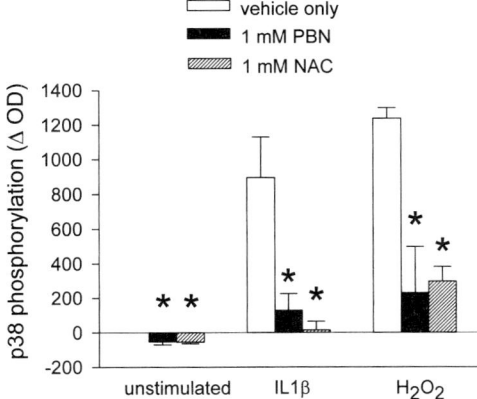

FIGURE 4. Data reconstructed from ref. 38 illustrating the activities of p38 in primary astrocytes 5 minutes after they were stimulated by 1L-1β or H_2O_2 and the suppression of p38 activation by PBN (0.5 mM) or N-acetylcysteine (NAC, 1mM).

H_2O_2.[38] FIGURE 4 presents some of these results. We have found that: A) p38 experiences first rapid activation after exposure to IL-1β or H_2O_2 and then it is subsequently shut down, and B) PBN as well as NAC inhibit p38 activation. It should be noted that our data also demonstrated that activation of p38 did not change the total amount of p38 present in astrocytes. Another important finding that these studies revealed (not shown) is that total phosphatase activity (measured by three different methods) varied inversely and in a correlative time-course manner with p38 activity. In other words, when p38 activity peaked there was a nadir in phosphatase activity.[38] This appears to be an important point for it demonstrates that phosphatase activity most likely governs signal transduction processes and, as will be discussed later in relation to the mechanistic basis of PBN action in signal transduction processes, the activity of phosphatases are oxidatively controlled. Thus PBN may be acting to indirectly control oxidation of phosphatases thereby influencing signal transduction processes.

p38 IS ACTIVATED NEAR PLAQUES IN ALZHEIMER'S BRAIN

A major step forward in the critical evaluation of the importance of neuro-inflammatory processes in Alzheimer's Disease (AD) is convincing proof that inflammatory processes are on-going in the AD brain in comparison to normal age-matched control brain. We were able to demonstrate that p38 was significantly activated in neurons surrounding plaques in the hippocampus and parahippocampal gyrus area of AD brain.[39] In comparison, none to very few p38 activated neurons were observed in the same brain regions of normal age-matched control brains.[39] The dually phosphorylated p38 antibody stained in a pronounced fashion classic features of the AD brain, including neuritic plaques, neuropil threads and neurofibrillary tangles. This data demonstrate that these areas are active in signal transduction processes as reflected by activated p38. We also demonstrated that aberrantly phosphorylated tau, another characteristic feature of AD, was not stained by the activated p38 antibody.[39] One of the consequences of p38 activation in the neurons surrounding plaques implicates that apoptotic processes are occurring in these cells.

BY-PRODUCTS OF NEURO-INFLAMMATION ARE INCREASED IN ALZHEIMER'S BRAIN

As noted earlier if neuro-inflammatory processes are activated in AD then by-products of these processes should be increased. Since p38 activation occurs in affected regions of the AD brain and since the p38 pathway controls the upregulation of iNOS one would expect to see increased nitric oxide being formed in the affected regions of the AD brain. Nitric oxide reacts very rapidly with superoxide to form peroxynitrite which reacts with tyrosine residues in protein to form 3-nitro-tyrosine adducts. We have developed an HPLC-electrochemical array detection method to analyze for 3-nitrotyrosine as well as for other oxidation products of tyrosine including dityrosine which is formed by free radical intermediates of tyrosine.[40] We have also developed a method of digesting protein to amino acid adduct residues so that dityrosine and 3-nitrotyrosine as well as some other oxidatively modified residues can be quantified and normalized to the total tyrosine content.[40] Application of this methodology to astrocytes activated by IL-1β has shown that iNOS is upregulated and that the 3-nitrotyrosine content of astrocyte protein is increased several fold by IL-1β stimulation.[40]

We have utilized HPLC-electrochemical array detection methodology to assess dityrosine and 3-nitrotyrosine content of the proteins from specific regions of AD brain versus age-matched control brain regions. Both of these residues are significantly increased in the AD brain.[41] Additionally the increased content of these residues is much higher in regions of the AD brain that has increased levels of plaques and tangles.[41] These residues are not increased in the cerebellum, a brain region that is not affected in AD. TABLE 2 presents a brief summary of our results. Up to a 7-fold increase is seen in 3-nitrotyrosine in the most affected region of AD, *i.e.* the hippocampus; whereas there is little or no increase in cerebellum. The other regions noted are intermediate in the effect that AD has on them, both neuro-pathologically as well as in the tyrosine adduct residues measured. TABLE 2 also demonstrates that the uric acid content of AD brain is significantly less than control brain. Uric acid has been shown to react with peroxynitrite and to protect neurons from this toxin.[42] Uric

TABLE 2. Neuro-inflammation by products in Alzheimer's brain. The numbers represent the mean ratio of analyte in Alzheimer's brain tissue to the corresponding level in normal brain tissue[a]

Brain Region	Product Analyzed		
	Dityrosine	3-Nitrotyrosine	Uric Acid
Hippocampus	4.6	7.8	0.52
Inferior parietal lobule	4.5	6.5	0.50
Superior and middle temporal gyri	2.7	5.2	0.42
Cerebellum	1.3	0.17	0.25

[a]HPLC-electrochemical array detection analysis of 11 Alzheimer's subjects and five normal subjects; mean age of both groups was 78 years old (38). The dityrosine and 3-nitrotryosine content was that present in the protein fraction of the tissue and normalized to the tyrosine content of the protein.

acid has also been shown to be an antioxidant. But the reason for its decrease in the AD brain is unknown and may not be related to its antioxidant property or its ability to protect from peroxynitrite toxicity.

MECHANISTIC BASIS OF THE PHARMALOGICAL ACTION OF PBN

Despite extensive effort by us as well as many others, a specific high affinity site mechanism basis of the action of PBN is not available. However, much advancement toward evaluating this "idealistic" goal has been accomplished in the last 2 to 3 years. Major facts that have helped delineate and helped focus the mechanism of PBN action into specific processes are as follows:

(a) Its bioavailability is an important factor, but its action as a classical antioxidant appears to be unimportant.
(b) Its pharmacological action does not appear to depend upon its mass action free radical spin-trapping activity.
(c) Its major activity appears to reside with its action in suppressing gene induction associated with oxidative damage.
(d) Its activity in the suppression of gene induction is due to its action in suppressing signal transduction processes.
(e) Its activity in suppressing age-enhanced susceptibility of brain to stroke resides in its ability to reverse the age-related propensity toward a neuro-inflammatory state and thus it suppresses gene induction brought on by the "smoldering" neuro-inflammation in the aged brain.

Even though these five summary statements regarding the action of PBN are important advancements, the question remains how does it suppress signal transduction processes? This question cannot be answered definitively now, but there are two likely events where its action are important. First, it has been shown that PBN very effectively interferes with oxidative processes which inactivate brain phosphatases.[38] This then allows the signal transduction mediated phosphorylation processes to be more active. Second, it has been shown that PBN very effectively interacts with respiring mitochondria to suppress their formation of H_2O_2 as a by-product.[43] Much available evidence indicates that H_2O_2 acts as a signaling molecule up-regulating genes in astrocytes.[38] These important points are indicated in a very simplified scheme of signal transduction processes shown in FIGURE 5. Perusal of FIGURE 5 indicates ROS flux from mitochondria responsible for the inactivation of phosphatases as well as the ROS acting as signaling molecules may be one and the same. Hence PBN suppression of ROS flux from mitochondria may then be suppressing both events. Clearly the processes involved are very dynamic and interrelated and this makes it very difficult to simplify.

If the primary action of PBN resides in altering signal transduction processes in the manner noted in FIGURE 5 then the ultimate specific high affinity site basis of its action may be an "idealistic" goal that is not realistic or misleading. To understand this point requires paradigm shifts in the understanding of the complexity and dynamic interactions involved. In other words, the specific high affinity site concept is viable when one considers specific inhibitors of specific enzymatic reactions. A rigorous study of the action of PBN in mitochondrial production of H_2O_2 reveals that

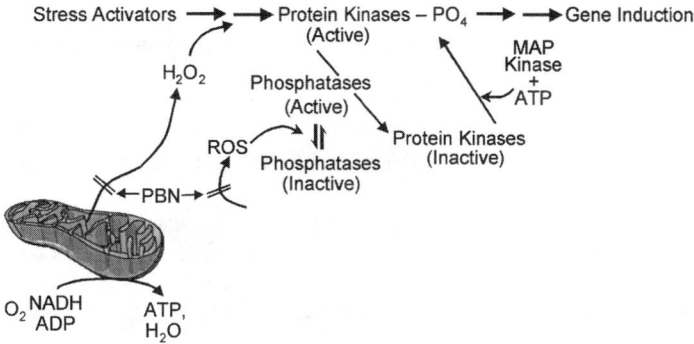

FIGURE 5. Scheme illustrating how ROS production by mitochondria influences both the redox state of phosphatases and the activation of signal transduction processes. PBN has been shown to suppress ROS production by mitochondria.[43]

the generation sites in the electron transport chain are "loose" in the sense that nature has evolved such that the electron transfer bleed serves a necessary and viable function, apparently in signaling systems. It should be noted also that there is more than one site of ROS production in the electron transfer chain and that PBN inhibits only at one site (complex 1) and that the mechanism of ROS formation at these different sites are not similar. The fact that both ROS sources as well as their actions are multiple is also a reason to consider that the need to have a specific high affinity site becomes an impediment to more advanced ways of thinking about complicated biological processes involving ROS and oxidative damage. Additionally, since the ROS are acting as essential signaling molecules then complete shut down of their formation would most likely be fatal. It follows then that only excessive flux of ROS is the condition of significant detriment to biological systems and hence agents that act to suppress excess production and/or oxidative damage may have valuable pharmacological activity. It appears then that agents such as PBN may be very useful in this regard.

ACKNOWLEDGMENTS

Research work was supported in part by NIH grant NS35747 and OCAST grant H97-067 and by a contract from Centaur Pharmaceuticals. The authors appreciate their colleagues including Charles Stewart, M. L. Maidt and Q. Pye for help with experiments and J. A. Schriewer for preparation of the manuscript.

REFERENCES

1. MICHAELIS, L. 1946. Fundamentals of oxidation and respiration. Am. Sci. **34:** 573–596.
2. GERSCHMAN, R., D.L. GILBERT, S.W. NYE, P. DWYER & W.O. FENN. 1954. Oxygen poisoning and x-irradiation: a mechanism in common. Science **119:** 623–626.

3. COMMONER, B., J. TOWNSEND & G.E. PAKE. 1954. Free radicals in biological materials. Nature **174:** 689–691.
4. KEILIN, D. 1966. The History of Cell Respiration and Cytochrome. Cambridge at the University Press
5. MCCORD, J.M. & I. FRIDOVICH. 1969. Superoxide dismutase: an enzymic function for erythrocuperin (Hemocuprein). J. Biol. Chem. **244:** 6049–6055.
6. LOSCHEN, G., L. FLOHE & B. CHANCE. 1971. Respiratory chain linked H_2O_2 production in pigeon heart mitochondria. FEBS Lett. **18:** 261–264.
7. ANZEN, E.G. 1971. Spin trapping. Acc. Chem. Res. **4:** 31–40.
8. BOLTON, J.R., R.K. CLAYTON & D.W. REED. 1969. An identification of the radical giving rise to light-induced electron spin resonance signal in photosynthetic bacteria. Photochem. Photobiol. **9:** 209–218.
9. POYER, J.L., R.A. FLOYD, P.B. MCCAY, E.G. JANZEN & E.R. DAVIS. 1978. Spin trapping of the trichloromethyl radical produced during enzymic NADPH oxidation in the presence of carbon tetrachloride or carbon bromotrichloromethane. Biochim. Biophys. Acta **539:** 402–409.
10. FLOYD, R.A., R. HENDERSON, J.J. WATSON & P.K. WONG. 1986. Use of salicylate with high pressure liquid chromatography and electrochemical detection (LCED) as a sensitive measure of hydroyxl free radicals in adriamycin treated rats. J. Free Rad. Biol. Med. **2:** 13–18.
11. CAO, W., J.M. CARNEY, A. DUCHON, R.A. FLOYD & M. CHEVION. 1988. Oxygen free radical involvement in ischemia and reperfusion injury to brain. Neurosci. Lett. **88:** 233–238.
12. FLOYD, R.A., J.J. WATSON, P.K. WONG, D.H. ALTMILLER & R.C. RICKARD. 1986. Hydroxyl free radical adduct of deoxyguanosine: sensitive detection and mechanisms of formation. Free Rad. Res. Commun. **1:** 163–172.
13. FLOYD, R.A., M.S. WEST, K.L. ENEFF, W.E. HOGSETT & D.T. TINGEY. 1988. Hydroxyl free radical mediated formation of 8-hydroxyguanine in isolated DNA. Arch. Biochem. Biophys. **262:** 266–272.
14. FLOYD, R.A., M.S. WEST, K.L. ENEFF & J.E. SCHNEIDER. 1989. Methylene blue plus light mediates 8-hydroxyguanine formation in DNA. Arch. Biochem. Biophys. **273:** 106–111.
15. FLOYD, R.A. 1990. Role of oxygen free radicals in carcinogenesis and brain ischemia. FASEB J. **4:** 2587–2597.
16. CLOUGH-HELFMAN, C. & J.W. PHILLIS. 1991. The free radical trapping agent N-*tert*-butyl-α-phenyl nitrone (PBN) attenuates cerebral ischaemic injury in gerbils. Free Rad. Res. Commun. **15:** 177–186.
17. ZHAO, Q., K. PAHLMARK, M.-I. SMITH & B.K. SIESJO. 1994. Delayed treatment with the spin trap α-phenyl-*N-tert*-butyl nitrone (PBN) reduces infarct size following transient middle cerebral artery occlusion in rats. Acta Physiol. Scand. **152:** 349–350.
18. COLTON, C.A. & D.L. GILBERT. 1987. Production of superoxide anions by a CNS macrophage, the microglia. Fed. Eur. Biochem. Soc. **223:** 284–288.
19. FLOYD, R.A. 1996. Protective action of nitrone based free radical traps against oxidative damage of the central nervous system. *In* Advances in Pharmacology. H. Sies, Ed. San Diego. Academic Press. **38:** 361–378.
20. HENSLEY, K., J.M. CARNEY, C.A. STEWART, T. TABATABAIE, Q. PYE & R.A. FLOYD. 1996. Nitrone-based free radical traps as neuroprotective agents in cerebral ischemia and other pathologies. *I.* Neuroprotective Agents and Cerebral Ischemia. A.R. Green & A.J. Cross, Eds. :299–317. London. Academic Press Ltd.
21. NAKAE, D., Y. KOTAKE, H. KISHIDA, K. L. HENSLEY, A. DENDA, Y. KOBAYASHI, W. KITAYAMA, T. TSUJIUCHI, C. SANG, C. A. STEWART, T. TABATABAIE, R.A. FLOYD & Y. KONISHI. 1998. Inhibition by phenyl *N-tert*-butyl nitrone on early phase carcinogenesis in the livers of rats fed a choline-deficient, L-amino acid-defined diet. 1998. Cancer Res. **58:** 4548–4551.
22. EDAMATSU, R., A. MORI & L. PACKER. 1995. The spin-trap N-*tert*-a-phenyl-butyl nitrone prolongs the life span of the senescence accelerated mouse. Biochem. Biophys. Res. Commun. **211:** 847–849.

23. SAITO, K., H. YOSHIOKA & R.G. CUTLER. 1998. A spin trap, N-*tert*-butyl-α-phenyl nitrone extends the life span of mice. Biosci. Biotechnol. Biochem. **62:** 792–794.
24. NOVELLI, G.P., P. ANGIOLINI, R. TANI, G. CONSALES & L. BORDI. 1985. Phenyl-T-butyl nitrone is active against traumatic shock in rats. Free Rad. Res. Commun. **1:** 321–327.
25. NOVELLI, G., P. ANGIOLINI, G. CANSALES, R. LIPPI & R. TANI. 1986. Anti-shock action of phenyl-t-butyl-nitrone, a spin trapper. *In* Oxygen Free Radicals in Shock. G.P. Novelli & F. Ursini, Eds. :119–124. Florence, Basel. Karger.
26. CARNEY, J.M., P.E. STARKE-REED, C.N. OLIVER, R.W. LANDRUM, M.S. CHEN, J.F. WU & R.A. FLOYD. 1991. Reversal of age-related increase in brain protein oxidation, decrease in enzyme activity, and loss in temporal and spacial memory by chronic administration of the spin-trapping compound N-*tert*-butyl-α-phenyl nitrone. Proc. Natl. Acad. Sci. USA **88:** 3633–3636.
27. CARNEY, J.M., M.S. KINDY, C.D. SMITH, K. WOOD, T. TATSUNO, J.F. WU, W. R LANDRUM & R.A. FLOYD. 1994. Gene expression and functional changes after acute ischemia: Age-related differences in outcome and mechanisms. *In* Cerebral ischemia and basic mechanisms. A. Hartmann, F. Yatsu & W. Kuschinsky, Eds. :301–311. Berlin, Heidelberg. Springer-Verlag.
28. IADECOLA, C. 1997. Bright and dark sides of nitric oxide in ischemic brain injury. Trends Neurosci. **20:** 132–139.
29. TABATABAIE, T., C. STEWART, Q. PYE, Y. KOTAKE & R.A. FLOYD. 1996. *In vivo* trapping of nitric oxide in the brain of neonatal rats treated with the HIV-1 envelope protein gp 120: Protective effects of α-phenyl-*tert*-butyl nitrone. Biochem. Biophys. Res. Commun. **221:** 386–390.
30. MIYAJIMA, T. & Y. KOTAKE. 1995. Spin trapping agent, phenyl N-*tert*-butyl nitrone, inhibits induction of nitric oxide synthase in endotoxin-induced shock in mice. Biochem. Biophys. Res. Commun. **215:** 114–121.
31. FLOYD, R.A. & J.M. CARNEY. 1995. Nitrone radicals traps (NRTs) protect in experimental neurodegenerative diseases. *In* Neuroprotective Approaches to the Treatment of Parkinson's Disease and Other Neurodegenerative Disorders. C.A. Chapman, C.W. Olanow, P. Jenner & M. Youssim, Eds. :69–90. London. Academic Press.
32. FLOYD, R.A. 1999. Neuroinflammatory processes are important in neurodegenerative diseases: an hypothesis to explain the increased formation of reactive oxygen and nitrogen species as major factors involved in neurodegenerative disease development. Free Rad. Biol. Med. **26:** In press.
33. SMITH, C.D., J.M. CARNEY, P.E. STARKE-REED, C.N. OLIVER, E.R. STADTMAN, R.A. FLOYD & W.R. MARKESBERY. 1991. Excess brain protein oxidation and enzyme dysfunction in normal aging and in Alzheimer's disease. Proc. Natl. Acad. Sci. USA **88:** 10540–10543.
34. GORDON, M.N., W.A. SCHREIER, X. OU, L.A. HOLCOMB & D.G. MORGAN. 1997. Exaggerated astrocyte reactivity after nigrostriatal deafferentation in the aged rat. J. Comp. Neur. **388:** 106–119.
35. SCHRECK, R., P. RIEBER & P.A. BAEUERLE. 1991. Reactive oxygen intermediates as apparently widely used messengers in the activation of the NF-κB transcription factor and HIV-1. EMBO J. **10:** 2247–2258.
36. WALTON, K.M., R. DiROCCO, B.A. BARTLETT, E. KOURY, V.R. MARCY, B. JARVIS, E.M. SCHAEFER & R.V. BHAT. 1998. Activation of p38^{MAPK} in microglia after ischemia. J. Neurochem. **70:** 1764–1767.
37. DA SILVA, J., B. PIERRAT, J.-L. MARY & W. LESSLAUER. 1997. Blockade of p38 mitogen-activated protein kinase pathway inhibits inducible nitric-oxide synthase expression in mouse astrocytes. J. Biol. Chem. **272:** 28373–28380.
38. ROBINSON, K.A., C.A. STEWART, Q.N. PYE, X. NGUYEN, L. KENNEY, S. SALZMAN, R.A. FLOYD & K. HENSLEY. 1999. Redox-sensitive protein phosphatase activity regulates the phosphorylation state of p38 protein kinase in primary astrocyte culture. J. Neurosci. Res. In press.

39. HENSLEY, K., R.A. FLOYD, N.-Y. ZHENG, R. NAEL, K.A. ROBINSON, X. NGUYEN, Q.N. PYE, C.A. STEWART, J. GEDDES, W.R. MARKESBERY, E. PATEL, G.V.W. JOHNSON & G. BING. 1999. p38 Kinase is activated in the Alzheimer's disease brain. J. Neurochem. In press.
40. HENSLEY, K., M.L. MAIDT, Q.N. PYE, C.A. STEWART, M. WACK, T. TABATABAIE & R.A. FLOYD. 1997. Quantitation of protein-bound 3-nitrotyrosine and 3,4-dihydroxyphenylalanine by high performance liquid chromatography with electrochemical array detection. Anal. Biochem. **251:** 187–195.
41. HENSLEY, K., M.L. MAIDT, Z. YU, W.R. MARKESBERY & R.A. FLOYD. 1998. Electrochemical analysis of protein nitrotyrosine and dityrosine in the Alzheimer brain indicates region-specific accumulation. J. Neurosci. **18:** 8126–8132.
42. MATTSON, M.P., Y. GOODMAN, H. LUO, W. FU & K. FURUKAWA. 1997. Activation of NF-κB protects hippocampal neurons against oxidative stress-induced apoptosis: evidence for induction of manganese superoxide dismutase and suppression of peroxynitrite production and protein tyrosine nitration. J. Neurosci. Res. **49:** 681–697.
43. HENSLEY, K., Q.N. PYE, M.L. MAIDT, C.A. STEWART, K.A. ROBINSON, F. JAFFREY & R.A. FLOYD. 1998. Interaction of α-phenyl-*N-tert*-butyl nitrone and alternative electron acceptors with complex I indicates a substrate reduction site upstream from the rotenone binding site. J. Neurochem. **71:** 254–2557.

Effects of Atypical Antioxidative Agents, S-nitrosoglutathione and Manganese, on Brain Lipid Peroxidation Induced by Iron Leaking from Tissue Disruption

PEKKA RAUHALA[a,b] AND CHUANG C. CHIUEH [a,c]

[a]*Unit on Neurodegeneration and Neuroprotection, Laboratory of Clinical Science, National Institute of Mental Health, NIH Clinical Center, Bethesda, Maryland 20892-1264, USA*

[b]*Institute of Biomedicine, Department of Pharmacology and Toxicology, Box 8, University of Helsinki, Helsinki, Finland*

ABSTRACT: A fluorescent assay of brain lipid peroxidation was used for screening new antioxidants for the prevention of neurodegeneration caused by free radicals. Incubation of rat brain homogenates led to a temperature-dependent increase in production of fluorescent adducts of peroxidized polyunsaturated fatty acids; it was inhibited completely by lowering the incubation temperature to 4°C. This tissue disruption-induced brain lipid peroxidation at 37°C was blocked by deferoxamine (IC_{50} = 0.3 µM) and EDTA; it was augmented by adding submicromolar iron and hemoglobin. Ferrous ion's pro-oxidative activities were five times more potent than ferric ion. Micromolar manganese completely inhibited lipid peroxidation, confirming earlier unexpected *in vivo* reports. Trolox and vitamin C suppressed brain lipid peroxidation with IC_{50} values of 20 and 500 µM, respectively. U-78517F was approximately 20 times more potent than Trolox. 17β-Estradiol, hydralazine, S-nitrosoglutathione and 3-hydroxybenzylhydrazine were as potent as Trolox. Melatonin, glutathione, α-lipoic acid and *l*-deprenyl were about 20 times less potent than Trolox. Surprisingly, N-tert-butyl-α-phenylnitrone was a weak antioxidant. Furthermore, this procedure can also detect pro-oxidative side effects of vitamin C, oxidized glutathione, penicillamine and Angeli's salt. The present results obtained from this selective fluorescent assay are consistent with earlier reports that iron complexes promote while manganese inhibits brain lipid peroxidation caused by cell disruption. S-Nitrosoglutathione, melatonin, 17β-estradiol, and manganese have been successfully tested in cell/animal models for their potential neuroprotective effects. In conclusion, monitoring fluorescent adducts of peroxidizing polyunsaturated fatty acids in brain homogenates is a simple, quantitative method for studying iron-dependent brain lipid peroxidation and for screening of potential neuroprotective antioxidants in both *in vitro* and *in vivo* preparations.

[c]Address for correspondence: Dr. Chuang C. Chiueh, Unit on Neurodegeneration and Neuroprotection, Laboratory of Clinical Science, National Institute of Mental Health, NIH Clinical Center, 10/3D-41, Bethesda, MD 20892-1264.
e-mail: chiueh@helix.nih.gov

INTRODUCTION

Increasing evidence suggests that oxidative stress may contribute to the pathogenesis of degenerative brain disorders.[1–5] Oxygen-centered radicals are inevitable by-products of biological redox reactions such as mitochondrial respiration and NADPH-dependent oxidase in phagocytes. They could also be generated by iron-containing proteins such as subtypes of cytochrome P450 and hemoglobin. In the presence of trace amounts of iron (i.e., ferrous citrate), both enzymatic and non-enzymatic oxidation of dopamine generates semiquinone radicals, hydrogen peroxide and hydroxyl radicals. These highly reactive hydroxyl radicals are generated from hydrogen peroxide through the Fenton's reaction catalyzed by copper and iron but not manganese.[6] In addition to Fenton's reaction, iron-oxygen complexes could also play a oxidative role in biological systems.[7] Reactive oxygen species may release excitatory amino acids (i.e., glutamate), trigger lipid peroxidation, protein oxidation, interfere with respiration or ATP formation in mitochondria, and cause DNA damage; leading to changes in membrane fluidity, disruption of intracellular ion homeostasis and progressive neuronal injury.[8,9] Oxidation of phospholipids may generate lipid radicals and reactive aldehyde species such as 4-hydroxynonenal which may induce apoptosis in differentiating cells and immature neurons.[10,11] Moreover, brain neurons are more vulnerable than non-neuronal cells to oxidative stress because they contain high levels of polyunsaturated fatty acids[d] (PUFA) and consume large amounts of oxygen. Therefore, there is an urgent need for the development of antioxidants to protect brain neurons, to prevent or slow down clinical deterioration in degenerative brain disorders and to improve quality of life.

During the lipid peroxidation chain reaction, peroxidizing PUFA decomposes to secondary products including alkoxyl and peroxyl lipid radicals and malondialdehyde species as well. Aldehyde species derived from peroxidizing brain lipids immediately react with primary amines such as amino acids and ethanolamines in brain phospholipids, leading to accumulation of fluorescent adducts of lipid peroxidation. The exact chemical structure of these dihydropyridine adducts is not fully understood.[12,13] However, previous *in vivo* studies indicate that the measurement of these stable fluorescent products of peroxidizing PUFA may be a reliable method for studying free radical induced oxidative stress and brain injury.[14–17] Moreover, the formation of fluorescent products of lipid peroxidation is directly correlated with the loss of PUFA in brain homogenates.[18] Therefore, we and others have proposed that these fluorescent adducts of PUFA may be used as a biological marker for monitoring lipid peroxidation in both *in vitro* and *in vivo* preparations. However, the usefulness of this fluorescent method for screening pro- and anti-oxidants in brain homogenates has not been systematically evaluated. In this study, this fluorescent method was used for comparing the antioxidative potency of typical (i.e., Trolox, vitamin C, U-78517F, glutathione or GSH) and atypical antioxidants (i.e., N-tert-butyl-α-phenylnitrone, melatonin, 17β-estradiol, manganese, and S-nitrosoglutathione) on tissue disruption-induced peroxidation of rat brain lipids at 37°C. Furthermore, we focused on the relative antioxidative potency of 17β-estradiol,

---

[d]ABBREVIATIONS: GSH, glutathione; GSNO, S-nitrosoglutathione; GSSG, oxidized glutathione; ·NO, nitric oxide; NO⁻, nitroxyl anion; NSD-1015, 3-Hydroxybenzylhydrazine; PUFA, polyunsaturated fatty acid

melatonin, *l*-deprenyl and *S*-nitrosoglutathione (GSNO) in brain homogenates, because these compounds have been proposed to protect brain neurons against oxidative stress.[16,19–28] We also investigated proposed pro- and anti-oxidative effects of metal ions (such as iron, copper, zinc and manganese).

METHODS

Fluorimetric Measurement of Brain Lipid Peroxidation

Male Sprague-Dawley rats (250–350 g, Taconic Farms, Germantown, NY) were decapitated and cortical brain samples were dissected and stored at −70°C.

Pooled brain samples were homogenized in ice-cold Ringer's solution (500 mg/ml) using an ultrasonic cell disrupter. Brain homogenates were then diluted to 50 mg/ml with ice-cold Ringer's solution. Brain homogenates (1 ml) were incubated at 37°C in a shaking water bath for 2–24 hours. In some experiments different incubation temperature (e.g., 0–4, 25, 31 or 42°C) were used. Drugs were added to homogenates at the beginning of incubation (co-treatment) or 2 h thereafter (post-treatment). Fluorescent products of brain lipid peroxidation, which consist of cross-linked adducts of primary amines with reactive aldehyde species derived from peroxidizing PUFA[12,13,29,30] were determined using a microassay procedure of Mohanakumar *et al.*[15] modified from Dillard and Tappel.[12] After incubation, a 200 µl aliquot was transferred to a tube containing 200 µl of chloroform and 100 µl of methanol. The mixture was vortexed and kept on ice for 15 min. After centrifugation (at 5,200 × *g* for 6 min) 150 µl from the chloroform extract was transferred to another tube containing 50 µl of methanol. The relative fluorescent intensities of the brain extracts (100 µl; corresponding to 3.75 mg of brain tissue) were measured in a microcuvette using a Perkin Elmer LS 50B spectrofluorometer (activation wavelength 356 nm and emission wavelength 426 nm; calibrated with quinine sulfate).

Chemicals

Ascorbic acid, deferoxamine mesylate, Na_2EDTA, D(−)-penicillamine hydrochloride, melatonin, 17β-estradiol, ferrous ammonium sulfate, ferric chloride, glutathione (GSH), oxidized glutathione (GSSG), bovine hemoglobin, hydralazine hydrochloride, manganese chloride, and zinc chloride were obtained from Sigma (St. Louis,MO). Trolox (6-hydroxy-2,5,7,8-tetramethylchroman-2-carboxylic acid), *N*-tert-butyl-α-phenylnitrone (PBN) and copper chloride (Cu^+ and Cu^{++}) were obtained from Aldrich Chemical Inc., Milwaukee, WI) and GSNO from Calbiochem (LaJolla, CA). Angeli's salt was purchased from Cayman Chemicals (Ann Arbor, MI). 3-Hydroxybenzylhydrazine (NSD-1015) was obtained from Research Biochemicals International (Natic, MA). α-Lipoic acid was provided in kind by Asta Medica (Frankfurt, Germany). U-78517F [2-{[4-(2,6-di(1-pyrrolidinyl)-4-pyrimidinyl)-1-piperazinyl]methyl}-3,4-dihydro-2,5,7,8-tetramethyl-2h-1-benzopynan-6-ol dihydrochloride] and *l*-deprenyl were kindly donated by Pharmacia-Upjohn Company (Kalamazoo, MI) and Chinoin (Budapest, Hungary), respectively. Melatonin, U-78517F, 17β-estradiol and α-lipoic acid were dissolved in dimethylsulfoxide and then diluted in Ringer's solution (final concentration <1% dimethylsulfoxide).

Proper amounts of dimethylsulfoxide were used in sham controls. All other drugs were dissolved and diluted in ice-cold Ringer's solution.

Statistical Analysis

Data are presented as the mean values ± SEM of indicated numbers of observations. Results were analyzed by one way analysis of variance and p values were calculated by using Newman-Keuls test (p values less than 0.05 were considered statistically significant).

RESULTS

Effects of Temperature on Brain Lipid Peroxidation Induced by Cell Disruption

Tissue disruption-induced brain lipid peroxidation was monitored for up to one-day incubation at 37°C using a fluorescent assay of primary amines cross-linked with aldehyde species derived from peroxidized brain PUFA. Maximal fluorescence was seen when brain extracts were excited with 356 nm and emission wavelength of 426 nm was used (FIG. 1). During the first four-hour incubation, samples incubated below 25°C emitted little fluorescence while samples incubated at higher temperature (e.g., 31, 37 and 42°C) emitted greater fluorescence intensity, corresponding to the temperature used (FIG. 2). The fluorescence intensity of the end products of lipid peroxidation increased further when the incubation period was increased from 4 to 24 h. However, there was no apparent increase in the peroxidation of brain lipids when the samples were incubated at 0–4°C (FIGS. 1 and 2), indicating that hypothermia may be a useful procedure for minimizing oxidative stress in the brain.

Pro- and Anti-oxidative Effects of Metal Ions

Both ferrous and ferric ions concentration-dependently (0 to 25 µM) increased lipid peroxidation (FIGS. 1 and 3). Iron had a maximal 4-fold increase between 25–125 µM iron concentrations following a 2-h incubation at 37°C. Ferrous ion was approximately 5 times more potent than ferric ion in augmenting brain lipid peroxidation. Moreover, brain lipid peroxidation caused by cell disruption and iron complexes were a time- and concentration-dependent process. The lipid peroxidation process continued during incubation of brain homogenates at 37°C up to 24 h.

At 1 µM concentration, only ferrous ion, but not copper and manganese caused a significant increase (162%) in lipid peroxidation of brain homogenates during a 2-h incubation at 37°C ($p < 0.05$, TABLE 1). At unphysiological extremely high concentrations (i.e. 100 µM), copper ions (Cu^+ and Cu^{++}) caused a small 80% increase while zinc induced only a 50% increase in brain lipid peroxidation. Surprisingly, in contrast to transition metals' pro-oxidant effect, manganese (Mn^{++}; 10 to 100 µM) completely suppressed brain lipid peroxidation, confirming prior *in vivo* results.[17]

Pro-oxidative Effects of Hemoglobin

Hemoglobin concentration dependently increased lipid peroxidation during a 2-h incubation in brain homogenates (FIG. 4). Similar to the pro-oxidant effect of iron,

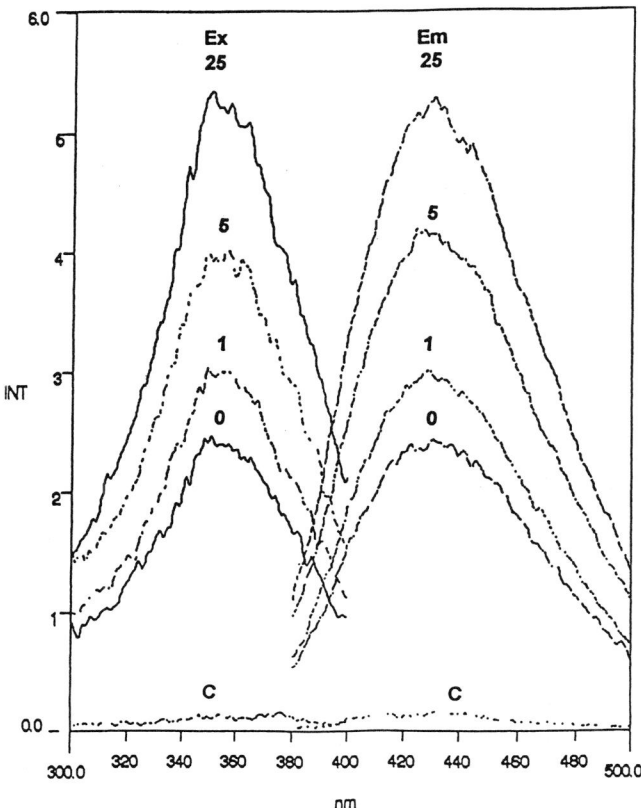

FIGURE 1. The excitation and emission spectra of fluorescent adducts in peroxidizing brain lipids catalyzed by ferrous ions. Rat brain homogenates (50 mg/ml) were incubated for 24 h at 37°C with ferrous ion (0 μM; 1 μM; 5 μM; 25 μM) in 1 ml Ringer's solution. One group of sample was incubated at 0–4°C (C). After incubation, peroxidized brain lipids and their fluorescent adducts were extracted and scanned for excitation (EX.) and emission (EM.) spectra using a spectrofluorometer. The maximal excitation and emission wavelength were peaked at 356 nm and 426 nm, respectively.

hemoglobin (0.33 to 33 μM) significantly and maximally increased 3-fold fluorescent products of oxidized brain lipids or PUFA ($p < 0.05$).

Anti- and Pro-oxidative Effects of Metal Chelators

Deferoxamine and EDTA, iron chelators were potent inhibitors of brain lipid peroxidation (FIG. 5A). The IC_{50} values for deferoxamine and EDTA were 0.3 and 5 μM, respectively. The antioxidant potency of these metal chelators declined with an increase in incubation time from 3 to 24 h. After a 3-h incubation the copper chelator penicillamine did not completely suppress lipid peroxidation even with

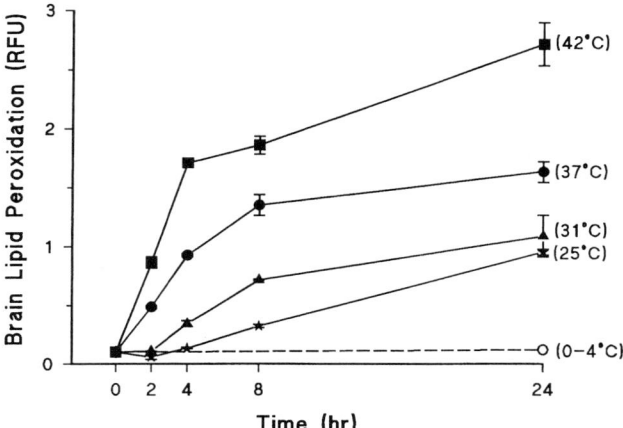

FIGURE 2. The effects of incubation temperature on brain lipid peroxidation. Rat brain homogenates in Ringer's solution (50 mg/ml) were incubated for up to 24 h. The incubation temperature was increased from 37°C to 42°C for hyperthermia experiments while it was decreased to 31°C or 25°C for hypothermia studies. Control samples were incubated at 0–4°C. After incubation, the relative intensity of fluorescent adducts of peroxidizing brain lipids in solvent extracts were measured (activation/emission wavelength: 356/426 nm). Each point represents the mean ± SEM of relative fluorescent intensities units (RFU) of fluorescent adducts of lipid peroxidation in 3.75 mg of brain tissue ($n = 3$–4).

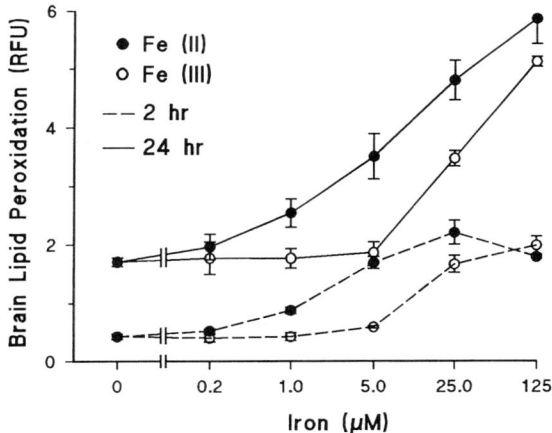

FIGURE 3. Comparison of the pro-oxidant effects of ferrous and ferric ion on brain lipid peroxidation. Micromolar concentrations of ferrous (●) and ferric ion (○) were added to 1 ml rat brain homogenates (50 mg/ml) and incubated at 37°C for 2 (--) or 24 (—) h. Fluorescent products of peroxidizing brain lipids were measured and the results depict the relative fluorescent intensity in 3.75 mg of brain tissue ($n = 3$–6).

TABLE 1. Effect of iron, copper, zinc and manganese on lipid peroxidation in brain homogenates[a]

	Concentration (μM)	Lipid Peroxidation (RFU)	Net Change (%)
Control	0	0.378 ± 0.014	0
Iron (II)	1	0.991 ± 0.046*	+162
	5	1.371 ± 0.020*	+363
Copper (I)	10	0.421 ± 0.041	+11
	100	0.691 ± 0.018*	+83
Copper (II)	10	0.437 ± 0.015	+16
	100	0.679 ± 0.054*	+80
Zinc (II)	10	0.420 ± 0.030	+11
	100	0.558 ± 0.045*	+48
Manganese (II)	10	0.012 ± 0.010*	−97
	100	0.031 ± 0.010*	−92

[a]Metal ions were added to rat brain homogenates (50 mg/ml) and incubated at 37°C for 2 h. Fluorescent products of peroxidizing brain lipids in 3.75 mg tissue were isolated and measured as relative fluorescence intensity (RFU at excitation/emission wavelength: 356/426 nm) using a spectrofluorometer. Control values show tissue disruption-induced increase of lipid peroxidation in the brain homogenates following incubation at 37°C for 2 h. Sham control samples were incubated at 0–4°C whose RFU were not significantly over that of reagent blank. $n = 3$. *$p < 0.05$ compared to control.

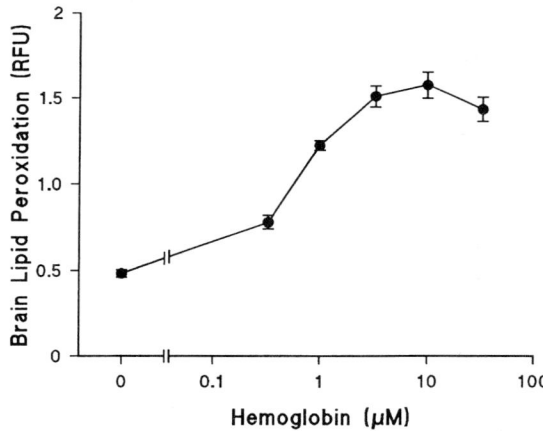

FIGURE 4. The effect of hemoglobin on brain lipid peroxidation. Hemoglobin (0 to 33 μM) was added to rat brain homogenates (50 mg/ml) and incubated at 37°C for 2 h. Fluorescent products of lipid peroxidation were measured as described in the METHODS. The ordinate represents the mean ± SEM of relative fluorescent intensity units (RFU) of peroxidized brain lipids in 3.75 mg of brain tissue (n = 3).

concentrations up to 800 µM. Furthermore, after a 24-h incubation period, these high concentrations of penicillamine were largely ineffective, even becoming pro-oxidative.

Anti- and Pro-oxidative Properties of Classic Antioxidants

Vitamin E analogs, such as Trolox and U-78517F significantly inhibited brain lipid peroxidation during a 3- to 24-h incubation period (FIG. 5B). Following a 3-h incubation IC_{50} values for U-78517F and Trolox were 1 µM and 20 µM, respectively; these values were not significantly altered following a prolonged incubation of up to 24 h. High concentrations of vitamin C produced an antioxidative effect on the peroxidation of brain lipids after both 3- and 24-h incubation periods, with IC_{50} values of 500 and 600 µM, respectively (FIG. 5B). However, low concentrations of ascorbate (below 200 µM) became a pro-oxidant which augmented tissue disruption-induced brain lipid peroxidation.

After a 3-h incubation with brain homogenates, GSH significantly suppressed lipid peroxidation with an IC_{50} value of 1.5 mM (FIG. 5C). After a 24-h incubation GSH's antioxidative activity significantly declines by two thirds (FIG. 5C). The oxidized GSH (GSSG) became a pro-oxidative agent which increased brain lipid peroxidation. α-Lipoic acid was as potent as the antioxidant GSH after a 3-h incubation but was more potent than GSH after incubation for 24 h.

Effects of GSNO and Angeli's Salt

GSNO, a nitric oxide (·NO) releasing compound suppressed lipid peroxidation with the same concentration range (IC_{50}: 20 µM) as Trolox and was more potent than GSH in the acute 3-h experiment (FIG. 5D). However, in the 24-h subchronic experiment 3- to 4-fold higher concentrations of GSNO were needed to produce a similar antioxidative effect as was seen in the 3-h acute experiment. Angeli's salt, a donor of nitroxyl anion (NO^-), induced a small pro-oxidant effect at 10–100 µM concentrations after a 3-h incubation (FIG. 5D). However, higher concentration of Angeli's salt (1 mM) suppressed lipid peroxidation. After a 24-h incubation period, both pro-oxidant and antioxidant effects of Angeli's salt were significantly attenuated.

Anti-oxidative Potency of Atypical Antioxidants

17β-Estradiol concentration-dependently (IC_{50}: 15 µM) suppressed brain lipid peroxidation induced by cell disruption (FIG. 5E). After a 3-h incubation 17β-estradiol was as potent as Trolox and about 25 and 50 times more potent than deprenyl and melatonin, respectively (FIG. 5E). The relative potency of 17β-estradiol and melatonin did not change after a 24-h incubation period. However, deprenyl's antioxidative action was further enhanced after increasing the incubation period from 3 to 24 h since N-demethylated metabolite is still active.

Hydrazine containing substances, such as hydralazine and NSD-1015 suppressed brain lipid peroxidation with IC_{50} of 15 and 40 µM, respectively (FIG. 5F). There was no significant difference between the potency of hydrazine compounds incubated for either 3 or 24 h.

N-tert-butyl-α-phenylnitrone, an electron paramangnetic resonance trapping agent and a radical scavenger, suppressed lipid peroxidation with millimolar concen-

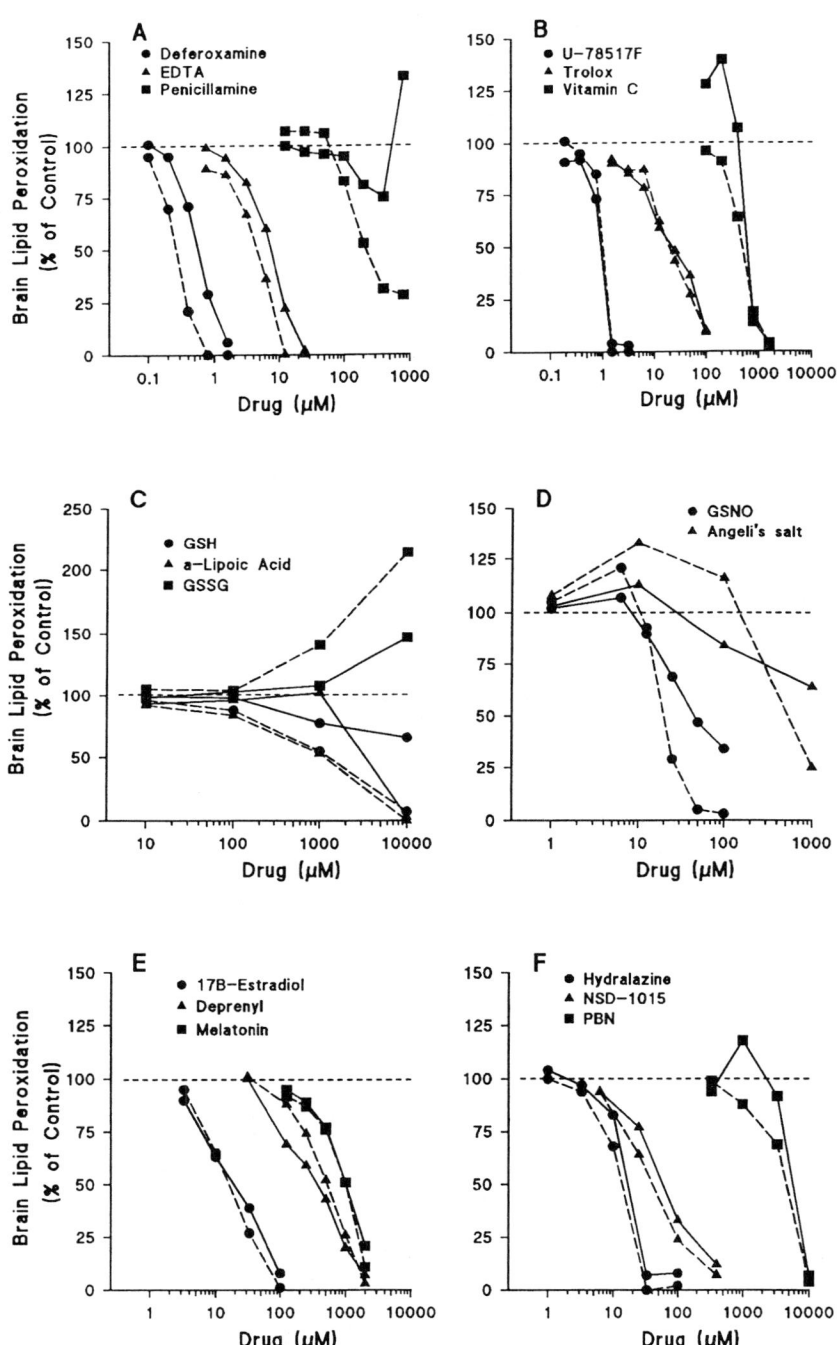

trations (1 to 10 mM) after a 3-h incubation period but only the highest dose (10 mM) significantly suppressed lipid peroxidation after a 24-h incubation (FIG. 5F).

Post-treatment Effects of Metal Chelators and Antioxidants on the Propagation of Brain Lipid Peroxidation

Iron chelators, deferoxamine (1.5 μM) and EDTA (25 μM) almost completely suppressed lipid peroxidation when added to brain homogenates at the beginning of incubation. These iron chelators also suppressed the propagation of lipid peroxidation for up to 24 h when they were added to brain homogenates 2 h after the beginning of incubation (FIG. 6A). Vitamin C (1 mM) and U-78517F (1.5 μM) effectively inhibited lipid peroxidation when added to brain homogenates either at the beginning of incubation or 2 h after initiation of peroxidation of PUFA (FIG. 6B). Therefore, these compounds could suppress brain lipid peroxidation at both the initiation and the propagation steps.

DISCUSSION

The present results demonstrate that the assay of fluorescent adducts of peroxidizing PUFA is a sensitive and reliable method for investigating oxidative stress or lipid peroxidation in brain homogenates. Moreover, the fluorescence spectra of peroxidized brain lipids after cell disruption was similar to those spectra seen after iron-induced peroxidation of PUFA (such as linolenic acid, arachidonic acid and docosahexanoic acid) in the presence of ethanolamine (data not shown). This is consistent with earlier reports that fluorescent products are cross-linked adducts of amino acids and primary amines with aldehyde species derived from peroxidizing PUFA or brain lipids.[13,29]

Brain lipid peroxidation caused by sonication or tissue disruption was completely blocked by micromolar iron chelators such as deferoxamine and EDTA, confirming that this is an iron-dependent lipid peroxidation chain reaction. It is noteworthy that

---

FIGURE 5. Effects of typical and atypical antioxidants on brain lipid peroxidation. Testing compounds, such as (**A**) metal chelators: deferoxamine (0 to 1.6 μM), EDTA (0 to 25 μM) and penicillamine (0 to 800 μM); (**B**) typical antioxidants: U-78517F (0 to 3.1 μM), Trolox (0 to 100 μM) and ascorbate (0 to 1600 μM); (**C**) thiol compounds: glutathione (GSH; 0 to 10 MM.), oxidized glutathione (GSSG; 0 to 10 MM.), and α-lipoic acid (0 to 10 mM); (**D**) nitric oxide and nitroxyl anion donors: S-nitrosoglutathione (GSNO; 0 to 100 μM) and Angeli's salt (0 to 1000 μM); (**E**) atypical antioxidants: 17β-estradiol (0 to 100 μM), l-deprenyl (0 to 2000 μM) and melatonin (0 to 2000 μM); (**F**) hydrazine and nitrone compounds: hydralazine (0 to 100 μM), 3-hydroxybenzylhydrazine (NSD-1015; 0 to 400 μM), and N-tert-butyl-α-phenyl-nitrone (PBN; 0 to 10 mM), were added to rat brain homogenates (50 mg/ml) immediately before the incubation. Samples were incubated at 37°C for a period of either 3 h (-----) or 24 h (———). After incubation fluorescent adducts of peroxidizing PUFA in brain samples were extracted using organic solvent. The relative fluorescent intensity units (RFU) in 3.75 mg of brain tissue were measured using a spectrofluorometer and used as an index for lipid peroxidation. Control values for different batches of brain lipid peroxidation following a 3-h or 24-h incubation at 37°C were between 0.6–0.9 and 1.9–2.8 RFU, respectively. Results depict the mean values and corresponding SEM was always less than 10% of the mean ($n = 3$–6).

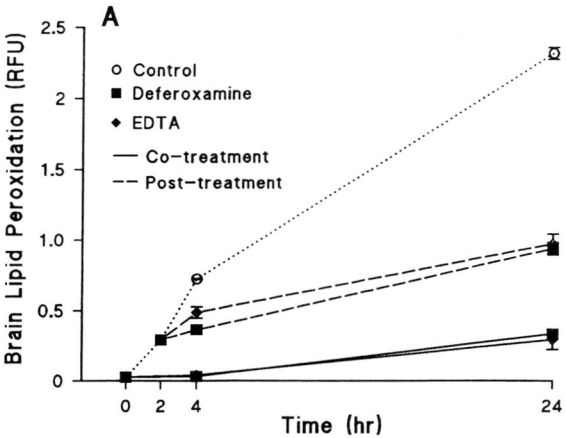

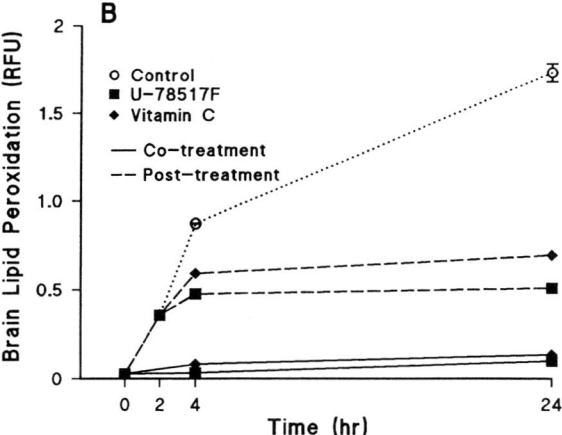

FIGURE 6. The effect of post-treatment with iron chelators and antioxidants on the propagation of brain lipid peroxidation. Control curve (.....) shows the time-dependent increase in lipid peroxidation of brain homogenates (50 mg/ml in Ringer's solution) following 24-h incubation at 37°C. Deferoxamine (1.5 µM), EDTA (25 µM), U-78517F (1.5 µM) and ascorbate (1 mM) were added just before (co-treatment, —) or 2 h after (post-treatment, -- --) the beginning of 37°C incubation. Zero time values are the mean results of brain samples incubated at 0–4°C. Results represent mean ± SEM (n = 4) of relative fluorescence intensity units (RFU) of fluorescent products of peroxidized brain lipids were measured as described in METHODS.

copper ions (Cu^+ and Cu^{++}) which catalyze the peroxidation of low-density lipoprotein in plasma were much less potent pro-oxidative metals than iron complexes in the generation of fluorescent adducts of peroxidized brain lipids. Furthermore, in the past, unphysiological high iron concentrations (i.e., >100 µM) were used for studying brain lipid peroxidation using a relatively non-specific thiobarbiturate procedure. Recently, a more relevant pro-oxidant effects of submicromolar iron complexes (<1 µM) on the initiation of brain lipid peroxidation has been demonstrated using the present method.[31] The present result further indicates that Fe^{++} ion was approximately 5 times more potent pro-oxidative than Fe^{+++} ion in the augmentation of brain lipid peroxidation caused by cell disruption. Moreover, similar to the effect of ferrous ion submicromolar concentrations of hemoglobin significantly increased lipid peroxidation in brain homogenates. This result suggests that hemoglobin may catalyze the propagation of lipid peroxidation in the injured brain where iron is released from the heme protein. These findings support a notion that low micromolar levels of free iron released during damage of brain tissue may initiate brain lipid peroxidation.[32] Therefore, iron-dependent cell disruption-induced lipid peroxidation may mimic oxidant stress/injury caused by head trauma and stroke.

Manganese (Mn^{++}), in contrast to iron's pro-oxidative effects, produced a significant antioxidative action in brain preparations. It completely prevented brain lipid peroxidation caused by cell disruption at 10 µM. Consistent with current *in vitro* findings, manganese has recently been shown to inhibit iron-induced oxidative stress and injury in the nigrostriatal dopaminergic neurons *in vivo*.[17] The present finding that brain neurons are prone to oxidative injury caused by iron complexes is in agreement with the hypothesis that oxidative stress caused by disruption of iron homeostasis in the brain may accelerate the progressive, neurodegenerative course in stroke, head trauma, Parkinson's disease and perhaps Alzheimer's disease. Furthermore, a selective accumulation of iron complexes in the basal ganglia may lead to a site-specific oxidative injury to the A9 nigrostriatal dopaminergic neurons more than the A10 dopaminergic neurons in Parkinson's disease.[4,33,34]

There are several findings showing that the present fluorescent assay of brain lipid peroxidation is sensitive and reliable screening method for new potential antioxidant compounds.

(i) A dual pro- and anti-oxidative effects of penicillamine and ascorbate were demonstrated by the present method which validates a practical use of this method for screening dual pro- and anti-oxidative effects of new drugs in brain samples.[35,36] Interestingly, the present results show that GSSG has pro-oxidative properties. It is possible that in the presence of trace amounts of iron GSSG is converted to reactive thiyl radical. Dual anti- and pro-oxidative effects may complicate the use of some of new antioxidants *in vivo*.

(ii) The results obtained from the present simple fluorescent method have verified the relative potency of known antioxidants, such as Trolox, vitamin C, α-lipoic acid and U-78517F, which were previously examined using different methods.[36–38]

(iii) It demonstrated the potential clinical use of hypothyroid procedure for minimizing brain lipid peroxidation caused by free radical generating compounds, head trauma and stroke.

(iv) In addition, the present method provided means for evaluating subchronic effects of atypical antioxidants for up to 24 h at 37°C. The comparison between 3- and 24-h experiments reveled that two times higher concentrations of EDTA and deferoxamine were needed to suppress brain lipid peroxidation when the incubation period was extended. The IC_{50} values of Trolox, U-78517F, 17β-estradiol and melatonin were not significantly altered by the incubation duration and in fact low concentrations of deprenyl were more potent during a 24-h incubation period than in 3-h experiments.

(v) Furthermore, this fluorescent assay can be used for measuring brain lipid peroxidation in both *in vitro* and *in vivo* preparations.[6,12–17,31,39]

Estrogen, 17β-estradiol, melatonin and deprenyl have been shown to protect brain neurons in both *in vitro* and *in vivo* preparations.[19,21,22,24–26,39–42] Similar to well-known antioxidants, the present results indicate that all these compounds suppressed lipid peroxidation in brain homogenates. It is obvious that these drugs have antioxidant properties in addition to previously known specific pharmacological actions. Surprisingly, 17β-estradiol was as potent antioxidant as Trolox. Interestingly, metabolites of 17β-estradiol have been shown to be more potent than 17β-estradiol against copper-induced oxidation of plasma low-density lipoproteins.[43] Moreover, melatonin and deprenyl were not potent antioxidants in brain homogenates *in vitro*. Besides the inhibition of lipid peroxidation, however, *l*-deprenyl and melatonin may modulate the expression of antioxidant defense genes,[44,45] which could greatly enhanced their antioxidative properties *in vivo*. The present and previous preclinical results indicate that at least a part of antioxidant properties of both deprenyl and estrogen may be due to their ability in the termination of brain lipid peroxidation and thus contribute to the beneficial effects of these drugs in slowing the progressive clinical deterioration in Parkinson's disease and/or Alzheimer's dementia.[20,23,27]

GSNO, an atypical antioxidant, suppressed lipid peroxidation with the same concentration range as Trolox. Furthermore, GSNO (IC_{50} = 20 μM) was 75 times more potent than GSH (IC_{50} = 1.5 mM) because it can slowly release micromolar concentration of ·NO. This is in agreement with previously demonstrated antioxidant properties of ·NO in both *in vivo* and *in vitro* preparations.[16,46–50] Furthermore, GSNO protects brain dopamine neurons against iron-induced oxidative stress/injury *in vivo*.[50] Since GSNO is formed *in vivo*,[51–53] it may be a part of the antioxidative cellular defense systems. To the best of our knowledge the effect of Angeli's salt or NO^- on lipid peroxidation have not been previously reported. In the present study this NO^- releasing compound produced a weak pro-oxidant effect at low micromolar levels and a strong antioxidant action at higher concentrations. The exact mechanism of pro-oxidant effects of Angeli's salt is not known. However, it has been suggested that the mixture of NO^- and oxygen may generate peroxynitrite,[54] a ·NO-derived oxidant, which may initiate significant lipid peroxidation *in vitro*[55] but is weak oxidant *in vivo*.[50] This discrepancy may be due to a fact that in the presence of physiological concentration of carbon dioxide peroxynitrite is rapidly decomposed to nitrates[56,57] which can be rapidly eliminated from the brain via the acid transporters. NO^- released from Angeli's salt is converted to ·NO by oxidizing agents,[58] leading to vasodilatation.[59]

Our preliminary results indicate that after the conversion of NO^- to $\cdot NO$ by copper, the pro-oxidant effects of Angeli's salt faded away and its antioxidant effects became more prominent (P. Rauhala & C.C. Chiueh, unpublished observation). It has been recently suggested that nitric oxide synthase may first generate NO^-, which is then converted to $\cdot NO$ in vivo.[60] Furthermore, increasing evidence supports a notion that $\cdot NO$ is a potent antioxidant which may potentiate the effects of carbon dioxide in the suppression of the peroxidative action of peroxynitrite. Recent in vivo studies clearly indicate that antioxidative and cellular modulatory effects of $\cdot NO$ are more prominent than a weak and undetectable pro-oxidative effect of peroxynitrite in the brain.[50] Thus, both $\cdot NO$ and GSNO protect against brain lipid peroxidation caused by tissue injury (cell disruption), iron complexes and peroxynitrite in vivo and in vitro. Interestingly, GSNO is approximately 100 times more potent than the classic antioxidant GSH, leading to a working hypothesis that $\cdot NO$ may scavenge peroxyl lipid radicals (LOO$\cdot$ + $\cdot NO$ → LOO-NO), convert thiyl radicals to a more potent antioxidant GSNO (GS$\cdot$ + $\cdot NO$ → GSNO) which may prove to be a part of antioxidative cellular defense system.[50,61]

In conclusion, this fluorimetric method is proven to be a reliable and sensitive method for investigating both anti- and pro-oxidant effects of compounds or atypical antioxidative drugs beyond their known pharmacological actions in both in vitro and in vivo preparations.

REFERENCES

1. HALLIWELL, B. 1992. Reactive oxygen species and the central nervous system. J. Neurochem. **59:** 1609–1623.
2. AMES, B.N., M.K. SHIGENAGA & T.M. HAGEN. 1993. Oxidants, antioxidants, and the degenerative diseases of aging. Proc. Natl. Acad. Sci. USA **90:** 7915–7922.
3. OLANOW, C.W. 1993. A radical hypothesis for neurodegeneration. Trends Neurosci. **16:** 439–444.
4. CHIUEH, C.C., H. MIYAKE & M.T. PENG. 1993. Role of dopamine autoxidation, hydroxyl radical generation, and calcium overload in underlying mechanisms involved in MPTP-induced parkinsonism. Adv. Neurol. **60:** 251–258.
5. SIMONIAN, N.A. & J.T. COYLE. 1996. Oxidative stress in neurodegenerative diseases. Annu. Rev. Pharmacol. Toxicol. **36:** 83–106.
6. SZIRAKI, I., P. RAUHALA, K.K. KOOH, P. VAN BERGEN & C.C. CHIUEH. 1999. Implications for atypical antioxidative properties of manganese in iron-induced lipid peroxidation and copper-dependent low density lipoprotein conjugation. Neurotoxicology, In press.
7. QIAN, Y. & G.R. BUETTNER. 1996. Iron-oxygen complexes may be more important than the Fenton reaction in initiating biological free radical oxidations: an EPR spin trapping study. The 3rd annual meeting of the oxygen society. Abstract 1–82.
8. HALL, E.D & J.M. BRAUGHLER. 1988. The role of oxygen radical-induced lipid peroxidation in acute central nervous system trauma. In Oxygen Radicals and Tissue Injury. B. Halliwell Ed. 92-98. FASEB, Bethesda.
9. FLOYD, R.A. & J.M. CARNEY. 1992. Free radical damage to protein and DNA: Mechanisms involved and relevant observations on brain undergoing oxidative stress. Ann. Neurol. **32:** S22–S27.
10. ESTERBAUER, H., R.J. SCHAUR & H. ZOLLNER. 1991. Chemistry and biochemistry of 4-hydroxynonenal, malonaldehyde and related aldehydes. Free Rad. Biol. Med. **11:** 81–128.

11. KRUMAN, I., A.J. BRUCE-KELLER, D. BREDESEN, G. WANG & M.P. MATTSON. 1997. Evidence that 4-hydroxynonenal mediates oxidative stress-induced neuronal apoptosis. J. Neurosci. **17:** 5089–5100.
12. DILLARD, C.J. & A.L. TAPPEL. 1973. Fluorescent products from reaction of peroxidizing polyunsaturated fatty acids with phosphatidyl ethanolamine and phenylalanine. Lipids **8:** 183–189.
13. SHIMASAKI, H. 1994. Assay of fluorescent lipid peroxidation products. Methods Enzymol. **233:** 338–346.
14. TRIGGS, W.J. & L.J. WILLMORE. 1984. In vivo lipid peroxidation in rat brain following intracortical Fe^{2+} injection. J. Neurochem. **42:** 976–980.
15. MOHANAKUMAR, K.P., A. DE BARTOLOMEIS, R.-M. WU, K.J. YEH, L. STERNBERGER, S.-Y. PENG, D.L. MURPHY & C.C. CHIUEH. 1994. Ferrous-citrate complex and nigral degeneration: Evidence for free-radical formation and lipid peroxidation. Ann. N.Y. Acad. Sci. **738:** 392–399.
16. RAUHALA, P., K.P. MOHANAKUMAR, I. SZIRAKI, A.M.-Y. LIN & C.C. CHIUEH. 1996. S-Nitrosothiols and nitric oxide, but not sodium nitroprusside, protect nigrostriatal dopamine neurons against iron-induced oxidative stress in vivo. Synapse **23:** 58–60.
17. SZIRAKI, I., K.P. MOHANAKUMAR, P. RAUHALA, H.G. KIM, K.J. YEH & C.C. CHIUEH. 1998. Manganese: A transition metal protects nigrostriatal neurons from oxidative stress in the iron-induced animal model of Parkinsonism. Neuroscience **85:** 1101–1111.
18. REHNCRONA, S., D.S. SMITH, B. ACKESSON, E. WESTERBERG, & B.K. SIESJO. 1980. Peroxidative changes in brain cortical fatty acids and phospholipid, as characterized during Fe^{2+} and ascorbic acid-stimulated lipid peroxidation in vitro. J. Neurochem. **34:** 1630–1638.
19. WU, R.-M., C.C. CHIUEH, A. PERT & D.L. MURPHY. 1993. Apparent antioxidant effect of *l*-deprenyl on hydroxyl radical formation and nigral injury elicited by MPP^+ in vivo. Eur. J. Pharmacol. **243:** 241–247.
20. THE PARKINSON STUDY GROUP. 1993. Effects of tocopherol and deprenyl on the progression of disability in early Parkinson's disease. New Engl. J. Med. **328:** 176–183.
21. GOODMAN, Y., J.A. BRUCE, B. CHENG & M.P. MATTSON. 1996. Estrogens attenuates and corticosterone exacerbates excitotoxicity, oxidative injury, and amyloid -peptide toxicity in hippocampal neurons. J. Neurochem. **66:** 1836–1844.
22. GIUSTI, P., M. LIPARTITI, D. FRANCESCHINI, N. SCHIAVO, M. FLOREANI & H. MANEV. 1996. Neuroprotection by melatonin from kainate-induced excitotoxicity in rats. FASEB J. **10:** 891–896.
23. TANG, M.X., D. JACOBS, Y. STERN, K. MARDER, P. SCHOFIELD, B. GURLAND, H. ANDREWS & R. MAYEUX. 1996. Effects of oestrogen during menopause on risk and age at onset of Alzheimer's disease. Lancet **348:** 429–432.
24. PAPPOLLA, M.A., M. SOS, R.A. OMAR, R.J. BICK, D.L. HICKSON-BICK, R.J. REITER, S. EFTHIMIOPOULOS & N.K. ROBAKIS. 1997. Melatonin prevents death of neuroblastoma cells exposed to the Alzheimer amyloid peptide. J. Neurosci. **17:** 1683–1690.
25. BEHL, C., T. SKUTELLA, F. LEZOUALCH, A. POST, M. WIDMANN, C.J. NEWTON & F. HOLSBOER. 1997. Neuroprotection against oxidative stress by estrogens: structure-activity relationship. Mol. Pharmacol. **51:** 535–541.
26. MYTILINEOU, C., P. RADCLIFFE, E.K. LEONARDI, P. WERNER & C.W. OLANOW. 1997. l-Deprenyl protects mesencephalic dopamine neurons from glutamate receptor-mediated toxicity in vitro. J. Neurochem. **68:** 33–39.
27. SANO, M., C. ERNESTO, R.G. THOMAS, M.R. KLAUBER, K. SCHAFER, M. GRUNDMAN, P. WOODBURY, J. GROWDON, C.W. COTMAN, E. PFEIFFER, L.S. SCHNEIDER & L.J. THAL. 1997. A controlled trial of selegiline, α-tocopherol, or both as treatment for Alzheimer's diseases. N. Engl. J. Med. **336:** 1216–1222.
28. BIRGE, S.J. 1997. The role of estrogen in the treatment of Alzheimer's disease. Neurology **48:** S36–S41.
29. KIKUGAWA, K., T.KATO, M. BEBBU & A. HAYASAKA. 1989. A fluorescent and cross-linked proteins formed by free radicals and aldehyde species generated during lipid peroxidation. Adv. Exp. Med. Biol. **266:** 345–356.

30. HALLIWELL, B. & S. CHIRICO. 1993. Lipid peroxidation: its mechanism, measurement, and significance. Am. J. Clin. Nutr. **57**(Suppl.): 715S–725S.
31. RAUHALA, P., I. SZIRAKI & C.C. CHIUEH. 1996. Peroxidation of brain lipids in vitro: nitric oxide versus hydroxyl radicals. Free Rad. Biol. Med. **21**: 391–394.
32. STOCKS, J., J.M.C. GUTTERIDGE, R.J. SHARP & T.L. DORMANDY. 1974. Assay using brain homogenates for measuring the antioxidant activity of biological fluids. Clin. Sci. Mol. Med. **47**: 215–222.
33. DEXTER, D.T., F.R. WELLS, A.J. LEES, F. AGID, Y. AGID, P. JENNER & C.D. MARSDEN. 1989. Increased nigral iron content and alterations in other metal ions occurring in brain in Parkinson's diseases. J. Neurochem. **52**: 1830–1836.
34. SOFIC, E., W. PAULUS, K. JELLINGER, P. RIEDERER & M.H.B. YOUDIM. 1991. Selective increase of iron in substantia nigra zona compacta of parkinsonian brains. J. Neurochem. **56**: 978–982.
35. CIUFFI, M., G. GENTILINI, S. FRANCHI-MICHELI & L. ZILETTI. 1992. D-Penicillamine affects lipid peroxidation and iron content in the rat brain cortex. Neurochem. Res. **17**: 1241–1246.
36. MILLER, D.M. & S.D. AUST. 1989. Studies of ascorbate-dependent, iron-catalyzed lipid peroxidation. Arch. Biochem. Biophys. **271**: 113–119.
37. HALL, E.D., J.M. BRAUGHLER, P.A. YONKERS, S.L. SMITH, K.L. LINSEMAN, E.D. MEANS, H.M., P.F. VON VOIGTLANDER, R.A. LAHTI & E.J. JACOBSEN. 1991. U-78517F: a potent inhibitor of lipid peroxidation with activity in experimental brain injury and ischemia. J. Pharmacol. Exp. Ther. **258**: 688–694.
38. NICKANDER, K.K., B.R. MCPHEE, P.A. LOW & H. TRITSCHLER. 1996. α-Lipoic acid: antioxidant potency against lipid peroxidation of neural tissue in vitro and implications for diabetic neuropathy. Free Rad. Biol. Med. **21**: 631–639.
39. WU, R.M., D.L. MURPHY & C.C. CHIUEH. 1995. Neuronal protective and rescue effects of deprenyl against MPP$^+$ dopaminergic neurotoxin. J. Neural Transm. Gen. Sec. **100**: 53–61.
40. IACOVITTI, L., N.D. STULL & K. JOHNSTON. 1997. Melatonin rescues dopamine neurons from cell death in tissue culture models of oxidative stress. Brain Res. **768**: 317–326.
41. CHIUEH, C.C., J.N. JOHANNESSEN, J.L. SUN, J.P. BACON & S.P. MARKEY. 1986. Reversible Neurotoxicity of MPTP in the nigrostriatal dopaminergic system of mice. In MPTP-Parkinsonian Syndrome Producing Neurotoxin. S.P. Markey, A.J. Trevor, N. Castagnoli & I.J. Kopin, Eds. :473–480. Academia Press, New York.
42. GRIDLEY, K.E., P.S. GREEN & J.W. SIMPKINS 1997. Low concentrations of estradiol reduce β-amyloid (25–35)-induced toxicity, lipid peroxidation and glucose utilization in human SK-N-SH neuroblastoma cells. Brain Res. **778**: 158–165.
43. SHWAERY, G.T., J.A. VITA & J.F. KEANEY. 1997. Antioxidant protection of LDL by physiological concentrations of β-estradiol: requirement for estradiol modification. Circulation **95**: 1378–1385.
44. TATTON, W.G. & R.M.E. CHALMERS-REDMAN. 1996. Modulation of gene expression rather than monoamine oxidase inhibition: (−)-Deprenyl-related compounds in controlling neurodegeneration. Neurology **47**(Suppl.): S171–S183.
45. ANTOLIN, I., C. RODRIGUEZ, R.M. SAINZ, J.C. MAYO, H. URIA, M.L. KOTLER, M.J. RODRIGUEZ-COLUNGA, D. TOLIVIA & A. MENENDEZ-PELAEZ. 1996. Neurohormone melatonin prevents cell damage: effects on gene expression for antioxidant enzymes. FASEB J. **10**: 882–890.
46. KANNER, J., L.S. HAREL & R. GRANIT. 1991. Nitric oxide as an antioxidant. Arch Biochem. Biophys. **289**: 130–136.
47. WINK, D.A., I. HANBAUER, M.C. KRISHNA, W. DEGRAFF, J.GAMSON & J.B. MITCHELL. 1993. Nitric oxide protects against cellular damage and cytotoxicity from reactive oxygen species. Proc. Natl. Acad. Sci. USA **90**: 9813–9817.
48. HOGG, N., B. KALYANARAMAN, J. JOSEPH, A. STRUCK & S. PARTHASARATHY. 1993. Inhibition of low-density lipoprotein oxidation by nitric oxide, potential role in atherogenesis. FEBS Lett. **334**: 170–174.

49. RUBBO, H., R. RADI, M. TRUJILLO, R. TELLERI, B. KALYANARAMAN, S. BARNES, M. KIRK & B.A. FREEMAN. 1994. Nitric oxide regulation of superoxide and peroxynitrite-dependent lipid peroxidation. J. Biol. Chem. **269:** 26066–26075.
50. RAUHALA, P., A. M.-Y. LIN & C.C. CHIUEH. 1998. Neuroprotection by S-nitrosoglutathione of brain dopamine neurons from oxidative stress. FASEB J. **12:** 165–173.
51. GASTON, B., J. REILLY, J.M. DRAZEN, J. FACKLER, P. RAMDEV, D. ARNELLE, M.E. MULLINS, D.J. SUGARBAKER, C. CHEE, D.J. SINGEL, J. LOSCALZO & J.S. STAMLER. 1993. Endogenous nitrogen oxides and bronchodilator S-nitrosothiols in human airways. Proc. Natl. Acad. Sci. USA **90:** 10957–10961.
52. KLUGE, I., U. GUTTECK-AMSLER, M. ZOLLINGER & K.Q. DO. 1997. S-Nitrosoglutathione in rat cerebellum: Identification and quantification by liquid chromatography-mass spectrometry. J. Neurochem. **69:** 2599–2607.
53. KIM, Y.-M., M.E. DE VERA, S.C. WATKINS & T.R. BILLIAR. 1997. Nitric oxide protects cultured rat hepatocytes from tumor necrosis factor-α induced apoptosis by inducing heat shock protein 70 expression. J. Biol. Chem. **272:** 1402–1411.
54. WINK, D.A. & M. FEELISCH. 1996. Formation and detection of nitroxyl and nitrous oxide. *In* Methods in Nitric Oxide Research. M. Feelisch & J.S. Stamler, Eds. Vol. 1: 403–412. New York. John Wiley & Sons Ltd.
55. RADI, R., J.S. BECKMAN, K.M. BUSH & B.A. FREEMAN. 1991. Peroxynitrite-induced membrane lipid peroxidation: the cytotoxic potential of superoxide and nitric oxide. Arch. Biochem. Biophys. **288:** 481–487.
56. LYMAR, S.V. & J.K. HURST. 1996. Carbon dioxide: physiological catalyst for peroxynitrite-mediated cellular damage or cellular protectant. Chem. Res. Toxicol. **9:** 845–850.
57. PRYOR, W.A., J.N. LEMERCIER, H. ZHANG, R.M. UPPU & G.L. SQUADRITO. 1997. The catalytic role of carbon dioxide in the decomposition of peroxynitrite. Free Rad. Biol. Med. **23:** 331–338.
58. FUKUTO, J.M., A.J. HOBBS & L.J. IGNARRO. 1993. Conversion of nitroxyl (HNO) to nitric oxide (NO·) in biological systems: the role of physiological oxidants and relevance to the biological activity of HNO. Biochem. Biophys. Res. Commun. **196:** 707–713.
59. FUKUTO, J.M., K. CHIANG, R. HSZIEH, P. WONG & G. CHAUDHURI. 1992. The pharmacological activity of nitroxyl: a potent vasodilator with activity similar to nitric oxide and/or endothelium-derived relaxing factor. J. Pharmacol. Exp. Ther. **263:** 546–551.
60. SCHMIDT, H.H.H., H. HOFMANN, U. SCHINDLER, Z.S. SHUTENKO, D.D. CUNNINGHAM & M. FEELISCH 1996. No NO from NO synthase. Proc. Natl. Acad. Sci. USA **93:** 14492–14497.
61. CHIUEH, C.C. & P. RAUHALA. 1999. The redox pathway of s-nitrosoglutathione, glutathione, and nitric oxide in cell to neuron communications. Free Rad. Res. **31:** 641–650.

Effect of MAO-B Inhibitors on MPP$^+$ Toxicity in Vivo

RUEY-MEEI WU,[a,b] RONG-CHI CHEN,[c] AND CHUANG C. CHIUEH[d]

[a]*Department of Neurology, National Taiwan University Hospital, College of Medicine, National Taiwan University, Taipei, Taiwan*

[c]*Department of Neurology, En-Chu-Kong Hospital, Taipei, Taiwan*

[d]*Unit on Neurotoxicology and Neuroprotection, Laboratory of Clinical Science, National Institute of Mental Health, NIH, Clinical Center 10/3D-41, Bethesda, Maryland, USA*

ABSTRACT: *l*-Deprenyl (Selegiline), a selective and irreversible type B monoamine oxidase inhibitor, has been used as an adjunct to levodopa therapy in Parkinson's disease. Recently, it is proposed as a putative neuroprotective agent in delaying the progression of cell death based on its capability of reducing the oxidative stress derived from the MAO-B dependent metabolism of dopamine, and blocking the development of MPTP-parkinsonism. However, a variety of experimental models suggest that *l*-deprenyl provides neuroprotection through multiple modes of mechanism other than the inhibition of MAO-B. We have previously shown that *l*-deprenyl protects midbrain dopamine neurons from MPP$^+$ toxicity by a novel antioxidant effect. In the present study we examined whether the protection against MPP$^+$ toxicity is also shared by other reversible or irreversible MAO-B inhibitors including (+)-deprenyl, Ro16-6491 and pargyline. Our data show that non of these MAO-B inhibitors changes the dopamine loss in the striatum induced by intranigral injection of MPP$^+$. Our result suggests that *l*-deprenyl may possess a unique neuroprotective action on nigral neuron against MPP$^+$ toxicity independent of the MAO-B inhibition.

INTRODUCTION

Monoamine oxidase (MAO) catalyzes the oxidative deamination of monoamine neurotransmitters and neuromodulators such as dopamine, noradrenaline, serotonin (5-hydroxytyramine), and β-phenylethylamine (PEA), as well as some exogenous bioactive monoamines. Two types of MAO are existing in mammalian tissues, MAO-A and MAO-B, with different substrate and inhibitors specificity.[1] MAO-A deaminates preferentially serotonin and is sensitive to selective inhibitors, such as clorgyline. MAO-B deaminates preferentially PEA and is sensitive to MAO-B inhibitors, such as *l*-deprenyl. Inhibitors of MAO-A have been proved to be effective antidepressant, while MAO-B blockers have been emphasized in the treatment of Parkinson's disease (PD).

---

[b]Address for correspondence: Dr. Ruey-Meei Wu, Department of Neurology, National Taiwan University Hospital, No. 7, Chuang-Shan South Road, Taipei, 100, Taiwan.
e-mail: rmwu@ha.mc.ntu.edu.tw

l-Deprenyl (Selegiline), first described in 1965 is a selective irreversible inhibitor of MAO-B.[2] As an adjunct to levodopa therapy, it has been shown to provide mild antiparkinsonian effect and reduce motor fluctuations in advanced PD.[3,4] In addition to the symptomatic effect, greater interest has focused on its role as a neuroprotective agent to prevent neurons from degeneration. Recent clinical trials have shown that early treatment with *l*-deprenyl delays the progression of disability necessitating the introduction of levodopa therapy and also the symptoms and signs.[5,6] The idea of *l*-deprenyl as a neuroprotective agent for patients with PD is based on the hypothesis that oxidative stress derived from the metabolism of dopamine,[7,8] and/or exposure to a selective neurotoxin, such as 1-methyl-4-phenyl-1,2,3,6-tetrahydropyridine (MPTP) may contribute to the cell death in PD patients.[9] MPTP, a neurotoxin causing selective destruction of dopaminergic neurons in the midbrain, has been shown to provide a striking pathological analogy to PD.[9–11] *l*-Deprenyl blocks MPTP toxicity by inhibiting the conversion of MPTP to MPP^+, the putative toxic metabolite,[12] and suppressing the oxidative stress secondary to the metabolism of dopamine.[7] However, recent laboratory studies suggest that *l*-deprenyl prevents neurodegeneration in various experimental models through a mechanism that does not depend on the inhibition of MAO-B activity. It has been shown that *l*-deprenyl enhances the activity of antioxidant defense enzymes such as superoxide dismutase and catalase in the striatum that in turn, may protect against MPP^+ toxicity.[13–15] Tatton and colleagues reported that *l*-deprenyl can rescue neurons from MPTP toxicity and enhance the survival of motor neurons after axotomy by inducing trophic-like substance.[16,17] Recently, they also demonstrated that *l*-deprenyl reduces apoptosis in cultured PC-12 cells through the enhancement of protein synthesis.[18,19] Our previous studies have shown that *l*-deprenyl can suppress the hydroxyl radical formation induced by MPP^+ in the striatum and thereby protect midbrain dopamine neurons against injury induced by MPP^+ *in vivo*.[20,21] Thus, a unique antioxidant effect of *l*-deprenyl, independent of the enzymatic inhibition of MAO was proposed to exert its neuroprotection.[22] In the present study, we further examined whether the protection on dopaminergic neurons against MPP^+ toxicity is shared by other MAO-B inhibitors such as (+)-deprenyl (a stereoisomer of *l*-deprenyl), RO16-6491, a moclobemide analog with reversible inhibition of MAO-B activity, and pargyline, or it is a pharmacological unique to *l*-deprenyl itself.

MATERIALS AND METHODS

Our previous study has shown that infusion of MPP^+ 4.2 nmol into the substantia nigra produced around 50% reduction of dopamine level in the ipsilateral striatum of control and *l*-deprenyl provided a complete protection on nigral neurons against this injury.[21] Thus, in this study, we used the same experimental paradigm to investigate the effects of tested MAO-B inhibitors on midbrain dopamine neurons against MPP^+ toxictiy. Sprague-Dawley rats (male, 250–350 g) were anesthetized with chloral hydrate (400 mg/kg, i.p.) and mounted in a stereotaxic apparatus (David Kopf Instruments) for intranigral infusion of drugs. MPP^+, *l*-deprenyl, (+)-deprenyl, Ro 16-6491 and pargyline (RBI, Natic, MA) were dissolved in Ringer solution for infusion, individually. MPP^+ 4.2 nmol/µL was infused into one side of the substantia nigra to produce a moderate nigral injury. The contralateral striatum served as a control. To investigate the protective effect of MAO-B inhibitors against MPP$^+$ toxicity, the other groups of animals received coadministration of MPP$^+$ (4.2 nmol) with equal mole (4.2 nmol) of l-deprenyl, (+)-deprenyl, Ro 16-6491 or pargyline, individually into the substantia nigra. In the control study, some of the animals received Ringer solution (as sham control) or the tested MAO-B inhibitors solution (1 µL) alone. The drug solution (1 µL) was stereotaxically infused into the substantia nigra of rat (Paxinos & Watson, 1986; coordinates A 3.2, R 1.8, H 1.8) at an infusion rate of 0.2 µL/min via a 30-gauge needle. The injection needle was held in place for additional 3 min after the infusion. The body temperature of the animals was maintained at 37°C using a thermo-regulator during the surgery. Four days after the surgery, rats were sacrificed by decapitation. Their brains were removed immediately and dissected to obtain tissue from caudate nucleus. Tissue samples were homogenized in 0.1 N perchloride acid. After centrifuge, the clear supernatant (5 µL) was injected into a high performance liquid chromatography (HPLC) coupled with electrochemical detector to assay the contents of dopamine and their metabolites.

Values were expressed as mean ± SEM. Statistical analysis was conducted by Student's t-test or ANOVA test for comparing the mean differences among groups, using $p < 0.05$ as significant.

RESULTS

Intranigral infusion of Ringer solution (1 µL), l-deprenyl (4.2 nmol in 1 µL Ringer solution), (+)-deprenyl (4.2 nmol/µL), Ro 16-6491(4.2 nmol/1 µL), or pargyline (4.2 nmol/ 1 µL) led to 9.6 ± 2.9% (N = 3), 11.3 ± 2.5% (N = 3), 10.2 ± 2.2% (N = 3), 10.0 ± 2.0% (N = 3), and 13.5 ± 2.5% (N = 3) reduction in dopamine levels in the ipsilateral striatum four days after surgery, respectively. There was no significant difference in the small decrease of striatal dopamine in the injected side induced by these agents (ANOVA test). Infusion of MPP$^+$ 4.2 nmol into the substantia nigra produced a significant reduction of dopamine levels to 53.4 ± 3.1% (N = 10) of control in the ipsilateral striatum four days after the intranigral injection procedure. Coadministration of l-deprenyl (4.2 nmol) with MPP$^+$ significantly attenuated the dopamine loss caused by MPP$^+$ to 88.0 ± 3.5% of control (N = 8; $p < 0.05$). However, infusion of (+)-deprenyl (4.2 nmol), Ro 16-6491 (4.2 nmol) or pargyline (4.2 nmol) along with MPP$^+$ into the substantia nigra did not significantly change the MPP$^+$-induced dopamine depletion (FIG. 1), and the corresponding dopamine levels in the ipsilateral striatum were 48.7 ± 3.6% (N = 8), 46.0 ± 3.2% (N = 10) and 52.3 ± 3.2% (N = 8) of control sides, respectively. The dopamine contents in the control sides of striatum were not significantly different among the experimental groups (range: 40–54 pmol/mg wet weight of tissue; ANOVA test).

DISCUSSION

Our present data showed that l-deprenyl provided significant protection on midbrain dopamine neurons against MPP$^+$ toxicity *in vivo*. None of the other MAO-B inhibitors such as (+)-deprenyl, Ro-16-6491 and pargyline conferred similar

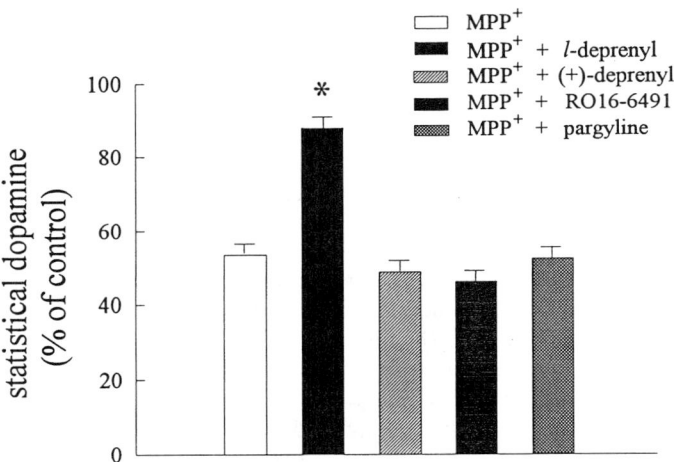

FIGURE 1. Effects of l-deprenyl, (+)-deprenyl, Ro16-6491 and pargyline on the reduction of dopamine level induced by intranigral injection of MPP^+. MPP^+ (4.2 nmol in 1 µL of Ringer solution) with or without MAO-B inhibitor was administered stereotaxically into one side of the substantia nigra of anesthetized rats. Four days after the intranigral injection, dopamine level in both sides of striatum were measured. Striatal dopamine content in the injected side was expressed as the percentage of control side. Values were expressed as mean ± SEM. $*p < 0.01$ as compared with the MPP^+ group.

protection. The MAO-B inhibitors (+)-deprenyl, Ro-166491 and pargyline investigated in this study possess some pharmacological properties different from those of l-deprenyl. (+)-Deprenyl is a stereoisomer of l-deprenyl. It is mostly metabolized to (+)-amphetamine and (+)-methamphetamine. l-Deprenyl is metabolized into (—)-desmethylselegiline (DES) and l-(—)-methamphetamine, which can both be converted to l-(—)-amphetamine in the liver.[23] The principal pharmacological action of l- or d-amphetamine is the release of catecholamine from the presynaptic neurons. On higher concentration, uptake of catecholamine may take place. The dopamine releasing effect of d-form amphetamine is 10 times potent than that of l-form.[24] Moreover, d-amphetamine is a more potent MAO-A inhibitor than l-amphetamine. Thus, the d-amphetamine is more potent in producing elevation of blood pressure and stereotypic locomotor behavior in rats.[23] Additionary, d-deprenyl and d-amphetamine have also been shown to exhibit a more potent inhibitory effect on amine uptake than l-derpenyl did.[25] In contrast to l-deprenyl and d-deprenyl, Ro 16-6491 is a reversible, highly potent and selective MAO-B inhibitor.[26] Also unlike both form of deprenyl, it is not metabolized to potentially active compounds such as amphetamine and methamphetamine, and without the inhibitory effect on dopamine uptake.[25] Its analog, lazabemide (Ro19-6327), a very potent and selective MAO-B inhibitor, has been shown to improve daily living activities in early Parkinson's disease.[27] Hence, The findings that d-deprenyl, Ro16-6491 and pargyline (a classic MAO-A and MAO-B inhibitor) did not change the dopamine depletion in the striatum induced by MPP^+ suggest that l-deprenyl may possess a unique neuroprotective effect on mesencephalic dopamine neurons against MPP$^+$ toxicity *in vivo*. The mechanism of neuroprotection is independent of the MAO-B inhibition and probably unrelated to other pharmacological properties of *l*-deprenyl, such as amphetamine effect and amine-uptake inhibition.

There is compelling evidence suggesting that the neuroprotective mechanisms of *l*-deprenyl against MPP$^+$ toxicity are not dependent on inhibition of MAO-B. Our previous studies have demonstrated that *l*-deprenyl may act like an antioxidant to suppress the generation of hydroxyl radical elicited by MPP$^+$ *in vivo* without altering MAO-B.[20] Moreover, supplementary treatment with *l*-deprenyl after single injection of MPP$^+$ in the substantia nigra can rescue midbrain dopamine neurons against severe injury induced by MPP$^+$ *in vivo*.[21] Tatton and Greenwood have also observed the rescue effect of *l*-deprenyl which protects dopaminergic neurons from MPTP even when it is administered 72 hours afterward and in a dose too small to block MAO activity.[16] Our recent studies have shown that intrastriatal infusion of MPP$^+$ increases fluorescent products of lipid peroxidation.[28] The potent inhibitor of lipid peroxidation U-78517F, and hydroxyl radical scavenger dimethylsulfoxide can protect nigral neurons against MPP$^+$ toxicity.[28] Accordingly, *l*-deprenyl may suppress hydroxyl radical formation and therefore cease the propagation of peroxidation of polyunsaturated fatty acids induced by MPP$^+$ *in vivo*. The findings that *l*-deprenyl suppressed the lipid peroxidation induced by NADPH and ascorbic acid in rat brain homogenate were consistent with this hypothesis.[29] Furthermore, the protection of *l*-deprenyl against MPTP toxicity may be also associated with the up-regulation of endogenous antioxidant enzymes and antiapoptotic molecules such as superoxidase dismutase, catalase, glutathione, Bcl-2 and BclX$_L$.[14,30] Over-expression of SOD or Bcl-2 in the transgenic mice were reported to protect against MPTP toxicity.[31,32] Thus, it appears that *l*-deprenyl can protect midbrain dopamine neurons against the oxidative injury induced by MPTP through multiple modes of action.

In addition to MPTP toxicity, *l*-deprenyl has been shown to provide neuroprotective effects against 6-OHDA and the noradrenergic neurotoxin DSP-4.[33,34] *l*-Deprenyl, but not other MAO-B inhibitors, has been reported to augment the CNTF gene expression[17] and reduce PC12 cell apoptosis.[18] More recently, Mytilineou and colleagues reported that *l*-deprenyl can prevent neurodegeneration induced by excitotoxicity or reduced glutathione secondary to administration of buthionine sulfoximine (BSO), in which the nonselective MAO-B inhibitor pargyline or dopamine uptake inhibitor mazindol did not diminish BSO toxicity.[35,36] Thus, it is clear from these laboratory data that *l*-deprenyl can exert its unique neuroprotection via the mechanisms more than MAO-B inhibition. To the extent, these laboratory data might also provide beneficial evidence for *l*-deprenyl as a neuroprotective agent in the treatment of Parkinson's disease.

In conclusion, the present study demonstrated that *l*-deprenyl, but not other MAO-B inhibitors such as (+)-deprenyl, Ro 16-6491 or pargyline, significantly attenuates the striatal dopamine loss caused by intranigral injection of MPP$^+$ *in vivo*. These findings suggest that *l*-deprenyl may protect against MPP$^+$ toxicity through a mechanism independent of MAO-B inhibition.

ACKNOWLEDGMENT

This study was supported from the grant of National Science Council, Taiwan, R.O.C. (NSC 84-2331-B002-249).

REFERENCES

1. JOHNSTON, J.P. 1968. Some observations upon a new inhibitor of monoamine oxidase in human brain. Biochem. Pharmacol. **17**: 1285–1297.
2. KNOLL, J., Z. ECSERI, K. KELEMEN, J. NIEVEL & B. KNOLL. 1965. Phenylisopropylmethylpropinylamine (E-250), a new spectrum psychic energizer. Arch. Int. Pharmacodyn. **155**: 154–164.
3. BIRKMAYER, W., P. RIEDERER, L. AMBROZI & M.B.H. YOUDIM. 1977. Implications of combined treatment with Madopar® and deprenyl in Parkinson's disease. Lancet **i**: 439–444.
4. GOLBE, L.I., A.N. LIEBERMAN, M.D. MUENTER, J.E. AHLSKOG, G. GOPINATHAN, A.N. NEOPHYTIDES, S-H. FOO & R.C. DUVOISIN. 1988. Deprenyl in the treatment of symptom fluctuations in advanced Parkinson's disease. Clin. Neuropharmacol. **2**: 45–55.
5. THE PARKINSON STUDY GROUP. 1989. Effect of deprenyl on the progression of disability in early Parkinson's disease. N. Engl. J. Med. **321**: 1364–1371.
6. OLANOW, C.W., R.A. HAUSER, L. GAUGER, T. MALAPIRA, S. KOLLER & J. HUBBLE. 1995. The effect of deprenyl and levodopa on the progression of Parkinson's disease. Ann. Neurol. **38**: 771–777.
7. COHEN, G. & M.B. SPINA. 1989. Deprenyl suppressed the oxidant stress associated with increased dopamine turnover. Ann. Neurol. **26**: 689–690.
8. FAHN, S. & G. COHEN. 1992. The oxidant stress hypothesis in Parkinson's disease: evidence supporting it. Ann. Neurol. **32**: 804–812.
9. LANGSTON, J.W., P. BALLARD, J.W. TETRUD & I. IRWIN. 1983. Chronic parkinsonism in humans due to a product of meperidine-analog synthesis. Science **219**: 979–980.
10. DAVIS, G.C., A.C. WILLIAMS, S.P. MARKEY, M.H. EBERT, E.D. CAINE, C.M. ERICHERT & I.J. KOPIN. 1979. Chronic parkinsonism secondary to intravenous injection of meperidine analogues. Psychiatry Res. **1**: 249–254.
11. BURNS, R.S., C.C. CHIUEH, S.P. MARKEY, M.H. EBERT, D.M. JACOBOWITZ & I.J. KOPIN. 1983. A primate model of parkinsonism: selective destruction of dopaminergic neurons in the pars compacta of the substantia nigra by N-methyl-4-phenyl-1,2,3,6-tetrahydropyridine. Proc. Natl. Acad. Sci. USA **80**: 4546–4550.
12. CHIBA, K., A. TREVOR & N. CASTAGNOLI, JR. 1984. Metabolism of the neurotoxic tertiary amine, MPTP, by brain monoamine oxidase. Biochem. Biophys. Res. Commun. **120**: 574–578.
13. KNOLL, J. 1989. The pharmacology of selegiline ((−)deprenyl). New aspects. Acta. Neurol. Scand. **126**: 83–91.
14. CARRILLO, M.C., S. KANAI, M. NOKUBO & K. KITANI. 1991. (−)-Deprenyl induces activities of both superoxide dismutase and catalase but not of glutathione peroxidase in the striatum of young male rats. Life Sci. **48**: 517–517.
15. VIZUETE, M. L., V. STEFFEN, A. AYALA, J. CANO & A. MACHADO. 1993. Protective effect of deprenyl against 1-methyl-4-phenylpyridinium neurotoxicity in rat striatum. Neurosci. Lett. **152**: 113–116.
16. TATTON, W.G. & C.E. GREENWOOD. 1991. Rescue of dying neurons: a new action for L-deprenyl in MPTP parkinsonism. J. Neurosci. Res. **30**: 666–672.
17. SENIUK, N.A., J.T. HENDERSON, W.G. TATTON & J.C. RODER. 1994. Increased CNTF gene expression in process-bearing astrocytes following injury is augmented by R(−)-deprenyl. J. Neurosci. Res. **37**: 278–286.
18. TATTON, W.G., W.Y.L. JU, D.P. HOLLAND, C. TAI & M. KWAN. 1994. (−)-L-Deprenyl rescues PC12 cell apoptosis by inducing new protein synthesis. J. Neurochem. **63**: 1572–1575.

19. WADIA, J.S., R.M.E. CHALMERS-REDMAN, W.J.H. JU, G.W. CARLILE, J.L. PHILLIPS, A.D. FRASER & W.G. TATTON. 1998. Mitochondrial membrane potential and nuclear changes in apoptosis caused by serum and nerve growth factor withdrawal: time course and modification by (−)-deprenyl. J. Neurosci. **18**: 932–947.
20. WU, R.-M, C.C. CHIUEH, A. PERT & D.L. MURPHY. 1993. Apparent antioxidant effect of l-deprenyl on hydroxyl radical formation and nigral injury elicited by MPP^+ in vivo. Eur. J. Pharmacol. **243**: 241–247.
21. WU, R.-M, D.L. MURPHY & C.C. CHIUEH. 1995. Neuroprotective and rescue effects of deprenyl against MPP^+ toxicity. J. Neural Transm. (Gen. Sect.) **100**: 53–61.
22. WU, R.-M, D.L. MURPHY & C.C. CHIUEH. 1996. Suppression of hydroxyl radical formation and protection of nigral neurons by l-deprenyl (selegiline). Ann. N.Y. Acad. Sci. **786**: 379–390.
23. HEINONEN E.H & R. LAMMINTAUSTS. 1991. A review of the pharmacology of selegiline. Acta. Neurol. Scand. **84** (Suppl. 136): 44–59.
24. FANG, J. & P.H. YU. 1994. Effect of L-Deprenyl, its structural analogues and some monoamine oxidase inhibitors on dopamine uptake. Neuropharmacol. **33**: 763–768.
25. CHIUEH, C.C. & K.E. MOORE. 1974. Relative potencies of d- and l-amphetamine on the release of dopamine from cat brain in vivo. Res. Commun. Chem. Path. Pharmacol. **7**: 189–199.
26. DA PRADA M., R. KETTLER, H.H. KELLER, A.M. CESURA, J.G. RICHARDS & M.L. SAURA. 1990. From moclobemide to Ro19-6327 and Ro 41-1049: the development of a new class of reversible, selective MAO-A and MAO-B inhibitors. J. Neural. Transm. **29**: 279–292.
27. THE PARKINSON STUDY GROUP. 1993. A controlled trial of lazabemide (Ro19-6327) in untreated Parkinson's disease. Ann. Neurol. **133**: 350–356.
28. WU, R.-M, D.L. MURPHY & C.C. CHIUEH. 1994. Protection of nigral neurons against MPP^+-induced oxidative injury by deprenyl (selegiline), U-78517F, and DMSO. New Trends Clin. Neuropharm. **8**: 187–188.
29. SZOKO, E., G. BATHORY, K. TEKES & K. MAGYAR. 1990. Effect of l-deprenyl on lipid peroxidation in rat brain homogenate. Eur. J. Pharmacol. **183**: 1549–1549.
30. TATTON, W.G., J.S. WADIA, W.Y.L. JU, R.M.E. CHALMERS-REDMAN & N.A. TATTON. 1996. (−)-Deprenyl reduces neuronal apoptosis and facilitates neuronal outgrowth by altering protein synthesis without inhibiting monoamine oxidase. J. Neural Transm. **48** (Suppl.): 45–59.
31. PRZEDBORSKI, S., V. KOSTIC, V. JACKSON-LEWIS, A.B. NAINI, S. SIMONETTI, S. FAHN, E. CARLSON, C.J. EPSTEIN & J.L. CADET. 1992. Transgenic mice with increase Cu/Zn-Superoxide dismutase activity are resistant to N-methyl-4-phenyl-1,2,3,6-tetrahydropyridine-induced neurotoxicity. J. Neurosci. **12**: 1658–1667.
32. YANG, L., R.T. MATTHEWS, J.B. SCHULZ, T. KLOCKGETHER, A.W. LIAO, J.C. MARTINOU, J.B. PENNEY, B.T. HYMAN & M.F. BEAL. 1998. 1-methyl-4-phenyl-1,2,3,6-tetrahydropyride neurotoxicity is attenuated in mice over-expressing bcl-2. J. Neurosci. **18**: 8145–8152.
33. SALONEN, T., A. HAAPALINNA, E. HEINONEN, J. SUHONEN & A. HERVONEN. 1996. Monoamine oxidase B-inhibitor selegiline protects young and aged rat peripheral sympathetic neurons against 6-hydroxydopamine-induced neurotoxicity. Acta Neuropathol. **91**: 466–474.
34. FINNEAGAN, K.T., J.J. SKRATT, I. IRWIN, L.E. DELANNEY & J.W. LANGSTON. 1990. Protection against DSP-4-induced neurotoxicity by deprenyl is not related to its inhibition of MAO B. Eur. J. Pharmacol. **184**: 119–126.
35. MYTILINEOU, C., P. RADCLIFFE, E.K. LEONARDI, P. WERNER & C.W. OLANOW. 1997. L-deprenyl protects mesencephalic dopamine neurons from glutamate receptor-mediated toxicity in vitro. J. Neurochem. **68**: 33–39.
36. MYTILINEOU, C., E.K. LEONARDI, P. RADCLIFFE, E.H. HEINONEN, S.K. HAN, P. WERNER, G. COHEN & C.W. OLANOW. 1998. Deprenyl and desmethylselegiline protect mesencephalic neurons from toxicity induced by glutathione depletion. J. Pharmacol. Exp. Therap. **284**: 700–706.

Neuroprotective Strategies in Parkinson's Disease Using the Models of 6-Hydroxydopamine and MPTP[a]

EDNA GRÜNBLATT, SILVIA MANDEL, AND MOUSSA B.H. YOUDIM[b]

Technion–Faculty of Medicine, Eve Topf and U.S. National Parkinson's Foundation, Centers for Neurodegenerative Diseases, Bruce Rappaport Family Research Institute and Department of Pharmacology, Haifa, Israel

ABSTRACT: The etiology of Parkinson's disease is not known. Nevertheless a significant body of biochemical data from human brain autopsy studies and those from animal models point to an on going process of oxidative stress in the substantia nigra which could initiate dopaminergic neurodegeneration. It is not known whether oxidative stress is a primary or secondary event. Nevertheless, oxidative stress as induced by neurotoxins 6-hydroxydopamine and MPTP (N-methyl-4-phenyl-1,2,3,6-tetrahydropyridine) has been used in animal models to investigate the process of neurodegeneration with intend to develop antioxidant neuroprotective drugs. It is apparent that in these animal models radical scavengers, iron chelators, dopamine agonists, nitric oxide synthase inhibitors and certain calcium channel antagonists do induce neuroprotection against such toxins if given prior to the insult. Furthermore, recent work from human and animal studies has provided also evidence for an inflammatory process. This expresses itself by proliferation of activated microglia in the substantia nigra, activation and translocation of transcription factors, NF κ-β and elevation of cytotoxic cytokines TNFα, IL1-β, and IL6. Both radical scavengers and iron chelators prevent LPS (lipopolysaccharide) and iron induced activation of NF κ-B. If an inflammatory response is involved in Parkinson's disease it would be logical to consider antioxidants and the newly developed non-steroid anti-inflammatory drugs such as COX2 (cyclo-oxygenase) inhibitors as a form of treatment. However to date there has been little or no success in the clinical treatment of neurodegenerative diseases per se (Parkinson's disease, ischemia etc.), where neurons die, while in animal models the same drugs produce neuroprotection. This may indicate that either the animal models employed are not reflective of the events in neurodegenerative diseases or that because neuronal death involves a cascade of events, a single neuroprotective drug would not be effective. Thus, consideration should be given to multi-neuroprotective drug therapy in Parkinson's disease, similar to the approach taken in AIDS and cancer therapy.

[a]This paper is dedicated to Dr. Daniel Gilbert, a true friend and an outstanding colleague.
[b]Address for correspondence: Prof. M.B.H. Youdim, Technion–Faculty of Medicine, Pharmacology Department, Efron St., POB 9649, Haifa 31096, Israel. Voice: +972-4-8295290; fax: +972-4-8513145.
e-mail. youdim@tx.technion.ac.il

INTRODUCTION

Parkinson's Disease (PD) is a neurodegenerative disorder involving the progressive degeneration of dopamine neurons arising in the substantia nigra and terminating with its terminals in the striatum. The disease is best described as a deficiency of striatal dopamine. The major problem concerning a better therapeutic approach to the treatment and prevention of the disease is the enigma of its underlying cause. This has remained obscure in spite of the many approaches and efforts made so far.[1,2] Nevertheless, in the past few years much has been learnt about the biochemical pathology of PD, specially in the substantia nigra pars compacta (SNPC) that gives hope not only to find the cause of the disease, but also to develop new preventive drugs that may either halt the progressive degeneration of the dopaminergic neurons, or may even reverse it. The current hypothesis concerning the pathogenesis of PD holds the belief that there is an on going selective oxidative stress (OS), that expresses itself with biochemical alterations compatible with this state.[1–8] Much of our knowledge about dopaminergic neurodegeneration has come from studies with two neurotoxins that produce animal models for OS and parkinsonism syndrome in rodents, primates and other species. Both neurotoxins, namely 6-hydroxydopamine (6-OHDA)[9] and MPTP (N-methyl-4-phenyl-1,2,3,6-tetrahydropyridine)[10–12] cause the degeneration of nigro-striatal dopamine neurons with the subsequent loss of striatal dopamine. Earlier studies with 6-OHDA had indicated that this neurotoxin is a highly reactive substance, which is readily autoxidized and oxidatively deaminated by monoamine oxidase to give rise to hydrogen peroxide and reactive oxygen species (ROS).[13] It was inferred that this neurotoxin exerts its neurodegenerative action via OS.[14,15] The consequence of OS is the initiation of ROS generation followed by brain membrane lipid peroxidation. The possibility that an endogenous toxin, similar to 6-OHDA or some other neurotoxin, may be formed in the brain and may be involved in the process of the neurodegeneration has been envisaged on many occasions.[9] Indeed, the discovery of MPTP and its neurodegenerative property led to the notion that an enviromental toxin similar to it might be responsible for the onset of PD. However MPTP is a synthetic substance and to date no such toxin has been identified in the enviroment or in the brain. MPTP, similar to 6-OHDA is thought to initiate its dopaminergic neurotoxicity via metabolism by monoamine oxidase (MAO), giving rise to its reactive metabolite MPP$^+$. This late is thought to begin the neurodegeneration process via ROS-induced OS and inhibition of mitochondrial complex I,[16–19] as it produces sustained dopamine oxidation, hydroxyl radical formation and membrane lipid peroxidation.[17,18,20] Furthermore, both toxins induce an inflammatory process that results in proliferation of reactive microglia in the SN-PC.[21] It is assumed that it is the reactive microglia, which is responsible for the induction of the inflammatory process, resulting from its ability to generate substantial amounts of ROS.[1] A similar picture is also seen in PD brains, where an increase of several folds in the inflammatory cytokines, IL1-β, IL2, IL6 and TNFα[22–24] was observed. The biochemical changes induced by the toxins are summarized in TABLE 1.

TABLE 1. Evidence for oxidative stress in 6-OHDA and MPTP-induced selective nigrostriatal degeneration

	6-OHDA	MPTP
Biochemical indices that increase		
Reactive microglia proliferation	28	21
Iron release	29	30, 31
Lipid peroxidation	32, 33	2, 25
Generation of ROS	34	2, 3
Biochemical indices that decrease		
Reduced glutathione (GSH)	34	35
Ratio of GSH to oxidized glutathione (GSSG)	34	36
Calcium-binding proteins	37	38, 39
Mitochondrial complex I (NADPH oxidase) activity	40	41

MPTP AND 6-OHDA–INDUCED NIGROSTRIATAL DEGENERATION

Earlier studies have indicated that there might be an increase in iron and a decrease in reduced glutathione (GSH) content in the substantia nigra of PD brains.[3,4,25] While the iron increases in the SNPC with the staging of the disease,[25] GSH decreases[3] and almost disappears in this region of the brain. The mechanisms by which these events occur are not known. Such findings are compatible with an on going OS state and have been observed with other diseases in other tissues such as the lung, heart and kidney, where OS have been implicated in pathological conditions (see ref. 26 for review). Tissue GSH is essential and responsible for the removal of hydrogen peroxide from the brain since it is the cofactor of the enzyme glutathione peroxidase, the only enzyme responsible for removal and disposition of hydrogen peroxide from the brain. In the presence of free tissue iron and low levels of GSH the highly reactive hydroxyl radical is formed by the classical Fenton chemistry.

Thus, if under abnormal conditions there is over production of hydrogen peroxide along with a lack of its disposition, OS can result. Indeed, our own extensive studies, supported by those from other laboratories, on the biochemistry of SN from PD patients, have shown neurochemical changes compatible with OS[1-5] and these changes have occurred specifically in the pars compacta. A closer examination of the biochemical pathology of parkinsonian SNPC shows a related similarity to that observed in the 6-OHDA and MPTP models. In both models there is exaggerated increase of iron in the SNPC, 2-fold and 5-fold increase of nigral iron in MPTP and 6-OHDA–treated animals, respectively,[27] which leads to increase in dopamine turnover, hydroxyl radical generation, lipid peroxidation, dopamine depletion and nigral neuron loss which is associated with significant motor dysfunction.[11]

NEUROPROTECTION IN MPTP AND IRON ANIMAL MODELS

The ability of the anti-parkinsonian drug, l-selegiline (l-deprenyl), an irreversible MAO-B inhibitor, to prevent MPTP-induced parkinsonism in mice, and non-human primates was the first example of neuroprotection for PD.[42] The mechanism of this process has been explained by the inhibition of MAO-B by l-selegiline, thus preventing the metabolism of the neurotoxin to the reactive metabolite, MPP^+, by the enzyme and the formation of hydrogen peroxide. This explanation appears to be too simple, as selegiline was shown to prevent as well to rescue cultured nigral neurons from the induced oxidative damage caused by the reactive metabolite MPP^+.[43] Moreover, other drugs not having MAO-B inhibitory action can also exert neuroprotection in this model. So far iron chelators (e.g., desferrioxamine),[31,44,45] antioxidants (vitamin E),[31,46,47] the dopamine agonists apomorphine[48,49,50] and bromocriptine,[51] GSH analogues[52] and nitric oxide synthase inhibitor (7NI but not L-NAME)[53,54] have been described to have the same effect. Although at first glance it does not seem to be a common mechanistic feature among these compounds, a closer examination reveals that they all converge to a possible participation of hydroxyl radicals and iron in MPTP-induce neurodegeneration. In normal circumstances, free iron is not found in the cell but rather stored and bound to ferritin. The nigra has the highest concentration of iron and ferritin in the brain. This is evident in the rodent, non-human primate and human brain.[25] However, when iron is released from its bound form, it can become toxic for the survival of the cell because of its redox state and promotion of reactive hydroxyl radical formation from hydrogen peroxide. Our own recent in vitro studies have clearly shown that MPTP (when metabolized by MAO-B to MPP^+) induces the release of iron from ferritin. *In vivo*, MPTP-treated mice and Green Monkeys show a highly significant increase of iron in SNPC.[55,56] Similar results were also reported for 6-OHDA.[27] Indeed, 6-OHDA similar to MPP^+ and nitric oxide, is a potent releaser of ferritin bound iron.[57,58] This may be the reason why iron chelator desferrioxamine and vitamin E are neuroprotective in MPTP-treated mice and 6-OHDA–treated rats.[59,60]

Intranigral iron injection to rats produces a relatively selective lesion of nigra dopamine neurons, which can be prevented by iron chelator, desferrioxamine.[59,61,62] However, intraventricular injection of iron up to 50 mM has no such effects. These results clearly point that iron does not easily cross the blood brain barrier, as has been shown previously.[63,64] Thus, it is assumed that the availability of free iron or its low molecular weight complex form is the factors probably related to neurodegeneration. However, ferrous citrate/iron and biologically active small molecular weight iron complexes, redoxically generate cytotoxic hydroxyl radicals, initiating membrane lipid peroxidation and cytotoxicity in submicromolar concentrations. It has been suggested on several occasions that nitric oxide may be involved in neurodegeneration as well as in the mechanism of action of MPTP-induced neurodegeneration. If nitric oxide is involved in neuropathology of neurodegeneration it would explain why nitric oxide synthase inhibitor, 7NI but not L-NAME is protective in the MPTP-induced neurodegeneration in the mice, despite the fact that both drugs equally inhibit NOS in the brain. However this is a disputed subject, since, 7NI have been shown to inhibit MAO and prevent the conversion of MPTP to MPP^+.[65]

It is more than a passing interest that in PD the increase of iron occurs in SNPC within the pigmented melanin-containing neurons, where iron is bound to neuromelanin.[66] The role of neuromelanin remains obscure, since human SNPC is the only one that contains neuromelanin. Never the less, recent studies have clearly shown that melanin is a two-edged sword. When not in a complex, melanin acts as a radical scavenger; by contrast, in the presence of divalent metals, especially iron, it is a very powerful ROS generator.[25,59]

FUTURE STRATEGIES FOR NEUROPROTECTION IN PARKINSON'S DISEASE

The current therapeutic approach to the treatment of PD is symptomatic and the most commonly is a dopamine replacement therapy with L-dopa or dopamine receptor agonists (bromocriptine, lisuride, pergolide and apomorphine) or increasing the availability of brain dopamine with inhibitors of dopamine metabolic enzymes (MAO-A and -B and catechol-o-methyltransferase). If neuroprotective therapy is to be realized in PD, a much better understanding of the biochemical pathology of the on going progressive dopaminergic degeneration is required.

In PD it is the melanin-containing pigmented dopamine neurons of SNPC which are most susceptible to neurodegeneration, where iron is accumulated in the dopamine neuron as a complex with melanin. Besides being cytotoxic in this condition, the other property of melanin-iron complex is its ability to promote an inflammatory process. Indeed evidence for inflammation in parkinsonian SNPC have come from identification of proliferation of reactive microglia[21,66] and increased inflammatory cytokines, IL1-β, IL2, IL6 and TNFα.[22,23,24] Thus, attention must be paid to these features if neuroprotection with drugs is going to be attempted. Although much can be learnt from the 6-OHDA and the MPTP animal models, both models have significant limitation. One is obviously the fact that there is very little melanin in mouse, rat or even in non-human primate SNPC. Thus, attention should be paid to how melanin is formed and accumulated in the human SNPC.

It is well recognized that catechols can readily be oxidized in the presence of divalent metals, especially iron and copper to give rise to semiquinones and superoxide. Semiquinone is thought to be involved in the formation of melanin via its polymerization. Analysis of melanin purified from normal human brain by mossbauer spectroscopy have indicated that it binds highly significant amounts of iron in a very similar manner to ferritin and hemosiderin[67] and is different from melanin synthesized *in vitro* from dopamine.[68] We are now analyzing the structural requirement of melanin isolated from PD brains and comparing it to that from control brains. We have suggested that in PD melanin could be different. Support for it has come from our recent studies where we have shown[58] that dopamine when incubated with iron and hydrogen peroxide *in vitro* generates significant amounts of 6-OHDA and 5-OHDA. Thus, theoretically 6-OHDA can be physiologically formed *in vivo* and can give rise to melanin with 6-OHDA and 5-OHDA structural backbone. Antioxidants and iron chelators do inhibit the autoxidation of dopamine to semiquinone, which leads to the formation of melanin.

The present hypothesis of etiology of PD is thus closely associated with an ongoing OS, the initiation of it is still not known. The animal models studies infer that iron chelators, radical scavengers, and GSH analogues could be a valid therapeutic approach to PD. However there are major therapeutic problems, since the nontoxic iron chelators capable of crossing the blood brain barrier are not available, nor are potent radical scavengers or therapeutically available nitric oxide inhibitors or glutathione analogues. Furthermore, an extend study with vitamin E (DATATOP). This has not prevented the search for other possible drugs. Some of the newer compounds that are being tested are shown in TABLE 2.

Many compounds have been and are being examined as radical scavengers and iron chelators.[60] One of the drugs so far tested by us, R-apomorphine (R-APO), the dopamine D1-D2 receptor agonist, has shown remarkable neuroprotective activity in models of PD.[49,50] Although clinically R-APO has several peripheral side effects, these can be managed and controlled with dopamine antagonist, domperidone. R-APO has also very poor pharmacokinetics and is readily oxidized *in vivo*. Even so, R-APO is the most effective anti-PD drug available, specially in fluctuating patients and in patients who loose the response to L-dopa. It is most often forgotten that R-APO is a catechol, such compounds bearing antioxidant properties at low concentrations, whereas being pro-oxidant at high concentrations. They also have the ability of chelating divalent metals. For these reasons we have examined the antioxidant, iron chelating and neuroprotective action of (R- and S-) APO enantiomers in several *in vitro* and *in vivo* models of neurodegeneration. We have demonstrated that *in vitro* R-APO is one of the most potent iron chelator and radical scavenger so far described, with IC_{50} 0.2–0.5 μM.[78] It inhibits iron-induced mitochondrial lipid peroxidation with similar potency. In cell culture experiments employing pheochromocytoma (PC12) cells, R-APO is a potent cytoprotective agent against iron, hydrogen peroxide and 6-OHDA–induced cell death with IC_{50} O.5–2.0 μM.[48] More recently we have demonstrated its neuroprotective properties *in vivo* using the MPTP model of PD.[49,50] In this condition, R-APO (5 and 10 mg/kg) prevents depletion of dopamine, nigral neuron loss, reduction in tyrosine hydroxylase and loss of reduced glutathione (GSH). The neuroprotective action of R-APO has been attributed to its iron chelating and radical scavenging properties and not to its dopamine receptor agonism. This is reasoned since its S- isomer, which is not a dopamine agonist, has similar iron chelating,

TABLE 2. Antioxidant and iron chelators as neuroprotective agents

Properties	Drugs
Iron chelators	desferrioxamine[59]
Radical scavengers	flavonoids,[69] green tea,[70,71] lipoic acid,[72] vitamins C and E[73,74]
Dopamine agonists	apomorphine,[48–50] bromocriptine[51]
Anti-inflammatory agents	aspirin, salicylic acid[75]
Monoamine oxidase B inhibitors	rasagiline,[76] L-deprenyl[77]

radical scavenging and neuroprotective actions. R-APO is also a reversible selective inhibitor of mouse brain MAO-A and -B. It may be assumed that its neuroprotective activity may be a combination of the properties described. We know of no other anti-PD drug so far described which has the broad neuroprotective spectrum of R-APO in the several models of cell death and PD. Thus, R-APO with its pharmacological shortcoming may be an ideal drug to examine neuroprotection in idiopathic parkinsonism.[60] Similarly, bromocriptine, a longer acting dopamine agonist, though not as effective as R-APO, has been reported to be effective against oxidative stress and striatal damage caused by MPTP and methamphetamine.[51]

GENE ALTERATIONS AND THERAPEUTIC APROCHES IN PD

PD is a neurodegenerative disorder involving the progressive degeneration of DA neurons arising in the SN and terminating with its terminals in the striatum. The major problem concerning a better therapeutic approach to the treatment and prevention of the disease is the enigma of its underlying cause. A number of recent studies have reported alterations in the expressions of various genes, such as a decrease in calcium-binding protein (28 kD calbindin-D) and D_3 receptor mRNA in the SN[79] and lymphocytes[80] from PD patients, respectively. Two gene alterations in the MPTP model for PD were observed. The first, an increase of glutamate decarboxylase mRNA in a subpopulation of neurons in the putamen of parkinsonian monkeys, which provides further evidence that striato-pallidal GABAergic neurons are hyperactive in MPTP-treated parkinsonian monkeys.[81] The second alteration, an increase in Bax mRNA expression (a cell death effector) in SN, with a concomitant increase in Bax immunoreactivity.[82] Furthermore, an increase in IL1-β mRNA was also reported in methamphetamine-treated rats.[83]

Previous studies showed significant increases in the translocation of NF κ-β to the nucleus, which is essential to its activation, in PD patients.[84,85] An increasing body of experimental evidence has indicated a direct signaling role of ROS in NF κ-β activation[86] and suppression of such activation by antioxidants.[87,88] NF κ-β induces transcription of cytokines such as TNFα and IL6.[89,90] Concomitant to these findings, it was observed that the protein levels of IL1-β, IL6, IL2 and IL4[22,24,91] and of tumor necrosis factor (TNF)α and transforming growth factor (TGF)α and β[22,23,92] are elevated in the brains of parkinsonian patients. Now days there is a great believe that the inflammatory process is one of the major processes involved in neurodegeneration, in which the cytotoxic cytokines seem to play a significant role. As it was shown in lipopolysaccharide-induced neurodegeneration in mice, there is an increase in IL1-β levels and a decrease in the binding to the IL1 receptor in the brain.[93]

As a consequence of the gene alterations that may be involved in the etiology of PD, there is a massive search for therapeutic strategies regarding gene regulation. It was previously shown for some neuroprotective drugs, that one of their mechanisms of action might involve gene regulation. L-deprenyl, for instance, an irreversible MAO-B inhibitor, reduced neuronal death independently of its MAO-B inhibition ability even after neurons have sustained seemingly lethal damage.[94] L-deprenyl also altered the synthesis of several proteins, such as the onco-proteins Bcl-2, which increases in the nigrostriatal dopaminergic regions in PD,[95] Bax and the scavenger proteins Cu/Zn-superoxide dismutase (SOD) and Mn-SOD.[94] In addition, l-deprenyl and melatonin, which exhibited neuroprotective properties in the MPTP-induced neurodegeneration model,[96] up-regulated GDNF (glial cell line–derived neurotrophic factor) gene expression at threshold doses lower than those needed for altering MAO-B activity and/or the antioxidant enzyme systems, respectively.[97] Melatonin also increased gene expression of antioxidant enzyme as glutathione peroxidase, Cu/Zn-SOD and Mn-SOD, in rat brains.[98] Another drug that caused differential gene expression is the D_2 agonist, U91356A, which reversed the MPTP-induced increase of D_2 and D_1 receptors mRNA.[99] This evidence provides strong hope for the development of new neuroprotective drugs, which exert their protective effect through their ability to modulate gene expression.

REFERENCES

1. YOUDIM, M.B.H. *et al.* 1997. Understanding Parkinson's disease. Sci. Am. **276:** 52–59.
2. JENNER, P. 1998. Oxidative mechanisms in nigral cell death in Parkinson's disease. Mov. Disord. Suppl. **13:** 24–34.
3. RIEDERER, P. *et al.* 1989. Transition metals, ferritin, glutathione, and ascorbic acid in parkinsonian brains. J. Neurochem. **52:** 515–520.
4. GERLACH, M. *et al.* 1994. Altered brain iron as a cause of neurodegeneration? J. Neurochem. **63:** 793–807.
5. YOUDIM, M.B.H. *et al.* 1993. The neurotoxicity of iron and nitric oxide. Relevance to the etiology of Parkinson's disease. Adv. Neurol. **60:** 259–266.
6. GOTZ, M.E. *et al.* 1994. Oxidative stress: free radical production in neural degeneration. Pharmacol. Therap. **63:** 37–122.
7. JENNER, P. *et al.* 1996. Oxidative stress and the pathogenesis of Parkinson's disease. Neurology Suppl. **47:** 161–170.
8. OLANOW, C.W. *et al.* 1996. Iron and neurodegeneration: prospects for neuroprotection. *In* Neurodegeneration and Neuroprotection in Parkinson's Disease. C.W. Olanow *et al.*, Eds. :55–69. London Academic Press.
9. KOSTRZEWA, R.M. *et al.* 1974. Pharmacological action of 6-hydroxydopamine. Pharmacol. Rev. **26:** 199–288.
10. BURNS, R.S. *et al.* 1983. A primate model of parkinsonism: selective destruction of dopaminergic neurons in the pars compacta of the substantia nigra by N-methyl-4-phenyl-1,2,3,6-tetrahydropyridine. Proc. Natl. Acad. Sci. USA **80:** 4546–4550.
11. CHIUEH, C.C. *et al.* 1993. Role of dopamine autoxidation, hydroxyl radical generation, and calcium overload in underlying mechanisms involved in MPTP-induced parkinsonism. Adv. Neurol. **60:** 251–258.
12. DAVIS, G.C. *et al.* 1979. Chronic parkinsonism secondary to intravenous injection of meperidine analogues. Psychiat. Res. **1:** 249–254.
13. COHEN, G. *et al.* 1974. The generation of hydrogen peroxide, superoxide radical and hydroxyl radical by 6-hydroxydopamine, dialuric acid, and related cytotoxic agents. J. Biochem. **249:** 2447–2452.
14. GLINKA, Y. *et al.* 1996. Nature of inhibition of mitochondrial respiratory complex I by 6-hydroxydopamine. J. Neurochem. **66:** 2004–2010.
15. GLINKA, Y, *et al.* 1997. Mechanism of 6-hydroxydopamine neurotoxicity. J. Neural Transm. Suppl. **50:** 55–66.
16. SEATON, T.A. *et al.* 1997. Free radical scavengers protect dopaminergic cell lines from apoptosis induced by complex I inhibitors. Brain Res. **777:** 110–118.
17. CHIUEH, C.C. *et al.* 1992. Intracranial microdialysis of salicylic acid to detect hydroxyl radical generation through dopamine autooxidation in the caudate nucleus: effects of MPP+. Free Rad. Biol. Med. **13:** 581–583.
18. CHIUEH, C.C. *et al.* 1992. Enhanced hydroxyl radical generation by 2′-methyl analog of MPTP: reversal by clorgyline and deprenyl. Synapse. **11:** 346–348.

19. SINGER, T.P. *et al.* 1987. Biochemical events in the development of parkinsonism induced by 1-methyl-4-phenyl-1,2,3,6-tetrahydropyridine. J. Neurochem. **49:** 1–8.
20. CHIUEH, C.C. *et al.* 1997. Free radicals and MPTP-induced selective destruction of substantia nigra compacta neurons. Adv. Pharmacol. **42:** 796–800.
21. MCGEER, P.L. *et al.* 1988 Expression of the histocompatibility glycoprotein HLA-DR in neurological disease. Acta Neuropathol. **76:** 550–557.
22. MOGI, M. *et al.* 1994. Interleukin-1β, interleukin-6, epidermal growth factor and transforming growth factor-α are elevated in the brain from parkinsonian patients. Neurosci. Lett. **180:** 147–150.
23. MOGI, M. *et al.* 1994. Tumor necrosis factor-α (TNF-α) increases both in the brain and in the cerebrospinal fluid from parkinsonian patients. Neurosci. Lett. **165:** 208–210.
24. MOGI, M. *et al.* 1996. Interleukin (IL)-1β, IL-2, IL-4, IL-6 and transforming growth factor-α levels are elevated in ventricular cerebrospinal fluid in juvenile parkinsonian and Parkinson's disease. Neurosci. Lett. **211:** 13–16.
25. YOUDIM, M.B.H. *et al.* 1993. The possible role of iron in the etiopathology of Parkinson's disease. Mov. Disord. **8:** 1–12.
26. RANDALL, B.L. 1992. Iron, aging, and human disease: historical background and new hypotheses. *In* Iron and Human Disease. B.L. Randall, Ed. :1–23. CRC Press.
27. OESTREICHER, E. *et al.* 1994. Degeneration of nigrostriatal dopaminergic neurons increases iron within the substantia nigra: a histochemical and neurochemical study. Brain Res. **660:** 8–18.
28. AKIYAMA, H. *et al.* 1989. Microglial response to 6-hydroxydopamine-induced substantia nigra lesions. Brain Res. **489:** 247–253.
29. LODE, H.N. *et al.* 1990. Release of iron from ferritin by 6-hydroxydopamine under anaerobic conditions. Free radical Res. Commun. **11:** 153–158.
30. YE, F.Q. *et al.* 1996. Basal ganglia iron content in Parkinson's disease measured with magnetic resonance. Mov. Disord. **11:** 243–249.
31. LAN, J. *et al.* 1997. Excessive iron accumulation in the brain: a possible potential risk of neurodegeneration in Parkinson's disease. J. Neural. Transm. **104:** 649–660.
32. KUMAR, R. *et al.* 1995. Free radical generated neurotoxicity of 6-hydroxydopamine. J. Neurochem. **64:** 1703–1707.
33. OGAWA, N. *et al.* 1994. Changes in lipid peroxidation, Cu/Zn-superoxide dismutase and its mRNA following an intracerebroventricular injection of 6-hydroxydopamine in mice. Brain Res. **646:** 337–340.
34. PERUMAL, A.S. *et al.* 1989. Regional effects of 6-hydroxydopamine on free radical scavengers in rat brain. Brain Res. **504:** 139–141.
35. PEARCE, R.K. *et al.* 1997. Alterations in the distribution of glutathione in the substantia nigra in Parkinson's disease. J. Neural. Transm. **104:** 661–677.
36. TOGHI, H. *et al.* 1995. Reduced and oxidized forms of glutathione and alpha-tocopherol in the cerebrospinal fluid of parkinsonian patients: comparison between before and after L-DOPA treatment. Neurosci. Lett. **184:** 21–24.
37. GERFEN, C.R. *et al.* 1987. The neostriatal mosaic: III. Biochemical and developmental dissociation of patch-matrix mesostriatal systems. J. Neurosci. **7:** 3935–3944.
38. GERMAN, D.C. *et al.* 1992.Midbrain dopaminergic cell loss in Parkinson's disease and MPTP-induced parkinsonism: sparing of calbindin-D28k-containing cells. Ann. N.Y. Acad. Sci. **648:** 42–62.
39. IACOPINO, A.M. *et al.* 1992. Calbindin-D28k-containing neurons in animal models of neurodegeneration: possible protection from excitotoxicity. Brain Res. Mol. Brain Res. **13:** 251–261.
40. GLINKA, Y.Y. *et al.* 1995. Inhibition of mitochondrial complexes I and IV by 6-hydroxydopamine. Eur. J. Pharmacol. **292:** 329–332.
41. MIZUNO, Y. *et al.* 1988. Inhibition of mitochondrial respiration by MPTP in mouse brain. Neurosci. Lett. **3:** 349–353.
42. HEIKKILA, R.E. *et al.* 1984. Protection against the dopaminergic neurotoxicity of 1-methyl-4-phenyl-1,2,3,6-tetrahydropyridine by monoamine oxidase inhibitors. Nature **311:** 467–469.
43. TATTON, W.G. 1993. Selegiline can mediate neuronal rescue rather than neuronal protection. Mov. Disord. **8:** S20–S30.

44. SANTIAGO, M. et al. 1997. Neuroprotective effect of the iron chelator desferrioxamine against MPP^+ toxicity on striatal dopaminergic terminals. J. Neurochem. **68:** 732–738.
45. MATARREDONA, E.R. et al. 1997. Involvement of iron in MPP+ toxicity in substantia nigra: protection by desferrioxamine. Brain Res. **773:** 76-81.
46. CADET, J.L. et al. 1989. Vitamin E attenuates the toxic effects of intrastriatal injection of 6-hydroxydopamine (6-OHDA) in rats: behavioral and biochemical evidence. Brain Res. **476:** 5–10.
47. PERUMAL, A.S. et al. 1992. Vitamin E attenuates the toxic effects of 6-hydroxydopamine on free radical scavenging systems in rat brain. Brain Res. Bull. **29:** 699–671.
48. GASSEN, M. et al. 1998. Apomorphine enantiomers protect pheochromocytoma (PC12) cells from oxidative stress induced by hydrogen peroxide and 6-hydroxydopamine. Mov. Disord. **13:** 242–248.
49. GRÜNBLATT, E. et al.1999. Potent neuroprotective and antioxidant activity of apomorphine in MPTP and 6-hydroxydopamine induced neurotoxicity. J. Neural Transm. (Suppl.) **55:** 57–70.
50. GRÜNBLATT, E. et al. 1999. Apomorphine protects against MPTP induced neurotoxicity in mice. Mov. Disord. **14:** 612–618.
51. MURALIKRISHNAN, D. et al. 1998. Neuroprotection by bromocriptine against 1-methyl-4-phenyl-1,2,3,6-tetrahydropyridine-induced neurotoxicity in mice. FASEB J. **12:** 905–012.
52. DI MONTE, D. et al. 1987. Increase efflux rather than oxidation is the mechanism of glutathione depletion by 1-methyl-4-phenyl-1,2,3,6-tetrahydropyridine (MPTP). Biochem. Biophys. Res. Commun. **148:** 153–160.
53. PRZEDBORSKI, S. et al. 1996. Role of neuronal nitric oxide in 1-methyl-4-phenyl-1,2,3,6-tetrahydropyridine (MPTP)-induced dopaminergic neurotoxicity. Proc. Natl. Acad. Sci. USA **93:** 4565–4571.
54. SCHULZ, J.B. et al. 1995. Role of nitric oxide in neurodegenerative diseases. Curr. Opin. Neurol. **8:** 480–486.
55. TEMLETT, J.A. et al. 1994. Increased iron in the substantia nigra compacta of the MPTP-lesioned hemiparkinsonian african green monkey: evidence from proton microprobe element microanalysis. J. Neurochem. **62:** 134–146.
56. MOCHIZUKI, H. et al. 1994. Iron accumulation in the substantia nigra of 1-methyl-4-phenyl-1,2,3,6-tetrahydropyridine (MPTP) induced hemiparkinsonism in monkeys. Neurosci. Lett. **168:** 251–253.
57. MONTEIRO, H.P. et al. 1989. 6-Hydroxydopamine releases iron from ferritin and promotes ferritin-dependent lipid peroxidation. Biochem. Pharmacol. **38:** 4177–4182.
58. LINERT, W. et al. 1996. Dopamine, 6-hydroxydopamine, iron, and dioxygen-their mutual interactions and possible implication in the development of Parkinson's disease. Biochem. Biophys. Acta **1316:** 160–168.
59. BEN-SHACHAR, D. et al. 1991. The iron chelator desferrioxamine (Desferal) retards 6-hydroxydopamine-induced generation of nigrostriatal dopamine. J. Neurochem. **56:** 1441–1444.
60. GASSEN, M. et al. 1997. The potential role of iron chelators in the treatment of Parkinson's disease and related neurological disorders. Pharmacol. Toxicol. **80:** 159–166.
61. BEN-SHACHAR, D. et al. 1991. Intranigral iron injection induces behavioral and biochemical "parkinsonism" in rats. J. Neurochem. **57:** 2133–2135.
62. SENGSTOCK, G.J. et al. 1994. Progressive changes in striatal dopaminergic markers, nigral volume, and rotational behavior following iron infusion into the rat substantia nigra. Exp. Neurol. **130:** 82–94.
63. BEN-SHACHAR, D. et al. 1992. Role of iron and iron chelation in dopaminergic-induced neurodegenaration: implication for Parkinson's disease. Ann. Neurol. Suppl. **32:** S105–S110.
64. LEENDER, S.K.L. et al. 1994. Blood to brain iron uptake in one rhesus monkeys using [Fe-52]-citrate and positron emission tomography (PET): influence of haloperidol. J. Neural Transm. Suppl. **43:** 123–132.

65. YOUDIM, M.B.H. et al. 1999. Neuroprotective strategies in Parkinson's disease and Huntington Chorea: MPTP and 3-NP induced neurodegeneration as models. *In* Mitochondrial Inhibition and Neurodegenerative Disorders. P.R. Sandberg, H. Nishino & C.V. Borlongan, Eds. NewYork. Humana Press. In Press.
66. JELLINGER, K. et al. 1990. Brain iron and ferritin in Parkinson's and Alzheimer's diseases, J. Neural. Transm. Park. Dis. Dement. Sect. **2:** 327–340.
67. GERLACH, M. et al. 1995. Mossbauer spectroscopic studies of purified human neuromelanin isolated from the substantia nigra. J. Neurochem. **65:** 923–926.
68. ZECCA, L. et al. 1992. The chemical characterization of melanin contained in substantia nigra of human brain. Biochem. Biophys. Acta. **1138:** 6–10.
69. KOSTYUK, V.A. et al. 1998. Antiradical and chelating effects in flavonoid protection against silica-induced cell injury. Arch, Biochem. Biophys. **355:** 43–48.
70. KATIYAR, S.K. et al. 1999. Polyphenolic antioxidant (-)epigallocatechin-3-gallate from green tea reduces UVB-induced inflammatory responses and infiltration of leukocytes in human skin. Photochem. Photobiol. **69:** 148–153.
71. PIETTA, P. et al. 1998. Relationship between rate and extent of catechin absorption and plasma antioxidant status. Biochem. Mol. Biol. Inter. **46:** 895–903.
72. BIEWENGA, G.P. et al. 1997. The pharmacology of the antioxidant lipoic acid. General Pharmacol. **29:** 315–331.
73. ROSLER, M. et al. 1998. Free radicals in Alzheimer's dementia: currently available therapeutic strategies. J. Neural. Transm. Supl. **54:** 211–219.
74. VATASSERY, G.T. 1998. Vitamin E and other endogenous antioxidants in the central nervous system. Geriatrics. **1:** 25–27.
75. AUBIN, N. et al. 1998. Aspirin and salicylate protect against MPTP-induced dopamine depletion in mice. J. Neurochem. **71:** 1635–1642.
76. FINBERG, J.P. et al. 1998. Increased survival of dopaminergic neurons by rasagiline, a monoamine oxidase B inhibitor. Neuroprotection **9:** 703–707.
77. LYYTINEN, J. et al. 1997. Simultaneous MAO-B and COMT inhibition in L-DOPA treated patients with Parkinson's disease. Mov. Disord. **12:** 497–505.
78. GASSEN, M. et al. 1996. Apomorphine is highly potent free radical scavenger in rat mitochondrial fraction. Eur. J. Pharmacol. **308:** 219–225.
79. IACOPINO, A.M et al. 1990. Specific reduction of calcium-binding protein (28-kilodalton calbindin-D) gene expression in aging and neurodegenerative diseases. Proc. Natl. Acad. Sci. USA **87:** 4078–4082.
80. NAGAI, Y. et al. 1996. Decrease of the D3 dopamine receptor mRNA expression in lymphocytes from patients with Parkinson's disease. Neurology **46:** 791–795.
81. SOGHOMONIAN, J.J. et al. 1997. Glutamate decarboxylase (GAD67 and GAD65) gene expression is increased in a subpopulation of neurons in the putamen of Parkinsonian monkeys. Synapse. **27:** 122–132.
82. HASSOUNA, I. et al. 1996. Increase in bax expression in substantia nigra following 1-methyl-4-phenyl-1,2,3,6-tetrahydropyridine (MPTP) treatment of mice. Neurosci. Lett. **204:** 85–88.
83. YAMAGUCHI, T. et al. 1991. Methamphetamine-induced expression of interleukin-1 beta mRNA in the rat hypothalamus. Neurosci. Lett. **128:** 90–92.
84. HUNOT, S. et al. 1997. Nuclear translocation of NF-κβ is increased in dopaminergic neurons of patients with Parkinson's disease. Proc. Natl. Acad. Sci. USA **94:** 7531–7536.
85. LIN, M. et al. 1997. Role of iron in NF-kappa β activation and cytokine gene expression by rat hepatic macrophages. Am. J. Phys. **272:** G1355–G1364.
86. SCHRECK, R. et al. 1991. Reactive oxygen intermediates as apparently widely used messengers in the activation of the NF-κβ transcription factor and HIV-1. EMBO J. **10:** 2247–2258.
87. SCHRECK, R. et al. 1992. Dithiocarbamates as potent inhibitors of nuclear factor κβ activation in intact cells. J. Exp. Med. **175:** 1181–1194.
88. SUZUKI, Y.T. et al. 1993. Inhibition of NF-κβ activation by vitamin E derivatives. Biochem. Biophys. Res. Commun. **193:** 277–283.

89. COLLART, M.A. et al. 1990. Regulation of tumor necrosis factor-α transcription in macrophages: involvement of four kβ-like motifs and of constitutive and inducible forms of NF-kB. Mol. Cell. Biol. **10:** 1498–1506.
90. SHIMIZU, H. et al. 1990. Involvement of a NF-κβ like transcription factor in the activation of the interleukin-6 gene by inflammatory lymphokines. Mol. Cell Biol. **10:** 561–568.
91. BLUM-DEGEN, D. et al. 1995. Interleukin-1β and interleukin-6 are elevated in the cerebrospinal fluid of Alzheimer's and de novo Parkinson's disease patients. Neurosci. Lett. **202:** 17–20.
92. MOGI, M. et al. 1995. Transforming growth factor-β1 levels are elevated in the striatum and in ventricular cerebrospinal fluid in Parkinson's disease. Neurosci. Lett. **193:** 129–132.
93. TAKAO, T. et al. 1993. Reciprocal Modulation of interleukin-1β (IL-1β) and IL-1 receptors by lipopolysaccharide (endotoxin) treatment in the mouse brain-endocrine-immune axis. Endocrinology. **132:** 1497–1504.
94. TATTON, W.G. et al. 1997. Apoptosis in neurodegenerative disorders: potential for therapy by modifying gene transcription. J. Neural Transm. Suppl. **49:** 245–268.
95. MOGI, M. et al. 1996. bcl-2 Protein is increased in the brain from parkinsonian patients. Neurosci. Lett. **215:** 137–139.
96. ACUNA-CASTROVIEJO, D. et al. 1997. Melatonin is protective against MPTP-induced striatal and hippocampal lesions. Life Sci. **60:** 23–29.
97. TANG, T.P. et al. 1998. Enhanced glial cell line-derived neurotrophic factor mRNA expression upon (−)-deprenyl and melatonin treatment. J. Neurosci. Res. **53:** 593–604.
98. KOTLER, M. et al. 1998. Melatonin increases gene expression for antioxidant enzymes in rat brain cortex. J. Pineal. Res. **24:** 83–89.
99. GOULET, M. et al. 1997. Continuous or pulsatile chronic D2 dopamine receptor agonist (U91356A) treatment of drug-naive 4-phenyl-1,2,3,6-tetrahydropyridine monkeys differentially regulates brain D1 and D2 receptor expression: in situ hybridization histochemical analysis. Neuroscience **79:** 497–507.

Neuroprotective Antioxidants from Marijuana[a]

A.J. HAMPSON,[b,c] M. GRIMALDI,[d] M. LOLIC,[e] D. WINK,[f] R. ROSENTHAL,[e] AND J. AXELROD[b]

[b]*Laboratory of Cellular and Molecular Regulation, NIMH, Bethesda, Maryland 20892, USA*

[d]*Laboratory of Adaptive Systems, NINDS, Bethesda, Maryland 20892, USA*

[e]*Department of Emergency Medicine, George Washington University, Washington DC 20052, USA*

[f]*Radiation Biology Branch, NCI, Bethesda, Maryland 20892, USA*

ABSTRACT: Cannabidiol and other cannabinoids were examined as neuroprotectants in rat cortical neuron cultures exposed to toxic levels of the neurotransmitter, glutamate. The psychotropic cannabinoid receptor agonist Δ^9-tetrahydrocannabinol (THC) and cannabidiol, (a non-psychoactive constituent of marijuana), both reduced NMDA, AMPA and kainate receptor mediated neurotoxicities. Neuroprotection was not affected by cannabinoid receptor antagonist, indicating a (cannabinoid) receptor-independent mechanism of action. Glutamate toxicity can be reduced by antioxidants. Using cyclic voltametry and a fenton reaction based system, it was demonstrated that Cannabidiol, THC and other cannabinoids are potent antioxidants. As evidence that cannabinoids can act as an antioxidants in neuronal cultures, cannabidiol was demonstrated to reduce hydroperoxide toxicity in neurons. In a head to head trial of the abilities of various antioxidants to prevent glutamate toxicity, cannabidiol was superior to both α-tocopherol and ascorbate in protective capacity. Recent preliminary studies in a rat model of focal cerebral ischemia suggest that cannabidiol may be at least as effective *in vivo* as seen in these *in vitro* studies.

INTRODUCTION

Cannabinoid components of marijuana are known to exert behavioral and psychotropic effects but also possess therapeutic properties including analgesia,[1] ocular hypotension,[2] and antiemesis.[3] This report examines another potential therapeutic role for cannabinoids as neuroprotectants and describes their mechanism of action in rat cortical neuronal cultures. During an ischemic episode, large quantities of the excitatory neurotransmitter, glutamate, are released in the brain. This event causes neuronal death by over-stimulation of NMDA[g] (NMDAr), AMPA and kainate type receptors, which massively increase intracellular calcium, resulting in metabolic

---

[a]*In vitro* data presented in this paper and FIGURES 1–6 were first published in Proceedings of the National Academy of Sciences (July 1998, **95**: 8268–8273).

[c]Address for correspondence: Dr. Aidan Hampson, Cortex Pharmaceuticals, Inc., 15231 Barranca Parkway, Irvine, CA 92618.
e-mail: aidan@codon.nih.gov

stress and production of toxic reactive oxygen species (ROS). Antioxidants such as α-tocopherol,[4,5] can prevent this toxicity by reducing the ROS formed during ischemic metabolism. Cannabinoids such as (-)Δ^9-tetrahydro-cannabinol (THC) and its psychoactive analogues have previously been suggested to reduce glutamate toxicity,[6] although this effect was apparently cannabinoid receptor mediated,[6,7] perhaps through inhibition of voltage sensitive calcium channels.[7–9] Our study examines cannabinoids as *in vitro* neuroprotectants, and focuses on the non-psychoactive cannabinoid, cannabidiol. As with THC, cannabidiol is a natural component of the marijuana plant, *Cannabis sativa*, although unlike THC, cannabidiol does not activate cannabinoid receptors in the brain and so is devoid of psychoactive effects. This presentation will demonstrate that cannabinoids are potent antioxidants which can protect neurons from ischemic injury without psychoactive side-effects.

CANNABIDIOL BLOCKS NMDA, AMPA, AND KAINATE RECEPTOR–MEDIATED NEUROTOXICITY

Glutamate neurotoxicity can be mediated either by NMDA, AMPA or kainate receptors. We therefore compared the ability of cannabinoids to prevent neurotoxicity mediated by all three types of glutamate receptors. To examine NMDAr mediated toxicity, rat cortical neuron cultures were exposed to glutamate for 10 min in a magnesium free medium. After this time, the culture media was replaced and the cells incubated for 20 hours at 37°C. In order to examine AMPA / kainate receptor mediated toxicity, neurons were incubated with glutamate for 20 hours in the presence of MK-801 (an NMDAr antagonist) and an agent to prevent receptor desensitization. To study AMPA or kainate receptors individually, glutamate was replaced with a specific receptor agonists (fluorowillardiine or 4-methyl-glutamate respectively). At the end of the incubation period, in both NMDA and AMPA / kainate models, toxicity was assessed by examination of lactate dehydrogenase (LDH) released into the media by dying cells.

Cannabidiol prevented cell death equally well with an EC50 of 2–4 μM in both NMDA and AMPA / kainate toxicity models (FIG. 1). Similar data was also observed when glutamate or AMPA-specific or kainate receptor specific ligands were used (data not shown). These results demonstrate cannabidiol protect equally regardless of whether toxicity is mediated by NMDA, AMPA or kainate receptors. This suggests glutamate receptors are probably not the site at which cannabidiol acts, protection is more likely to be due to a mechanism that occurs downstream of the initial glutamate receptor activation event.

NEUROPROTECTION BY TETRAHYDROCANNABINOL

Unlike cannabidiol, THC is a ligand for the brain cannabinoid receptor[10] and this action has been proposed to explain THC's ability to protect neurons from NMDAr

[g]ABBREVIATIONS: AMPA, 2-amino-3-(4-butyl-3-hydroxyisoxazol-5-yl)propionic acid; BHT, Butylhydroxy- toluene; NMDA, N-methyl-D-aspartate; NMDAr, NMDA receptors; ROS, reactive oxygen species; THC, -Δ^9-tetrahydrocannabinol.

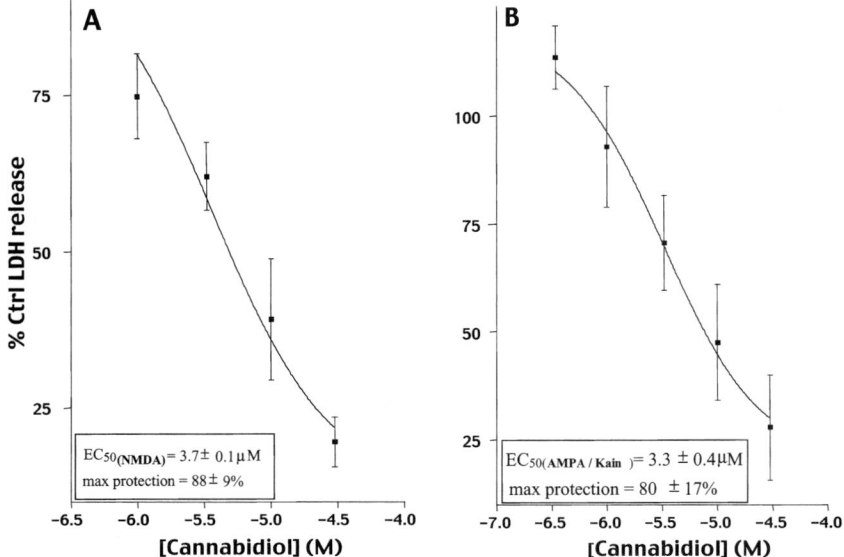

FIGURE 1. Effect of cannabidiol on NMDAr (**A**) and AMPA/kainate receptor (**B**) mediated neurotoxicity. Data shown represents mean values ± SEM from a single experiment with four replicates. Each experiment was repeated on at least four occasions with essentially the same results. Cannabinoids were present during (and, in the case of NMDAr mediated toxicity, after) the glutamate exposure periods. See text for further experimental details.

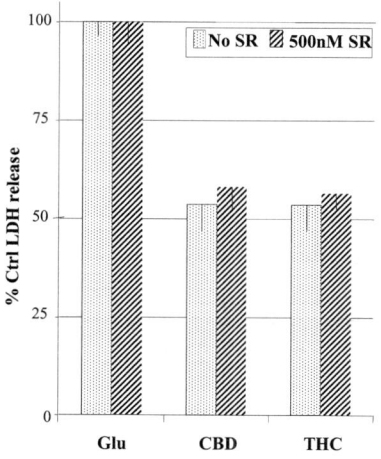

FIGURE 2. Effect of THC, cannabidiol and cannabinoid receptor antagonist on glutamate induced neurotoxicity. Neurons exposed to glutamate in an AMPA/ kainate receptor toxicity model, were incubated with 10 μM cannabidiol or THC in the presence or absence of SR141716A (500 nM). See text for experimental details. Data represents mean values ± SEM from four experiments each with three replicates.

toxicity *in vitro*.[6] However, THC and cannabidiol were similarly protective in AMPA/kainate receptor toxicity assays, suggesting that cannabinoid neuroprotection does not involve cannabinoid receptor activation. This was confirmed using the cannabinoid receptor antagonist, SR-141716A (FIG. 2). Neither THC or cannabidiol neuroprotection was affected by cannabinoid receptor antagonist indicating their action is not cannabinoid receptor–mediated.

CANNABINOIDS AS ANTIOXIDANTS

Easily oxidizable compounds such as glutathione, ascorbate and α-tocopherol, are used by living cells as disposable antioxidants which protect vital membranes and proteins from ROS damage. This type of ROS damage has previously been demonstrated to be a factor in glutamate neurotoxicity.[4,5] To investigate whether cannabinoids might possess antioxidant abilities and so protect neurons by absorbing the ROS formed following glutamate receptor activity, the antioxidant properties of cannabidiol and other cannabinoids were assessed by both cyclic voltametry and in a Fenton reaction system (iron catalyzed ROS generation). Cyclic voltametry, which examines the ability of a compound to accept or donate electrons under a variable voltage potential, was used to measure the oxidation potentials of both natural and synthetic cannabinoids (FIG. 3). All of the cannabinoids tested (cannabidiol, cannabinol, THC, nabilone, HU-211 and levanantrodol), yielded electron donation profiles similar to that of the known antioxidant, butylhydroxy-toluene (BHT). Anandamide, an endogenous cannabinoid receptor ligand that is structurally unrelated to cannabinoids and is not a good electron donor, was included as a negative control. The antioxidant properties of cannabinoids were also examined in a Fenton reaction system using the lipid and water-soluble compound, *Tert*-butyl hydroperoxide as a substrate.

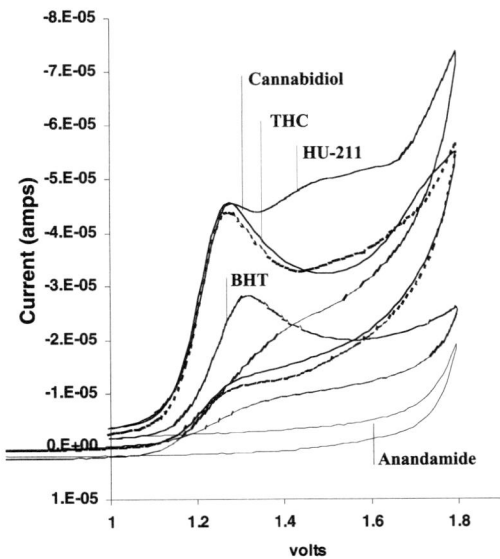

FIGURE 3. A comparison of the oxidation potentials of cannabinoids and the antioxidant, BHT. The oxidation profiles of (750 µM) BHT, cannabinoids and anandamide, were compared by cyclic voltametry. Anandamide, a cannabinoid receptor ligand with a non-cannabinoid structure was used as a non-responsive control. Experiments were repeated three times with essentially the same results. See text for experimental details.

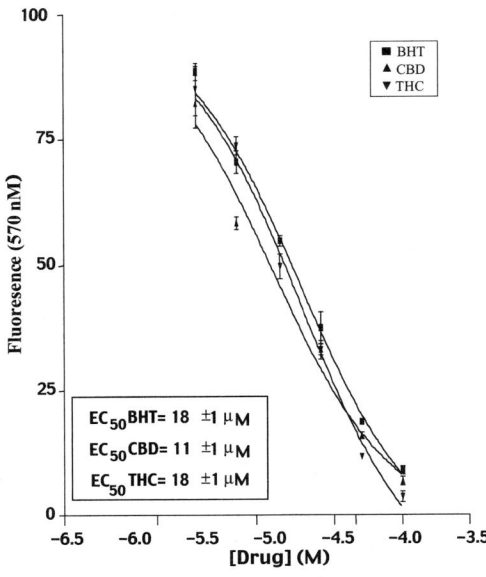

FIGURE 4. Effect of cannabidiol and THC on dihydrorhodamine oxidation. Cannabinoids were compared with BHT for their ability to prevent t-butyl hydroperoxide induced oxidation of dihydrorhodamine. See text for experimental details. Data represent mean values ± SEM from a single experiment with three replicates. This experiment was repeated four times with essentially the same results.

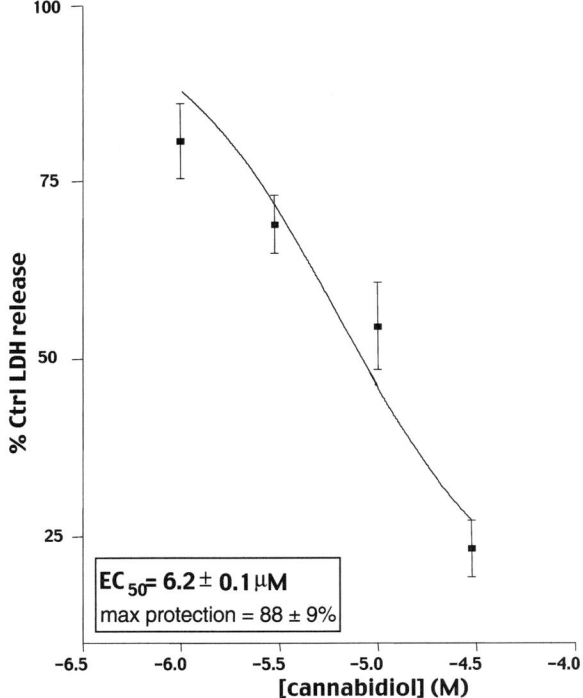

FIGURE 5. The effect of cannabidiol on oxidative toxicity in neuronal cultures. Toxicity was induced by addition of 250 μM *T*-butyl hydroperoxide in the presence or absence of cannabidiol. Each experiment represents the mean of four replicates, repeated on three occasions.

Dihydrorhodamine, an oxidation sensitive fluorescent dye, served as the target (and indicator) of oxidation in this reaction. Cannabidiol and THC both prevented dihydrorhodamine oxidation in a concentration dependent manner similar to that of the antioxidant, BHT (FIG. 4).

CANNABINOIDS PREVENT OXIDANT TOXICITY IN NEURONAL CULTURES

The ability of cannabinoids to prevent ROS toxicity in cultured neuron preparations was also examined. (FIG. 5). Tertbutyl hydroperoxide was again used as the oxidant, because its solubility in both aqueous and organic solvents, facilitates oxidation in both cytosolic and membrane delimited cellular compartments. As previously shown with glutamate toxicity studies, cannabidiol protected neuron cultures well against hydroperoxide toxicity (in a dose dependent manner), so that 30 µM cannabidiol was able to rescue 75% of neurons from 250 µM peroxide (a dose calculated to have maximal lethal effect).

CANNABIDIOL IS A POWERFUL ANTIOXIDANT IN NEURONAL CULTURES

The protective capacity of cannabidiol was compared with more familiar antioxidants in an AMPA / kainate toxicity model where neurons were exposed to both glutamate and equal concentrations (5 µM) of cannabidiol, α-tocopherol, BHT or ascorbate (FIG. 6). While all of the antioxidants attenuated glutamate toxicity to varying degrees, cannabidiol was 30–50% more protective than either α-tocopherol or ascorbate.

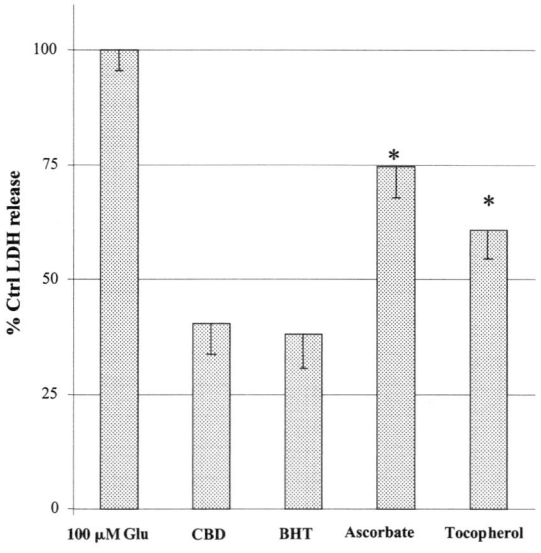

FIGURE 6. Comparison of antioxidants and cannabidiol for their ability to prevent glutamate toxicity in neurons. The effects of cannabidiol, BHT, ascorbate and α-tocopherol (10 µM) were examined in a model of AMPA/kainate receptor dependent toxicity. All drugs were present throughout the glutamate exposure period. Each experiment represents the mean of four replicates, repeated on three occasions. See text for further experimental details. Significant differences between cannabidiol and other antioxidants are indicated with an asterisk.

PRELIMINARY *IN VIVO* STUDIES OF CANNABIDIOL

The efficacy of cannabidiol as an antiischemic agent has recently been examined in a rat stroke model. In anesthetized Wistar rats, a suture was fed through the carotid artery up into the middle cerebral artery (MCA). The suture prevented blood flow and was left in place for 90 minutes, after which time it was removed. The animals were allowed to recover for 48 hours and then a six point battery of neurological tests was performed. After these tests, the animals were sacrificed and their brains were fixed, sliced and the area of infarct calculated by computer imaging. At the onset of ischemia, either 5 mg/kg of cannabidiol or vehicle was intravenously administered to the animals using a "blinded" protocol. A second 20 mg/kg dose was administered by intra-peritoneal injection 12 hours after surgery. Forty eight hours after surgery the animals were sacrificed and their brains perfused with a 2% solution of triphenyltetrazolium chloride. The samples were then fixed, sliced and the infarct volume calculated by computer imaging. Representative images of brain slices taken from a typical control and cannabidiol treated animal are presented in FIGURE 7. In this study cannabidiol reduced infarct size by 60% by comparison with vehicle treated animals. Behavioral parameters were also significantly improved ($p = 0.016$, $n = 7$) by cannabidiol treatment although the drug had no significant effect on blood pressure, glucose levels, blood gases or rectal temperature.

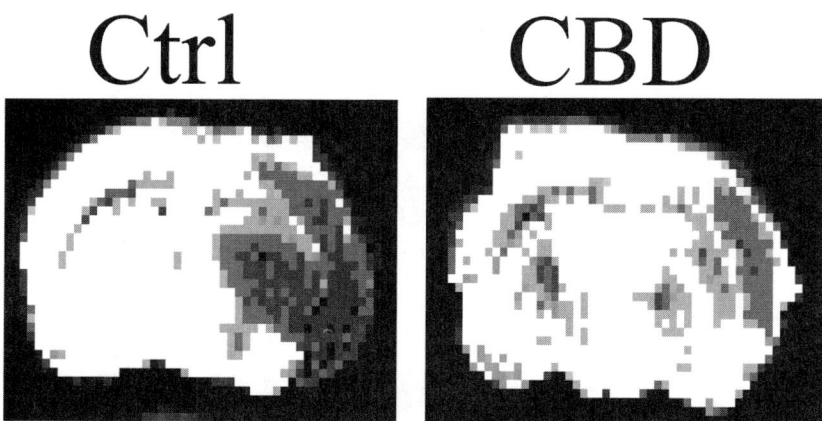

FIGURE 7. Cannabidiol as a neuroprotective agent in an MCA occlusion model of focal ischemia. The effects of cannabidiol were examined in Wistar rats subjected to 90 min of focal ischemia induced by occlusion of the middle cerebral artery (MCA). Forty-eight hours after surgery the animals were sacrificed and their brains perfused with a 2% solution of triphenyltetrazolium chloride. The samples were then fixed, sliced and the infarct volume calculated by computer imaging. The left-hand image is representative of brain taken from a rat treated with a vehicle control. The right image represents an animal that received 5 mg/kg (i.v.) cannabidiol (CBD) immediately prior to onset of ischemia.

DISCUSSION

The non-psychoactive marijuana constituent, cannabidiol, can prevent both glutamate neurotoxicity and ROS induced cell death. Tetrahydrocannabinol the psychoactive principle of *Cannabis*, also blocks neurotoxicity with a potency similar to that of cannabidiol. This neuroprotection was unaffected by cannabinoid receptor antagonist, which demonstrates that cannabinoids have effects that are independent of their involvement with cannabinoid receptors, the system responsible for cannabinoid psychoactivity.[11]

Cannabidiol and THC were equally potent at blocking glutamate toxicity regardless of which glutamate receptor mediated the toxicity. This suggests that either, cannabidiol and THC antagonize three different glutamate receptors with a similar affinity or more likely, cannabinoids protect by an action downstream of the initial receptor activation event. Cannabidiol, THC and other structurally related cannabinoids were all demonstrated to be antioxidants by cyclic voltametry. Using a glutamate neuronal toxicity model cannabidiol was demonstrated to be significantly more protective than either of the antioxidant vitamins, α-tocopherol or ascorbate and comparable to the industrial antioxidant, BHT. However, unlike BHT, cannabinoids do not appear to be tumor promotors.[12,13]

These properties of cannabinoids suggest they may have a therapeutic role as neuroprotectants, and the particular properties of cannabidiol make it a good candidate for such development. The lack of psychoactivity associated with cannabidiol allows it to be administered in higher doses than would be possible with psychotropic cannabinoids such as THC. It is hoped that therapeutics developed from non-psychoactive cannabinoids may also avoid the toxic side effects associated with clinical use of other promising antiischemic agents such as NMDAr antagonists.[14]

Preliminary studies from a currently ongoing study using a model of focal cerebral ischemia suggest that cannabidiol may well prove to be a good protective agent *in vivo*. Our studies in rats have indicated that 5 mg/kg of cannabidiol (iv) reduced both infarct volume and neurological impairment by 50–60%. While it is difficult to extrapolate drug doses of given in rodent studies to humans it is worth noting that psychoactive cannabinoids produce strong physiological responses in rats when administered in the 5–20 mg/kg range (ip[15]) while humans experience psychoactive effects and make physiological responses in the 10–60 mg/70kg range (po[15]). This means that owing to enhanced the metabolism of rodents, 5–20× the amount of cannabinoid must be administered comparative to a human dosage in order to achieve a similar effect. If one assumes that this ratio would also hold true for cannabidiol when used as a neuroprotectant, one might expect to achieve results similar to those seen in our rat ischemia studies at a human dose of less than 1 mg/kg. Previous clinical studies using cannabidiol in humans have already demonstrated that cannabidiol has a low toxicity, even when chronically administered to humans[16] or given in large acute doses of 10 mg/kg/day.[17] From these admittedly crude calculations, it is hoped that cannabidiol may one day reach clinical trials, and if so, it may be expected that the required dose will be low, thereby reducing the chances of toxic side effects. (See Proc. Natl. Acad. Sci., July 1998, **95:** 8268–8273 for the first publication of the *in vitro* data presented here.)

REFERENCES

1. WELCH, S.P. & D.L. STEVENS. 1992. Antinociceptive activity of intrathecally administered cannabinoids alone, and in combination with morphine, in mice. J. Pharmacol. Exp. Ther. **262**(1): 8–10.
2. MERRITT, J.C., W.J. CRAWFORD, P.C. ALEXANDER, A.L. ANDUZE & S.S. GELBART. 1980. Effect of marihuana on intraocular and blood pressure in glaucoma Ophthalmol. **87**(3): 222–228.
3. ABRAHAMOV, A. & R. MECHOULAM. 1995 An efficient new cannabinoid antiemetic in pediatric oncology Life Sci. **56**(23-24): 2097–2102.
4. CIANI, E., L. GRONENG, M. VOLTATTORNI, V. ROLSETH, A. CONTESTABILE & R.E. PAULSEN. 1996. Inhibition of free radical production or free radical scavenging from the excitotoxic cell death mediated by glutamate in cultures of cerebellar granule neurons. Brain Res. **728**: 1–6.
5. MACGREGOR, D.G., M.J. HIGGINS, P.A. JONES, W.L. MAXWELL, M.W. WATSON, D.I. GRAHAM & T.W. STONE. 1996. Ascorbate attenuates the systemic kainate-induced neurotoxicity in the rat hippocampus Brain Res **727**(1–2): 133–144.
6. SKAPER, S.D., A. BURIANI, R. DAL TOSO, I.L. PETRELL, L. ROMANELLO, L. FACCI & A. LEON. 1996. The aliamide, palmitoylethanolamide and cannabinoids, but not anandamide, are protective in a delayed, postglutamate paradigm of excitotoxic death in cerebellar granular neurons. Proc. Natl. Acad. Sci. USA **93**: 3984–3989.
7. HAMPSON, A.J., L.M. BORNHEIM, M. SCANZIANI, C.S. YOST, A.T. GRAY, B.M. HANSEN, D.J. LEONOUDAKIS & P.E. BICKLER. 1998. Dual Effects of Anandamide on NMDA Receptor-Mediated Responses and Neurotransmission J. Neurochem. **70**: 671–676.
8. TWITCHELL, W., S. BROWN & K. MACKIE. 1997. Cannabinoids inhibit N- and P/Q-type calcium channels in cultured rat hippocampal neurons J. Neurophysiol. **78** (1):43–50.
9. ESHHAR, N., S. STRIEM, R. KOHEN, O. TIROSH & A. BIEGON. 1995. Neuroprotective and antioxidant activities of HU-211, a novel NMDA receptor antagonist. Eur. J. Pharmacol. **283**(1–3):19-29
10. DEVANE, W.A., F.A. DYSARZ, M.R. JOHNSON, L.S. MELVIN & A.C. HOWLETT. 1988. Determination and characterization of a cannabinoid receptor in rat. Mol. Pharmacol. **34**(5): 605–613.
11. LEDENT, C., O. VALVERDE, G. COSSU, F. PETITET, J.F. AUBERT, F. BESLOT, G.A. BÖHME, A. IMPERATO, T. PEDRAZZINI,. B.P. ROQUES, G. VASSART, W. FRATTA & M. PARMENTIER. 1999. Unresponsiveness to cannabinoids and reduced addictive effects of opiates in CB1 receptor knockout mice. Science **283**: 401–404
12. THOMPSON, J.A., J.L. BOLTON & A.M. MALKINSON. 1991. Relationship between the metabolism of butylated hydroxytoluene (BHT) and lung tumor promotion in mice Exp. Lung Res. **172**(2): 439–453.
13. LINDENSCHMIDT, R.C., A.F. TRYKA, M.E. GOAD & H.P. WITSCHI. 1986. The effects of dietary butylated hydroxytoluene on liver and colon tumor development in mice. Toxicology **38**(2): 151–160.
14. AUER, R.N. 1994. Assessing structural-changes in the brain to evaluate neurotoxicological effects of NMDA receptor antagonists. Psychopharmacol Bull. **30**(4): 585–591.
15. NEUMEYER, J.L. & R.A. SHAGOURY. 1971. Chemistry and pharmacology of marijuana. J. Pharm. Sci. **60**(10):1433–1457.
16. CUNHA, J.M., E.A. CARLINI, A.E. PEREIRA, O.L. RAMOS, C. PIMENTEL, R. GAGLIARDI, W.L. SANVITO, N. LANDER & R. MECHOULAM. 1980. Chronic administration of cannabidiol to healthy volunteers and epileptic patients. Pharmacology **21**(3): 175–185.
17. CONSROE, P., J. LAGUNA, J. ALLENDER, S. SNIDER, L. STERN, K.R. SANDY, K. KENNEDY & K. SCHRAM. 1991. Controlled clinical trial of cannabidiol in Huntington's disease. Pharmacol. Biochem. Behav. **40**(3): 701–708.

A Positive-Feedback Model for the Loss of Acetylcholine in Alzheimer's Disease

GERALD EHRENSTEIN,[a,b] ZYGMUNT GALDZICKI,[c] AND G. DAVID LANGE[d]

[a]*Biophysics Section, National Institute of Neurological Disorders and Stroke, Bethesda, Maryland 20892, USA*

[c]*Department of Physiology, Uniformed Services University of the Health Sciences, 4301 Jones Bridge Road, Bethesda, Maryland 20814, USA*

[d]*Instrumentation and Computer Section, National Institute of Neurological Disorders and Stroke, Bethesda, Maryland 20892, USA*

ABSTRACT: We describe a two-component positive-feedback system that could account for the large reduction of acetylcholine that is characteristic of patients with Alzheimer's disease (AD). One component is β-amyloid–induced apoptosis of cholinergic cells, leading to a decrease in acetylcholine. The other component is an increase in the concentration of β-amyloid in response to a decrease in acetylcholine. We describe each mechanism with a differential equation, and then solve the two equations numerically. The solution provides a description of the time course of the reduction of acetylcholine in AD patients that is consistent with epidemiological data. This model may also provide an explanation for the significant, but lesser, decrease of other neurotransmitters that is characteristic of AD.

INTRODUCTION

Although there is strong evidence that β-amyloid plays an important role in Alzheimer's disease (AD),[1,2] it is not yet clear what that role is. There are several mechanisms whereby β-amyloid could affect the concentration of neurotransmitters. In this paper, we consider one of these mechanisms, β-amyloid-induced apoptosis,[1,3] as part of a positive feedback loop that would result in a significant loss of acetylcholine (ACh).

We previously considered β-amyloid-induced leakage of choline[4,5] out of cholinergic neurons as a possible mechanism for the reduced ACh concentration that correlates with AD.[6,7] Since the choline concentration is rate-limiting for the production of ACh, choline leakage would cause a reduction in ACh concentration. We then modeled the positive feedback[8] that would be generated as a result of this reduction of ACh concentration, whose basic cause is an increase in β-amyloid concentration, and the increase in β-amyloid concentration that has been shown to be caused by the reduction of ACh concentration. One difficulty with our previous approach is that there is not yet any experimental information regarding the magnitude of choline leakage in a relevant system. Accordingly, we had to use an arbitrary value for the

[b]Address for correspondence: Voice: 301-496-3206; fax: 301-496-8765.
e-mail: gerry@helix.nih.gov

magnitude of choline leakage in our calculations. Interestingly, with an appropriate choice for this value, we were able to obtain results that are consistent with epidemiological data, thus demonstrating the potential value of a positive feedback model involving loss of ACh and increase of β-amyloid.

We then sought an alternative mechanism for the reduction of ACh concentration by β-amyloid for which there is experimental data on the magnitude of the effect. Interestingly, such a mechanism has been reported. It had been shown previously that β-amyloid induces apoptosis in cultured neurons.[1] It had also been shown that β-amyloid is much more toxic to PC12 cell mutants that express p75 than to PC12 mutants that do not express p75 and that this toxicity is significantly enhanced by transfecting the latter with p75.[9] Recent experiments have shown that both β-amyloid and p75 have a role in causing cell death.[3] These experiments not only demonstrated directly that binding of β-amyloid to p75 receptors causes apoptosis, but also provided data on the magnitude of the effect.[3] Accordingly, the mechanism we consider in this paper is β-amyloid–induced apoptosis.

In addition to a large decrease in the concentration of ACh, AD patients also experience significant, but lesser, decreases in the concentrations of other neurotransmitters.[10] A possible explanation for this is that β-amyloid-induced apoptosis affects all secreting cells that contain p75. As a result of the death of these secreting cells, there would be a decrease in the concentration of the various neurotransmitters that would be proportional to the relative concentration of p75 in cells secreting each neurotransmitter. ACh concentration would be most affected because the relative concentration of p75 is highest in cholinergic neurons of the basal forebrain.[11,12]

Our treatment includes a mathematical description of the binding of β-amyloid to p75 and the resultant apoptosis, but does not address the intermediate steps between binding and apoptosis. Although the details of these intermediate steps are not yet known, it is likely that free radicals play a crucial role. Evidence that binding to p75 could lead to formation of free radicals was obtained in experiments showing colocalization of p75 and nitric oxide synthase in rat brains.[13] Evidence that free radicals can cause apoptosis has been found both directly and indirectly. Ever since the pioneering work of Gerschman, Gilbert, and their collaborators,[14] it has been clear that free radicals can have a devastating effect on cell viability. Recent work shows that free radicals can cause apoptosis in both neurons[15] and glial cells.[16] Furthermore, antioxidants that destroy excess free radicals, such as the bcl-2 gene product, have been shown to prevent apoptosis.[17]

DESCRIPTION OF POSITIVE FEEDBACK MODEL

β-amyloid, which is a product of the breakdown of the amyloid precursor protein (APP), binds to p75 in cholinergic cells, leading to apoptosis of these cells[3] and a consequent decrease of ACh. A decrease in ACh concentration has two effects relevant to the concentration of β-amyloid. It causes an increase in the synthesis of APP in cerebral cortex, as reflected by increased levels of APP mRNA,[18] and it favors the processing of APP by means of the β-amyloid pathway.[19,20] Both effects tend to increase the concentration of β-amyloid in response to a decrease in ACh. Thus, the proposed decrease in ACh caused by an increase in β-amyloid and the resultant increase in apoptosis would lead to a further increase in β-amyloid and hence a further decrease in ACh, resulting in a positive feedback loop. Because of this positive feedback, eventually there would be a significant increase in the concentration of β-amyloid and a significant decrease in the concentration of ACh. The increase in β-amyloid would also cause a decrease in the concentration of other neurotransmitters.

MATHEMATICAL DESCRIPTION

Let a denote the concentration of ACh; b, the concentration of β-amyloid; p, the concentration of unbound p75 receptors in cholinergic neurons; p_t, the total concentration of p75 receptors in cholinergic neurons; and let p_b denote the concentration of p75 bound to β-amyloid.

The loss of cholinergic neurons by apoptosis is caused by the binding of β-amyloid to p75 receptors. The rate of apoptosis, which is the rate at which cholinergic cells are lost, is proportional to p_b. Assuming first-order binding of β-amyloid to p75,

$$\frac{da}{dt} = -k'p_b = -\frac{k'pb}{k_b}, \tag{1}$$

where k_b is the dissociation constant for binding of β-amyloid to p75 and k' is a constant of proportionality. The concentration of available receptors, p, can be expressed in terms of the total concentration of receptors, p_t:

$$p = \frac{p_t}{1 + b/k_b}. \tag{2}$$

Combining **(1)** and **(2)**,

$$\frac{da}{dt} = \frac{-k'p_t b}{k_b + b}. \tag{3}$$

Thus, a and p_t are each proportional to the number of cholinergic cells, and hence are proportional to each other:

$$p_t = k''a. \tag{4}$$

Combining **(3)** and **(4)**,

$$\frac{da}{dt} = \frac{-k'k''ab}{k_b + b}. \tag{5}$$

In order to simplify comparison of this equation for loss of ACh with the comparable equation in our previous model, let $k'k'' = k_1 k_b$. Then

$$\frac{da}{dt} = \frac{-k_1 k_b ab}{k_b + b}. \tag{6}$$

Equation **(6)**, which describes the loss of ACh as a result of apoptosis of cholinergic neurons, is one of a pair of equations that represent the positive feedback model. The other equation describes the increase in β-amyloid in response to a decrease in ACh. As indicated previously,[8] the increase in β-amyloid can be described by the following equation:

$$\frac{db}{dt} = k_2 - k_3 a - k_4 b. \quad (7)$$

Equation (7) describes the increase in the concentration of β-amyloid resulting from its formation from the breakdown of APP (represented by the term k_2), the effect of ACh on its rate of formation[18,19,20] (represented by the term $-k_3 a$), and the decrease in the concentration of β-amyloid resulting from enzymatic breakdown and absorption into neuronal membranes (represented by the term $-k_4 b$).

The simultaneous solution of Equations (6) and (7) provides a description of the time course of the decline in ACh concentration and the increase in β-amyloid concentration. These equations were solved numerically by use of the Mathematica program. Two of the constants in these equations can be estimated from the data presented by Yaar et al.[3] The dissociation constant k_b was measured directly, and is about 0.025 μM. To calculate k_1, consider Equation (6) for small values of b:

$$\frac{da/dt}{a} = -k_1 b.$$

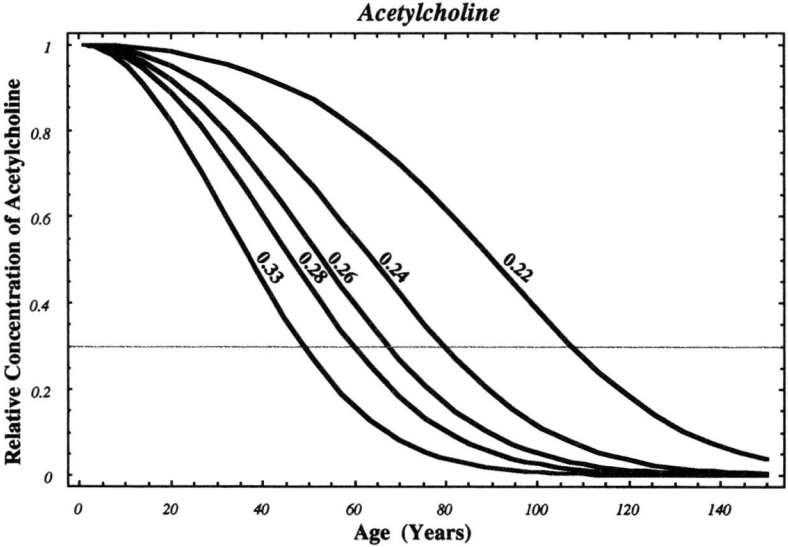

FIGURE 1. Calculated time course of the decrease in ACh concentration for several values of k_2, the rate of β-amyloid production. Each curve is labeled with the value of k_2 in units of nM/yr. Initial value of $a = 50$ nM; $k_1 = 9.0$ μM^{-1} yr^{-1}; $k_3 = 0.0042$ yr^{-1}; $k_4 = 0.01$ yr^{-1}. The horizontal line, which corresponds to 30% of the initial ACh concentration, is a typical ACh concentration at the onset of AD.

For this condition, the fractional rate of decrease of a is proportional to b. Thus, the value of k_1 can be estimated from the limiting slope for small b of the curve showing the rate of decrease of a as a function of b. From the data on melanocytes by Yaar et al.[3], we estimate the limiting slope to be about -9 μM^{-1} yr^{-1}. Thus, $k_1 = 9$ μM^{-1} yr^{-1}. Interestingly, this experimentally based value for k_1 is quite close to the value for the corresponding constant k_1 that we chose in our previous model[8] to give results consistent with epidemiological data.

The assumed initial value for b is zero. Also, the same initial value of a and the same values for k_3 and k_4 were used in this calculation as in our previous model. These values are indicated in the legend of FIGURE 1. Since the numerical solutions were found to be very sensitive to the value of k_2 (the rate of β-amyloid production that would occur if no ACh or β-amyloid were present), solutions were obtained for a range of k_2 values.

SIGNIFICANCE OF THE PREDICTIONS OF THE POSITIVE FEEDBACK MODEL

The time course of the decline in ACh concentration according to the positive feedback model described above is shown in FIGURE 1. It can be seen that ACh declines relatively slowly at first, and more rapidly at a later age. Measurement of the CSF ACh concentration in AD patients and in age-matched controls shows that the decline in CSF ACh concentration approximately parallels the severity of dementia[6,7] and that a reasonable estimate of the ACh concentration at the onset of AD is 30% of normal.[7] Accordingly, a horizontal line representing 30% of the initial ACh concentration is drawn in FIGURE 1. The intersections of this line with the curves in FIGURE 1 provide rough estimates of the predicted ages at which the ACh concentration decreases to the level characteristic of the onset of AD, and thus of the predicted ages of onset of AD, itself. This correspondence is based simply on the correlation between the loss of ACh and the onset of AD.

For the uppermost curve in FIGURE 1, where $k_2 = 0.22$ nM/yr, the age of onset is about 107 years. Thus, this curve pertains to an individual who is very unlikely to develop AD. At the other extreme in FIGURE 1, where $k_2 = 0.33$ nM/yr, 50% higher than the value for the uppermost curve, the age of onset is about 49 years. This might correspond to an individual with Down's syndrome, since the presence of an extra 21st chromosome would be expected to cause a 50% increase in APP,[21] and hence a 50% increase in k_2. Indeed, it is well documented that although AD is usually a disease of old age, individuals with Down's syndrome develop AD in middle age.[22–24]

The dependence of the age of onset of AD on the value of k_2 for the positive feedback model described above is shown in FIGURE 2, where it can be seen that the age of onset declines rapidly for small increases in the value of k_2. For example, an increase in k_2 of 10% lowers the age of onset from 107 to 78 and an increase of 20% lowers it to 65.

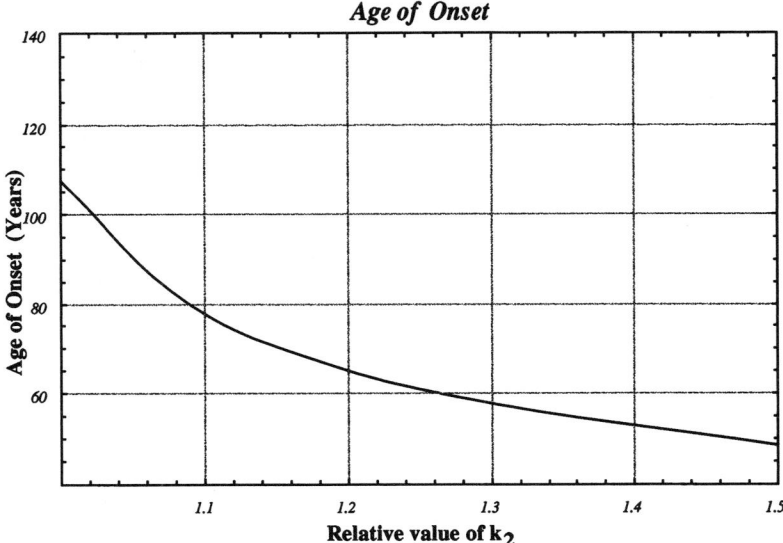

FIGURE 2. Dependence of calculated age of onset on k_2. Age of onset is determined by the intersection of each time course curve with the horizontal line in FIGURE 1.

DISCUSSION

Positive feedback between a decrease in ACh concentration and an increase in β-amyloid concentration requires two interacting mechanisms—one mechanism whereby β-amyloid causes a loss of ACh and one mechanism whereby ACh loss causes an increase in β-amyloid. In the present paper, we invoke the same mechanism for the increase in β-amyloid that we used in a previous attempt to model this type of positive feedback.[8] Therefore, one of the differential equations is the same in both treatments. For the loss of ACh, however, we invoke a completely different mechanism in this paper than we did in our previous treatment. Here we consider β-amyloid–induced apoptosis, whereas we previously considered β-amyloid–induced choline leakage. Although these two mechanisms are very different, the equations describing them are rather similar. The main difference in the form of the two equations is that apoptosis is based on binding, and hence the amount of apoptosis tends to saturate at relatively high concentrations of β-amyloid. By contrast, choline leakage was assumed to be linear, since the experiments describing β-amyloid–induced choline leakage showed a linear dependence of leakage on β-amyloid concentration.

In view of the general similarity between the two sets of equations, it is not surprising that the overall results shown in FIGURES 1 and 2 are rather similar to the equivalent figures presented in the previous treatment. In both models, FIGURES 1 and 2 provide a rationale for the well-known observations that AD is a disease of old age and that there is a wide range in the age of onset. A major advantage of the present treatment is that we can use experimental data to estimate the constants that describe the magnitude of the apoptosis leading to the loss of ACh. The fact that FIGURE 1 is consistent with epidemiological data on AD indicates that the rate of apoptosis that has been found experimentally can account for the observed loss of ACh in AD.

The present treatment can also explain several important observations regarding the loss of neurotransmitters in AD. The neurotransmitter most affected in AD is ACh.[10] However, not all cholinergic neurons are affected. According to the present treatment, only those neurons expressing p75 would undergo apoptosis. Interestingly, cholinergic neurons of the basal forebrain, which are particularly predisposed to degeneration in AD,[25,26] express the highest levels of p75 of any neurons tested,[27] whereas cholinergic neurons of the pontomesencephelon, which do not undergo degeneration in AD,[28] do not express p75.[27] Thus, our model provides a rationale for the spatial variability of ACh loss.

Although the positive feedback loop we have described affects only ACh, the model predicts that the increased β-amyloid concentration brought about by the positive feedback would also increase apoptosis of any neurons that contain p75. As a result, there would be a decrease in the concentration of other neurotransmitters that is proportional to the relative concentration of p75 in the relevant secreting cells. Although earlier experiments indicated that p75 was limited to certain cholinergic neurons,[29] more sensitive methods have demonstrated that lower concentrations of p75 are present in many areas of both the rat brain[30] and the human brain,[12] including the hippocampus, caudate, cerebellum, and olfactory bulb. Since p75 mRNA was found in areas of the brain that do not contain, and are not innervated by, cholinergic neurons, the model also provides a rationale for the observed decrease in concentration of neurotransmitters other than ACh.[10]

Finally, a positive-feedback model can provide a possible rationale for the fact that acetylcholinesterase inhibitors have had only limited success as therapeutic agents in AD. Because of the putative positive feedback, the loss of ACh would cause an increase in β-amyloid and a consequent loss of more ACh. Thus, even if the acetylcholinesterase inhibitors do increase the ACh concentration, they would be working in a system with a relatively large concentration of β-amyloid and with increasing ACh deficits. For a positive-feedback system, the best way to cope with this difficulty is to start treatment at a very early stage of the disease.[8] Since it is currently difficult to detect AD at a very early stage, a first step might be to provide acetylcholinesterase inhibitors or agents that prevent β-amyloid from binding to p75 to asymptomatic individuals who have a very high probability of developing AD, such as individuals with mutations known to cause AD or individuals with Down's syndrome.

REFERENCES

1. LOO, D.T., A. COPANI, C.J. PIKE, E.R. WHITTEMORE, A.J. WALENCEWICZ & C.W. COTMAN. 1993. Apoptosis is induced by beta-amyloid in cultured central nervous system neurons. Proc. Natl. Acad. Sci. USA **90:** 7951–7955.
2. HYMAN, B.T. & R.D. TERRY. 1994. Apolipoprotein E, A beta, and Alzheimer disease. J. Neuropathol. Exp. Neurol. **53:** 427–428.

3. YAAR, M., S. ZHAI, P.F. PILCH, S.M. DOYLE, P.B. EISENHAUER, R.E. FINE & B.A. GILCHREST. 1997. Binding of beta-amyloid to the p75 neurotrophin receptor induces apoptosis. A possible mechanism for Alzheimer's disease. J. Clin. Invest. **100:** 2333–2340.
4. GALDZICKI, Z., R. FUKUYAMA, K.C. WADHWANI, S.I. RAPOPORT & G. EHRENSTEIN. 1994. beta-Amyloid increases choline conductance of PC12 cells: possible mechanism of toxicity in Alzheimer's disease. Brain Res. **646:** 332–336.
5. ALLEN, D.D., Z. GALDZICKI, S.K. BRINING, R. FUKUYAMA, S.I. RAPOPORT & Q.R. SMITH. 1997. Beta-amyloid induced increase in choline flux across PC12 cell membranes. Neurosci. Lett. **234:** 71–73.
6. DAVIS, K.L., J.Y.-K. HSIEH, M.I. LEVY, T.B. HORVATH, B.M. DAVIS & R.C. MOHS. 1982. Cerebrospinal fluid acetylcholine, choline, and senile dementia of the Alzheimer's type. Psychopharmacol. Bull. **18:** 193–195.
7. TOHGI, H., T. ABE, K. HASHIGUCHI, M. SAHEKI & S. TAKAHASHI. 1994. Remarkable reduction in acetylcholine concentration in the cerebrospinal fluid from patients with Alzheimer type dementia. Neurosci. Lett. **177:** 139–142.
8. EHRENSTEIN, G., Z. GALDZICKI & G.D. LANGE. 1997. The choline-leakage hypothesis for the loss of acetylcholine in Alzheimer's disease. Biophys. J. **73:** 1276–1280.
9. RABIZADEH, S., C.M. BITLER, L.L. BUTCHER & D.E. BREDESEN. 1994. Expression of the low-affinity nerve growth factor receptor enhances beta-amyloid peptide toxicity. Proc. Natl. Acad. Sci. USA **91:** 3060–3063.
10. BOWEN, D.M. & A.N. DAVISON. 1986. Biochemical studies of nerve cells and energy metabolism in Alzheimer's disease. Br. Med. Bull. **42:** 75–80.
11. HEFTI, F., J. HARTIKKA, A. SALVATIERRA, W.J. WEINER & D.C. MASH. 1986. Localization of nerve growth factor receptors in cholinergic neurons of the human basal forebrain. Neurosci. Lett. **69:** 1–7.
12. GOEDERT, M., A. FINE, D. DAWBARN, G.K. WILCOCK & M.V. CHAO. 1989. Nerve growth factor receptor mRNA distribution in human brain: normal levels in basal forebrain in Alzheimer's disease. Mol. Brain Res. **5:** 1–7.
13. PENG, Z.C., S. CHEN, G. BERTINI, H.H. SCHMIDT & M. BENTIVOGLIO. 1994. Co-localization of nitric oxide synthase and NGF receptor in neurons in the medial septal and diagonal band nuclei of the rat. Neurosci. Lett. **166:** 153–156.
14. GERSCHMAN, R., D.L. GILBERT, S.W. NYE & W.O. FENN. 1954. Oxygen poisoning and x-irradiation: a mechanism in common. Science **119:** 623–626.
15. ESTEVEZ, A.G., N. SPEAR, S.M. MANUEL, R. RADI, C.E. HENDERSON, L. BARBEITO & J.S. BECKMAN. 1998. Nitric oxide and superoxide contribute to motor neuron apoptosis induced by trophic factor deprivation. J Neurosci **18:** 923–931.
16. KITAMURA, Y., T. OTA, Y. MATSUOKA, I. TOOYAMA, H. KIMURA, S. SHIMOHAMA, Y. NOMURA, H.P. GEBICKE & T. TANIGUCHI. 1999. Hydrogen peroxide-induced apoptosis mediated by p53 protein in glial cells. Glia **25:** 154–164.
17. HOCKENBERY, D.M., Z.N. OLTVAI, X.M. YIN, C.L. MILLIMAN & S.J. KORSMEYER. 1993. Bcl-2 functions in an antioxidant pathway to prevent apoptosis. Cell **75:** 241–251.
18. WALLACE, W., S.T. AHLERS, J. GOTLIB, V. BRAGIN, J. SUGAR, R. GLUCK, P.A. SHEA, K.L. DAVIS & V. HAROUTUNIAN. 1993. Amyloid precursor protein in the cerebral cortex is rapidly and persistently induced by loss of subcortical innervation. Proc. Natl. Acad. Sci. USA **90:** 8712–8716.
19. HUNG, A.Y., C. HAASS, R.M. NITSCH, W.Q. QIU, M. CITRON, R.J. WURTMAN, J.H. GROWDON & D.J. SELKOE. 1993. Activation of protein kinase C inhibits cellular production of the amyloid beta-protein. J. Biol. Chem. **268:** 22959–22962.
20. BUXBAUM, J.D., A.A. RUEFLI, C.A. PARKER, A.M. CYPESS & P. GREENGARD. 1994. Calcium regulates processing of the Alzheimer amyloid protein precursor in a protein kinase C-independent manner. Proc. Natl. Acad. Sci. USA **91:** 4489–4493.
21. RUMBLE, B., R. RETALLACK, C. HILBICH, G. SIMMS, G. MULTHAUP, R. MARTINS, A. HOCKEY, P. MONTGOMERY, K. BEYREUTHER & C.L. MASTERS. 1989. Amyloid A4 protein and its precursor in Down's syndrome and Alzheimer's disease [see comments]. N. Engl. J. Med. **320:** 1446–1452.

22. WISNIEWSKI, K.E., H.M. WISNIEWSKI & G.Y. WEN. 1985. Occurrence of neuropathological changes and dementia of Alzheimer's disease in Down's syndrome. Ann. Neurol. **17:** 278–282.
23. MANN, D.M. 1988. The pathological association between Down syndrome and Alzheimer disease. Mech. Ageing Dev. **43:** 99–136.
24. LAI, F. & R.S. WILLIAMS. 1989. A prospective study of Alzheimer disease in Down syndrome. Arch. Neurol. **46:** 849–853.
25. BARTUS, R.T., R.L. DEAN, III, B. BEER & A.S. LIPPA. 1982. The cholinergic hypothesis of geriatric memory dysfunction. Science **217:** 408–414.
26. WINBLAD, B., E. MESSAMORE, C. O'NEILL & R. COWBURN. 1993. Biochemical pathology and treatment strategies in Alzheimer's disease: emphasis on the cholinergic system. Acta Neurol. Scand. Suppl. **149:** 4–6.
27. WOOLF, N.J., E. GOULD & L.L. BUTCHER. 1989. Nerve growth factor receptor is associated with cholinergic neurons of the basal forebrain but not the pontomesencephalon. Neuroscience **30:** 143–152.
28. WOOLF, N.J., R.W. JACOBS & L.L. BUTCHER. 1989. The pontomesencephalotegmental cholinergic system does not degenerate in Alzheimer's disease. Neurosci. Lett. **96:** 277–282.
29. BUCK, C.R., H.J. MARTINEZ, I.B. BLACK & M.V. CHAO. 1987. Developmentally regulated expression of the nerve growth factor receptor gene in the periphery and brain. Proc. Natl. Acad. Sci. USA **84:** 3060–3063.
30. BUCK, C.R., H.J. MARTINEZ, M.V. CHAO & I.B. BLACK. 1988. Differential expression of the nerve growth factor receptor gene in multiple brain areas. Dev. Brain Res. **44:** 259–268.

Microglial Contribution to Oxidative Stress in Alzheimer's Disease

CAROL A. COLTON,[a,b] OLGA N. CHERNYSHEV,[a] DANIEL L. GILBERT,[c] AND MICHAEL P. VITEK[d]

[a]*Department of Physiology, Georgetown University Medical School, Washington, DC 20007, USA*

[c]*Unit on Reactive Oxygen Species, Biophysics Section, NINDS, National Institutes of Health, Bethesda, Maryland 20892, USA*

[d]*Department of Neurology, Duke University Medical Center, Durham, North Carolina*

ABSTRACT: Microglia are the CNS macrophage and are a primary cellular component of plaques in Alzheimer's disease (AD) that may contribute to the oxidative stress associated with chronic neurodegeneration. We now report that superoxide anion production in microglia or macrophages from 3 different species is increased by long term exposure (24 hours) to Aβ peptides. Since Aβ competes for the uptake of opsonized latex beads and for the production of superoxide anion by opsonized zymosan, a likely site of action are membrane receptors associated with the uptake of opsonized particles or fibers. The neurotoxic fibrillar peptides Aβ (1–42) and human amylin increase radical production whereas a non-toxic, non-fibrillar peptide, rat amylin, does not. We also report that the effect of Aβ peptides on superoxide anion production is not associated with a concomitant increase in nitric oxide (NO) production in either human monocyte derived macrophages (MDM) or hamster microglia from primary cultures. Since NO is known to protect membrane lipids and scavenge superoxide anion, the lack of Aβ-mediated induction of NO production in human microglia and macrophages may be as deleterious as the overproduction of superoxide anion induced by chronic exposure to Aβ peptides.

INTRODUCTION

A key feature of Alzheimer's disease (AD) is the presence of microglia, the CNS macrophage, within amyloid-containing plaques. The role of these cells in plaque formation remains controversial.[1–7] However, the ability of microglia to produce and release a variety of cytoactive factors increases the likelihood that microglia contribute to the neurodegenerative process.[1,3,4,8–10] One of the most invariant features of microglial activation is the production of reactive oxygen and reactive nitrogen species, *i.e.*, superoxide anion and nitric oxide (NO).[11] Macrophages are well known to express a membrane bound NADPH oxidase associated with the respiratory burst and an inducible nitric oxide synthase (iNOS). When activated, these enzymes and

---

[b]Address for correspondence: Carol Colton, Ph.D, Department of Physiology, Georgetown University Medical Center, 3900 Reservoir Rd. NW, Washington DC 20007. Voice: 202-687-1508; fax: 202-687-7407.
  e-mail: glia01@aol.com

their products contribute to the generation of oxidative and/or nitrosative stress in tissues.[12–17] This response is initiated by a variety of events, including exposure of the cell to bacteria, viruses, foreign proteins, cellular debris and abnormal cellular components.[12]

An increasing number of studies support an important role for oxidative stress in AD. Markers of oxidative stress such as oxidized proteins, DNA and lipids are readily observed in autopsy samples of AD brain and suggest that oxidative processes contribute to the neuropathophysiology.[18–22] The exact source of the oxidative stress is unclear and multiple biochemical pathways have been implicated, ranging from the leak of superoxide anion or hydrogen peroxide from damaged mitochondria to the inactivation of key enzymes such as creatine kinase.[18] Since chronic inflammation is a principal feature of AD, production of reactive oxygen species (ROS) by microglia also contributes to any potential oxidative stress. Klegeris et al.[23] and Van Muiswinkel et al.[24] have shown that Aβ peptides, fragments of amyloid precursor protein (APP), increase superoxide anion production in rat peritoneal macrophages and in cultured rat microglia. Interestingly, Aβ peptides in synergy with other factors such as lipopolysacharide (LPS) and gamma-interferon (γIFN) have also been reported to increase NO in a mouse microglial cell line.[25–27]

Rat or mouse macrophages, however, may not be a suitable model of human disease because of the large differences in response of human macrophages compared to rat or mouse cells. For example, human phagocytic cells are well known to have different regulatory mechanisms for nitric oxide production compared to rat or mouse microglia.[28–31] To explore this further, we have examined the production of superoxide anion and NO in response to Aβ peptides using primary cultures of human monocyte derived macrophages (MDM). We also report results on primary cultures of hamster microglia that have been previously shown to demonstrate a profile of activation that is reminiscent of human macrophages more than rat or mouse.[28]

METHODS

Materials

Polyinosinic: polycytidylic acid (PIC) -potassium salt, cytochrome C (horse heart), 12-phorbol-13-myristate acetate (PMA), rat amylin, human amylin and recombinant γ-interferon (γIFN) were purchased from Sigma Chemical Co (St. Louis, MO). PIC was diluted into RNAase-free distilled water to a concentration of 5 mg/ml. Stock solutions were aliquoted and stored at −20°C. Cytochrome C was prepared by diluting into Earles' balanced salt solution (EBSS) and used immediately. PMA was diluted into dimethylsulfoxide (DMSO) and stored at −80°C. Aβ (1–40), Aβ (1–42) and Aβ (25–35) were purchased from BACHEM (Torrence, CA), diluted into sterile phosphate buffered saline and allowed to aggregate at room temperature for 48 hours. Stock solutions prepared in this manner were then frozen at −80°C until used. Each stock solution was tested for toxicity to N2a cells, a rat neuronal cell line, using the MTT assay as described below. Rat and human amylin were also tested for toxicity. Superoxide dismutase (SOD) and lipopolysaccharide (LPS) were obtained from Calbiochem (LaJolla, CA). Culture supplements for the human monocyte derived macrophages included recombinant human M–Colony Stimulating factor

(CSF) (Genzyme-R & D, Minneapolis, MN) and human AB serum (ABi Technologies, Rockville, MD). Dulbecco's modified eagle medium (DMEM), EBSS, fetal bovine serum (FBS) and other tissue culture supplies were purchased from GIBCO-BRL (Rockville, MD).

Cell Cultures

Fresh human monocytes were obtained from the laboratory of Dr. Deborah Webb (Center for Biologics Evaluation and Research, NIH) and were converted to macrophages. Briefly, monocytes were elutriated, and the cells were maintained in 75 cm^2 flasks at 37°C in a 5% CO_2 humidified atmosphere for 5-10 days in DMEM containing 20% FBS, 10% human AB serum, and 20 ηg/ml human M-CSF. After 5–10 days in culture the monocytes undergo morphologic changes which include spreading and the formation of branches. These cells are commonly termed monocyte-derived macrophages (MDM).

Primary cultures of neonatal hamster microglia were prepared in a similar fashion as rat microglia[11] and were grown for 12 to 14 days at 37°C in a humidified 95% air, 5% CO_2 atmosphere in DMEM supplemented with 25 µg/ml gentamycin, 10% FBS and 2 mM glutamine. Microglia were removed from the underlying astrocyte layer by shaking and were plated at a density of 40,000 cells /well into 96 well dishes. Cells were allowed to adhere at 37°C for 30 minutes and then washed with EBSS 2 times to remove non-adherent cells. Prepared in this manner, the cultures contained greater than 90% microglia as identified with GS-1 lectin staining.

RAW (264.7) cells, a rodent macrophage cell line, were purchased from American Type Tissue Culture (ATTC) and were grown to confluence in a humidified 5% CO_2, 95% air atmosphere at 37°C in DMEM with 10% FBS, 10 mM HEPES, 2 mM glutamine and 25 µg/ml penicillin-streptomycin. RAW cells were removed from the flask using trypsin EDTA, the cells spun at $400 \times g$ to pellet and then plated into 96 well dishes at a density of 40,000 cells/well.

Superoxide Anion Assay

After plating into 96 well dishes, cells were allowed to recover overnight in normal media. The following day the media was replaced with serum-free media containing the various treatment agents in the presence or absence of Aβ peptides. Cells were treated for 24 hours after which the serum-free media was replaced with EBSS containing cytochrome C in the presence and absence of 300 U/ml SOD. The level of superoxide anion in the cell supernatants was measured directly in the 96 well plate using the cytochrome C reduction assay as previously described.[11] The difference in OD between SOD-free and SOD containing wells was taken as the level of superoxide anion-specific reduction. Values were normalized to mg protein using the BCA protein assay (Pierce, Rockford, IL) with bovine serum albumin as the standard. Data are presented as nmoles/mg protein.

Nitric Oxide Assay

Nitric oxide production was determined from the supernatant levels of nitrite using the Griess reaction.[32] Cells were plated as described above and the media in each well was replaced with serum-free media in the presence and absence of putative activators of the inducible nitric oxide synthase (iNOS). The supernatants were then collected at 24 hours and assayed for nitrite by the addition of Griess reagents. Optical density (OD) was read at 550 nm and nitrite levels were obtained from a standard curve using sodium nitrite as the standard.

Phagocytosis

The uptake of fluorescent latex beads was used as an indicator of phagocytosis. Briefly, 1 micron diameter fluorescent beads (Molecular Probes, Inc., Eugene, OR) were either coated with purified bovine serum albumin as a non-specific control or were opsonized by equilibrating the beads for 30 minutes at 37°C with 100% goat serum. The beads (in a ratio of 100 beads per cell) were then added to untreated or Aβ pretreated microglia plated onto glass coverslips in 24 well dishes and the uptake followed over 2 hours. The level of fluorescence was detected using a fluorescent plate reader (Cytofluor, PerSeptives Biosystems, Bedford, MA) and normalized to mg DNA. DNA content was determined using the DAPI dye (Molecular Probes, Inc., Eugene, OR) with calf thymus DNA as a standard.[33] To determine background fluorescence, treated and untreated microglia from each group were killed with 4% paraformaldehyde prior to exposure to the latex beads and a value of fluorescence obtained. This value was subtracted from the fluorescence of live treated and untreated microglia.

Statistics

Data points represent the average value ± SEM for at least 3 wells assayed per experimental condition for at least 3 different culture groups. Statistical significance was determined using a paired Student's t-test or a one way analysis of variance (ANOVA) with the Bonferroni post-test.

Western Blot

In some cases the cells were grown in 35 mm dishes or flasks until confluent and then either remained untreated or treated with the appropriate activating agents. After 12 hours, cells were washed with Hanks balanced salt solution (HBSS), scraped into lysis buffer, equal protein samples loaded onto a 4 to 15% gradient polyacrylamide gel and the proteins separated using SDS-PAGE. Proteins were transferred onto nitrocellulose membrane, blocked for 2 hours in Blotto (5% non-fat dry milk in TBS and immunostained overnight at 4°C for inducible nitric oxide synthase (iNOS) using a polyclonal iNOS antibody (anti-universal NOS, PA1-039, Affinity Bioreagents, Inc., Golden, CO) which detects human and rodent iNOS. RAW cell lysate served as a positive control for iNOS. The blots were then washed and immunoreactive bands detected using enhanced chemiluminesence (ECL™) per the manufacturer's instructions (Amersham Life Sciences, Arlington Heights, IL).

Toxicity Assay

The reduction of 3-[4,5-dimethylthiazol-2-yl]-2,5-diphenyltetrazolium bromide (MTT) was used as an indicator of cellular toxicity. The assay was carried out in standard fashion[34] using the Cell Proliferation Kit I (MTT) (Boehringer Mannheim,

Indianapolis, IN). Human amylin, Aβ (1–42), Aβ (1–40) and Aβ (25–35) were toxic to N2A cells after 24 hours of treatment. No change in survival was seen in microglia treated for up to 72 hours with Aβ (1–40) or Aβ (25–35).

RESULTS

The Effect of Aβ peptides on NO Production

NO production was analyzed by measuring the level of nitrite, a stable NO endproduct, in the cell supernatants from cultures that were stimulated with agents known to induce NOS. Because human MDM and hamster microglia do not generate measurable levels of nitrite in response to lipopolysaccharide (LPS), we used the double stranded polyribonucleotide, polyinosinic acid:polycytidylic acid (PIC) to stimulate production of NO in our study. PIC is a viral mimetic which has been shown to activate macrophages from all species studied.[12,30,35,36] Hamster microglia, human MDM and mouse macrophage clonal cells (RAW cells) were treated for 24 hours in the presence and absence of Aβ peptides alone and with Aβ peptides plus PIC. Neither Aβ (1–40) nor Aβ (25–35) alone had an effect on nitrite production in any of the cells studied (data not shown). As shown in FIGURE 1 for the mouse macrophage cell line, treatment with varying concentrations of PIC for 24 hours generated a dose-dependent increase in NO production. However, simultaneous treatment with 10 μM Aβ (25–35) did not alter the level of nitrite produced by PIC. The lack of an effect of Aβ peptides on NO production was not due to interaction of NO with superoxide anion, thus reducing measurable levels of NO, because addition of 300 U/ml of superoxide dismutase to remove superoxide anion did not alter the supernatant level of NO. Aβ peptides also did not affect NO production in hamster microglia and in human MDM as previously reported by our lab.[37]

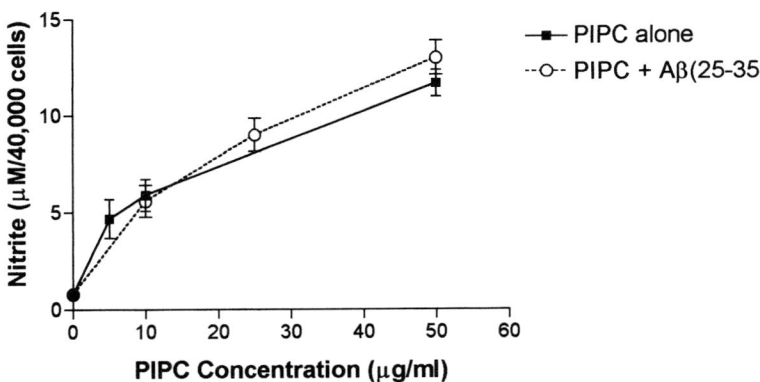

FIGURE 1. Aβ (25–35) does not affect PIC-induced NO production. Supernatant nitrite levels were measured in RAW cells (40,000 cells/well) exposed for 24 hours to varying concentrations of PIC in the presence and absence of 10 μM Aβ (25–35). $n = 20$ wells assayed

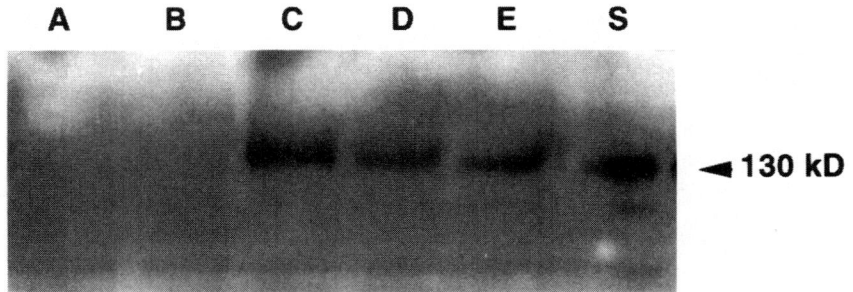

FIGURE 2. Expression of iNOS in RAW cells in response to Aβ (1–40). Cells were either untreated (*Lane A*), treated for 24 h with 10 μM Aβ (1–40) (*Lane B*), 1 μg/ml LPS as a positive control (*Lane C*), 50 μg/ml PIC (*Lane D*) and the combination of 50 μg/ml PIC plus 10 μM Aβ (1–40) (*Lane E*). *Lane S* = cell lysate standard for iNOS.

To confirm the lack of an effect of Aβ peptides on NO production, the expression of iNOS protein was examined. Using Western blot analysis, induction of iNOS in the mouse macrophagic cell line was seen in cells treated with LPS or with PIC alone (FIG. 2, Lanes C and D). Inducible NOS expression was not evident in untreated cells or in cells treated with 10 μM Aβ alone (lanes A and B). In addition, no apparent enhancement of iNOS expression was seen in cells treated with Aβ (1–40) plus PIC (lane E).

The Effect of Aβ Peptides on Superoxide Anion Production

The effect of Aβ peptides on microglial superoxide anion production was examined in cultured neonatal hamster microglia treated with varying concentrations of Aβ (1–40) for 24 hours. The microglia were then washed to remove Aβ in the media and superoxide anion production of the pre-treated cells was assayed. A significant dose-dependent increase in superoxide anion production was seen compared to untreated controls, with a maximal change of about 4 fold at 10 μM Aβ (1–40) (FIG. 3A).

The ability of Aβ (1–40) to modulate superoxide anion production induced by other agents was also studied. For these studies, microglia were either untreated or pretreated with varying concentrations of Aβ (1–40) for 24 hours, the cells were washed to remove Aβ in the media and PMA-stimulated superoxide anion production was then determined using the cytochrome C reduction assay. PMA is a phorbol ester used to initiate activation of the NADPH oxidase in monocytic-lineage cells.[12,13] A 7.5 fold increase in superoxide anion production was induced by 2 hours of stimulation with PMA compared to resting (non-stimulated) cells (FIG. 3B). Pretreatment of microglia with Aβ (1–40) for 24 hours further increased PMA-stimulated levels of superoxide anion. As shown in FIGURE 3B, 10 μM Aβ (1–40) significantly increased PMA-stimulated production by 80 ± 19% ($p < 0.004$). Aβ peptides also affected superoxide anion production induced by LPS. Microglia pre-treated with Aβ (1–40) plus LPS (1 μg/ml) for 24 hours demonstrated a significantly greater

($p < 0.02$) amount of superoxide anion production compared to LPS alone (FIG. 3B). However, no additional effect on superoxide anion production was seen when opsonized zymosan (OPZ) was used to stimulate microglial superoxide anion production. Phagocytosis of OPZ is associated with the generation of superoxide anion in monocytes, neutrophils and macrophages[11,12] and generated about a 3-fold increase in superoxide anion production compared to non-stimulated microglia. In microglia pre-treated with Aβ (1–40) for 24 hours, washed to remove remaining Aβ and then stimulated with 0.1 mg/ml OPZ, superoxide production induced by OPZ was not significantly different from untreated OPZ-stimulated microglia.

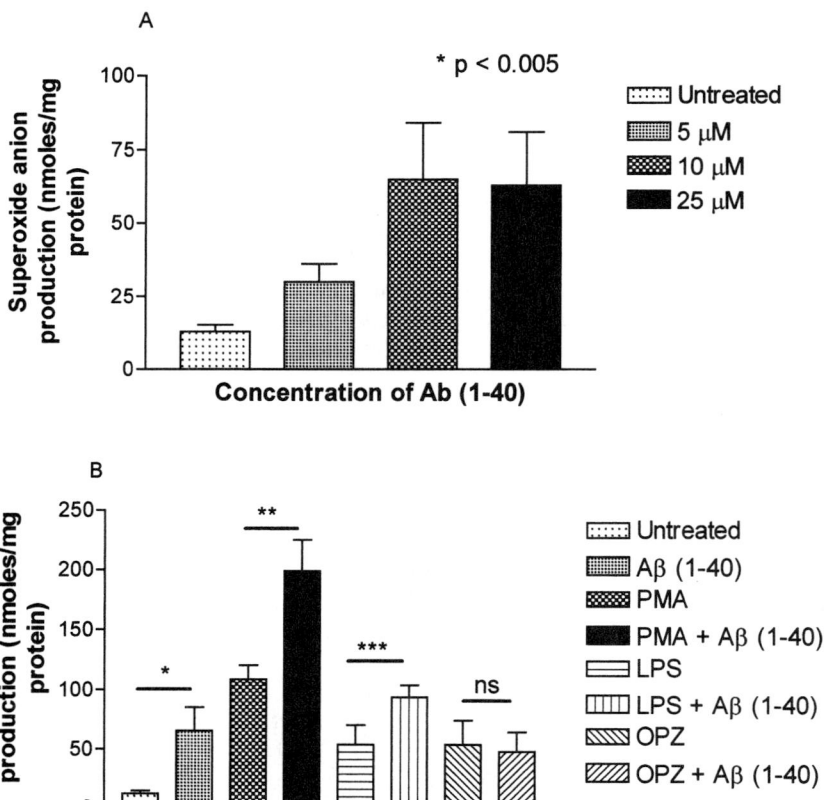

FIGURE 3. A: Effect of Aβ (1–40) on superoxide anion production in hamster microglia. Superoxide anion production was measured in microglia pretreated for 24 hours with varying concentrations of Aβ (1–40). $*p < 0.005$ as determined by ANOVA $n \geq 12$ wells assayed per experimental condition. **B:** Modulation of stimulated superoxide anion production by Aβ (1–40). Following pretreatment with 10 μM Aβ (1–40), superoxide anion production in hamster microglia was determined for PMA (0.05 μg/ml)-stimulated, LPS (1 μg/ml)-stimulated or OPZ (0.1 mg/ml)-stimulated conditions. $*p < 0.002$; $**p < 0.004$; $***p < 0.04$ compared to the appropriate control condition for at least 12 wells assayed per experimental condition.

TABLE 1. Phagocytosis of coated beads by neonatal hamster microglia in culture

	Untreated	Aβ (1–40)
Albumin-coated beads	81.3 ± 26 (6)	97.0 ± 12.5 (6)
Opsonized beads	95.2 ± 26 (6)	46.0 ± 6.4* (6)

Microglia were pretreated for 24 hours with either 5 µM Aβ (1–40) or remained untreated and then assayed for uptake of beads. Fluorescent beads were coated as indicated and uptake was expressed as the average value (± SEM) of arbitrary fluorescent units/ mg DNA obtained after 1 hour of equilibration at 37°C. n is at least duplicate coverslips assayed for a minimum of 3 different litter groups. *$p < 0.05$.

Because an additive effect on superoxide anion production was seen with PMA and LPS-induction but not OPZ-induction, it is likely that Aβ peptide competes at the membrane site(s) involved with the binding and uptake of OPZ. We further examined the interaction of Aβ peptide with OPZ-binding sites by determining the effect of Aβ peptide on the uptake of opsonized beads. Phagocytosis of opsonized particles was determined in microglia pretreated with 10 µM Aβ (1–40) for 24 hours. These microglia were then washed to remove Aβ in the media and exposed for 1 hour to fluorescent latex beads which had been coated with albumin as a non-specific protein control or opsonized with serum. As shown in TABLE 1, Aβ (1–40) significantly inhibited the uptake of fluorescently labeled beads that had been opsonized but did not affect beads which had been albumin coated.

Dependence of the Aβ Effect on the Fibrillar Nature of Aβ Peptides

Superoxide anion production is induced in macrophages during the uptake of a variety of particles, including fibrillar molecules such as asbestos.[38–40] To determine if Aβ-mediated production of superoxide anion is dependent on the fibrillar nature of the peptide, we examined the production of superoxide anion by hamster microglia in

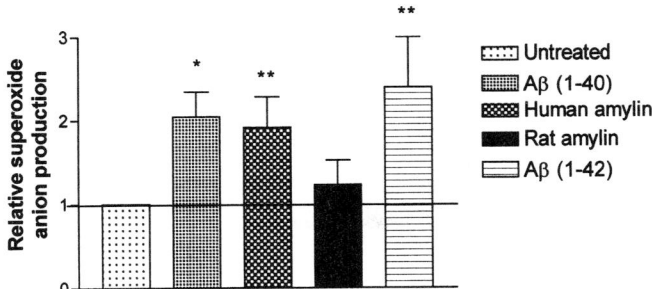

FIGURE 4. Response of hamster microglia to fibrillar and non-fibrillar proteins. Superoxide anion production was compared in microglia pretreated for 24 h with 5 µM of the following peptides: Aβ (1–40), Aβ (1–42), rat amylin and human amylin. Data are presented as the relative change in superoxide anion production of each experimental condition to the untreated control. *$p < 0.002$; **$p < 0.02$ compared to the untreated control; $n \geq 12$ wells assayed per experimental condition.

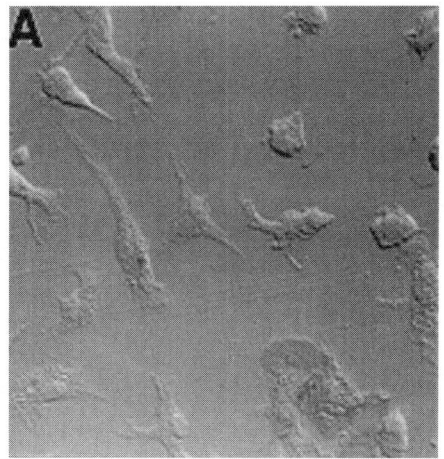
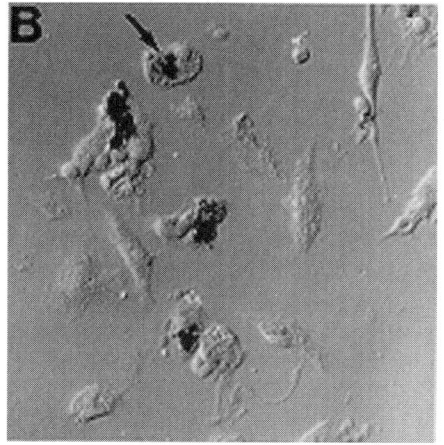
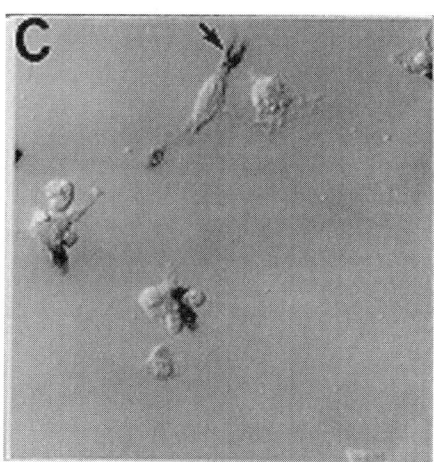

FIGURE 5. Photomicrograph of NBT staining in microglia treated with Aβ (1–40). Microglia were either untreated (*Panel A*) or treated for 6 hours with 0.1 mg/ml OPZ (*Panel B*) or 5 μM Aβ (1–40) (*Panel C*). At 3 hours of treatment, NBT (5 μg/ml) was added to detect superoxide anion production and cells were fixed at 6 hours.

response to fibrillar and non-fibrillar peptides similar to Aβ (1–40). Pretreatment with 5 μM rat amylin, a non-fibrillary and non-toxic peptide,[41,42] produced no change in superoxide anion levels compared to untreated controls (FIG. 4). Pretreatment with either 5 μM Aβ (1–42) or human amylin, a non-Aβ peptide but one which forms a toxic, fibrillar structure, produced a similar effect as Aβ (1–40) on superoxide anion production. In this case, superoxide anion production increased by 140 ± 60% and 92 ± 34% with Aβ (1–42) or human amylin pretreatment, respectively, compared to 105 ± 30% for 5 μM Aβ (1–40). Human amylin, like Aβ (1–40), was tested for toxicity to a rodent neuroblastoma cell line and was found to be toxic at the doses used in these experiments (data not shown).

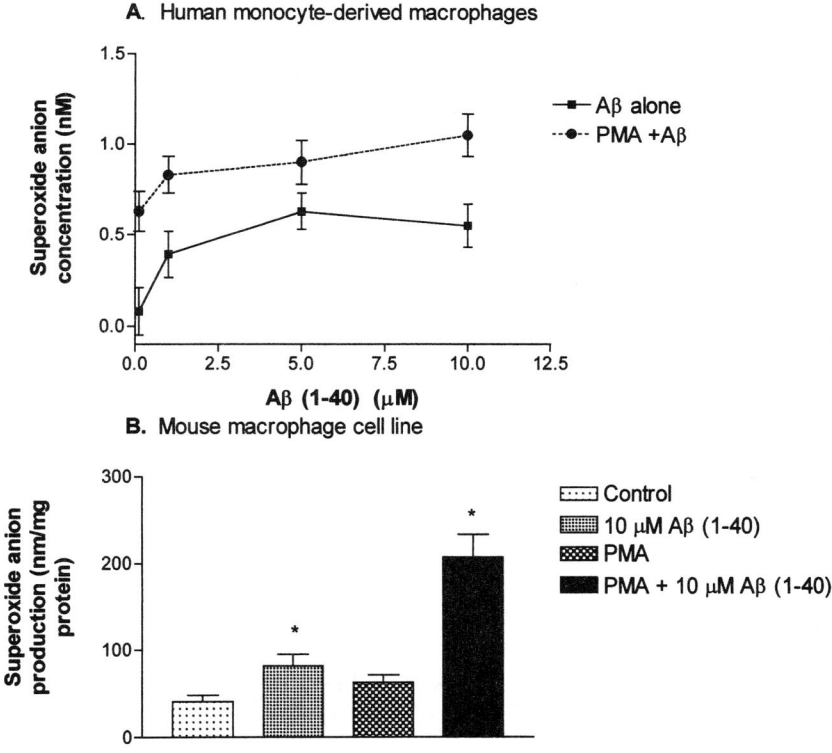

FIGURE 6. Superoxide anion production in human MDM and in a mouse macrophage cell line. A- Human MDM were treated for 24 h with varying doses of Aβ (1–40). Non-stimulated and PMA-stimulated superoxide anion production was then obtained. The data represent a typical dose response relationship and is obtained from 1 of 4 individuals. $n = 4$ wells assayed. **B:** Superoxide anion production was measured under non-stimulated and PMA-stimulated conditions in cells from a mouse macrophage cell line after 24 h of treatment with 10 μM Aβ (1–40). $n \geq 12$ wells assayed per experimental condition. $*p < 0.01$ compared to either untreated, non-stimulated control or to untreated, PMA-stimulated control.

Cellular Localization of Aβ Peptide-Induced Superoxide Anion Production

To further examine the cellular mechanisms underlying the response of the microglia to Aβ peptides, we utilized the deposition of formazan, an insoluble product of the interaction of superoxide anion and nitroblue tetrazolium (NBT).[43] Microglia were treated with Aβ (1–40) or remained untreated for 3 hours, followed by the addition of NBT to the incubation media for an additional 3 hours. Microglia were also treated with opsonized zymosan as a positive control for the localized production of superoxide anion. The cells were then fixed and viewed with Hoffman optics for the presence of blue-black precipitates. As shown in FIGURE 5, untreated microglia rarely demonstrated formazan deposition (Panel A) while cells treated with opsonized zymosan demonstrated a characteristic pattern of vesicular NBT deposition inside the cell. These deposits are associated with the uptake of the zymosan particles into the cell (arrow, Panel B). A different pattern of formazan deposition was seen in microglia treated with Aβ (1–40). In this case, clustered regions of formazan deposition were associated with the cell membrane (arrow, Panel C). The response to Aβ peptide was highly localized since superoxide anion could not be detected in the culture media over the 6 hours of exposure using the cytochrome C assay (data not shown).

The Effect of Aβ (1–40) on Superoxide Anion Production in Human Monocyte–Derived Macrophages and Clonal Mouse Macrophages

Because of the large species variations recently found in the regulation and production of reactive nitrogen intermediates,[28] we also examined the effect of Aβ (1–40) on superoxide anion production in cultured human monocyte-derived macrophages (MDM) and cells from a mouse macrophage line. The same treatment protocol as that described for the hamster microglia was used. Pretreatment with 5 μM Aβ (1–40) for 24 hours produced an increase in superoxide anion levels in the presence and absence of PMA in both human and mouse macrophages (FIG. 6).

DISCUSSION

The effect of Aβ (1–40) on production of oxyradicals appears to be a general phenomenon that affects most, if not all, cells exposed to Aβ peptides.[23,34,44–47] Behl et al.[45] and others [47–49] have measured intracellular changes in reactive oxygen species in neuronal cell lines or primary neuronal cultures using fluorescent dyes such as dichlorofluorescein. Hydrogen peroxide (H_2O_2) was identified as the reactive oxygen species generated, but the non-specificity of the dye used in these studies and the ubiquitous presence of transition metals (hence conversion of superoxide anion to H_2O_2 and OH· via Haber-Weiss/Fenton processes) do not rule out the possibility that superoxide anion production was a primary site of action. Our data confirm that long term (24 h) exposure to Aβ peptides induces superoxide anion production in each of the macrophagic cells studied, including primary cultures of microglia obtained from hamster cortex, human monocyte derived macrophages and a clonal mouse macrophage cell line. This direct effect of Aβ (1–40) is long-lasting and generates lower levels of superoxide anion than that associated with agents known to activate the "respiratory burst" NADPH oxidase such as PMA. An effect of Aβ (1–40) on

PMA-stimulated oxyradical production in macrophages has also been demonstrated.[9,23,50] Our data show that the direct effect of Aβ peptide on superoxide anion production appears to be additive with PMA or with LPS-stimulated production of superoxide anion. This is distinct from the typical synergistic effect of "priming" agents such as LPS or γ interferon on PMA-stimulated superoxide levels.[13,51] Synergism results in a greater response than that predicted for either agent alone or in combination and suggests a confluence of pathways. The presence of an additive response suggests that Aβ peptide has a site of action which is different than the sites for either PMA or LPS.

One potential site for Aβ action is the scavenger receptor described by El Khoury et al.[52] Microglia, like other tissue macrophages, are phagocytes and by both *in vivo* and *in vitro* studies have been shown to ingest Aβ peptides and amyloid from cerebral deposits.[53–56] Phagocytosis is associated with activation of the NADPH oxidase and the consequent production of superoxide anion. A clear example of this activation follows exposure of microglia and other macrophagic cells to opsonized particles with subsequent superoxide anion release. Opsonization promotes particle uptake and radical production through the concerted action of multiple receptors including the C3b, Fc and scavenger receptors.[57,58] Our data indicate that Aβ peptide competes for the uptake sites of opsonized particles, suggesting that like other fibers such as asbestos, Aβ (1–40) induces oxidative stress via activation of an NADPH oxidase. The lack of an effect of non-fibrillar, non-toxic peptides such as rat amylin[41,42] further supports this idea. Aβ (1–42) which more readily forms fibers,[42,59] and human amylin have a similar effect as Aβ (1–40) on superoxide anion production. It is important to note that these fibrillar peptides are also neurotoxic.[41,42,59,60]

Our data further demonstrate that NO is not produced in hamster microglia or human MDM in response to Aβ peptides. Both of these cell types are "low-output" NO systems and respond primarily to viral mimetics like PIC or to agents which activate CD23.[29,30] Thus, treatment with Aβ peptides does not directly increase or synergistically modulate NO levels. A similar result was found in the RAW cells, a mouse macrophage cell line. In this case, treatment with Aβ peptides did not induce iNOS or modify the production of NO induced by PIC. This is in contrast to published data by Meda et al.[26] and others[25,27] who have demonstrated NO production in mouse microglia from primary cultures and in a mouse microglial cell line. The differences in our results may be explained, in part, by the different induction agent used (PIC vs. LPS) and by potential differences in rodent clonal cell lines. The data underscore, however, the fact that NO production in response to Aβ peptides is more variable than superoxide anion production.

The precise role of microglial superoxide anion production in chronic neurodegenerative diseases such as Alzheimer's remains to be clearly defined. Footprints of oxidative damage are seen in AD and at least part of the disease process in AD involves the creation of pro-oxidant conditions.[18–22,61–63] Since the generation of a pro-oxidant environment can be achieved by either a failure of antioxidant protection mechanisms or an increased production of ROS, the mechanisms involved in ROS production in the brain and how they are regulated become critical issues. Microglia are clearly an important ROS source and the respiratory burst NADPH oxidase generates enough superoxide anion to be toxic to adjacent cells.[64] In addition, the inability of the microglia to fully ingest and dispose of the Aβ fibrils leads to "frustrated phagocytosis" and the accumulation of fibrils at the cell surface. This area could then act both as a nidus for further aggregation of the fibrils and as a site for localized oxyradical production. Oxidation is known to promote Aβ aggregation[1,65,66] and a local increase in superoxide anion would enhance this process. Under these conditions, a lack of NO production may further compound the membrane damage and oxidative imbalance since NO reduces oxidative stress by inhibiting lipid peroxidation and by scavenging superoxide anion.[67–70] Contrary to common belief, then, the lack of NO production in human microglia and macrophages may be as deleterious as the overproduction of superoxide anion. Thus, the lack of NO production in human macrophages coupled with the chronic resting release of superoxide anion induced by Aβ (1–40) may promote oxidative stress and enhance the overall disease process in AD.

ACKNOWLEDGMENTS

This work was supported in part by a grant from the Alzheimer's Association (PRG- 95-077) to CAC, by NIH NS36718 to CAC and by NIH AG 12851, 15609,15732 to MPV.

REFERENCES

1. COLTON, C. 1996. Products of the activated microglia: their role in chronic neurodegenerative disease. *In* Topical Issues in Microglial Research. E Ling, C Tan, Eds. :255-278. Kent Ridge, Singapore. Singapore Neuroscience Assoc.
2. CRAS, P., M. KAWAI, S. SIEDLAK, *et al.* 1990. Neuronal and microglial involvement in β-amyloid protein deposition in Alzheimer's disease. Am. J. Pathol. 37: 241–246.
3. AKIYAMA, H. 1994. Inflammatory response in Alzheimer's disease. Tohoku J. Exp. Med. 174: 295–303.
4. ITAGAKI, S., P. MCGEER, H. AKIYAMA, S. ZHU & D. SELKOE. 1989. Relationship of microglia and astrocytes to amyloid deposits of Alzheimer's disease. J. Neuroimmunol. 24: 173–182.
5. MATTIACE, L., P. DAVIES, S. YEN & D. DICKSON. 1990. Microglia in cerebellar plaques in Alzheimer's disease. Acta Neuropathol. 80: 493–498.
6. OHGAMI, T., T. KITAMOTO, R. SHIN, Y. KANEKO, K. OGOMOR & J. TATEISHI. 1991. Increased senile plaques without microglia in Alzheimer's disease. Acta Neuropathol. 82: 242–247.
7. PERLMUTTER, L., E. BARRON & H. CHUI. 1990. Morphologic association between microglia and senile plaque amyloid in Alzheimer's disease. Neurosci. Lett. 119: 32–36.
8. CHAO, C., S. HU, W. SHENG & P. PETERSON. 1995. Tumor necrosis factor-alpha production by human fetal microgial cells: regulation by other cytokines. Dev. Neurosci. 17: 97–105.
9. KLEGERIS, A. & P. MCGEER. 1997. Beta-amyloid protein enhances macrophage production of oxygen free radicals and glutamate. J. Neurosci. Res. 49: 229–235.
10. LONDON, J. & D. BIEGEL. 1996. Neurocytopathic effects of beta-amyloid stimulated monocytes: a potential mechanism for central nervous system damage in Alzheimer disease. Proc. Natl. Acad. Sci. 93: 4147–4152.
11. COLTON, C. & D. GILBERT. 1987. Production of superoxide anions by a CNS macrophage, the microglia. Federation of European Biochemical Societies Letters 223: 284–288.

12. ADAMS, D. & T. HAMILTON. 1984. The cell biology of macrophage activation. Ann. Rev. Immunol. **2:** 283–318.
13. BADWEY, J., J. DING, P. HEYWORTH & J. ROBINSON. 1991. Products of inflammatory cells synergistically enhance superoxide production by phagocytic leukocytes. *In* Cell to Cell Interactions in the Release of Inflammatory Mediators. P. Wong & C. Serhan, Eds. :19–33. New York. Plenum Press.
14. MOREL, F., J. DOUSSIERE & P. VIGNAIS. 1991.The superoxide-generating oxidase of phagocytic cells. Eur. J. Biochem. **251:** 523–546.
15. GRIOT, C, T. BURGE, M. VANDEVELDE & E. PETERHAUS. 1989. Antibody-induced generation of reactive oxygen radicals by brain macrophages in canine distemper encephalitis: a mechanism for bystander demyelination. Acta Neuropathol. **78:** 396–403.
16. WEISS, S. 1989. Tissue destruction by neutrophils. New Eng. J. Med. **320:** 363–376.
17. ZIELASEK, J., M. TAUSCH, K. TOYKA & H. HARTUNG. 1992. Production of nitrite by neonatal rat microglia cells/brain macrophages. Cell. Immunol. **141:** 111–120.
18. BOWLING, A. & F. BEAL. 1995. Bioenergetic and oxidative stress in neurodegenerative diseases. Life Sci. **56:** 1151–1171.
19. COHEN, G. & P. WERNER. 1994. Free radicals, oxidative stress and neurodegeneration. *In* Neurodegenerative Diseases. D. Calne, Ed. :139-161. W.B. Saunders. Philadelphia, PA.
20. SMITH, C., J. CARNEY & P. STARKE-REED. 1991. Excess brain protein oxidation and enzyme dysfunction in normal aging and Alzheimer disease. Proc. Natl. Acad. Sci. **88:** 10540–10543.
21. VITEK, M., K. BHATTACHARYA, M. GLENDENING *et al.* 1994. Advanced glycation end products contribute to amyloidosis in Alzheimer disease. Proc. Natl. Acad. Sci. **91:** 4766–4770.
22. BUTTERFIELD, D. 1996. Alzheimer's disease: A disorder of oxidative stress. Alzheimer's Dis. Rev. **1:** 68–70.
23. KLEGERIS, A., D. WALKER & P. MCGEER. 1994. Activation of macrophages by Alzheimer beta amyloid peptide. Biochem. Biophys. Res. Commun. **199:** 984–991.
24. VAN MUISWINKEL, F. & R. VEERHUIS. 1996. Amyloid β protein primes cultured rat microglial cells for an enhanced phorbol 12-myristate 13-acetate induced respiratory burst activity. J. Neurochem. **66:** 2468–2476.
25. II, M., M. SUNAMOTO, K. OHNISHI & Y. ICHIMORI. 1996. β-amyloid protein dependent nitric oxide production from microglial cells and neurotoxicity. Brain Res. **720:** 93–100.
26. MEDA, L., M. CASSATELLA, G. SZENDREL, *et al.* 1995. Activation of microglial cells by β-amyloid protein and interferon-γ. Nature **374:** 647–650.
27. GOODWIN, J., E. UEMURA & J.E. CUNNICK. 1995. Microglial release of nitric oxide by the synergistic action of β-amyloid and IFN-γ. Brain Res. **692:** 207–214.
28. COLTON, C., S. WILT, D. GILBERT, O. CHERNYSHEV, J. SNELL & M. DUBOIS-DALCQ. 1996. Species differences in the generation of reactive oxygen species by microglia. Molec. Chem. Neuropathol. **28:** 15–20.
29. DENIS, M. 1996. Human monocytes/macrophages: NO or no NO? J. Leukocyte Biol. **55:** 682–684.
30. SNELL, J., O. CHERNYSHEV, D. GILBERT & C.A. COLTON. 1997. Polyribonucleotides induce nitric oxide production by human monocyte-derived macrophages. J. Leukocyte Biol. **62:** 369–373
31. SCHNEEMANN, M., G. SCHOEDON, S. HOFER, N. BLAU, L. GUERRERO & A. SCHAFFNER. 1993. Nitric oxide synthase is not a constituent of the antimicrobial armature of human mononuclear phagocytes. J. Infect. Dis. **167:** 1358–1363.
32. GREEN, L., D. WAGNER, J. GLOGOWSKI, P. SKIPPER, J. WISHNOK & S. TANNENBAUM. 1982. Analysis of nitrate, nitrite, and [15N] nitrate in biological fluids. Anal. Biochem. **126:** 131–138.
33. MITCHEN, J., D.R. BLETZINGER, G. RAGO &. G. WILDING. 1995. Use of a DNA microfluorometric assay to measure proliferative response of mink lung cells to purified TGF beta and to TGF beta activity found in prostate conditioned medium. In Vitro Cell Dev. Biol. Anim. **31:** 692–697.

34. SHEARMAN, M., I. RAGAN & L. IVERSEN. 1994. Inhibition of PC12 cell redox activity is specific, early indicator of the mechanism of β-amyloid-mediated cell death. Proc. Natl. Acad. Sci. **91:** 1470–1474.
35. LANDOLFO, S., M. GARIGLIO, G. GRIBAUDO & G. GAROTTA. 1994. Double-stranded RNAs as gene activators. Prog. Mol. Subcell. Biol. **14:** 15–27.
36. LAKE, F., E. DEMPSEY, J. SPAHN & D. RICHES. 1994. Involvement of protein kinase C in macrophage activation by poly (I.C). Am. J. Physiol. **266:** C134–C142.
37. VITEK, M., J. SNELL, O. CHERNYSHEV & C. COLTON. 1997. Modulation of nitric oxide production in human macrophages by apolipoprotein E and amyloid beta peptide. Biophys. Biochem. Res. Commun. **240:** 391–394.
38. BRANCHAUD, R., L. GARANT & A. KANE. 1993. Pathogenesis of mesothelial reactions to asbestos fibers. Pathobiology **61:** 154–63.
39. HILL, I., P. BESWICK & K. DONALDSON. 1995. Differential release of superoxide anion by macrophages treated with long and short fiber amosite asbestos is a consequence of differential affinity for opsonin. Occup. Environ. Med. **52:** 92–96.
40. HILL, I.M., P. BESWICK & K. DONALDSON. 1996. Enhancement of the macrophage oxidative burst by immunoglobulin coating of respirable fibers: fiber specific differences between asbestos and man-made fibers. Exp. Lung Res. **22:** 133–148.
41. LORENZO, A. & B.YANKER. 1996. Amyloid fibril toxicity in Alzheimer's disease and diabetes. Ann. N.Y. Acad Sci. **777:** 89–95.
42. GIULIAN, D., L. HAVERKAMP, J. YU et al. 1996. Specific domains of β-amyloid from Alzheimer plaque elicit neuron killing in human microglia. J. Neurosci. **16:** 6021–6037.
43. HALLIWELL, B. & J. GUTTERIDGE. 1989. Free Radicals in Biology and Medicine, 2nd edition Oxford, UK. Clarendon Press.
44. SCHUBERT, D., C. BEHL, R. LESLEY, et al. 1995. Amyloid peptides are toxic via a common oxidative mechanism. Proc. Natl. Acad. Sci. USA **92:** 1989–1993.
45. BEHL, C., J. DAVIS, R. LESLEY & D. SCHUBERT. 1994. Hydrogen peroxide mediates amyloid β protein toxicity. Cell **77:** 817–827.
46. THOMAS, T., G. THOMAS, C. MCLENDON, T. SUTTON & M. MULLAN. 1996. β-amyloid-mediated vasoactivity and vascular endothelial damage. Nature **380:** 168–171.
47. HARRIS, M., K. HENSLEY, A. BUTTERFIELD, R. LEEDLE & J. CARNEY. 1995. Direct evidence of oxidative injury produced by the Alzheimer's β-amyloid peptide (1–40) in cultured hippocampal neurons. Exp. Neurol. **131:** 193–202.
48. MECOCCI, P., M. MACGARVEY & M.F. BEAL. 1994. Oxidative damage to mitochondrial DNA is increased in Alzheimer's disease. Ann. Neurol. **36:** 747–751.
49. MATTSON, M. & Y. GOODMAN. 1995. Different amyloidogenic peptides share a similar mechanism of neurotoxicity involving reactive oxygen species and calcium. Brain Res. **676:** 219–224.
50. VAN MUISWINKEL, F., R.VEERHUIS & P. EIKELENBOOM. 1996. Amyloid β protein primes cultured rat microglial cells for an enhanced phorbol 12-myristate 13-acetate induced respiratory burst activity. J. Neurochem. **66:** 2468–2476.
51. ADAMS, D. 1992. Regulation of macrophage function by interferon-γ. In Interferon. S. Baron, D. Coppenhaver & F. Dianzani, Eds. :341–351. University of Texas Medical Branch. Galveston, TX.
52. EL KHOURY, J., S. HICKMAN, C. THOMAS & L. CAO. 1996. Scavenger receptor-mediated adhesion of microglia to beta amyloid fibrils. Nature **382:** 716–719.
53. WISNIEWSKI H., M. BARCIKOWSKA & E. KIDA. 1991. Phagocytosis of β/A4 amyloid fibrils of the neuritic neocortical plaques. Acta Neuropathol. **81:** 588–590.
54. PARESCE, D.M., H. CHUNG & F.R. MAXFIELD. 1997. Slow degradation of aggregates of the Alzheimer's disease amyloid β-protein by microglial cells. J. Biol. Chem. **272:** 29390–29397.
55. SHIGEMATSU, K., P. MCGEER, D. WALKER, T. ISHII & E. MCGEER. 1992. Reactive microglia/macrophages phagocytose amyloid precursor protein produced by neurons following neural damage. J. Neurosci. Res. **31:** 443–453.
56. WELDON, D., S. ROGERS, J. GHILARDI et al. 1998. Fibrillar beta-amyloid induces microglial phagocytosis, expression of inducible nitric oxide synthase and loss of a select population of neurons in the rat CNS in vivo. J. Neurosci. **18:** 2161–2173.

57. ABSOLOM, D. 1986. Basic methods for the study of phagocytosis. *In* Methods in Enzymology. S Colowick & N. Kaplan,Eds. :95–180. New York. Academic Press, Inc.
58. IAMAMICHI, T., H. SATO, S. IWAKI, T. NAKAMURA & J. KOYAMA. 1990. Different abilities of two types of Fc gamma receptor on guinea pig macrophages to trigger the intracellular Ca^{2+} mobilization and O_2^- generation. Mol. Immunol. **27:** 829–838.
59. SEILHEIMER, B., B. BOHRMANN, L. BONDOLFI, F. MULLER, D. STUBER & H. DOBELI. 1997. The toxicity of the Alzheimer's beta-amyloid peptide correlates with a distinct fiber morphology. J. Struct. Biol. **119:** 59–71.
60. MAY, P., L. BOGGS & K. FUSON. 1993. Neurotoxicity of human amylin in rat primary hippocampal cultures: similarity to Alzheimer's disease amyloid beta neurotoxicity. J. Neurochem. **61:** 2330–2333.
61. SMITH, M., S. TANEDA, P. RICHEY, *et al.* 1994. Advanced Maillard reaction end products are associated with Alzheimer disease pathology. Proc. Natl. Acad. Sci. **91:** 5710–5714.
62. JEANDEL, C., M. NICOLAS, F. DUBOIS, F. NABET-BELLEVILLE, F. PENIN & G. CUNY. 1989. Lipid peroxidation and free radical scavengers in Alzheimer's disease. Gerontology **35:** 275–282.
63. SUBBARAO, K., S. RICHARDSON & L. ANG. 1990. Autopsy samples of Alzheimer's cortex show increased peroxidation in vitro. J. Neurochem. **55:** 342–345.
64. THERY, C., B. CHAMAK & M. MALLAT. 1991. Cytotoxic effect of brain macrophages on developing neurons. Eur. J. Neurosci. **3:** 1155–1164.
65. DYRKS, T., E. DYRKS, T. HARTMANN, C. MASTERS & K. BEYREUTHER. 1992. Amyloidogenicity of βA4 and βA4-bearing amyloid protein precursor fragments by metal-catalyzed oxidation. J. Biol. Chem. **267:** 18210–18217.
66. HENSLEY, K., J. CARNEY, M. MATTSON, *et al.* 1994. A model for β-amyloid aggregation and neurotoxicity based on free radical generation by the peptide: relevance to Alzheimer disease. Proc. Natl. Acad. Sci. USA **91:** 3270–3274.
67. WINK, D., J. COOK, M. KRISHNA, *et al.* 1996. Nitric oxide protects against alkyl peroxide-mediated cytotoxicity: further insights into the role nitric oxide plays in oxidative stress. Arch. Biochem. Biophys. **319:** 402–407.
68. MILES, A., D. BOHLE, P. GLASSBRENNER, B. HANSERT, D. WINK & M. GRISHAM. 1996. Modulation of superoxide-dependent oxidation and hydroxylation reactions by nitric oxide. J. Biol. Chem. **271:** 40–47.
69. RUBBO, H., R. RADI, M. TRUJILLO, *et al.* 1994. Nitric oxide regulation of superoxide and peroxynitrite-dependent lipid peroxidation. J. Biol. Chem. **269:** 26066–26075.
70. WINK, D., I. HANBAUER, M. GRISHAM, *et al.* 1996. Chemical biology of nitric oxide: regulation and protective and toxic mechanisms. Curr. Topics Cell. Reg. **34:** 159-187.

Antioxidant Status and Human Health

Use of Cyclic Voltammetry for the Evaluation of the Antioxidant Capacity of Plasma and of Edible Plants

SHLOMIT CHEVION[a] AND MORDECHAI CHEVION

The Hebrew University of Jerusalem, P.O.Box 12272, Jerusalem 91120, Israel and Nestle Research Center, CH-1000 Lausanne 26, Switzerland

ABSTRACT: The low molecular weight antioxidants (LMWA) play a major role in protecting biological systems against reactive oxygen-derived species (ROS), and reflect the antioxidant capacity of the system. The cyclic voltammetry (CV) has been conveniently used and validated for the quantitation of the antioxidant capacity of the LMWA of blood plasma, tissue homogenates, and plant extracts. The CV tracing provides the biological oxidation potential (E and $E_{1/2}$ which relate to the nature of the molecule(s)), the intensity of the anodic current wave (I_a), and its area S (both relate to the concentration of the molecule(s)). The components of the first anodic wave of plasma were identified by comparison with HPLC-electrochemical detection. CV together with another plasma parameter R, which reflects the level of oxidized ascorbate, were used for the evaluation of the antioxidant status and the oxidative stress in healthy subjects and in chronic (diabetes mellitus) and acute patients (subjected to total body irradiation prior to bone marrow transplantation).These methodologies could be widely employed for rapid evaluation of subjects, in health and disease, for monitoring of their response to treatment and nutritional supplementation, and for screening of specific populations.

INTRODUCTION

Living cells, including those of man, animals and plants are continuously exposed to a variety of challenges which exert oxidative stress. These could stem from endogenous sources through normal physiological processes, such as mitochondrial respiration. Alternatively, they could result from exogenous sources, such as exposure to pollutants and ionizing irradiation.

Oxidative stress will be manifested in a biological system following either an increased exposure to oxidants or a decrease in antioxidant capacity of the system, or both. Oxidative stress is often associated with or leads to the generation of reactive oxygen species (ROS),[1] amongst which free radicals are of particular interest.[2] ROS are strongly implicated in the pathophysiology of disease, including diabetes, atherosclerosis, neoplasia and aging.[3]

---

[a]Author for correspondence: Dr. Shlomit Chevion, The Hebrew University of Jerusalem, P.O. Box 12272, Jerusalem, Israel IL-91120. Voice: +972-2-675-8160; fax: +972-2-641-5848.
e-mail: chevion@cc.huji.ac.il

Cells are equipped with several defense systems, acting through a variety of mechanisms. These allow the cells to withstand and overcome the continuously imposed challenge. These mechanisms include prevention of the induction of damage,[1] induction of repair mechanisms,[3] and directly acting against deleterious metabolites.[4] It is assumed that the antioxidant defense systems have developed concurrently with the increase of oxygen concentration during evolution.[1] These defense systems can be categorized into protection via enzymatic activities and protection through low molecular weight antioxidants (LMWA).[5] The LMWA family consists of many compounds, each of which acts as a direct chemical scavenger neutralizing ROS component(s),[4] or indirectly, through transition metal chelation.[6] LMWA are small molecules that can often penetrate into cells, accumulate (at high concentrations) at specific compartments near where oxidative damage might occur, and be regenerated by the cell.[1]

In human tissues, the cellular LMWA are obtained from various sources: glutathione, NADH and carnosine[7] are synthesized by the cells, uric acid (UA)[8] and bilirubin[9] are waste products of the cellular metabolism, while ascorbic acid (AA)[10] tocopherols and polyphenols are antioxidants obtained from the diet.

Edible plants constitute a major nutritional source for LMWA in mammals, including man. Consumption of fresh fruits and vegetables has been shown to be associated with lower incidence and lower mortality rates of cancer, heart diseases and atherosclerosis, brain degenerative diseases, aging and other pathologies.[11–18]

Plasma is an available tissue, which is often used for the evaluation of (free radical–induced) damage. Plasma contains critical targets for oxidative damage such as lipoproteins, and LDL in particular. It contains also important antioxidants, such as AA and UA. Plasma represents a biologically relevant milieu, since it reflects the integrated antioxidant status which arises from both body tissues (synthesis and reservoirs) and nutrition. Hence, in order to evaluate the interrelationship between disease, diet, free radicals and vitamin supplementation, methods based on the characterization of plasma parameters are considered bona fide representatives of the antioxidant status of the whole organism.

The total antioxidant status of plasma reflects the integrated capacity of the organism to cope with and to combat the oxidative insult. As part of the antioxidant reservoir is obtained through the diet, there is a high interest in the assessing of the total antioxidant capacity of plasma and the antioxidant capacity of edible plants. In parallel, it is important to evaluate the individual contribution of each of the plant components to the overall (total) antioxidant capacity of the organism.

Several methods for measuring the total antioxidant capacity of a biological sample, including plasma and plants, have been proposed and recently reviewed.[19–22] These are related to the capacity of a sample to compete for and scavenge a specific ROS. They provide useful information, which is not sufficient for the evaluation of the overall antioxidant profile of the biological fluid or tissue homogenate, or the plant extract.

We have suggested the use of the cyclic voltammetry methodology (CV) as an instrumental tool for the evaluation of the total antioxidant capacity of the LMWA of human (and animal) plasma, other body fluids and homogenated tissues.[23–27] Likewise, CV had been used for the evaluation of the total antioxidant capacity of edible plants.[28] The CV measurement provides information concerning the integrated antioxidant capacity, which arises from the LMWA, without the specific determination of the contribution of each individual component.

The CV method is based on the measurement of the reductive potential of a given compound, and/or a mixture of compounds. The CV tracing indicates the ability of the compound to donate electron(s). Most of the LMWA are reducing agents, which quench ROS through donation of electron(s) to the ROS, neutralizing its activity. Therefore, evaluation of the overall reducing power of a biological fluid, tissue homogenate or plant extract would reflect the antioxidant activity of its LMWA. At the same time this method allows the measurement of the combined concentration equivalent of these reducing agents. Hence, the total antioxidant capacity of the sample is a function of two sets of parameters: the total antioxidant reducing power and the total concentration of these reducing agents. Together these are equivalent to the evaluation of the total antioxidant activity. The integrated value extracted from the CV tracing, thus, represents the total antioxidant capacity of the sample, *without* the necessity to measure the specific antioxidant ability of each of its components.

The CV tracing of the sample (plasma or other body fluids, tissue homogenate or plant extract) is constituted of anodic wave(s). The oxidation potential of an anodic wave is characterized by $E_{1/2}$, the potential at half height of the anodic wave. $E_{1/2}$ is a compound-specific value, which correlates with the ability of the specific compound to donate electron(s). The anodic wave is also characterized by its current height (intensity), I_a, which correlates with the concentration of the component. We have further proposed to use the area of the anodic wave (S; related to the charge), rather than I_a, as a better parameter reflecting the total antioxidant capacity of the sample.[28] This provides a marked advantage in some cases, in particular when an anodic wave contains more than a single component.

The CV methodology allows the monitoring of changes, even small ones, in the antioxidant capacity of a sample. For plasma, it will provide information about the *in vivo* exposure of the subject to oxidative stress (prior to blood collection), and about the antioxidant pool of the subject, when the sample is exposed to a challenge, *in vitro*. CV tracing of plasma can also be used for monitoring the antioxidant status of patients, and the success (or failure) of the treatment they receive. Clinical examples include diabetes and cancer (see below). Likewise, CV allows the monitoring of the total antioxidant capacity of edible plants, and the monitoring of the quality of a food product during its shelf life (see below).

THE CYCLIC VOLTAMMETRY METHODOLOGY

The Principle

Cyclic voltammetry (CV) is a methodology extensively used in electrochemistry, to determine redox properties of molecules in solution. Experimentally, the potential of a working electrode is linearly scanned (vs. a reference electrode—typically Ag| AgCl) from an initial value to a final value, and back, while recording of the anodic current (I_a) (FIG. 1). FIGURE 2 shows the current obtained when the potential excitation signal (typically -0.45 V to $+1.35$ V) is applied to a platinum (working) electrode immersed in potassium ferrocyanide, which is often used as the standard, particularly, when old analog instruments are used. As the potential is scanned positively (lower

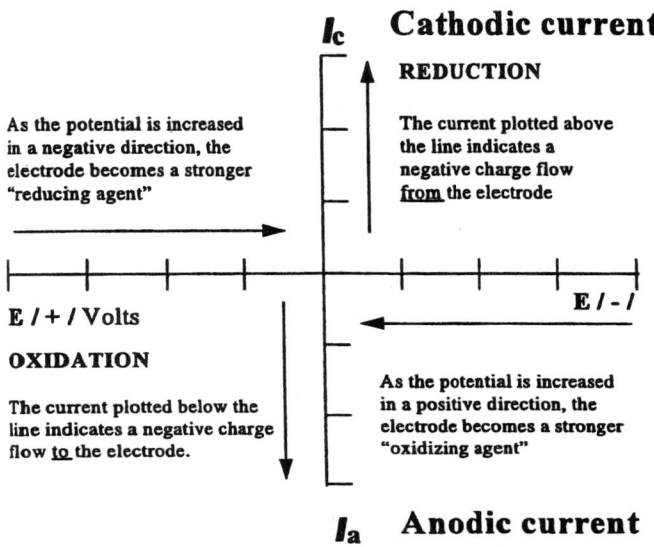

FIGURE 1. Current–Potential (I–E) axes for voltammetric techniques.

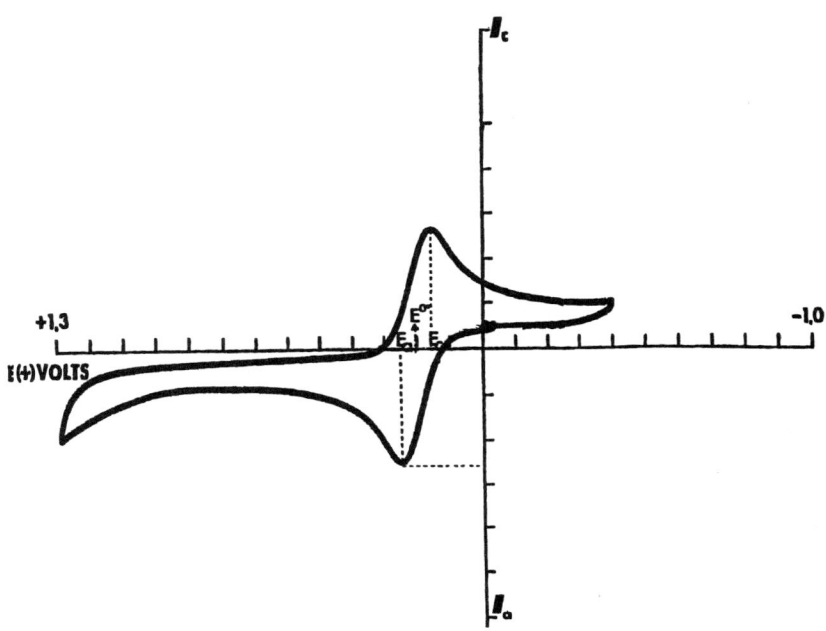

FIGURE 2. Analysis of CV tracing of potassium ferrocyanide.

part of the trace), the electrode becomes sufficiently oxidizing to oxidize Fe^{+2} to Fe^{+3}, and gives rise to the recorded anodic current (I_a). I_a increases rapidly until [Fe^{+2}] at the anodic surface approaches zero, causing the current to peak. At 1.3V the scan direction is switched to negative (upper part of scan), giving rise to analogous cathodic current (I_c), completing the (first) cycle. Thus, CV is a method capable of rapidly generating a new species during the forward scan and then monitoring its fate on the reverse scan.

The important parameters of a CV scan are the magnitude of the peak current (I) and the peak potential (E). These are illustrated in FIGURE 2. The peak current can be described by the Randles-Sevcik equation: $I = 2.69 n^{3/2} A D^{1/2} C V^{1/2}$, where I = peak current, amperes; n = electron stoichiometry, equiv/mol; A = electrode area, cm^2; D = diffusion coefficient, cm^2/sec; C = concentration, $mole/cm^3$; and V = scan rate, V/sec.

A redox couple in which both species are stable and rapidly exchange electrons with the working electrode is termed an electrochemically reversible couple. The formal reduction potential ($E^{\circ\prime}$) for a reversible couple is centered between E_a-anodic and E_c-cathodic of the peak: $E^{\circ\prime} = (E_a + E_c)/2$ (FIG. 2).

The number of electrons transferred (n) during the electrode reaction for a reversible couple can be determined from the separation between the anodic and cathodic peak potentials from: $E_a - E_c = 0.058/n$

The Instrument

Analyses using cyclic voltammetry (CV) were carried out on three different electrochemical analyzers, an analog instrument (BAS) and two digital Electrochemical Analyzers BAS 100A and BAS 27. A modified low volume cell, containing ~0.5–2 ml sample of plasma or tissue, or plant homogenate or extract, or a standard solution (ferrocyanide, ascorbate, thioctic acid, trolox), was used. A three electrode system was used throughout the study: the reference electrode (Ag| AgCl), the working electrode (a glassy carbon disc (BAS) 3.2 mm in diameter), and a counter electrode (platinum wire). CV tracings were recorded typically at a range from –0.5 to +1.5 volts versus the reference electrode, at a scan rate of 100 mV/sec or 400 mV/sec.

The Tracing

It was our goal to determine the total (integrated) reducing capacity of biological samples, including plasma, mammalian tissues and plant samples, as a measure of their antioxidant capacity. An increasing potential was applied to the working electrode, the CV tracing was recorded, and allowed the evaluation of the reducing equivalents present in the sample.

The Dependency of I_a and S on the Concentration

The CV tracings in FIGURE 3 show the dependency of the peak current of the anodic wave (I_a), as well as the area under the anodic peak (S), as a function of the concentration of commonly used standards—ascorbate (AA) and trolox. Both parameters well obey a linear relationship with the concentration.

For pure AA and UA in phosphate buffer $E_{1/2}$ was 380 and 420 mV, respectively. This is in accord with their antioxidant capacity. AA, which is a more potent antioxidant,

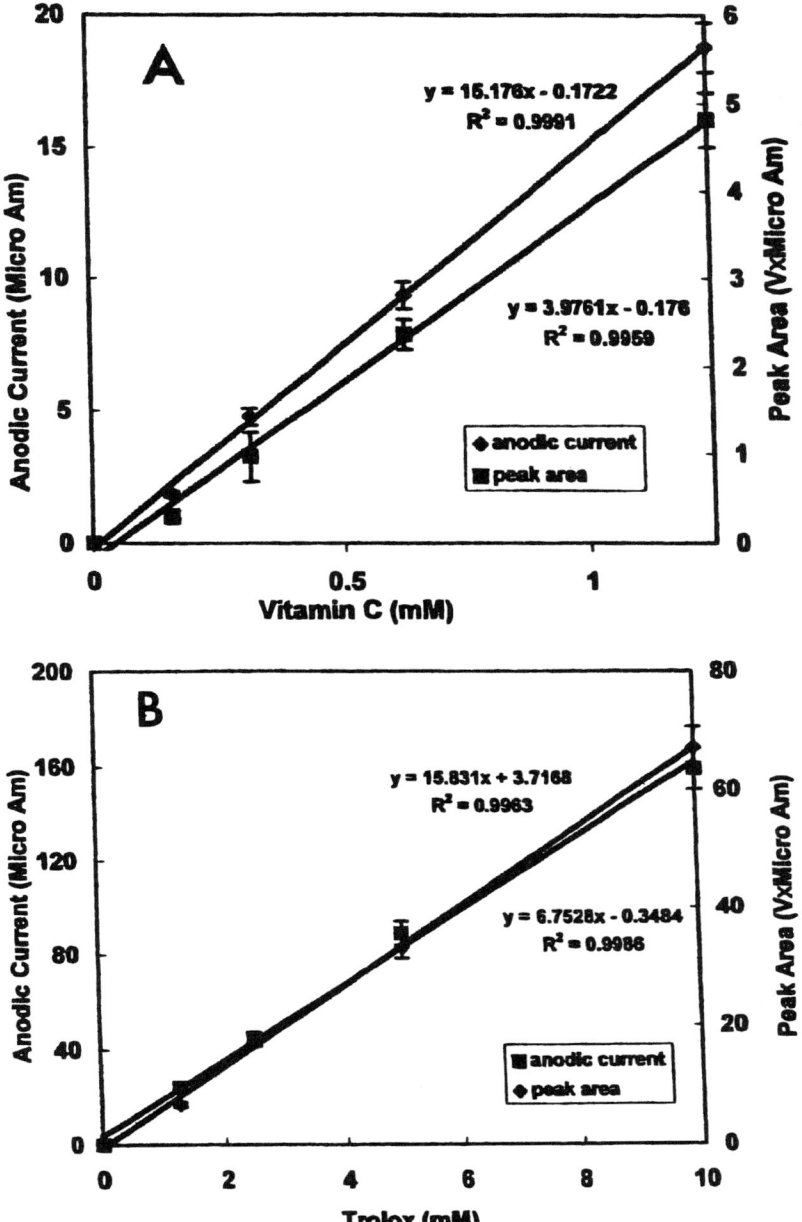

FIGURE 3. The anodic current I_a and the area under the anodic wave, S, of AA and trolox *versus* their concentrations. **A:** Anodic current amplitude (I_a) and peak area S for increasing concentrations of ascorbate (0–1.25 mM) in water/acetic acid/acetonitrile mixture (40/30/30), pH 1.9. **B:** I_a and S for increasing concentrations (0–10 mM) of trolox in phosphate buffer 0.2M, pH 7.4.

donates its electrons at a lower oxidation potential. The maximal change in the oxidation potential of the first anodic wave, assuming that AA and UA are its sole or major components, could be 40 mV, the difference between the oxidation potentials of AA and UA.

As an anodic wave of a biological sample often represents more than a single component, each of which could donate electron/s around the same potential, we have proposed[28] that S could prove more useful than I_a for monitoring changes in these components. An example is the anodic wave of plasma, around $E_{1/2}$ = ~400 mV. We have shown[23,24,26] that this anodic peak is comprised of AA and uric acid (UA). The changes in the concentration of one of these components, particularly AA, which is the minor component, could be better monitored through the changes in S, rather than by I_a. As the typical potentials, $E_{1/2}$, for AA (380 mV) and UA (420 mV) are similar but different, the change in the concentration of AA will significantly affect the width of the anodic peak, and S, while only marginally affecting I_a.

Higher CV scan rates give rise to higher I_a; I_a is proportional to the (scan rate)$^{1/2}$ (according to Randles-Sevcik equation) and as has been shown experimentally for these compounds.[28] Concentrations of AA, as low as 5–10 mM, can be measured with precision (signal to noise > 5) using scan rate of 400 mV/sec. I_a and S values can also be used for quantitation of relatively high concentrations of AA (verified up to 10 mM).

Phosphate buffer (0.2 M; pH 7.4) and solvent systems used, including water/acetic acid (70/30, and water/acetic acid/acetonitrile (40/30/30, pH 1.9) did not show any anodic wave.

STUDIES OF HUMAN PLASMA IN HEALTH AND DISEASE STATES

CV Tracing of Human Plasma

FIGURE 4 shows a typical CV tracings of human plasma from a healthy subject. Two major peaks were identified: at above 500 and 1000 mV (characterized by $E_{1/2}$ values of ~400 and ~900 mV). Also, we have shown[24] that the *first anodic wave* consists of AA and UA. This was verified by several line of evidence: reconstitution of the peak from the two individual components, decrease of the peak by oxidizing AA, decrease of the peak following treatment with uricase, and its complete disappearance following the removal of both components. The tracing of such a peak can be deconvoluted into two (or more) anodic current waves.[26]

Thioctic acid (TA, lipoic acid) is a natural cofactor of α-keto dehydrogenases. It was shown to be a candidate component for the second anodic wave.[25] The anodic current I_a was shown to linearly depend on the concentration of TA, both in buffer and when added in plasma. Also, diabetic patients who ingest TA daily were found to show higher anodic currents of the *second wave*, than normal controls, or patients who do not ingest this drug. The detailed composition of this anodic wave still awaits further investigation. Leakage of NADPH from red blood cells has been considered and ruled out as a contributor to the second anodic wave. NADPH demonstrates a relative low CV response, and this second wave is present in all samples, including those which were carefully handled and did not show any other sign of leakage or hemolysis.

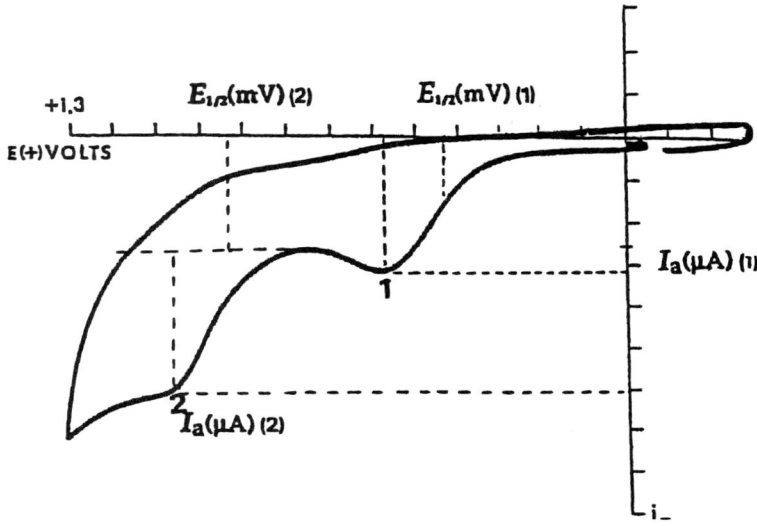

FIGURE 4. CV tracing of plasma from a healthy human subject. The range of the voltage is from −0.3 to 1.3 V, at a rate of 100 mV/sec, *versus* Ag|AgCl reference electrode.

Changes of the pH often lead to alterations in the redox properties of the solution, and cause a horizontal potential (voltage) shift.[24,28] Thus, it is important to characterize the appropriate oxidation potential according to the pH of the sample. This should be of concern when de-proteination (by acid or organic solvent) is used, and for plant samples, which are often acidic.

An important concern is that the working electrode has to be intensively polished before every measurement in order to remove biological residues from its surface, and to maintain its sensitivity.

Special attention should be given during the collection of plasma samples. The CV tracing of plasma collected in Heparin shows the two anodic waves specified above, at $E_{1/2}$ = ~400 and ~920 mV. EDTA, which is often used as an anticoagulant in blood samples, exhibits its specific anodic wave at around 900mV, and thus, interferes with the second anodic wave of the plasma.[23,24]

Changes in the CV Tracing of Plasma Subjected to Oxidative Stress

Oxidative stress was exerted on plasma, *in vitro*.[23,24] It was expected that challenges by reactive oxygen–derived species (ROS) will lead to the decrease of the antioxidant capacity of the plasma, and this in turn, will be reflected by a parallel decrease in I_a of the CV anodic waves. Plasma samples were exposed to:

(a) *Peroxyl radicals*: incubation of plasma samples at 37°C with AAPH led to a reciprocal time dependencies of I_a on the concentration (0–200 mM) of AAPH and the duration of incubation (0–3 h). In general, a linear dependency was found between log I_a and the extent of the different oxidative stresses.[25]

(b) *Copper ions*: incubation of plasma samples with copper(II) sulfate for 1h showed a reciprocal time dependencies between I_a and [Cu(II)] (0–400 μM), and I_a and the duration of incubation (0–3 h).

(c) *Ionizing irradiation*: plasma samples were subjected to ionizing irradiation (0, 1200, 100,000 and 1,000,000 cGy).

TABLE 1 shows the marked decrease in the two anodic waves as determined by CV. It also shows the selective decreases in the levels of AA and UA, the components of the first anodic wave, as measured by HPLC coupled with electrochemical detection. The effect of AAPH caused a reduction of 92% in the intensity of the first anodic current wave, I_a, and corresponding decreases of 96% in [AA] and 83% in [UA]. Exposure to copper ions results in a 31% reduction in I_a of the first wave, and 94% in [AA]; no change in [UA] was induced. Exposure to irradiation resulted in a 79% reduction in the I_a of the first wave, concurrent with a 32% decrease in [AA] and 92% in [UA]. No significant changes in $E_{1/2}$ of the CV tracing were found.

We have systematically characterized the use of anodic current parameters, I_a and S, rather than $E_{1/2}$ alone, as the major parameters determining the antioxidant capacity of plasma LMWA, as reflected by CV.[23,24,28]

We did not find any changes in the oxidation potential ($E_{1/2}$) of the anodic waves following a broad variety of treatments. The observed changes of the CV tracings found were in I_a, reflecting analogous changes in the concentration of the LMWA. While changes in $E_{1/2}$ are not expected, this has been the main claim of a series of CV studies.[29–31] This claim, that the oxidation potential of a sample is dependent on the treatment it had received, was based on measurements using an analog instrument, which could have led to inaccurate determinations of $E_{1/2}$, as can be identified in the published CV tracings. The authors would probably have offered other explanation, if they had one.

It was expected that in clinical disorder and/or under various types of oxidative stress, the changes in the antioxidant capacity will be reflected by equivalent changes in the anodic current I_a, and not in $E_{1/2}$. Thus, groups of patients suffering from a pathology, which is likely to cause oxidative stress, will show a reduction in I_a, but

TABLE 1. Changes in levels of LMWA following oxidative stress as evaluated by the anodic current, I_a, of the CV and by HPLC-ECD

Inducer of oxidative stress	Concentration	Percent change by CV		Percent change by HPLC	
		at 420 mV	at 900 mV	AA	UA
AAPH	200 mM	−92 ± 2	−48 ± 3	−96 ± 3	−83 ± 2
CuSO$_4$	400 μM	−31 ± 2	−31 ± 2	−94 ± 2	no change
γ-irradiation	1,000,000 rad	−79 ± 3	−27 ± 2	−32 ± 3	−92 ± 4

Stress with AAPH, CuSO$_4$, H$_2$O$_2$ and γ-irradiation was exerted on human plasma samples as described above. The CV tracings were recorded and analyzed and the changes were evaluated from the anodic current heights at around 400 and 900 mV. Ascorbate (AA) and urate (UA) were analyzed by HPLC-ECD using an amperometric detector (with electrode potential of +0.5 V versus Ag|AgCl reference electrode).

no change in $E_{1/2}$. Also, enrichment of the level of antioxidants or other changes in diet, are expected to be reflected by I_a of the CV tracing.

The Antioxidant Status of Healthy Subjects

To determine the effects of vitamin C on cardiovascular risk factors, the effect of dietary enrichment with vitamin C was studied in 40 young healthy Yeshiva students consuming a high-saturated fatty acid diet.[26,18] After a month run-in period consuming 50 mg/d AA (Lo-C), half were randomized to receive 500 mg/day AA (Hi-C) for a further 2 months, while the other half remained on the original diet (Lo-C). Plasma AA increased from 13 to 52 mmol/L ($p < 0.01$) on the Hi-C diet. At the same time we monitored an additional plasma parameter of these students, R, which we consider as a marker of the oxidant stress.

$$R = [DHAA]/\{[AA] + [DHAA]\},$$

the ratio between the oxidized form of AA—dehydro-ascorbate (DHAA)—and the total AA, significantly decreased in this group, indicating a better antioxidant status. In the Lo-C diet group the level of AA, which decreased during the run-in period from 21 to 13 mmol/L, remained in this level, while the ratio R remained high. An increase in the lag period of the *in vitro* oxidation of LDL was observed, and correlated well with the plasma AA concentrations. The changes of the antioxidant capacity of the plasma by CV showed a high positive correlation between AA levels and the I_a of the first anodic wave. Since the level of UA in plasma is up to 10 times higher than the level of AA, and since they are both components of the first anodic wave, we aimed to eliminate the contribution of UA by treatment of the plasma by uricase, and then recognize the difference. Addition of excess cysteine to the plasma, following the removal of UA, allows the determination of the total AA and DHAA. Subsequently, R can be determined by CV and yield results, which are similar to the values obtained by HPLC-ECD.

Antioxidant Status of Patients

Two groups of patients were studied. The first consisted of patient subjected to total body irradiation (TBI) prior to bone marrow transplantation (BMT),[26,27] and the other consisted of patients suffering from diabetes mellitus (DM).[25,26]

Total Body Irradiation

Total body irradiation (TBI) is a routine preconditioning procedure for the treatment of leukemia and aplastic anemia, prior to bone marrow transplantation. Ionizing irradiation generates ROS that react with cellular components, inactivating bone marrow cells. Plasma of 14 patients undergoing TBI prior to BMT was examined and evaluated by CV for the antioxidant status of these patients. In addition, the levels of AA and UA were determined by HPLC-ECD, and R was calculated. The antioxidant capacity (as reflected by I_a of the CV tracings) was lower by 36% ($p < 0.02$) than normal matched healthy controls; 4 months after the TBI treatment I_a recovered to a level 22% higher than before the treatment ($p < 0.05$). Both, AA and UA, decreased following irradiation by 84% ($p < 0.02$) and 24% ($p < 0.05$), respectively, but returned to a level of 21% and 320% after 4 months compared to baseline values. The changes in [UA] were probably affected by Allopurinol (a xanthine oxidase inhibitor), given as a routine pre-transplant therapy until day −1. R value was 45%, while the range for normal healthy controls = 13.2 ± 1.5%. R increased by 69% following TBI.

It is interesting to note that in order to induce, *in vitro*, in plasma, a decrease of the antioxidant capacity (as indicated by I_a), which is comparable to the *in vivo* decrease similar to that demonstrated for the patients, a 1000-fold higher dose of irradiation was required. TBI alters antioxidant homeostasis and dramatically enhances the damage inflicted on the cells. Employing CV measurements has led to a better understanding of the balance between oxidative stress and antioxidant utilization, and to the reconsideration of the routine use of Allopurinol as a pretreatment for TBI patients, and the prescription of antioxidant support before and/or after TBI treatment.

Diabetes mellitus

Ninety-two diabetes mellitus (DM) patients were included in a study. Their blood samples were analyzed by HPLC for the levels of AA, DHAA and UA, and by CV for their antioxidant capacity (by I_a). The patients were divided into groups according to the severity of their nephropathy, or according to the severity of their total complication count. Severe DM patients who were under treatment with thioctic acid (TA) showed higher level of the second anodic wave current, than those not taking the drug. The level of AA was found to be lower with increasing albuminuria and with their complication score by over 60%. Severe patients without TA had significantly higher levels of their R values ($p < 0.02$), while for the group taking TA the R values was not different than non-diabetic controls. Generally, when comparing the two groups of patients with the same number of complications, those who did not receive TA had significantly lower [AA] levels combined with a higher R value then those taking TA. Thus, based on the use of these parameters it could be demonstrated that the antioxidant status of these patients is significantly improved by treatment with TA. This positive correlation between TA and AA, and the negative correlation between AA and R values strongly indicate that TA or other antioxidant/s should be considered as drugs for improving antioxidant status of DM patients.

STUDIES OF EDIBLE PLANTS

We proposed the adaptation of the CV for the evaluation of the total antioxidant capacity of edible plants.[28] Different vegetables and fruits, which are often consumed in the U.S. diet, have been studied. The LMWA were extracted from these edible plants either with water, or aqueous acetic acid (30%), or water/acetic acid/acetonitrile mixture (40/30/30). It was found that it is impossible to fully extract the LMWA in a single extraction, and several steps (typically 3–5) were required. The ratio between the amounts of LMWA extracted in two consecutive steps varied for the different plants used. In all the samples this ratio was always > 2, allowing a complete extraction (> 95%). The 'extraction solution' *itself*, which contains a combination of acetic acid and acetonitrile, did not show any anodic wave in the relevant range, and proved to be the most effective for extraction. The extraction of ascorbate by water alone was unsuccessful, and with acetic acid the efficiency was low. As the extraction with the combination was effective, it is reasonable to assume that the physical homogenization of a plant, with this 'extraction solution' causes a disruption of cellular and sub-cellular structures releasing the ascorbate (and possibly other LMWA). Additional experiments showed that the recovery rate of ascorbate was 93%. Samples, weighing 50–100 g were used, and the total content of LMWA was evaluated from the area of the anodic wave—S—of the CV tracing, and translated into equivalents of ascorbate in 100 g of edible plant.

CV Tracing of Edible Plants

The CV tracings of extracts of edible plants, revealed up to three anodic current waves. Strawberry showed one broad anodic wave (around $E = 750$ mV as automatically calculated by the Instrument); broccoli, cauliflower, potato, and tomato showed two anodic waves (around $E = 600$ and 1200 mV), and corn showed three anodic waves (around $E = 600$, 900, and 1200 mV; TABLE 2). The biological oxidation potential, E, of vitamin C is dependent on the pH. At pH = 2 and with acetonitrile (30%), the E value of ascorbate was found around 600 mV, while its $E_{1/2} = 480$ mV (higher than the value obtained at pH 7.4). FIGURE 5 shows a typical CV tracing of a plant sample (cauliflower).

The Antioxidant Capacity of the Plants

FIGURE 6A shows the CV tracings of extracts of cauliflower and depicts each of the 5 steps that were needed to accomplish complete extraction with the 'extraction solution.' FIGURE 6B depicts S multiplied by the volume of the 'extraction solution' for every step i $(S_i \times V_i)$ versus the number of steps (N) needed for a complete extraction (> 95%) of the LMWA of cauliflower. This procedure was accomplished by only two steps for strawberry, and by 4–5 steps for broccoli, cauliflower, tomato, potato and corn.

TABLE 2. Combined LMWA, $\Sigma_i(S_i \times V_i)$ for each peak (the sum of the products of the area of the anodic current wave multiplied by the volume of the extract of the edible plants tested)

Plant	N	Peak 1		Peak 2		Peak 3	
		E (mV)	$\sum_i S_i V_i$	E (mV)	$\sum_i S_i V_i$	E (mV)	$\sum_i S_i V_i$
Broccoli	5	654 ± 12	954	1261 ± 13	385		
Cauliflower	5	568 ± 31	840	1255 ± 15	534		
Strawberry	2	749 ± 27	1934				
Tomato	4	588 ± 63	186	1017 ± 30	74		
Potato	4	612 ± 16	620	1225 ± 16	777		
Corn	4	615 ± 11	7.4	1279 ± 12	135	962 ± 14	3.3

$i = 1–N$, the # of extraction steps; E, oxidation potential; S, area of anodic current wave; V, volume of sample. V was normalized to 100 g plant equivalent. The product was calculated from the area, S, of the CV tracing and the volume of the extract, for each step. The CV instrument calculated automatically the values of the oxidation potentials E and S, for each anodic current wave (peak).

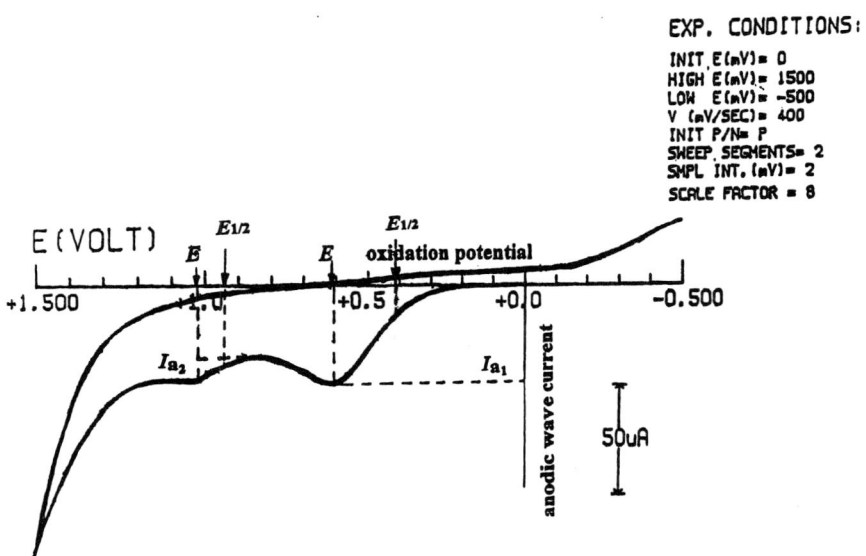

FIGURE 5. CV tracing of a typical plant sample (cauliflower) The range of voltage is −0.5 to 1.5 V, at a scan rate of 400 mV/sec, *versus* Ag|Agcl reference electrode.

TABLE 2 shows the content of LMWA for the plants used. S_i represents the area of the anodic wave in the i-th extraction step, and V_i represents the volume of the 'extraction solution' (calculated) per 100 g of plant. $\Sigma_i (S_i \times V_i)$ (i =1...N) represents the combined amount of LMWA for a given peak. The consecutive values of $S \times V$ represent a declining function which could be well-fitted to a polynomial function of the third power (see FIG. 6B).

TABLE 3 shows the sum of the values of $\Sigma i (S_i \times V_i)$ (i = 1 − N) for the three anodic waves (j) identified, $(\Sigma_j \Sigma_i (S_i \times V_i)$ (i = 1 − N; j = 1 − 3)). These sums are equivalent to the total level of antioxidants of the plant and are expressed in (area)×(volume) units per 100 g of sample. Using ascorbate standard, the sums could be converted and were expressed as vitamin C equivalent units (mg of vitamin C), which represented the actual vitamin C as well as the other LMWA in the plant extract. For comparison, the last column of TABLE 3 depicts the actual levels of vitamin C in these edible plants, as published by USDA.[32] The total antioxidant capacity of each plant, expressed as vitamin C equivalent units, was higher than the vitamin C levels in the plant, as previously reported. The difference between the last two columns of TABLE 3 could reflect the capacity of the other LMWA (except ascorbate), in each plant.

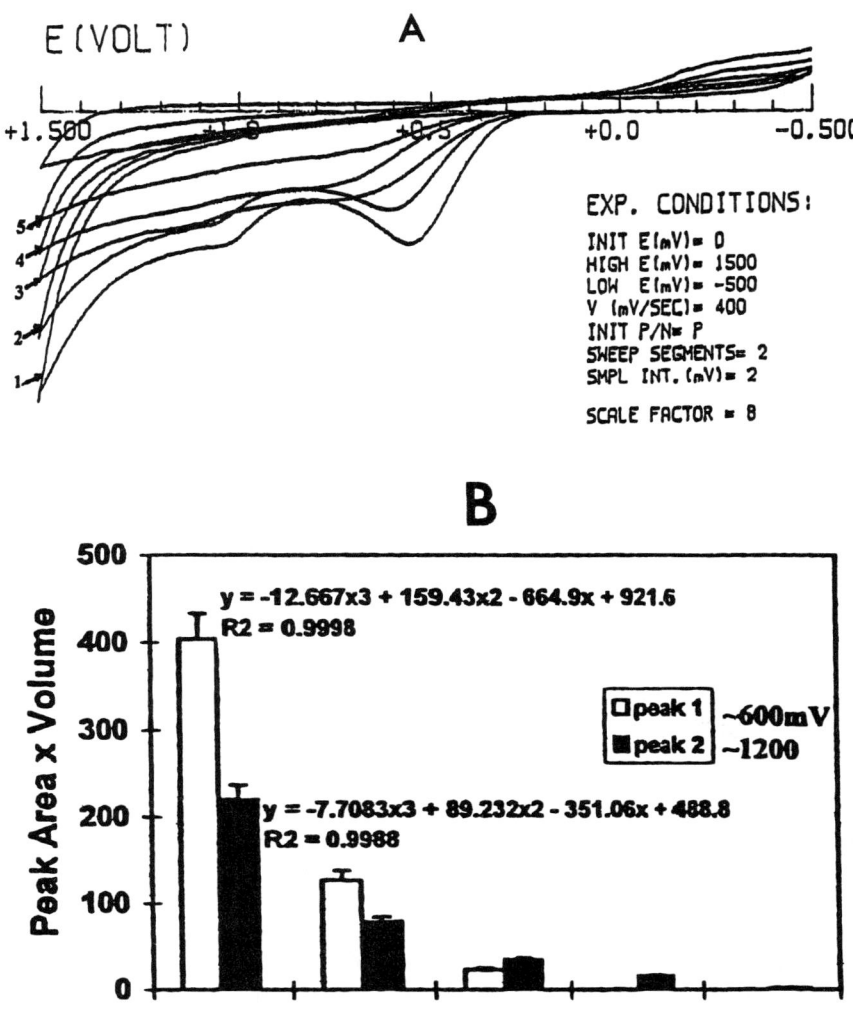

FIGURE 6. The extraction steps of cauliflower and their evaluation by CV. **A:** CV of the 5 consecutive steps needed to complete the extraction of cauliflower (sample weight 66 g, peak 1: $E = 568$mV; peak 2: $E = 1255$mV). The # denotes the number of the extraction step. **B:** $S \times V$, is the product of the area of the anodic current wave multiplied by the extraction volume, of the supernatant fraction for every extraction step, *versus* the number N for all the steps which were needed to complete the extraction process.

TABLE 3. The total amounts of LMWA in the edible plants tested, expressed as equivalents of AA per 100 g plant sample, as evaluated according to $\sum_j \sum_i S_{ij} V_{ij}$ ($i = 1-N; j = 1-3$) from the anodic waves

Plant	N	$\sum_j \sum_i S_{ij} V_{ij}$ $j = 1-3; i = 1-N$	Vit C (mg) equivalent units	Vit C (mg) USDA*
broccoli	5	1339	117.45	93.2
cauliflower	5	1374	120.52	46.4
strawberry	2	1934	169.64	56.7
tomato	4	260	22.8	19.1
potato	4	1397	122.54	19.7**
corn	4	148	12.98	6.8

$i = 1-N$; the number of extraction steps.
$j = 1-3$; the number of peaks in the CV tracings.
Vitamin C equivalent units (mg of vitamin C per 100 g sample) represent the actual vitamin C in addition to the other LMWA, in the plant extract.
*USDA Nutrient Database for Standard Reference, 12.[32] The actual vitamin C content in each of the edible plants.
**USDA value represents the content of vitamin C in flesh and skin of raw potato, while we have used only the flesh of raw potato.

CONCLUSIONS

We have shown several line of evidence that the total reducing power of human plasma of patients, as measured by the CV, well correlates with the severity of the disease. Also, CV parameters well correlate with the severity of an acute treatment the subject receives. This is based on that plasma, *ex vivo*, represents the *in vivo* antioxidant status (capacity) prior to the collection of the blood.

Enrichment of the diet of healthy subjects with an antioxidant, like ascorbate or thioctic acid, led to concomitant changes in I_a and R; an increase of the antioxidant capacity of the plasma, expressed as an increase of the intensity of the anodic wave current, I_a, and a decrease in the oxidative stress as reflected by a decrease in R.

It has been repeatedly confirmed that human diet, which is rich in fruits and vegetables, provides a marked health benefit, through lowering the incidence of cancer, heart disease and other pathologies. It has further been proposed that this effect stems from the increased nutritional supply of vitamins (and minerals). The vitamins include ascorbate (vitamin C), as well as vitamin E (the tocopherols), carotenoids, polyphenols, and others. Among those, the water soluble ones are represented in the family of LMWA, reported directly by the CV tracing. The lipid soluble antioxidants can also be visualized by the CV methodology (see below).

The CV is an efficient instrumental tool for evaluating the total (integrated) antioxidant capacity of LMWA in human plasma and animal tissues[23-27] as well as in edible plants (fruits and vegetables).[28]

The total reducing power of the sample is provided by the CV tracing, without the necessity to measure the specific antioxidant capacity of each component alone.

CV does not require the labor-intensive characterization of the activity of each component against a specific ROS, which in turn, serves as the basis for other methodologies measuring the total antioxidant capacity.

The preparation of a sample for CV measurement is simple and rapid, and does not require advanced procedures.

The CV methodology allows a rapid screening of many samples and preparations, and is especially suitable for screening studies.

The sensitivity of CV is sufficient for determining physiological concentrations of antioxidants.

The area (S) under the anodic current (related to the charge), rather than the anodic current itself (I_a) should be used for the evaluation of the total antioxidant capacity of a system, when the anodic wave is comprised of several components with similar, but different $E_{1/2}$ values. The CV of such a combination is represented by a broader anodic wave, rather than by an increase in I_a.

S, rather than I_a, is associated with the monitoring of the changes in the concentration of a specific LMWA component, within a wave, as a result of stress. The changes for individual components, within a single anodic current wave, could be different. The changes in S would better represent the residual antioxidant capacity, and will allow to better quanitate the loss in the specific component.

CV tracings can be obtained in aqueous medium, as well as in organic solvents like acetonitrile, or water/acetonitrile mixtures, provided that there is/are redox active component/s, and enough electrolytes in the solution to support redox reaction(s) on the electrode surface. This modification allows the quantitation of lipid soluble components as well.

The combination of acetic acid, acetonitrile and water is optimal for the quantitative extraction of antioxidants (both water and lipid soluble), and for quantitative evaluation of the antioxidant capacity.

In most cases, both blood plasma and edible plants extracts show anodic waves at similar oxidation potentials. This is in accord with that plants serve as sources for some antioxidant components of the human body, and that there is some similarity between the redox biochemical pathways of plants and mammalian tissues.

Individual components of the first anodic wave in human plasma have been identified. The complete identification and assignment of the individual component/s of the second wave in human plasma, as well as in other tissues, is of marked importance and awaits further investigation.

REFERENCES

1. HALLIWELL, B. & J.M.C. GUTTERIDGE. 1989. Protection against oxidants in biological systems: the superoxide theory of oxygen toxicity. In Free Radicals in Biology and Medicine. Oxford. Clarendon Press, pp. 86–179.
2. GUTTERIDGE, J.M.C. 1993. Free radicals in disease processes: a complication of cause and consequence. Free Rad. Res. Commun. **19:** 141–158.
3. AMES, B.M., M.K. SHIGENAGA & T.M. HAGEN. 1993. Oxidants, antioxidants and the degenerative diseases of aging. Proc. Natl. Acad. Sci. USA **90:** 7915–7922.
4. HALLIWELL, B. 1994. Free radicals and antioxidants: a personal view. Nutr. Rev. **52:** 253–263.

5. HALLIWELL, B. 1990. How to characterize a biological antioxidant. Free Rad. Res. Commun. **9:** 1–32.
6. CHEVION, M. 1988. A site-specific mechanism for free radical induced biological damage: The essential role of redox active transition metals. Free Rad. Biol. Med. **5:** 27–37.
7. CHANCE, P.A., H. SIES & A. BOVERIS. 1979. A hydroperoxide metabolism in mammalian organs. Physiol. Rev. **59:** 527–605.
8. AMES, B.M., R. CATHCART, E. SCHWIERS & P. HOCHSTEIN. 1981. Uric acid produces an antioxidant defense in humans against oxidant and radical-caused aging and cancer: a hypothesis. Proc. Natl. Acad. Sci. USA **73:** 6858–6862.
9. STOCKER, R., Y. YAMAMOTO, A. MCDONAGH, A.N. GLAZER & B.N. AMES. 1987. Bilirubin is an antioxidant of possible physiological importance. Science **235:** 1043–1045.
10. FREI, B., L. ENGLAND & B.N. AMES. 1989. Ascorbate is an outstanding antioxidant in human blood plasma. Proc. Natl. Acad. Sci. USA **86:** 6377–6381.
11. DOLL, R. 1990. An overview of the epidemiological evidence linking diet and cancer. Proc. Nutr. Soc. **49:** 119–131.
12. DRAGSTED, L.O., M. STRUBE & J.C. LARSEN. 1993. Cancer protective factors in fruits and vegetables: biochemical and biological background. Pharmacol. Toxicol. **72** (Suppl. 1)**:** 116–135.
13. WILLETT, C.W. 1994. Diet and health: what should we eat? Science **264:** 532–537.
14. WILLETT, C.W. 1994. Micronutrients and cancer risk. Am. J. Clin. Nutr. **59:** 162S–165S.
15. LA VECCHIA C., A. DECARLI & R. PAGANO. 1998. Vegetable consumption and risk of chronic disease. Epidemiology **9:** 208–210, 1998.
16. GEY, K.F. 1990. The antioxidant hypothesis of cardiovascular disease: epidemiology and mechanisms. Biochem. Soc. Trans. **18:** 1041–1045.
17. POSNER, B.M., M. FRANZ & P. QUATROMONI. 1994. The Interhealth Steering Committee. Nutrition and the global risk for chronic diseases: the interhealth nutrition initiative. Nutr Rev. **52:** 201–207.
18. HARATS, D., S. CHEVION, M. NAHIR, Y. NORMAN, O. SAGEE & E.M. BERRY. 1998. Citrus fruit supplementation reduces lipoprotein oxidation in young men ingesting a diet high in saturated fat: presumptive evidence for an interaction between vitamins C and E in vivo. Am. J. Clin. Nutr. **67:** 240–245.
19. RICE-EVANS, C. & N.J. MILLER. 1994. Total antioxidant status in plasma and body fluids. Methods Enzymol. **234:** 279–293.
20. WANG, H., G. CAO & R.I. PRIOR. 1996. Total antioxidant capacity of fruits. J. Agric. Food Chem. **44:** 701–705.
21. CAO, G., E. SOFIC & R.L. PRIOR. 1996. Antioxidant capacity of tea and common vegetables. J. Agric. Food Chem. **44:** 3426–3431.
22. CAO, G. & R.L. PRIOR. 1998. Comparison of different analytical methods for assessing total antioxidant capacity of human serum. Clin. Chem. **44:** (6) 1309–1315.
23. CHEVION, S., E.M. BERRY, N. KITROSSKY & R. KOHEN. 1995. Evaluation of plasma low molecular weight antioxidant capacity by cyclic voltammetry. International meeting on free radicals in health and disease. October 1995, Istanbul, Turkey. [abstract, no. 300]
24. CHEVION, S., E.M. BERRY, N. KITROSSKY & R. KOHEN. 1997. Evaluation of plasma low molecular weight antioxidant capacity by cyclic voltammetry. Free Rad. Biol. Med. **22:** 411–421.
25. CHEVION, S., M. HOFMANN, R. ZIEGLER, M. CHEVION & P.P. NAWROTH. 1997. The antioxidant properties of thioctic acid: characterization by cyclic voltammetry. Biochem. Mol. Biol. Int. **41:** 317–327.
26. CHEVION, S. 1998. The Antioxidant Status of Plasma in Health and Disease. Ph.D Thesis, The Hebrew University of Jerusalem, Jerusalem, Israel.
27. CHEVION, S., R. OR & E.M. BERRY. 1999. The antioxidant status of patients subjected to total body irradiation. Biochem. Mol. Biol. Int. **47**(6): 1019–1027.

28. CHEVION, S., M. CHEVION, P.B. CHOCK & G.R. BEECHER. 1999. The antioxidant capacity of edible plants: extraction protocol and direct evaluation by cyclic voltammetry. J. Med. Food. **2:** 1–110.
29. KOHEN, R., O. TIROSH & K. KOPOLOVICH. 1992. The reductive capacity index of saliva obtained from donors of various ages. Exp. Gerontol. **27:** 161–168.
30. KOHEN, R., O. TIROSH & R. GORODETSKY. 1992. The biological reductive capacity of tissues is decreased following exposure to oxidative stress: a cyclic voltammetry study of irradiated rats. Free Rad. Res. Commun. **17:** 239–248.
31. KOHEN, R. 1993. The use of cyclic voltammetry for the evaluation of oxidative damage in biological samples. J. Pharmacol. Toxicol. Methods **29:** 185–193.
32. USDA. USDA Nutrient Database for Standard Reference, 12. http://www.nal.usda.gov/fnic/foodcomp

Antioxidants in Nutrition

SLOBODAN V. JOVANOVIC[a] AND MICHAEL G. SIMIC[b,c]

[a]*Helix International, 381 Viewmount Drive, Nepean, Ontario, Canada K2E 7RG*
[b]*Gaithersburg, Maryland 20877, USA*

ABSTRACT: The harmful effects of oxidative processes in living organisms, in addition to chemical and biochemical media, can be reduced by antioxidants. The efficacy of an antioxidant depends on its reduction potential and kinetics of elimination of diverse free radicals. Redox potentials and reaction rate constants of selected gallocatechins and flavonoids were measured by pulse radiolysis and laser photolysis. The reduction potentials of the flavonoids studied were in the range of 0.33 V (quercetin) and 0.75 V (kaempferol). The rate constants of the superoxide radical with 15 flavonoids ranged from 10^5–10^7 $M^{-1} s^{-1}$. Singlet oxygen quenching by flavonoids was also very rapid (from 10^5 to 10^8 $M^{-1} s^{-1}$). These studies may be crucial in optimizing health and increasing longevity by reducing oxidative stress and biological damage.

INTRODUCTION

The intrinsic processes of conception, growth, and maintenance of living organisms are inextricably associated with deleterious oxidative processes. In most living organisms, oxygen is not only necessary for life but also a major cause of damaging processes and death.

To control oxidative processes, i.e. to reduce if not prevent their harmful effects, diverse antioxidants are ingested into organisms with foods. The average human lifespan is more than 82 years in Iceland and Japan but in some other countries may be less than 50 years. Nutrition is believed to be an important factor.[1–8] Consequently, optimal nutrition may be of primary importance in healthier and longer life.

There is no universal antioxidant. Hence, a complete understanding of the kinetics and reaction mechanisms of damaging processes, as well as their prevention, is necessary.

Efficient reduction of free radical processes by antioxidants can be achieved if the kinetics of inactivation of free radicals is sufficiently fast. The kinetics of these reactions is studied by fast time resolved techniques, such as pulse radiolysis and laser flash photolysis. Scavenging and inactivation of free radicals as well as repair of biological radicals often occur in the micro- (10^{-6} s) and millisecond (10^{-3} s) ranges.[2, 9,10] The kinetic parameters must be measured at physiological pH (in the pH range from 3 to 8).

The efficacy of antioxidants depends on their oxidation potentials (more appropriately defined as the reduction potentials of one-electron deficient daughter radicals). One-electron reduction potentials of antioxidant radicals are conveniently determined from equilibrium kinetics of electron transfer reactions, which are studied by pulse radiolysis and laser flash photolysis. Although the reduction potential is the major factor in determining the antioxidant capacity, the overall ability to

---

[c]*Author for correspondence: Dr. Michael G. Simic, 9404 Bac Pl., Gaithersburg, MD 20877.*

act as an antioxidant is also determined by other physicochemical characteristics. For example, in spite of inferior oxidation potential ($E_7 = 0.51$ V)[11] theaflavin reacts much faster with the superoxide radical, $k = 1 \times 10^7$ M^{-1} s^{-1}, than $k = 7.3 \times 10^5$ M^{-1} s^{-1} of epigallocatechin gallate with $E_7 = 0.43$ V.[12]

One of the major properties of antioxidants is their polarity and consequently their distribution between polar and non-polar media in the body. For example, vitamin E is strictly located in non-polar cell membranes. It would minimally protect cellular components outside of the membrane.

Another important factor is the bioavailability and pharmacokinetics of various antioxidants. Many antioxidants are readily absorbed through the gut and persist for some time in the blood either unchanged or as sulfated, methylated, conjugated or broken down metabolites.[13–15] However, the antioxidants which are not as readily absorbed, such as quercetin, may still have a very significant role in chemoprevention of colorectal carcinogenesis.

Despite the obvious importance of bioavailability and distribution of antioxidants in humans for their antioxidant and chemopreventive action, there are only a few studies on this subject. Considerably more needs to be done to understand the mechanisms of nutritional antioxidants *in vivo*.

We shall discuss here gallocatechins and flavonoids as examples of the complexities associated with antioxidants in biological systems.

FLAVONOIDS AS ANTIOXIDANTS

Antioxidants are a broad class of chemicals and biochemicals capable of preventing oxidation. Oxidation is a chain reaction initiated by free radical generators, such as ionizing and UV radiation, Fenton chemistry, or simply molecular oxygen. Free radicals, typically peroxyl, propagate chain oxidation, leading to massive chemical changes. A typical free radical chain reaction is presented below:

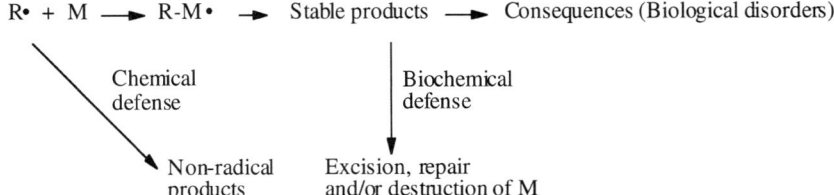

To avert the chain oxidation and minimize oxidative damage, antioxidants can intervene at several points along the chain of events,

1. inhibit initiation by blocking cellular free radical generators,
2. scavenge peroxyl radicals, converting them to hydroperoxides, which are then disposed of by glutathione peroxidase,
3. repair long-lived biological radicals before they are converted into stable products,
4. inhibit cellular expression of the mutagenic lesion,
5. induce apoptosis of damaged cells, and
6. induce and assist enzymatic antioxidants and detoxifying agents.

Unlike more specific and efficient antioxidants (such as vitamin E or A), which protect only cell membranes,[10] flavonoids and gallocatechins have many other antioxidant functions. First, they are fair to excellent free radical and oxy species interceptors in aqueous media.[9,16–19] Then, gallocatechins are very efficient iron chelators,[20] thus preventing redox cycling of iron and chain initiation. They can also inhibit enzymatic radical generators,[21,22] proliferation of cancer cells[23] and induce apoptosis of cancer cells.[16,17,24]

REDUCTION POTENTIALS OF FLAVONOID RADICALS

Flavonoids and gallocatechins, which are plant polyhydroxybenzenes, are fair to excellent electron donors. One-electron reduction potentials of flavonoid radicals at pH 7 (TABLE 1), which are equal to the oxidation potentials of parent flavonoids, range from 0.33 V for quercetin to 0.72 V for hesperidin.

On the basis of lower reduction potentials, all flavonoids may fully restitute any DNA base ($E_7 = 1.29$ V[25] for the most oxidizable base, guanosine) or protein amino acid radical ($E_7 = 0.9$ V[26] for the most vulnerable, tyrosine).

Alkyl peroxyl radicals with high $E_7 = 1.06$ V[27] are inactivated to hydroperoxides by electron transfer from lower oxidation potential flavonoids.

TABLE 1. Reduction potentials of flavonoid radicals at pH 7, versus NHE

Flavonoid radicals from	E_7, V[a]
Catechin	0.57
Taxifolin	0.5
Hesperidin	0.72
Theaflavin	0.51[b]
Epigallocatechin	0.42
Epigallocatechin gallate	0.43
Galangin	(0.62)
Rutin	0.6
Kaempferol	(0.75)
Morin	(0.6)
Fisetin	(0.33)
Luteolin	(0.6)
Quercetagetin	(0.33)
Quercetin	0.33
Myricetin	(0.36)

[a]From Reference 9.
[b]From Reference 11.

The oxidation potential of a parent flavonoid, which comprises a benzopyrene and one or several phenol rings, equals the lowest oxidation potential of a component ring.[28] For example, the oxidation potential of hesperidin ($E_7 = 0.72$ V) with the 4'-hydroxy-3'-methoxybenzene as the B ring, is similar to that of 2-methoxy-4-methylphenol, $E_7 = 0.68$ V.[28]

Reduction potentials of flavonoid radicals in neutral media are higher than $E_7 = 0.28$ V[29] of ascorbate. Consequently, flavonoid radicals may oxidize and deplete ascorbate *in vivo*.

Some flavonoids, epigallocatechin ($E_7 = 0.42$ V), epigallocatechin gallate ($E_7 = 0.43$ V)[12] and quercetin ($E_7 = 0.33$ V),[28] may repair vitamin E radical ($E_7 = 0.48$ V),[29] thus maintaining desirable concentration of this important physiological antioxidant.

INACTIVATION OF SUPEROXIDE RADICAL AND SINGLET OXYGEN

Flavonoids and gallocatechins are very efficient scavengers of the superoxide radical (TABLE 2).

The reactivities of flavonoids and gallocatechins with the superoxide radical are high, as expected from their excellent electron donating abilities. The highest rate constants were measured for theaflavins, k $\sim 10^7$ M^{-1} s^{-1},[11] very probably because of favorable charge separation within the benzocycloheptenone moiety in these

TABLE 2. Rate constants[a] for the reactions of the superoxide radical with flavonoids

Flavonoid, F	pH	$k(O_2^{-} + F)$, M^{-1} s^{-1}
Catechin	7	6.4×10^4
Epicatechin	7	6.8×10^4
Epigallocatechin	7	4.1×10^5
Epicatechin gallate	7	4.3×10^5
Epigallocatechin gallate	7	7.3×10^5
Theaflavin	7	1.0×10^{7b}
Catechin	10	1.8×10^4
Fisetin	10	1.3×10^4
Quercetin	10	4.7×10^4
Rutin	10	5.1×10^4
Hesperidin	10	2.8×10^4
Hesperetin	10	5.9×10^3
Kaempferol	10	2.4×10^3
Morin	10	1.6×10^3
Galangin	10	8.8×10^2

[a]From Reference 9.
[b]From Reference 11.

black tea derivatives. However, it should be emphasized that even the poorest electron donor, galangin, with the inactivation rate constant of $k = 8.8 \times 10^2$ M^{-1} s^{-1},[30] still efficiently scavenges superoxide radical.

The mechanism of the inactivation of the superoxide radical by flavonoids results in the formation of flavonoid phenoxyl radical[9,11,30] and hydrogen peroxide, as shown below:

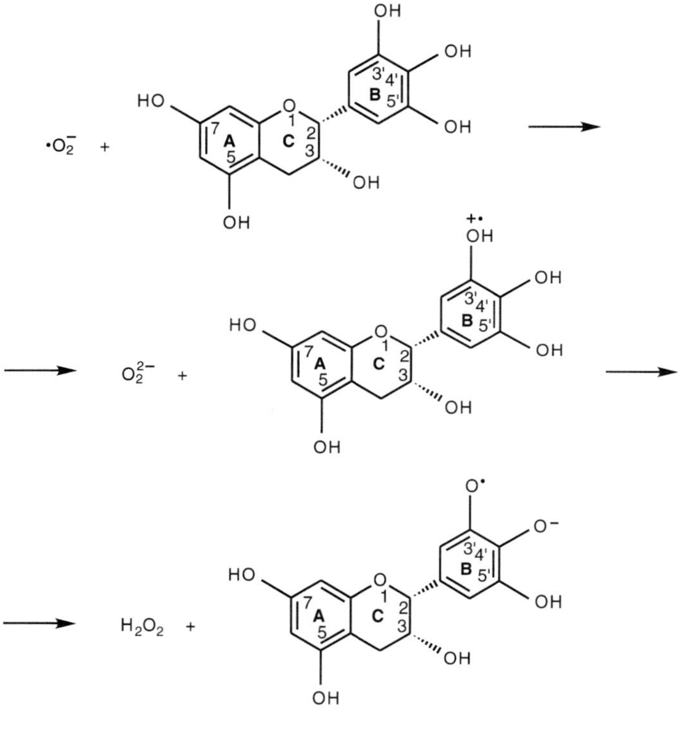

[Reaction scheme showing $\cdot O_2^-$ + flavonoid → O_2^{2-} + flavonoid radical cation → H_2O_2 + flavonoid phenoxyl radical]

This mechanism has been proven by gamma radiolysis measurements in the pH range from 7 to 10.[30]

The rate constants of the inactivation of the superoxide radical by flavonoids are low, $k \sim 10^2$-10^7 M^{-1} s^{-1}, in comparison with typical radical reactions ($k \sim 10^9$ M^{-1} s^{-1}). However, this is enough for very efficient *in vivo* scavenging of the superoxide radical. Given that the reactivities of the superoxide with the biological molecules are of the order of ~1 s^{-1} or less,[26] even at ~10^{-6} M concentrations[31] flavonoids would favorably compete for the superoxide ($k = 10^{-6}$ M × 10^7 M^{-1} s^{-1} = 10 s^{-1}).

Singlet oxygen, $^1O_2$, is a by-product of biological oxidation reactions and a component of photochemical smog. It may oxidize various biomolecules causing cell death and mutations.[6,32,33] Flavonoids are efficient scavengers of singlet oxygen (TABLE 3), very probably by a chemical reaction.

The rate constants of the reactions of $^1O_2$ with flavonoids, $k = 10^4 - 10^8$ M^{-1} s^{-1} are comparable to that of vitamin E, $k = 5 \times 10^8$ M^{-1} s^{-1}.[34] Higher rate constants were measured for flavonoids with lower oxidation potentials, e.g. k ($^1O_2$ + epigal-

TABLE 3. Singlet oxygen quenching by flavonoids[a]

Flavonoid, F	Solvent	$k(^1O_2 + F)_t$, M^{-1} s^{-1}
Quercetin	CD_3OD	2.4×10^6
Galangin	CD_3OD	1.2×10^6
Kaempferol	CD_3OD	7.1×10^5
Fisetin	CD_3OD	3.1×10^6
Fisetin	$CCl_3:CH_3OH=1:3$	1.9×10^8
Rutin	CD_3OD	1.6×10^6
Troxerutin	CD_3OD	7×10^4
Luteolin	CD_3OD	1.3×10^6
Chrysin	CD_3OD	2.4×10^5
Tangeretin	CD_3OD	2.4×10^5
Taxifolin	CD_3OD	1.1×10^6
Eriodyctol	CD_3OD	1.4×10^6
Naringenin	CD_3OD	5×10^4
Catechin	CD_3OD	5.8×10^6
Catechin	CH_3CN	1.1×10^8
Epicatechin	CH_3CN	9.6×10^7
Epigallocatechin	CH_3CN	1.1×10^8
Epicatechin gallate	CH_3CN	2.2×10^8
Epigallocatechin gallate	CH_3CN	2.2×10^8

[a]Total quenching (physical + chemical), taken from Reference 9.

locatechin gallate) = 2.2×10^8 M^{-1} s^{-1},[12] indicating that a polar transition state is involved in the quenching.

FLAVONOIDS AND GALLOCATECHINS AS PHYSIOLOGICAL ANTIOXIDANTS.

In order to function as physiological antioxidants flavonoids and gallocatechins have to be present at the right place in the right concentrations at the right time. A healthy human diet contains about 1 g of flavonoids from food and from 0 to 2 g of catechins and gallocatechins from tea. Approximately 80% of flavonoids and catechins is absorbed in the gut.[4,14,31,35–40] They are metabolized in the liver, where the B ring is sulfated or methylated, or they are conjugated to gluconorides or broken down to benzoic acids.[39–41] Most of the ingested flavonoids and catechins, about 90%, are excreted unchanged or metabolized.

Approximately 3% of ingested epigallocatechin gallate enters various rat organs.[42]

A very rough picture of distribution of flavonoids and gallocatechins is as follows: 20–40% stays in the colon, 80–60% enters the circulation and is cleared in the urine, ~3% enters the cells of various organs (mostly GI tract). The peak concentration in plasma and organs occurs in the rat at 0.5 h and at 24 h upon ingestion.[42]

The concentrations in various biological fluids and organs (of the order of 1 µmol/l in plasma[31]) are generally sufficient for any antioxidant or chemopreventive action of flavonoids or gallocatechins. It is interesting to note that high levels of tea polyphenols in tissues can be maintained by a repeated and frequent consumption of tea.[38]

There is little doubt that flavonoids have beneficial antioxidant and chemopreventive action in humans. However, what exactly is the mechanism(s) of this action is far from being fully understood.

The *in vitro* studies have conclusively shown that flavonoids and gallocatechins can scavenge damaging reactive oxygen (ROS) and nitrogen (NOS) species. The scavenging reaction generates a peroxide, which is harmful and has to be neutralized by catalase or glutathione peroxidase, and a flavonoid phenoxyl radical. Flavonoid phenoxyls are mildly oxidizing, resonance stabilized radicals, usually single negatively charged at pH 7.[9,11] Such radicals can be reduced by an appropriate electron donor, such as ascorbate, to regenerate the parent flavonoid. Flavonoid phenoxyls can react with each other, whereby various quinones and other products can be generated.[43–45] Little is known about the metabolism and clearance of stable products of flavonoid phenoxyl radicals.

Kinetic and mechanistic studies of certain gallocatechins and flavonoids may serve as models for the study of many other antioxidants, with the ultimate goal of determining essential nutritional antioxidant requirements for optimal health and increased longevity.

REFERENCES

1. HELZLSOUER, K.J., G. BLOCK, J. BLUMBERG, *et al.* 1994. Summary of the round table discussion on strategies for cancer prevention: diet, food, additives, supplements, and drugs. Cancer Res. (Suppl.) **54**: 2044s.
2. SIMIC, M.G. & S.V. JOVANOVIC. 1994. Inactivation of oxygen radicals by dietary phenolic compounds in anticarcinogenesis. Food Phytochemicals for Cancer Prevention II. Washington, D. C.: American Chemical Society.
3. STAVRIC, B. 1994. Role of Chemopreventers in human diet. Clin. Biochem. **27**: 319.
4. STAVRIC, B. & T.I. MATULA. 1992. Flavonoids in foods: their significance for nutrition and health. *In* Lipid-Soluble Antioxidants: Biochemistry and Clinical Applications. A.S.H. Ong & L. Packer, Eds. :274. Basel, Switzerland. Birkhäuser Verlag.
5. SIMIC, M.G. & D.S. BERGTOLD. 1991. Dietary modulation of DNA damage in human. Mutat. Res. **250**: 17.
6. SIMIC, M.G. 1994. DNA markers of oxidative processes in vivo: Relevance to carcinogenesis and anticarcinogenesis. Cancer Res. (suppl.) **54**: 1918s.
7. GRANT, W.B. 1997. Dietary links to Alzheimer's disease. Alz. Dis. Rev. **2**: 42.
8. SMITH, M.A., G.J. PETOT & G. PERRY. 1997. Commentary: diet and oxidative stress: A novel synthesis of epidemiological data on Alzheimer's disease. Alz. Dis. Rev. **2**: 58.
9. JOVANOVIC, S.V., S. STEENKEN, M.G. SIMIC & Y. HARA. 1998. Antioxidant properties of flavonoids: reduction potentials and electron transfer reactions of flavonoid radicals. *In* Flavonoids in Health and Disease. C.A. Rice-Evans & L. Packer, Eds. :137. New York. Marcel Dekker, Inc.
10. VON SONNTAG, C. 1987. The Chemical Basis of Radiation Biology. London. Taylor and Francis.

11. JOVANOVIC, S.V., Y. HARA, S. STEENKEN & M.G. SIMIC. 1997. Antioxidant potential of theaflavins. A pulse radiolysis study. J. Am. Chem. Soc. **119:** 5337.
12. JOVANOVIC, S.V., Y. HARA, S. STEENKEN & M.G. SIMIC. 1995. Antioxidant potential of gallocatechins. A pulse radiolysis and laser photolysis study. J. Am. Chem. Soc. **117:** 9881.
13. HARBORNE, J.B. & T.J. MABRY, Eds. 1982. The Flavonoids: Advances in Research. London. Chapman and Hall.
14. GRIFFITHS, L.A. 1982. Mammalian metabolism of flavonoids. In The Flavonoids: Advances in Research. J.B. Harborne & T.J. Mabry, Eds. :681. London. Chapman and Hall.
15. HARBORNE, J.B. 1988. The flavonoids: recent advances. In Plant Pigments. T.W. Goodwin, Ed. :299. London. Academic Press.
16. WEISBURGER, J.H., A. RIVENSON, C. ALIAGA, et al. 1998. Effect of green tea extracts, polyphenols, and epigallocatechin gallate on azoxymethane-induced colon cancer. Proc. Soc. Exp. Biol. Med. **217:** 104.
17. STEELE, V.E., S. SHARMA, C.W. BOONE, et al. 1996. Chemopreventive efficacy of black and green tea extracts in in vitro assays. Proc. Annu. Meet. Am. Assoc. Cancer Res. **37:** 1859.
18. OHSHIMA, H., Y. YOSHIE, S. AURIOL & I. GILIBERT. 1998. Antioxidant and pro-oxidant actions of flavonoids: effects on DNA damage induced by nitric oxide, peroxynitrite and nitroxyl anion. Free Rad. Biol. Med. **25:** 1057.
19. ISHIKAWA, T., M. SUZUKAWA, T. ITO, et al. 1997. Effect of tea flavonoid supplementation on the susceptibility of low-density lipoprotein to oxidative modification. Am. J. Clin. Nutr. **66:** 261.
20. JOVANOVIC, S.V., M.G. SIMIC, S. STEENKEN & Y. HARA. 1998. Iron complexes of gallocatechins. Antioxidant action or iron regulation? J. Chem. Soc. Perkin 2 **11:** 2365.
21. LIN, Y.L. & J.K. LIN. 1997. (−)-Epigallocatechin-3-gallate blocks the induction of nitric oxide synthase by down-regulating lipopolysaccharide-induced activity of transcription factor nuclear factor kappa B. Mol. Pharmacol. **52:** 465.
22. AUCAMP, J., A. GASPAR, Y. HARA & Z. APOSTOLIDES. 1997. Inhibition of xanthine oxidase by catechins from tea (*Camellia sinesis*). Anticancer Res. **17:** 4381.
23. LU, L.H., S.S. LEE & H.C. HUANG. 1998. Epigallocatechin suppression of proliferation of vascular smooth muscle cells: correlation with c-jun and JNK. Br. J. Pharmacol. **124:** 1227.
24. AHMAD, N., D.K. FEYES, A.L. NIEMINEN, R. AGARWAL & H. MUKHTAR. 1997. Green tea constituent epigallocatechin-3-gallate and induction of apoptosis and cell cycle arrest in human carcinoma cells. J. Natl. Cancer Inst. **89:** 1881.
25. STEENKEN, S. & S.V. JOVANOVIC. 1997. How easily oxidizable is DNA? One-electron reduction potentials of adenosine and guanosine radicals in aqueous solution. J. Am. Chem. Soc. **119:** 617.
26. JOVANOVIC, S.V. & L. JOSIMOVIC. 1992. Radiation chemistry of fatty and amino acids. In The Chemistry of Acid Derivatives, Suppl. B. S. Patai, Ed. :1199. London. John Wiley and Sons.
27. JOVANOVIC, S.V., I. JANKOVIC & L. JOSIMOVIC. 1992. Electron transfer reactions of alkyl peroxy radicals. J. Am. Chem. Soc. **114:** 9018.
28. JOVANOVIC, S.V., S. STEENKEN, Y. HARA & M.G. SIMIC. 1996. Reduction potentials of flavonoid phenoxyl radicals. Which ring is responsible for antioxidant activity? J. Chem. Soc. Perkin 2. 2497.
29. STEENKEN, S. & P. NETA. 1982. One-electron redox potentials of phenols. Hydroxy- and aminophenols and related compounds of biological interest. J. Phys. Chem. **86:** 3661.
30. JOVANOVIC, S.V., S. STEENKEN, M. TOSIC, B. MARJANOVIC & M.G. SIMIC. 1994. Flavonoids as antioxidants. J. Am. Chem. Soc. **116:** 4846.
31. PAGANGA, G. & C.A. RICE-EVANS. 1997. The identification of flavonoids as glycosides in human plasma. FEBS Lett. **401:** 78.
32. GORMAN, A.A. & M.A.J. RODGERS. 1989. Singlet oxygen. In Handbook of Organic Photochemistry. J.C. Scaiano, Ed. :229. Boca Raton, FL. CRC Press.

33. DEVASAGAYAM, T.P.A., S. STEENKEN, M.S.W. OBENDORF, W.A. SCHULZ & H. SIES. 1991. Formation of 8-hydroxy(deoxy)guanosine and generation of strand breaks at guanine residues in DNA by singlet oxygen. Biochemistry **30:** 6283.
34. GORMAN, A.A., I.R. GOULD, I. HAMBLETT & M.C. STANDEN. 1984. reversible exciplex formation between singlet oxygen, $^1\Delta_g$, and vitamin E. Solvent and temperature effects. J. Am. Chem. Soc. **106:** 6956.
35. CHEN, L., M.J. LEE, H. LI & C.S. YANG. Absorption, distribution, elimination of tea polyphenols in rats. Drug Metab. Dispos. **25:** 1045.
36. NAKAGAWA, K. & T. MIYAZAWA. 1997. Absorption and distribution of tea catechin, (−)-epigallocatechin-3-gallate, in the rat. J. Nutr. Sci. Vitaminol. (Tokyo) **43:** 679.
37. NAKAGAWA, K., S. OKUDA & T. MIYAZAWA. 1997. Dose-dependent incorporation of tea catechins, (−)-epigallocatechin-3-gallate and (−)-epigallocatechin, into human plasma. Biosci. Biotechnol. Biochem. **61:** 1981.
38. SUGANUMA, M., S. OKABE, M. ONIYAMA, Y. TADA, H. ITO & H. FUJIKI. 1998. Wide distribution of [^{3}H](−)-epigallocatechin gallate, a cancer preventive tea polyphenol, in mouse tissue. Carcinogenesis **19:** 1771.
39. PIETTA, P.G., P. SIMONETTI, C. GARDANA, A. BRUSAMOLINO, P. MORAZZONI & E. BOMBARDELLI. 1998. Catechin metabolites after intake of green tea infusions. Biofactors **8:** 111.
40. HE, Y.H. & C. KIES. 1994. Green and black tea consumption by humans: impact on polyphenol concentration in feces, blood and urine. Plant Foods Human Nutr. **46:** 221.
41. WERMIELLE, M., E. TURIN & L.A. GRIFFITHS. 1983. Identification of the major urinary metabolites of (+)-catechin and 3-O-methyl-(+)-catechin in man. Eur. J. Drug Metabol. Pharmacokin. **8:** 77.
42. KOHRI, T., C.C. CONAWAY, D. DESAI, S. AMIN & Y. HARA. 1998. Uptake, tissue distribution and excretion of green tea polyphenol EGCG in F344 rats. AACR Meeting, March 1998.
43. DEEBLE, D.J., B.J. PARSONS, G.O. PHILLIPS, H.-P. SCHUCHMANN & C. VON SONNTAG. 1988. Pulse radiolysis of pyrogallol. Int. J. Radiat. Biol. **54:** 179.
44. YASUDA, H. & T. ARAKAWA. Deodorizing mechanism of (−)-epigallocatechin gallate against methyl mercaptan. Biosci. Biotechnol. Biochem. **59:** 1232.
45. SAWAI, Y. & K. SAKATA. 1998. NMR analytical approach to clarify the antioxidative molecular mechanism of catechin using 1,1-diphenyl-2-picrylhydrazyl. J. Agric. Food Chem. **46:** 111.

Free Radical Intermediates in Sonodynamic Therapy

VLADIMÍR MIŠÍK[a] AND PETER RIESZ[b,c]

[a]*Institute of Experimental Pharmacology, Slovak Academy of Sciences, 84216 Bratislava, Slovak Republic*

[b]*Radiation Biology Branch, National Cancer Institute, NIH, Bethesda, Maryland 20892-1002, USA*

ABSTRACT: Current understanding of the mechanism of sonodynamic action (i.e. the ultrasound-dependent enhancement of the cytotoxic action of certain drugs - sonosensitizers) with potential applications for cancer therapy is presented. The experimental evidence suggests that sonosensitization is due to the chemical activation of sonosensitizers inside or in the close vicinity of hot collapsing cavitation bubbles to form sensitizer-derived free radicals either by direct pyrolysis or due to reactions with ·H and ·OH radicals, formed by pyrolysis of water. These free radicals (mostly carbon-centered) react with oxygen to form peroxyl and alkoxyl radicals. Unlike ·OH and ·H, which are also formed by pyrolysis inside cavitation bubbles, the reactivity of alkoxyl and peroxyl radicals with organic components dissolved in biological media is lower and hence have higher probability of reaching critical cellular sites. Sonodynamic therapy appears to be a promising modality for cancer treatment since ultrasound can penetrate deep within the tissue and can be focused in a small region of tumor to chemically activate relatively non-toxic molecules (e.g. porphyrins) thus minimizing undesirable side effects.

INTRODUCTION

Sonodynamic therapy is a promising new modality for cancer treatment based on the synergistic effect on cell killing by the combination of a drug (a sonosensitizer) and ultrasound.[1–11] The effectiveness of sonodynamic therapy has been demonstrated in cell studies and in tumor-bearing animals (TABLE 1). The mechanism of this drug-dependent sonosensitization is unknown, but it seems likely that various mechanisms of sonosensitization operate for different classes of sonosensitizers. Here we outline our hypothesis supported by experimental data that free radicals produced from sonosensitizers by ultrasound play a role in sonodynamic action.

[c]Address for correspondence: Dr. Peter Riesz, National Cancer Institute, NIH, Building 10, Room B3-B69, Bethesda, MD 20892-1002. Voice: 301-496-4036; fax: 301-480-2238.
e-mail: sono@helix.nih.gov

TABLE 1. Synergistic effect of drugs and ultrasound in cancer research

Compound	Experimental system	Reference
Nitrogen mustard	Inoculation of mice with mouse leukemia L1210 cells	[1]
Daunomycin	Rats bearing Yoshida sarcoma	[2]
Adriamycin	Rats bearing Yoshida sarcoma	[2]
	Fibrosarcoma (RIF-1) or melanoma (B-16) bearing mice	[3]
	V79 chinese hamster fibroblast cells	[4]
	CHO and MCF-7 WT cells	[5]
	Uterine cervical squamous cell carcinoma implanted in the cheek pouch of Syrian hamster	[5]
Diaziquone	CHO and MCF-7 WT cells	[5]
	Uterine cervical squamous cell carcinoma implanted in the cheek pouch of Syrian hamster	[5]
Hematoporphyrin	Mice bearing sarcoma 180	[6]
Photofrin	HL-60 cells	[7]
	Adult T-cell leukemia cells	[8]
Ga-porphyrin ATX-70	Isolated sarcoma 180 cells	[9]
Various porphyrins	Murine leukemia L1210 cells	[10]
DMF, DMSO Methylformamide	HL-60 human promyelocytic leukemia cells	[11]

EXPERIMENTAL

Chemicals

The nitroso spin trap 3,5-dibromo-4-nitrosobenzene sulfonic acid, sodium salt (DBNBS) was obtained from OMRF Spin Trap Source, Oklahoma. Hematoporphyrin IX (HP), *meso*-tetra (4-sulfonatophenyl) porphine (MTSPP), protoporphyrin IX (PPIX) were acquired from Porphyrin Products, Logan, UT. The gallium porphyrin photosensitizer ATX-70 (7,12-bis(1-decyloxyethyl)-3,8,13,17-tetramethyl-porphyrin-2,18-bispropyonylasparginic acid gallium(III) salt, purity > 95%) was a generous gift from Dr. Isao Sakata, Toyo Hakka Kogyo Ltd. (Okayama, Japan).

Ultrasound Exposure

The details of experimental conditions will be specified as appropriate. In most experiments, samples (0.8 ml) containing the spin trap were fixed in the center of the Bransonic 1200 47 kHz sonication bath (temperature of the coupling water 20°C) in a Pyrex test-tube and bubbled with argon or other gases before and during exposure to ultrasound. For comparison, sonolysis of 0.8 ml of aqueous argon-bubbled ferrous sulfate Fricke dosimeter solution[12] for 5 minutes in this experimental set-up gave an absorbance of 0.536 ± 0.050 at 302 nm in a 1 cm cell.

EPR Measurements

Immediately after sonication the samples were transferred to EPR quartz flat cells and the acquisition of the spectrum started, typically within 1 minute after the end of sonication. A Varian E-9 X-band spectrometer with 100-kHz modulation frequency and a microwave power of 20 mW was used to record the spectra. The EPR software EPRDAP, written by Dr. Kuppusamy (U.S. EPR, Inc., Clarksville, MD), was used for acquisition, analysis and simulation of EPR data.

RESULTS AND DISCUSSION

The biological effects of ultrasound (pulsed or continuous) are due to one or a combination of several following factors (FIG. 1):

(i) thermal effects—due to the absorption and dissipation of ultrasound energy:[13] applications include hyperthermia treatment and physical therapy;

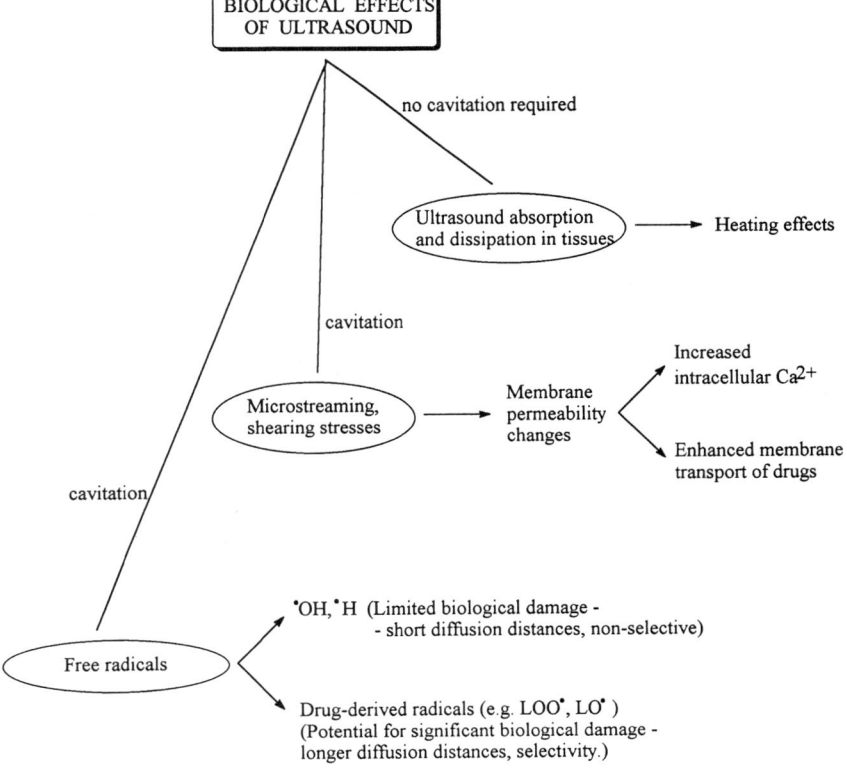

FIGURE 1. Overview of biological effects of ultrasound that may play a role in ultrasound-dependent activation of drugs.

(ii) cell membrane permeability changes and/or cell membrane rupture—due to shearing forces resulting from stable cavitation (oscillations of pre-existing gas bodies in response to varying acoustic pressure) or inertial cavitation (bubbles grow and implode violently after reaching a critical size): applications include cell membrane disruption and sonophoresis, which allows transmembrane transport of large molecules;[14,15]

(iii) free radical effects—due to the formation of free radicals by pyrolysis of molecules present inside collapsing cavitation microbubbles.

The mechanisms of free radical formation by ultrasound will be discussed in greater detail in the following paragraph.

Pressure changes associated with ultrasound traveling through liquids force microbubbles that are present in the medium to oscillate at the frequency of the sound field. During this process, the bubble size increases by a process known as rectified diffusion until it reaches resonant radius.[16] At this point the bubble rapidly expands and collapses in less than a half cycle of the sound frequency. During this violent compression phase extremely high temperatures and pressures are formed in the bubble interior.[17–19] Chemical effects and a number of biological effects of ultrasound are mediated by this process of bubble formation, growth and collapse, called cavitation.[20,21] The bubble interior represents a combustion-chemistry microreactor in which pyrolysis of the molecules present inside the bubbles (such as gases dissolved in the liquid and vapors of the liquid) takes place to produce free radicals and new molecules formed by reactions of these radicals.[20] Assuming adiabatic collapse of a spherical bubble, the final temperature of the gas within the bubble is a function of γ, which is the ratio of the heat capacity at constant pressure (C_p) and the heat capacity at constant volume (C_V). Additionally, high heat conductivity of the gas/vapor will enhance heat transport from the bubble interior to the liquid, thus resulting in lowering the temperature of the collapse. The heat carried away from the bubble interior produces heating of the thin shell of liquid adjacent to the bubble. These temperatures may also be sufficient to pyrolyse molecules that accumulate at the gas-liquid interface, thus producing free radicals.[22] Free radicals produced in the hot regions (such as ·OH radicals and ·H atoms in aqueous solutions) that do not recombine or disproportionate, will react with other molecules dissolved in the bulk of the solution by reactions known from radiation chemistry.[20]

The genotoxic potential of ·OH radicals is known from radiation biology and is used to kill cancer cells in radiation therapy. About 2/3 of the X-ray damage to DNA in mammalian cells (for sparsely ionizing radiation such as X-rays) is due primarily to ·OH radicals.[23] It seems likely that this damage includes some contribution from ·H atoms and hydrated electrons.[24]

In aqueous sonochemistry ·OH radicals are the radical species primarily responsible for reactions in the bulk of the liquid at ambient temperatures and pressures, with a smaller contribution from ·H atoms. The reason for this disparity is not immediately apparent, since equal amounts of ·H and ·OH are produced initially by water pyrolysis inside cavitation bubbles:

$$H_2O \rightarrow \cdot H + \cdot OH$$

The following factors may be involved:

(i) while a highly thermally stable end-product, H_2 (bond dissociation energy, $E_a = 402$ kJ), is formed by recombination of ·H atoms, a relatively labile end-product of recombination of ·OH radicals, H_2O_2, may undergo O-O bond homolysis ($E_a = 190$ kJ) in the interfacial region of the cavitation bubble, to regenerate ·OH radicals;

(ii) if the temperature is high and the number of water molecules inside cavitation bubbles exceeds the number of radicals, a considerable part of ·H atoms will be converted to ·OH radicals (·H + H_2O → H_2 + ·OH) before they can enter the solution;[25]

(iii) in the high temperature region inside cavitation bubbles ·H atoms may be converted to ·OH radicals by the following mechanism: ·OH + ·OH → ·O· + H_2O (disproportionation at high temperature), followed by ·O· + ·H → ·OH.

Although cavitation-producing ultrasound can generate ·OH radicals and ·H atoms, there is only limited evidence about the genotoxic potential of ultrasound. The reason for this apparent paradox (genotoxicity if radicals are produced by radiation, limited or no genotoxicity if these radicals are produced by ultrasound) is the site of their production: unlike ionizing radiation and photodynamic exposure, where free radicals and singlet oxygen respectively can be produced intracellularly, exposure of cells to ultrasound results in extracellular production of free radicals. There is no conclusive evidence so far that cavitation can also occur intracellularly, but such an event would lead to immediate cell destruction because the resonant size of the cavitation bubbles, in the low MHz range of ultrasound, is comparable to the size of cells.[26] Hence, there would be no role for intracellular free radical damage, since such cells would simply be ruptured by the rapidly expanding bubbles within. Therefore, if free radicals are involved in biological damage by ultrasound, the resulting damage is likely to depend on extracellularly produced reactive intermediates reaching the cell membrane, or even penetrating it, to destroy important intracellular targets.

Studies on the Mechanism of Sonosensitization of N,N-Dimethylformamide

Recently, Jeffers *et al.* demonstrated that N,N-dimethyl formamide (DMF, FIG. 2), a widely used polar solvent, exhibits synergistic killing of HL-60 human promyelocytic leukemia cells when combined with ultrasound.[11] Anti-tumor properties of DMF in the absence of ultrasound have been reported,[27] but its potential use for cancer therapy was limited due to the hepatotoxic effect of DMF. Jeffers *et al.* demonstrated the cell killing by DMF and ultrasound at DMF concentrations (0.11 M), which by themselves were non-toxic, thus improving the prospects for therapeutic application of DMF for locally intense chemotherapy.[11] Other polar solutes, N-methylformamide (MMF, FIG. 2) and dimethylsulfoxide (DMSO, FIG. 2), also showed a moderate toxicity when sensitized by ultrasound, but their effects were weaker compared to DMF.[11] Using a careful experimental design, Jeffers *et al.* eliminated the bulk solution heating effect of ultrasound. The observed cytotoxic effect of DMF required the presence of acoustic cavitation, but was not due to the enhancement of cell susceptibility to shear forces, which are associated with oscillating bubbles, or to the changes of other sonomechanical parameters by DMF. These data suggest that sonochemical activation of these solutes plays a role. Therefore, Jeffers *et al.*

DMF $H_3C-N(CH_3)-CHO$

MMF $H_3C-NH-CHO$

DMSO $H_3C-S(=O)-CH_3$

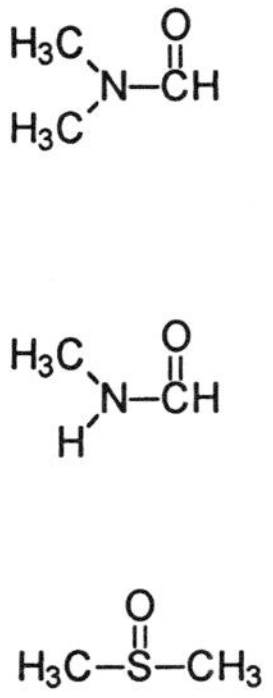

FIGURE 2. Chemical structures of low-molecular weight sonosensitizers, DMF, MMF, and DMSO.

concluded that unknown short-lived intermediates produced from these solutes by ultrasound are responsible for the toxicity, since no toxic effect was detected when sonicated solutions containing these solutes were subsequently added to the cells.[11]

We have investigated the short-lived intermediates proposed to be responsible for ultrasound mediated HL-60 cell killing by DMF, MMF, and DMSO. Using the spin trap 3,5-dibromo-4-nitrosobenzene sulfonate (DBNBS) in nitrogen-saturated aqueous solutions of DMF, MMF, or DMSO exposed to 47 kHz ultrasound, we detected formation of ·CH_3 and ·$CH_2N(CH_3)CHO$ radical adducts for DMF, mostly ·CH_2NHCHO adducts for MMF, and ·CH_3 adducts for DMSO.[28] These radicals were formed either by reactions of the solutes with ultrasound-generated ·H and ·OH radicals (such as ·CH_2R-type radicals in DMF and MMF, and ·CH_3 radicals in DMSO), or by direct pyrolysis of the weak bonds in the solute molecules (e.g. ·CH_3 radicals from DMF). In air-saturated sonicated solutions these carbon centered radicals were converted to the corresponding peroxyl radicals and spin trapped with 5,5-dimethyl-1-pyrroline-N-oxide (DMPO); ·$OOCH_2N(CH_3)CHO$ radicals were identified in DMF, ·$OOCH_2NHCHO$ radicals in MMF and ·$OOCH_3$ radicals in DMSO solutions.[28] We suggest that these radical species by virtue of their longer lifetimes and higher selectivity, compared to ·OH radicals which are also formed in sonicated solutions, are the species responsible for sonodynamic cell killing by the combined effect of ultrasound with DMF, MMF, or DMSO.

Investigation of the Mechanisms of Sonodynamic Action of Porphyrins

As discussed earlier, porphyrins are another class of sonosensitizers, however, the mechanism of their action is obscure. Umemura *et al.* have proposed that the sonoluminescent light that is produced during cavitational collapse of microbubbles is responsible for the photoexcitation of the sensitizer, with subsequent formation of singlet oxygen, a known reactive toxic species.[9] They suggested that the enhancement of sonochemical yields of TMP-NO nitroxide (2,2,6,6-tetramethyl-4-piperidone-N-oxyl) formed from the secondary amine TMP (2,2,6,6-tetramethyl-4-piperidone hydrochloride) in aqueous solutions of a gallium porphyrin derivative ATX-70 (FIG. 3) exposed to ultrasound was evidence for the formation of singlet oxygen in the system. We have investigated the aqueous sonochemical reactions of ATX-70 using 47 kHz ultrasound.[29] The experiments were carried out in the

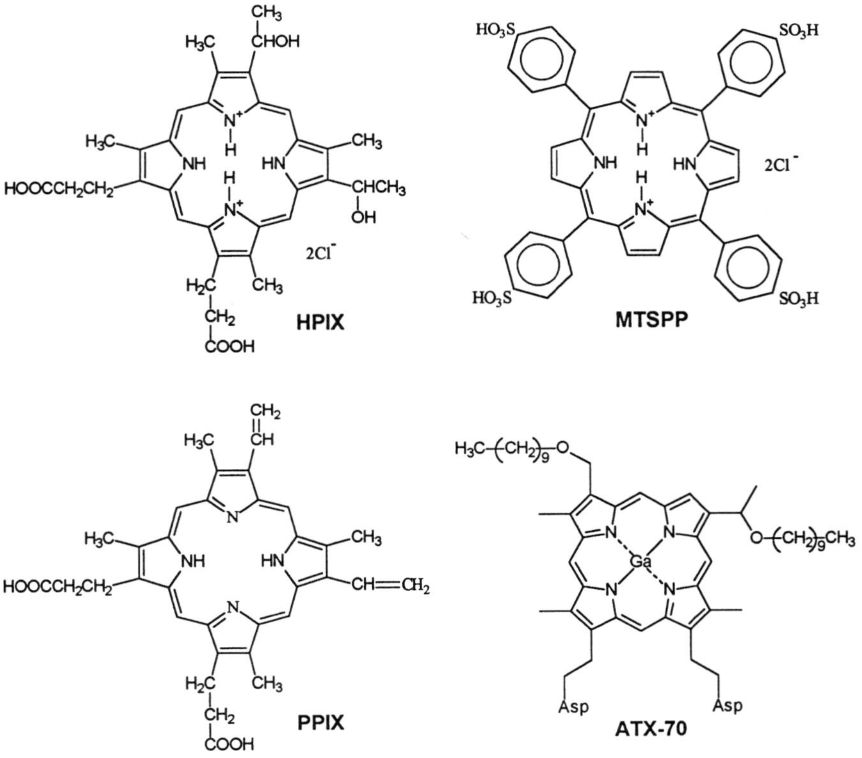

FIGURE 3. Chemical structures of selected porphyrins.

presence of TMP, which reacts with singlet oxygen or ·OH radicals to give the EPR-observable nitroxide TMP-NO. Our results show that the surfactant cetyltrimethylammonium bromide (CTAB) can mimic the ATX-70-induced increase of the TMP-NO signal, but it fails to reproduce the behavior of ATX-70 in D_2O: while the yields of TMP-NO in the presence of ATX-70 increase in D_2O, the opposite effect was found with the surfactant CTAB. However, our data show that the increased TMP-NO yields in D_2O are paralleled by an increased concentration of ATX-70-dimer, a form that is inactive in the photochemical generation of singlet oxygen. Our finding that the ATX-70-dependent enhancement of the TMP-NO signal was highest at ~20% O_2, both in N_2/O_2 and argon/O_2 mixtures, and decreased with increasing oxygen concentration, is not compatible with the singlet oxygen mechanism. Finally, our results on the temperature dependence of the ATX-70-induced TMP-NO formation are not consistent with the photochemical excitation of ATX-70 by sonoluminescent light: the ATX-70-dependent enhancement of the TMP-NO signal increased with temperature in the range 10-25°C, while the intensity of sonoluminiscence of aqueous solutions both in multiple bubble fields and in single bubble experiments is known to decrease with increasing temperature.[30,31]

Recently we have studied the effect of Ga-porphyrin ATX-70 on the sonosensitization in human leukemia HL-525 cells.[32] We found that low concentrations (< 10 µM) of gallium porphyrin ATX-70 significantly enhanced cellular toxicity in HL-525 cells exposed to 47 kHz ultrasound. The mechanism of this ATX-70–dependent sonosensitization is unknown, but we have established the requirement of extracellular localization of ATX-70 molecules for sonosensitization. Short-lived toxic intermediates produced from ATX-70 by ultrasound are implicated in the mechanism, since no cytotoxicity was found when medium containing ATX-70 was sonicated and subsequently added to the cells.

Radical intermediates from porphyrins resisted identification for some time. However, recently we were able to spin trap porphyrin-derived radicals from hematoporphyrin IX (HP), *meso*-tetra (4-sulfonatophenyl) porphine (MTSP),

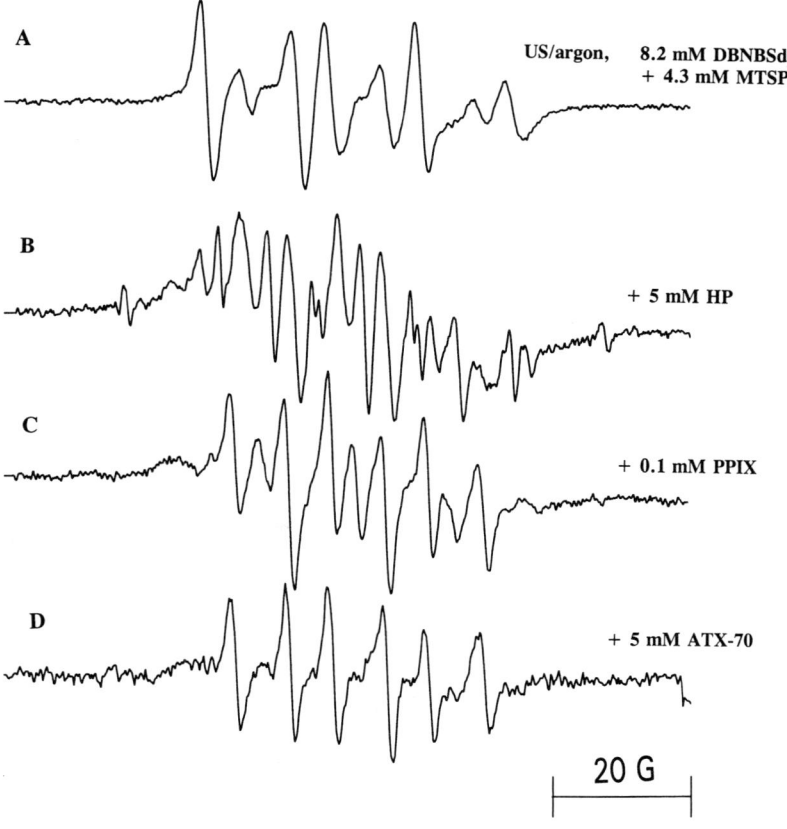

FIGURE 4. EPR spectra of porphyrin-derived radical adducts of DBNBS, formed by 10 minutes of 47kHz ultrasound exposure in argon-saturated aqueous solutions of porphyrins containing 8.2 mM DBNBS. See TABLE 2 for identification of the EPR spectra. The following concentrations of the porphyrins were used to obtain the spectra: **A:** 4.3 mM MTSPP; **B:** 5 mM HP; **C:** 0.1 mM PPIX; **D:** 5 mM ATX-70.

protoporphyrin IX (PPIX), and ATX-70 (their structures are shown in FIG. 3). Porphyrins were dissolved in 0.1 N NaOH and pH was adjusted to neutral with 1N HCl. The spin trap 3,5-dibromo-4-nitrosobenzene sulfonate (DBNBS, 8.2 mM) was added to the solutions of the porphyrins, samples were partially deoxygenated by 3 minute pre-bubbling with argon, followed by 10 minutes of ultrasound exposure with simultaneous argon-bubbling. EPR spectra of the spin trapped porphyrin-derived radicals are shown in FIGURE 4. Identification of the types of the spin adducts based on successful simulations (not shown) of the experimental spectra and a comparison with the literature data,[33] is shown in TABLE 2.

Water Soluble Azo-Compounds—a Promising New Class of Sonosensitizers

Water-soluble azo-compounds (R-N=N-R') are known to be thermally labile and decompose to form carbon centered radicals and ultimately peroxyl radicals (ROO·) if the decomposition is carried out in the presence of a dissolved oxygen:[34]

$$R\text{-}N=N\text{-}R' \xrightarrow{heat} \cdot R + \cdot R' + N_2$$
$$\cdot R + O_2 \to ROO\cdot$$

These radicals have a known cytotoxic potential and, therefore, water soluble azo-compounds have been proposed to be useful sensitizers for hyperthermia treatment.[35] We have investigated the feasibility of using focused ultrasound for site-specific decomposition of azo-compounds by studying the formation of free radicals during the decomposition of several water-soluble azo compounds (FIG. 5) by 47 kHz ultrasound in aqueous solutions.[36] Using the spin trap 3,5-dibromo-4-nitrosobenzene sulfonate (DBNBS) tertiary carbon-centered radicals from 2,2'-azo-

TABLE 2. DBNBS spin adducts of porphyrin-derived radicals formed by ultrasound exposure[a]

Porphyrin	Type of DBNBS adduct (hfc)	
HP	·CH_3	$a_N = 14.4$ G; $3a_H = 13.5$ G; $2a_H^m = 0.8$ G
	·CH_2R	$a_N = 14.2$ G; $2a_H = 9.6$ G
	·CHR_1R_2	$a_N = 13.9$ G; $a_H = 7.2$ G
	·C-tert	$a_N = 13.6$ G
MTSPP	·CHR_1R_2	$a_N = 18.2$ G; $a_H = 13.3$ G
	·C-tert	$a_N = 17.3$ G
PPIX	·CH_2R	$a_N = 13.2$ G; $2a_H = 11.8$ G
	·CHR_1R_2	$a_N = 14.06$ G; $a_H = 7.8$ G
	·C-tert	$a_N = 13.9$ G
ATX-70	·CHR_1R_2	$a_N = 14.3$ G; $a_H = 8.2$ G

[a]Samples containing 8.2 mM spin trap DBNBS were partially deoxygenated by 3 min pre-bubbling with argon, followed by 10 min of 47 kHz ultrasound exposure with simultaneous argon-bubbling.

bis (N,N'-dimethyleneisobutyramidine) dihydrochloride (VA-044), 2-(carbamoylazo)-isobutyronitrile (V-30), and 2,2'-azobis (2-amidinopropane) dihydrochloride (AAPH) and ·CH$_3$ radicals from 1,1'-azobis (N,N'-dimethylformamide) (ADMF) were detected in argon-saturated solutions and the corresponding oxygen-centered radicals from VA-044, V-30, and AAPH were identified using the spin trap 5,5'-dimethyl-1-pyrroline-N-oxide (DMPO) in aerated sonicated solutions. No free radicals from 4,4'-dihydroxyazobenzene-3,3'-dicarboxylic acid, disodium salt (DHAB) could be found in either system. While VA-044 and AAPH could also be readily decomposed by heat (42.5°C and 80°C), V-30 decomposition only occurred in the ultrasound-exposed solutions. The most likely mechanism of decomposition of azo compounds by ultrasound is their thermolysis in the heated shell of the liquid surrounding cavitating bubbles driven by ultrasound and/or by pyrolysis inside these bubbles. Experiments using scavengers of ·OH and ·H, which are produced by sonolysis in aqueous solutions, demonstrated that these radicals are not involved in the ultrasound-mediated radical production from the azo compounds. Owing to the known cytotoxic potential of free radicals produced from azo compounds,[34] the use of these compounds as ultrasound sensitizers appears to be a promising approach for sonodynamic cell killing.

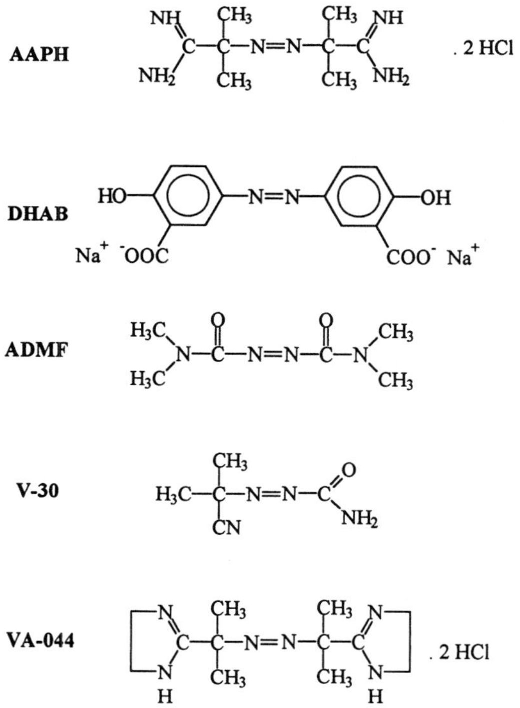

FIGURE 5. Chemical structures of selected water-soluble azo-compounds.

SUMMARY

Here we propose that the synergistic action of certain compounds (sonosensitizers) with ultrasound for cancer treatment (i.e. sonodynamic action) is due to the activation of these compounds by ultrasound. The proposed mechanism of sonosensitization is outlined in FIGURE 6. We propose that sonosensitization is due to the chemical activation of sonosensitizers inside or in the close vicinity of hot collapsing cavitation bubbles to form sensitizer-derived free radicals. These free radicals (mostly carbon-centered) react with oxygen to form peroxyl and alkoxyl radicals. Unlike ·OH and ·H, which are also formed by pyrolysis inside cavitation bubbles, the reactivity of alkoxyl and peroxyl radicals with organic components dissolved in biological media is lower and hence they have a higher probability of reaching critical cellular sites. Among the likely consequences of the reactions of these radicals with membrane phospholipids is initiation of lipid peroxidation, which has deleterious consequences in cell membranes (including loss of fluidity, a decrease in electrical resistance, a depression of protein mobility in the membrane and increased phospholipid exchange between bilayers.[37] The breakdown of cellular membranes can cause inactivation of membrane-bound enzymes and loss of decompartmentalization, events that are catastrophic to the normal functioning of cells.[37] Also indirect effects of lipid peroxidation due to production of aldehydes (e.g. malonaldehyde has been shown to cross-link and aggregate membrane proteins).[37]

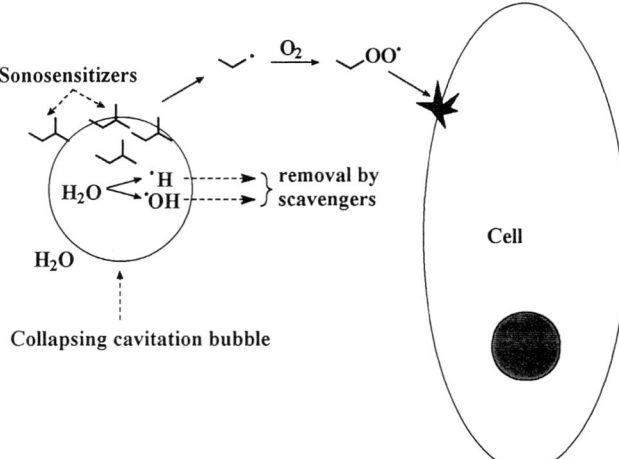

FIGURE 6. Working hypothesis of the mechanism of sonosensitization: sonosensitizer undergoes pyrolysis inside collapsing cavitation bubbles or in the heated gas-liquid interface, forming free radical intermediates. These intermediates (possibly carbon-centered radicals) react with dissolved O_2 to form peroxyl radicals, capable of attacking critical cellular sites due to their ability to diffuse significant distances. In contrast, ·OH radicals and ·H atoms which are also formed during cavitational collapse in aqueous sonochemistry, are unable to cause significant cellular damage, due to their extremely high reactivities and hence short diffusion distances.

Based on the proposed mechanism a suitable sonosensitizer should be able to be activated by ultrasound to free radical intermediates capable of causing significant biological damage. The following classes of compounds fit this description:

- small molecules with significant vapor pressure, which may accumulate inside cavitation bubbles (e.g. DMF, DMSO);

- surfactants (e.g. porphyrins), which accumulate at the gas-liquid interphase of the cavitation bubbles; and

- thermally labile molecules (e.g. azo-compounds) capable of thermal activation either from ultrasound-dependent heating of the bulk liquid, or due to pyrolysis inside or in the immediate vicinity of the collapsing cavitation bubbles.

Formation of secondary radicals due to reactions with ·OH and ·H formed by water pyrolysis could probably not represent a significant mode of sonosensitization *in vivo*, due to the rapid removal of ·H and ·OH by millimolar concentrations of ascorbate and thiols (e.g. glutathione, cysteine) in biological systems.

Thus, sonodynamic therapy seems to be a promising modality for cancer treatment since ultrasound can penetrate deep within the tissue and can be focused into a small region of tumor to activate relatively non-toxic molecules (e.g. porphyrins) or reduce the effective dose (e.g. DMF) thus reducing undesirable side effects.

REFERENCES

1. KREMKAU, F.W., J.S. KAUFMANN, M.M. WALKER, P.G. BURCH & C.L. SPURR. 1976. Ultrasonic enhancement of nitrogen mustard cytotoxicity in mouse leukemia. Cancer **37:** 1643–1647.
2. YUMITA, N., A. OKUMURA, R. NISHIGAKI, K. UMEMURA & S. UMEMURA. 1987. The combination treatment of ultrasound and antitumor drugs on Yoshida Sarcoma. Jpn. J. Hyperthermic Oncol. **3:** 175–182.
3. SAAD, A.H. & G.M. HAHN. 1989. Ultrasound enhanced drug toxicity on chinese hamster ovary cells in vitro. Cancer Res. **49:** 5931–5934.
4. LOVEROCK, B.P., G. TER HAAR, M.G. ORMEROD & P.R. IMRIE. 1990. The effect of ultrasound on the toxicity of adriamycin. Br. J. Radiol. **63:** 542–546.
5. HARRISON, G.H., E.K. BALCER-KUBICZEK & H.A. EDDY. 1991. Potentiation of chemotherapy by low-levels of ultrasound. Int. J. Radiat. Biol. **59:** 1453–1466.
6. YUMITA, N., R. NISHIGAKI, K. UMEMURA & S. UMEMURA. 1990. Synergistic effect of ultrasound and hematoporphyrin on sarcoma 180. Jpn. J. Cancer Res. **81:** 304–308.
7. TACHIBANA, K., N. KIMURA, M. OKUMURA, H. EGUCHI & S. TACHIBANA. 1993. Enhancement of cell killing of HL-60 cells by ultrasound in the presence of the photosensitizing drug Photofrin II. Cancer Lett. **72:** 195–199.
8. TACHIBANA, K., T. UCHIDA, S. HISANO & E. MORIOKA. 1997. Eliminating adult T-cell leukaemia cells with ultrasound. Lancet **349:** 325.
9. UMEMURA, S., N. YUMITA & R. NISHIGAKI. 1993. Enhancement of ultrasonically induced cell damage by a gallium-porphyrin complex, ATX-70. Jpn. J. Cancer Res. **84:** 582–588.
10. KESSEL, D., R. JEFFERS, J.B. FOWLKES & C. CAIN. 1994. Porphyrin-induced enhancement of ultrasound cytotoxicity. Int. J. Radiat. Biol. **66:** 221–228.
11. JEFFERS, R.J., R.Q. FENG, J.B. FOWLKES, J.W. HUNT, D. KESSEL & C.A. CAIN. 1995. Dimethylformamide as an enhancer of cavitation-induced cell lysis *in vitro*. J. Acoust. Soc. Am. **97:** 669–676.

12. SPINKS, J.W.T. & R.J. WOODS. 1976. An Introduction to Radiation Chemistry, 2nd edition. New York. J. Wiley and Sons.
13. AIUM. 1993. Bioeffects and Safety of Diagnostic Ultrasound. Laurel, MD. American Institute of Ultrasound in Medicine.
14. MITRAGOTRI, S., D. BLANKSCHTEIN & R. LANGER. 1995. Ultrasound-mediated transdermal protein delivery. Science **269**: 850–853.
15. MALGHANI, M.S., J. YANG & J. WU. 1998. Generation of bilayer defects induced by ultrasound. J. Acoust. Soc. Am. **103**: 1682–1685.
16. FLYNN, H.G. 1964. Physics of acoustic cavitation in liquids. In Physical Acoustic. Principles and Methods, Vol. I, part B. W.P. Mason, Ed. :58-172. New York. Academic Press.
17. FLINT, E.B. & K.S. SUSLICK. 1991. The temperature of cavitation. Science **253**: 1397–1399.
18. SUSLICK, K.S. & K.A. KEMPER. 1994. Pressure measurements during acoustic cavitation by sonoluminiscence. In Bubble Dynamics and Interface Phenomena. J.R. Blake, Ed. :311–320. Dordrecht, Netherlands. Kluwer Academic Publishers.
19. MIŠÍK, V., N. MIYOSHI & P. RIESZ. 1995. An EPR spin trapping study of the sonolysis of H_2O/D_2O mixtures—probing the temperatures of cavitation regions. J. Phys. Chem. **99**: 3605-3611.
20. RIESZ, P. & T. KONDO. 1993. Free radical formation induced by ultrasound and its biological implications. Free Rad. Biol. Med. **13**: 247–270.
21. BARNETT, S.B., Ed. 1998. Consensus Report of the World Federation for Ultrasound in Medicine. Ultrasound Med. Biol. **24** (Suppl. 1): S29–S34.
22. ALEGRIA, A.E., Y. LION, T. KONDO & P. RIESZ. 1989. Sonolysis of aqueous surfactant solutions. Probing the interfacial region of cavitation bubbles by spin trapping. J. Phys. Chem. **93**: 4908–4913.
23. HALL, E. 1993. Radiation Biology for the Radiologist, 4th edition. Philadelphia. J.B. Lippincott Company.
24. VON SONNTAG, C. 1987. The Chemical Basis of Radiation Biology. Philadelphia. Taylor and Francis.
25. MARK, G., A. TAUBER, R. LAUPERT, H.-P. SCHUCHMANN, D. SCHULZ, A. MUES & C. VON SONNTAG. 1998. OH-radical formation by ultrasound in aqueous solutions—Part II: Terephthalate and Fricke dosimetry and the influence of various conditions on the sonolytic yield. Ultrasonics Sonochem. **5**: 41–52.
26. MILLER, M.W., D.L. MILLER & A.A. BRAYMAN. 1996. A review of in vitro bioeffects of inertial ultrasonic cavitation from a mechanistic perspective. Ultrasound Med. Biol. **22**: 1131–1154.
27. VAN DONGEN, G., B.J.M. BRAAKHUIS, A. LEYVA, H.R. HENDRIKS, B.B.A. KIPP, M. BAGNAY & G.B. SNOW. 1989. Anti-tumor and differentiation-inducing activity of N,N-dimethylformamide (DMF) in head-and-neck cancer xenografts. Int. J. Cancer **43**: 285–292.
28. MIŠÍK, V. & P. RIESZ. 1996. Peroxyl radical formation in aqueous solutions of N,N-dimethylformamide, N-methylformamide, and dimethylsulfoxide by ultrasound: implications for sonosensitized cell killing. Free Rad. Biol. Med. **20**: 129–138.
29. MIYOSHI, N., V. MIŠÍK, M. FUKUDA & P. RIESZ. 1995. Effect of gallium-porphyrin analogue ATX-70 on nitroxide formation from a cyclic secondary amine by ultrasound: on the mechanism of sonodynamic activation. Radiat. Res. **143**: 194–202.
30. DIDENKO, Y.T., D.N. NASTICH, S.P. PUGACH, Y.A. POLOVINKA & V.I. KVOCHKA. 1994. The effect of bulk solution temperature on the intensity and spectra of water sonoluminescence. Ultrasonics **32**: 71–76.
31. BARBER, B.P., C.C. WU, R. LOFSTEDT, P.H. ROBERTS & S.J. PUTTERMAN. 1994. Sensitivity of sonoluminescence to experimental parameters. Phys. Rev. Lett. **72**: 1380–1383.
32. MIYOSHI, N., V. MIŠÍK & P. RIESZ. 1997. Sonodynamic toxicity of gallium-porphyrin analogue ATX-70 in human leukemia cells. Radiat. Res. **148**: 43–47.
33. LI, A.S.W., K.B. CUMMINGS, H.P. ROETHLING, G.R. BUETTNER & C.F. CHIGNELL. 1988. A spin trapping database implemented on the IBM PC/AT. J. Magn. Reson. **79**: 140–142.

34. NIKI, E. 1990. Free radical initiators as source of water- or lipid- soluble peroxyl radicals. *In* Methods in Enzymology, Vol. 186. L. Packer & A.N. Glazer, Eds. :100–108. New York. Academic Press.
35. KRISHNA, M.C., M.W. DEWHIRST, H.S. FRIEDMAN, J.A. COOK, W. DEGRAAF, A. SAMUNI, A. RUSSO & J.B. MITCHELL. 1994. Hyperthermic sensitization by the radical initiator 2,2'-azobis (2-amidinopropane) dihydrochloride (AAPH). I. *In vitro* studies. Int. J. Hypertherm. **10:** 271–281.
36. MIŠÍK, V., N. MIYOSHI & P. RIESZ. 1996. EPR spin trapping study of the decomposition of azo compounds in aqueous solutions by ultrasound: potential for use as sonodynamic sensitizers for cell killing. Free Rad. Res. **25:** 13–22.
37. HALLIWELL, B. & J.M.C. GUTTERIDGE. 1990. Role of free radicals and catalytic metal ions in human disease: An overview. *In* Methods in Enzymology, Vol. 186, Part B. L. Packer & A.N. Glazer, Eds. :1–85. San Diego, CA. Academic Press.

Glucose Deprivation-Induced Oxidative Stress in Human Tumor Cells

A Fundamental Defect in Metabolism?

DOUGLAS R. SPITZ,[a,b] JULIA E. SIM,[a] LISA A. RIDNOUR,[a] SANDRA S. GALOFORO,[c] AND YONG J. LEE[c]

[a]*Section of Cancer Biology, Radiation Oncology Center, Washington University School of Medicine, St. Louis, Missouri 63108, USA*

[c]*Department of Radiation Oncology, William Beaumont Hospital, Royal Oak, Michigan 48073, USA*

ABSTRACT: Recently, glucose deprivation-induced oxidative stress has been shown to cause cytotoxicity, activation of signal transduction (i.e., ERK1, ERK2, JNK, and Lyn kinase), and increased expression of genes associated with malignancy (i.e., bFGF and c-Myc) in MCF-7/ADR human breast cancer cells. These results have led to the proposal that intracellular oxidation/reduction reactions involving hydroperoxides and thiols may provide a mechanistic link between metabolism, signal transduction, and gene expression in these human tumor cells. The current study shows that several other transformed human cell types appear to be more susceptible to glucose deprivation-induced cytotoxicity and oxidative stress than untransformed human cell types. In a matched pair of normal and SV40-transformed human fibroblasts the cytotoxic process is shown to be dependent upon ambient O_2 concentration. A theoretical model to explain the results is presented and implications to unifying modern theories of cancer are discussed.

INTRODUCTION

For 70 years it has been noted that cells that have undergone neoplastic transformation (cancer cells) demonstrate altered metabolism when compared to untransformed (normal) cells.[1-4] The most pronounced and almost universal metabolic disruptions appear to involve metabolism of glucose and the loss of regulation between glycolytic metabolism and respiration.[1-4] In general it has been found that cancer cells exhibit increased glycolysis and pentose phosphate cycle activity, while demonstrating only slightly reduced rates of respiration.[1-4] Initially these metabolic differences were thought to arise as a result of "damage" to the respiratory mechanism and tumor cells were thought to compensate for this defect by increasing glycolysis.[1] However, studies of the mechanism involved in these metabolic changes focused on ATP production and energy metabolism,[1] which turned out not to be a

---

[b]Address for correspondence:Dr. Douglas R. Spitz, B180 Medical Laboratories, Free Radical and Radiation Biology Program, The University of Iowa, Iowa City, IA 52242.
e-mail: douglas-spitz@uiowa.edu

fruitful line of investigation for understanding carcinogenesis or designing cancer therapies.

Recently, it was discovered that glucose deprivation causes cytotoxicity in the MCF-7/ADR human multidrug-resistant breast carcinoma cell line.[5–8] Glucose deprivation-induced cytotoxicity in this model system was found to be preceded by the rapid activation of several signal transduction pathways (within 10 min) including extracellular regulated protein kinases (ERK1/ERK2), Lyn kinase (a *src* family kinase), and c-Jun N-terminal kinase (JNK).[7,8] Likewise the activities of MEK, Raf, Ras, and PKC were found to increase rapidly in glucose-deprived MCF-7/ADR cells.[5,7,8] In addition glucose-deprivation of MCF-7/ADR caused an increase in the DNA-binding activity of the AP-1 transcription factor as well as an increase in the expression of cellular homologues of oncogenes (c-Fos, c-Jun, c-Myc) and the angiogenic factor, basic fibroblast growth factor (bFGF).[5,6] Finally, it has been shown that over-expression of the mitochondrial protein, Bcl-2, protects MCF-7/ADR from glucose deprivation-induced cytotoxicity, suggesting that mitochondrial metabolism might be involved in the process that caused cytotoxicity.[6] These results show that removal of glucose from these human tumor cells results in cytotoxicity as well as activation of signal transduction pathways and increased expression of genes thought to be involved with neoplastic transformation. These results support the hypothesis that alterations in glycolytic metabolism could be linked by some process to signal transduction and gene expression associated with the malignant phenotype.

Since these original observations, studies to determine if oxidation/reduction reactions mediate glucose deprivation-induced cytotoxicity as well as the process linking glycolytic metabolism to alterations in signal transduction and gene expression were accomplished.[9,10] The fact that ERK1/ERK2 are members of the mitogen-activated protein kinase (MAPK) family which had been previously shown to be activated by oxidants (such as H_2O_2) and suppressed by reductants,[11,12] lead to the hypothesis that reductants such as N-acetylcysteine (NAC) could suppress activation of MAPKs induced by glucose deprivation in MCF-7/ADR. This hypothesis was supported by the findings that MAPK activation during glucose deprivation could be inhibited by treatment with the thiol antioxidant NAC.[9] These studies went on to show that glucose deprivation-induced cytotoxicity, activation of Lyn kinase and JNK as well as increases in steady state levels of mRNA coding for bFGF and c-Myc could all be inhibited by treatment with 1 mM NAC.[9,10]

The fact that a thiol-containing antioxidant was capable of inhibiting glucose deprivation-induced activation of signal transduction, increased expression of genes thought to be involved with maintenance of the malignant phenotype, and cytotoxicity lead to the hypothesis that oxidative stress was responsible for the effects seen during glucose deprivation in MCF-7/ADR. This hypothesis was supported by experiments showing that glucose deprivation resulted in the stimulation of glutathione (GSH) synthesis (2-fold), an increase in steady state levels of intracellular oxidized glutathione (GSSG) content (3- to 10-fold), and an increase in intracellular prooxidant production (2- to 4-fold).[9,10] A causal link between glucose deprivation-induced oxidative stress and glucose deprivation-induced cytotoxicity, activation of signal transduction, and increased gene expression was supported by the fact that NAC not only inhibited changes in signal transduction and gene expression, but entered the cells in the reduced form and inhibited parameters indicative of oxidative stress and cytotoxicity.[9,10] Finally, the hypothesis that metabolism of O_2 to hydroperoxides was the source of increased prooxidant production was supported by the fact that pyruvate, an intracellular hydroperoxide scavenger,[13] inhibited increased prooxidant production as well as cytotoxicity during glucose deprivation.[9]

After the formation of glucose 6-phosphate (via hexokinase) the major pathways of glucose metabolism include glycolysis and the pentose phosphate cycle.[4] Glycolysis results in the formation of pyruvate and the pentose phosphate pathway results in the formation of NADPH.[4] Pyruvate, in addition to being a substrate for the formation of acetyl-CoA and energy metabolism via the tricarboxylic acid (TCA) cycle and mitochondrial oxidative phosphorylation, has been shown to scavenge H_2O_2 and other hydroperoxides.[13] NADPH, by virtue of being the source of reducing equivalents for the glutathione/glutathione peroxidase/glutathione reductase system, has also been shown to participate in the metabolic decomposition of H_2O_2 and organic hydroperoxides.[14] Therefore, in addition to its well known role in energy production, glucose metabolism appears to be integrally related to the metabolic detoxification of intracellular hydroperoxides formed as byproducts of oxidative metabolism. In fact, increasing glucose concentrations in tissue culture media has been shown to render CHO cells resistant to H_2O_2-induced cytotoxicity.[15] Because glucose metabolism appears to be involved with the detoxification of intracellular hydroperoxides, we propose the theoretical model shown in FIGURE 1, to explain the results observed during glucose deprivation in MCF-7/ADR cells.

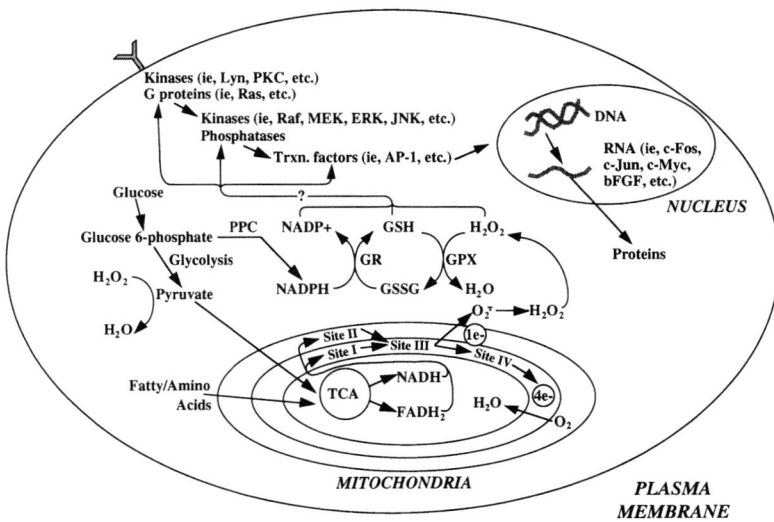

FIGURE 1. Theoretical model outlining of pathways contributing to the observed effects of glucose deprivation-induced oxidative stress in MCF-7/ADR. TCA = Tricarboxylic Acid Cycle, PPC = Pentose Phosphate Cycle; GR = Glutathione Reductase; GPX = Glutathione Peroxidase, ? = possibly thioredoxin, glutaredoxin, Ref-1, or direct oxidation/reduction.

During glucose deprivation steady state levels of intracellular prooxidants (presumably hydroperoxides) appear to increase immediately.[10] This suggests that hydroperoxides are being produced by ongoing metabolic processes and that the metabolic decomposition of these prooxidants is compromised by the removal of glucose probably via a decrease in intracellular NADPH and pyruvate (FIG. 1). We hypothesize that prooxidant production is occurring via mitochondrial electron transport chain activity because in the absence of glucose, fatty acids and amino acids would provide alternative substrates for the TCA cycle leading to the production of NADH and $FADH_2$ as the source of electrons for mitochondrial ATP production.[4] During mitochondrial respiration, O_2 acts as the terminal acceptor of electrons, with the 4-electron reduction of O_2 yielding H_2O. However, there exists a finite probability (which differs in different tissues) that 1-electron reduction of O_2 to yield superoxide will occur,[16] probably at Site I (NADH-dehydrogenase) or Site III (ubiquinone-cytochrome b) of the electron transport chain.[16] Superoxide then rapidly dismutes to form H_2O_2. It has been estimated during normal respiration 1–4% of O_2 consumption results in superoxide and hydrogen peroxide production.[16] Therefore, decreased peroxide scavenging (by pyruvate and NADPH dependent pathways) during glucose deprivation, could result in increased steady state levels of prooxidants (i.e., H_2O_2) produced as by products of mitochondrial electron transport chain activity (FIG. 1).

Increased steady state levels of prooxidants during glucose deprivation cause oxidative stress and cytotoxicity as evidenced by the accumulation of GSSG and increased clonogenic cell killing, which are inhibited by the thiol antioxidant (NAC) as well as a peroxide scavenger (pyruvate [9,10]). The cells appear to respond to glucose deprivation-induced oxidative stress by increasing the synthesis of glutathione[9,10] but in the absence of substrates necessary for the regeneration of NADPH, glutathione cannot be maintained in the reduced state.[10] In support of this notion glutamate rescues MCF-7/ADR cells from glucose deprivation-induced cytotoxicity and suppresses prooxidant production, presumably via the capacity of glutamate dehydrogenase to generate NADPH during the conversion of glutamate to α-ketoglutarate.[4,9] Because the thiol reductant (NAC) inhibits: 1) increased prooxidant production, 2) accumulation of oxidized glutathione, 3) increased synthesis of glutathione, and 4) activation of signal transduction and gene expression, we hypothesize that oxidation reactions involving thiols and hydroperoxides are causally involved with the intracellular signaling processes activated during glucose deprivation (FIG. 1).

Glucose deprivation-induced signal transduction cascades are believed to originate in either a receptor-dependent or a receptor-independent fashion. The former case is mediated through an interaction of cellular membrane receptors or G protein coupled receptors with protein kinases. The latter case is mediated through activation of non-receptor protein kinases. Activation of up-stream molecules (i.e., Lyn kinase, PKC, Ras) leads to the activation of a series of intermediary kinases (i.e., Raf, MEK) that ultimately activate cytosolic kinases (i.e., ERKs and JNKs). These cytosolic kinases then phosphorylate substrates (i.e., c-Fos and c-Jun) leading to the activation of transcription factors (i.e., AP-1) which bind target genes leading to the synthesis of new gene products (FIG. 1). Signal transduction cascades are believed to stand poised ready to sense intra- or extra-cellular stimuli and when activated are believed to redirect cellular function to respond to the stimulus. Our results, using MCF-7/ADR cells, suggest that alterations in oxidative metabolism caused by the removal of glucose trigger a number of these signaling cascades (i.e. Lyn kinase, PKC, Ras, Raf, MEK, ERKs, and JNK) resulting in the activation of transcription factors (i.e., AP-1) and increased expression of cellular homologues of oncogenes (i.e., c-Myc, c-Fos, and c-Jun) as well as angiogenic factors (i.e. bFGF) (FIG. 1).

At present it is unknown how glucose deprivation-induced oxidative stress activates signal transduction and gene expression, but a consideration of the literature reveals three distinct possibilities. First, it is well known that proteins involved in signaling cascades (i.e., Lyn kinase, Ras, AP-1) contain critical sulfhydryl residues that are sensitive to oxidation/reduction reactions and appear to modulate the transmission of signals via alterations in the activity of the signaling proteins.[17–20] It is therefore possible that direct oxidation reactions occurring at critical sulfhydryl residues could result in changes in activity of these signaling cascades.[17,20] Secondly, it is known that alterations in the steady state of oxidized/reduced ratios of $NADP^+$/NADPH and GSH/GSSG can alter the oxidation reduction status of thiol residues in thioredoxin and glutaredoxin that are capable of transmitting these signals to oxidation/reduction sensitive sites on kinases and transcription factors involved in signaling cascades.[20,21] Finally, it is also known that proteins [i.e., redox factor-1 (Ref-1)] containing critical sulfhydryl residues can interact with thioredoxin to transmit redox signals to thiol containing transcription factors such as AP-1, thereby altering their DNA-binding activity.[20] We hypothesize that these types of alterations in thiol redox status occurring on kinases, phosphatases, G-proteins, and/or transcription factors "sense" changes in intracellular oxidation/reduction reactions during glucose deprivation of MCF-7/ADR cells and result in the activation of signal transduction cascades leading to increases in gene expression (FIG. 1). In this way alterations in the flow of electrons through redox-sensitive signaling circuitry could represent one means by which alterations in electron transport chain activity are coupled to alterations in gene expression necessary to redirect cellular function.

While the model in FIGURE 1 is partially supported by the data gathered using the MCF-7/ADR human breast carcinoma cells, the generality of these results to other normal and tumor cell lines remains largely unknown.[10] Furthermore, it is unclear if alterations in glycolytic metabolism that have been noted for many years in transformed cells[1–4] relate to their susceptibility to glucose deprivation–induced cytotoxicity and oxidative stress.

The studies presented here were designed to determine:

1. if glucose deprivation induces cytotoxicity and oxidative stress in a variety of transformed human cell types;
2. if untransformed and transformed human cells are differentially susceptible to glucose deprivation–induced cytotoxicity and oxidative stress; and
3. if glucose deprivation–induced cytotoxicity and oxidative stress are dependent upon O_2-metabolism.

To accomplish these goals two matched pairs of SV40-transformed and untransformed human fibroblasts as well as HT29 human colon carcinoma cells and aortic smooth muscle cells were deprived of glucose. GSSG content was used as an index of oxidative stress and clonogenic cell survival was used as an endpoint indicative of cytotoxicity. To begin to determine if the metabolism of O_2 was involved in the cytotoxic process, selected experiments were done in 21% O_2 vs. 4% O_2.

MATERIALS AND METHODS

Cells and Culture Conditions

IMR90 (rep# I90 PO4) untransformed (normal) human fibroblasts and their SV40-transformed counterparts [designated IMR90 SV40 (rep#AG02804C)] were obtained from the Coriell Institute and maintained in Eagle's MEM supplemented with 1 × MEM vitamins, 1 × MEM essential amino acids, 2 × non-essential amino acids and 20% fetal bovine serum (Hyclone). GM00037F untransformed (normal) human fibroblasts and their SV40-transformed counterparts (GM00637G) were obtained from the Coriell Institute and maintained in Eagle's MEM with 10% fetal bovine serum. HT29 human colon carcinoma cells were obtained from ATCC and maintained in McCoy's 5A media supplemented with 10% fetal bovine serum. Untransformed human aorta smooth muscle cells were obtained from ATCC (#CRL-1999) and maintained in F12K media supplemented with 2 mM glutamine, 0.01 mg/ml insulin, 0.01 mg/ml transferrin, 10 ng/ml sodium selenite, 0.02 mg/ml endothelial cell growth supplement, and 10% fetal bovine serum. All stock cultures were maintained in 5% CO_2 and air in a humidified 37°C incubator in the absence of antibiotics.

Glucose Deprivation Conditions and Cell Survival Experiments

For experiments, cells were plated in 60 mm tissue culture dishes and grown for 2 to 4 days until each dish contained $1-2 \times 10^6$ cells in the presence of antibiotics (penicillin/streptomycin). At the beginning of each experiment the cells were rinsed with phosphate buffered saline (PBS) to remove glucose and placed in media containing all other additives except glucose and dialyzed fetal bovine serum as previously described.[9,10] Control cultures were treated identically except glucose was added at the normal concentration found in media (1 g/L). Cells were then placed in an incubator and harvested at the times indicated. For experiments in less than ambient O_2 (21% O_2 vs. 4% O_2), a Heraeus Instruments model 6000 (BB6220) tri-gas incubator was utilized. The desired O_2 concentrations were verified using both the internal detector in the incubator as well as external gas analyzers calibrated to certified gas standards. Following exposure to glucose deprivation, clonogenic cell survival was assayed as previously described[9] and surviving fraction of glucose deprived cells was normalized to the respective control plus glucose.

Biochemical Analysis

Following treatment cells were scrape-harvested in PBS at 4°C, centrifuged, the PBS discarded and the cell pellets frozen at −80°C. Samples were thawed and homogenates were prepared as described.[9,10] Total glutathione (GSH + GSSG), glutathione (GSH), and glutathione disulfide (GSSG) were determined using a spectrophotometric recycling assay and normalized per mg protein as previously described.[9,10]

RESULTS

FIGURE 2 shows the results of experiments where normal and transformed cells were exposed to 48 hours of glucose deprivation and clonogenic cell survival was assayed. IMR90 and GM00037F normal human fibroblasts deprived of glucose for 48 hours demonstrated no clonogenic cell killing, relative to their respective control (FIG. 2). The untransformed human aortic smooth muscle cells deprived of glucose for 48 hours demonstrated a slight (20%) reduction in clonogenic cell survival (relative to their control), but this did not reach statistical significance ($p > 0.05$, FIG. 2). In contrast, the transformed cells (IMR90 SV40, GM00637G, and HT29) demonstrated significant ($p < 0.05$) clonogenic cell killing following 48 hours of glucose deprivation (FIG. 2). The glucose-deprived IMR90 SV40 demonstrated ≈50% cell killing, while GM00637G as well as HT29 demonstrated > 90% cell killing, relative to their respective controls (FIG. 2). These results indicate that, while differences exist between the transformed cell lines, they are in general more susceptible to glucose deprivation–induced cytotoxicity than the untransformed cells tested.

Glutathione analysis following 48 hours of glucose deprivation in untransformed cells (IMR90, GM00037F, and aortic smooth muscle) showed no change in total glutathione content (GSH + GSSG) (FIG. 3A). Furthermore, GSSG was below the limit

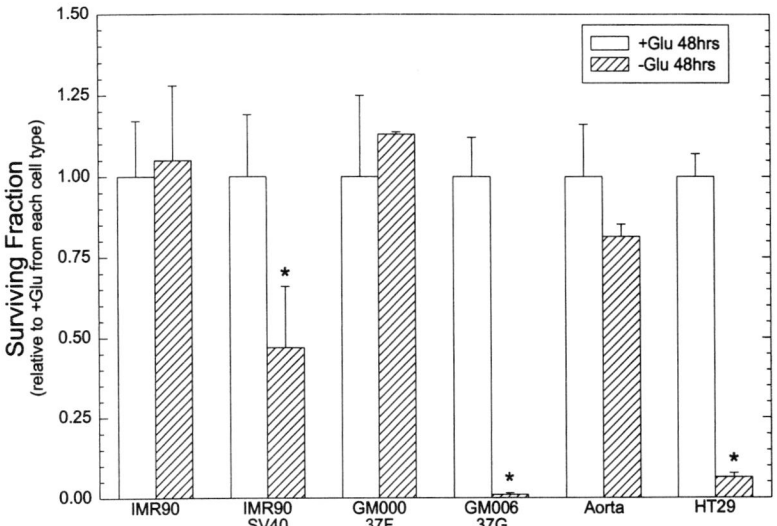

FIGURE 2. Clonogenic cell survival data showing the differential susceptibility of untransformed (IMR90, GM00037F, Aorta) and transformed (IMR90 SV40, GM00637G, HT29) human cell types to glucose deprivation-induced cytotoxicity. Errors represent ± 1 standard deviation (SD). Asterisks indicate significant differences between +Glu and −Glu from each cell line ($p < 0.05$, paired t-test). Data from IMR90 and IMR90 SV40 represent the mean of at least 12 determinations from 3 separately treated dishes in 2 separate experiments. Data from all other cell types are the mean of at least 4 determinations from each treatment group in one experiment.

of detection in the glucose-deprived untransformed cells (FIG. 3B). In contrast, glucose deprivation of transformed cell types (IMR90 SV40, GM00637G, and HCT) resulted in increased total glutathione, GSSG, and GSH (FIG. 3A,B,C). The fact that total and GSSG increased in the glucose-deprived transformed cells, supports the previously proposed hypothesis that these cells were experiencing oxidative stress and responding to that stress by increasing the synthesis of glutathione.[9,10] However, it appears that in the absence of glucose, the transformed cells were unable to maintain the newly synthesized glutathione in the reduced state, as evidenced by the accumulation of GSSG. Furthermore, during glucose deprivation, GSSG accumulated

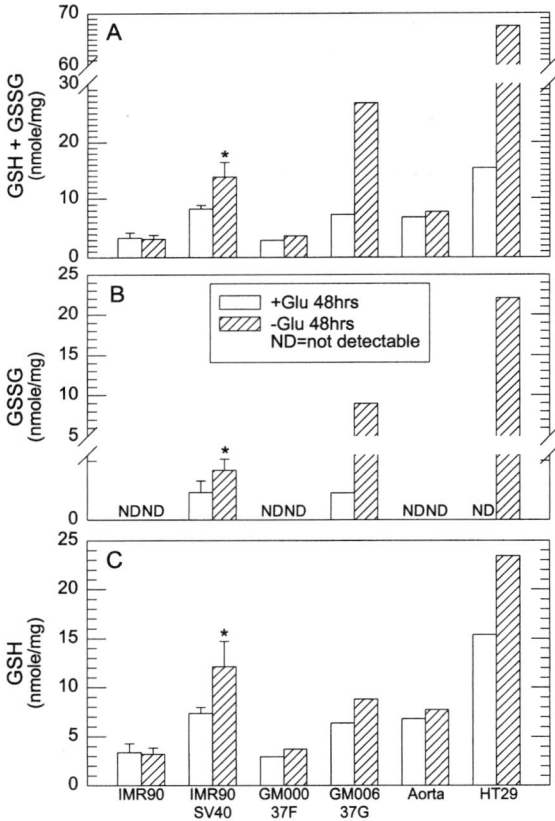

FIGURE 3. Glutathione analysis of untransformed (IMR90, GM00037F, Aorta) and transformed (IMR90 SV40, GM00637G, HT29) cell types exposed to glucose deprivation in FIGURE 2. Data points from IMR90 and IMR90 SV40 are the mean of 6 determinations done in 2 separate experiments (errors = ±1 SD, asterisks indicate significant differences between +Glu and −Glu, paired t-test, $p < 0.05$). Data from other cell types are single determinations in one experiment.

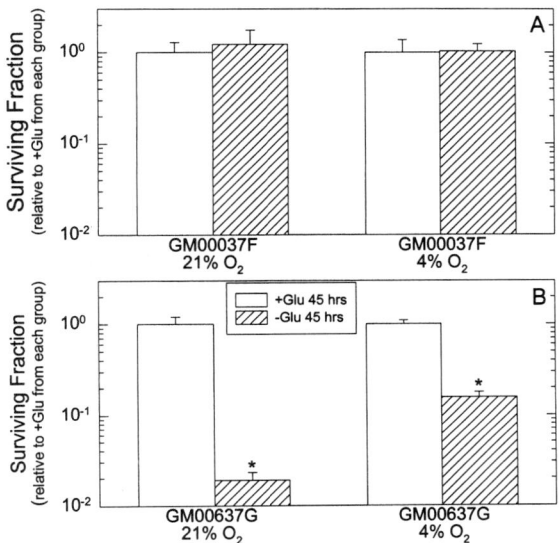

FIGURE 4. Clonogenic cell survival data showing the effect of O_2 concentration (21% vs. 4%) on glucose deprivation-induced cytotoxicity in untransformed (**A**) vs. SV40-transformed (**B**) human fibroblasts. Errors represent ± 1 SD. Asterisks indicate significantly different from all other treatment groups within a cell type ($p < 0.05$, paired t-tests). Data points are the mean of 4 determinations from each treatment group in one experiment.

to a greater extent in the transformed cell types which were the most susceptible to glucose deprivation–induced cytotoxicity (GM00637G and HT29 vs. IMR90 SV40, FIGS. 2 and 3). These results support the hypothesis that glucose deprivation–induced oxidative stress occurred in the transformed cells and contributed to the cytotoxicity noted in FIGURE 2.

In an attempt to determine if glucose deprivation–induced cytotoxicity was dependent on the metabolism of O_2, experiments were performed in 4% vs. 21% O_2 (FIG. 4). Confirming the results shown in FIGURE 2, glucose deprivation caused no significant clonogenic cell killing in the untransformed GM00037F human fibroblasts at either 21% O_2 or 4% O_2 (FIG. 4A). Again confirming the results shown in FIGURE 2, the glucose-deprived SV40-transformed cells (GM00637G) demonstrated significant clonogenic cell killing, with only 2% survival after 45 hours in 21% O_2 (FIG. 4B). When the SV40-transformed cells were glucose-deprived in the presence of 4% O_2, clonogenic cell killing was significantly less than was seen at 21% O_2 (15% survival vs. 2% survival, respectively; FIG. 4B). These results confirm that glucose deprivation was selectively cytotoxic to SV40-transformed human fibroblasts and show that the process leading to cytotoxicity was dependent upon ambient O_2 concentration. These results support the hypothesis that glucose deprivation–induced cytotoxicity is mediated by a process involving the metabolism of O_2.

DISCUSSION

The results in the current report extend the previous observations using MCF-7/ADR human breast carcinoma cells into three potentially important areas. First, it appears that glucose deprivation–induced cytotoxicity and oxidative stress occurs in several human transformed cell lines in addition to MCF-7/ADR (IMR90 SV40, GM00637G, and HT29). These transformed cells were derived from spontaneous transformation *in vivo* (MCF-7/ADR and HT29) as well as viral transformation events mediated by SV40 *in vitro* (IMR90 SV40 and GM00637G). These data allow for the speculation that susceptibility to glucose deprivation–induced cytotoxicity and oxidative stress is not limited to a single type of transformation event or transformed human cell type. This opens the possibility that glucose deprivation–induced cytotoxicity and oxidative stress could represent a general phenomenon common to several different types of cancer cells. Secondly, transformed cells appear to be more susceptible to glucose deprivation–induced cytotoxicity and oxidative stress than non-transformed cells (including two matched sets of normal and virally transformed fibroblasts). These results allow for the speculation that normal cells may be less susceptible to glucose deprivation-induced cytotoxicity and oxidative stress than cancer cells. This also supports the speculation there may be a fundamental defect in oxidative metabolism in cancer cells that could be exploited to gain a therapeutic advantage when trying to kill cancer cells while sparing normal tissues. Finally, it appears in one transformed cell line (GM00637G), glucose deprivation–induced cytotoxicity is dependent upon the metabolism of O_2. This result is consistent with the speculation that a defect in the transformed cell's respiratory mechanism may give rise to the reactive specie(s) responsible for glucose deprivation–induced cytotoxicity and oxidative stress. The data presented here are consistent with the metabolic pathways outlined in the model shown in FIGURE 1. However, the effect of glucose deprivation on signal transduction and gene expression in the vast majority of transformed human cell lines remains unknown.

Perhaps the most thought provoking and speculative considerations that can be derived from the recent work with glucose deprivation (refs. 5–10, FIGS. 2–4) pertain to a historical perspective on understanding of the origin of cancer cells. Warburg suggested that cancer was a metabolic disease is which respiration was "damaged" and glycolysis was increased to compensate for defective respiration.[1] Warburg hypothesized that this defect in metabolism gives rise to the cancerous phenotype. Oberley *et al.*[22] proposed that cancer cells had a defect in the metabolism of superoxide (produced as a byproduct of mitochondrial respiration) that could trigger immortalization, uncontrolled cell proliferation, and the development of the malignant phenotype. During this same period of time it was also discovered that environmental insults and chemicals that lead to increases in prooxidant production could act as both initiators and promoters of the carcinogenic process.[23,24] In addition it was shown that antioxidant enzymes and vitamins could inhibit both the initiation and promotion of carcinogenesis.[23–25] Finally it was suggested that tumor cells produce large amounts of prooxidants (presumably hydroperoxides)[26] and that prooxidants appear to be involved with both stimulation of cell proliferation during mitogenesis as well as the generation of mutations and genomic instability, which are hallmarks of the transformed phenotype.[27–29] Therefore, it appears that transformed cells demonstrate metabolic abnormalities involving glucose and respiration as well as alterations in intracellular oxidation/reduction reactions towards a more prooxidant environment. Furthermore, prooxidants appear to accelerate mutagenesis as well as carcinogenesis, and antioxidants appear to be anticarcinogens.

More recently the theory that cancer is a genetic disease has gained widespread support.[30,31] In this model carcinogenesis is thought to be a multi-step process in which genetic mutations gradually accumulate over time, eventually resulting in immortalization, the loss of control of cell proliferation, and progression to the malignant phenotype. Mutations in, and altered expression of, cellular homologues of oncogenes (i.e., Ras, c-Fos, c-Jun, and c-Myc) associated with growth and development as well as tumor suppressor genes (i.e., p53) have been suggested to contribute to the process of neoplastic transformation.[30–34] Many of these genes have also been suggested to be involved in signal transduction pathways (*src* family kinases, Ras, etc.). In addition several of the mutations associated with conversion of non-oncogenic genes into their oncogenic counterparts appear to involve changes in oxidation/reduction sensitive regulatory sites (particularly thiol residues) on the respective proteins (i.e., v-Fos, v-Jun, and mutant p53).[34–36] Furthermore, changes in the redox status of thiols have long been associated with the process of cell division and cancerous liver tissue has been shown to have an altered pattern of reduced thiols when compared to normal liver (reviewed in ref. 37). In addition it has been suggested that genes coding for proteins that block the process of programmed cell death (apoptosis) (i.e. Bcl-2), may permit cells, carrying a potentially oncogenic defect and destined to die via programmed cell death, to survive and continue down the path to neoplastic transformation.[38] Bcl-2 has also been suggested to regulate the intracellular localization of respiratory proteins and thiols as well as protecting cells from oxidative stress and glucose deprivation–induced cytotoxicity.[6,38–41] Therefore, it appears that a common thread linking metabolic and genetic theories of cancer involves glucose metabolism, respiration, mutations in redox regulated proteins governing growth and development, and intracellular oxidation/reduction reactions involving hydroperoxides and thiols.

Given this interesting historical commonality between theories of cancer and the model for explaining recent results obtained with glucose deprivation (FIG. 1), it is now possible to propose a scenario incorporating critical aspects of both genetic and metabolic theories of cancer. All O_2 metabolizing cells are thought to produce a relatively constant low level of prooxidants (i.e., superoxide, hydrogen peroxide, etc.) as by products of electron transport chain activity that is balanced by cellular antioxidant capacity to maintain a viable non equilibrium steady state environment that is predominantly reducing. NADPH and the glutathione/glutathione peroxidase/glutathione reductase system are thought to represent major sources of reducing equivalents necessary to maintain this environment in a reduced state. If cancer cells have a defective respiratory mechanism (either increased prooxidant production and/or decreased cellular antioxidant capacity) that leads to increases in steady state levels of hydroperoxide production, they may exist in a relatively prooxidant intracellular environment (with respect to normal cells). Metabolism of glucose could be up regulated to produce more pyruvate and NADPH to compensate for this defect thereby rescuing the transformed cell from a respiratory-dependent cell death. Although not lethal (because of increased metabolism of glucose), the ensuing prooxidant environment produced by defective respiration may be capable of stimulating seemingly uncontrolled cell division via aberrant activation of redox regulated signal transduction pathways and/or cell cycle regulatory proteins as well as inhibition of processes involved with differentiation (i.e., DNA methylation). In addition this prooxidant environment may also contribute to mutagenesis, genomic instability, and cellular heterogeneity, which are hallmarks of the progression to the malignant phenotype. Likewise, genetic mutations that result in the loss of redox control of cell cycle regulatory and/or signal transduction proteins that govern growth and development may cause these proteins to become activated or inactivated at inappropriate redox potentials leading to uncontrolled cell growth and the inability to differentiate. In addition mutations in redox regulated tumor suppressor genes (i.e., p53) may predispose cells to undergo malignant transformation more readily (without a preexisting defect in respiration) or escape cell death in the presence of defective in respiration. In fact, apoptosis mediated by p53 has recently been suggested to involve oxidative stress in cancer cells.[42] Finally, over expression of proteins (such as Bcl-2), which could rescue cells from a lethal respiratory defect, might allow the defective cells to continue progressing towards malignant transformation. Cancer might therefore represent a constellation of metabolic and/or genetic diseases where the common theme is uncoupling of normal cellular processes that govern cell growth and development caused by the inappropriate flow of electrons from metabolic oxidation/reduction reactions to redox sensitive proteins governing signal transduction and gene expression.

While the current data linking glucose derivation–induced cytotoxicity, activation of signal transduction, and gene expression to metabolic oxidative stress in human tumor cells opens many theoretical possibilities with potentially far reaching implications to the understanding of malignancy, the generality and predictive power of the results remains largely unknown. Many future studies will be required to test the validity of these concepts.

ACKNOWLEDGMENTS

The authors thank Dr. Daniel Gilbert for his inspirational career achievements in the field of Free Radical Biology and providing forums for new ideas. The authors thank Lori A. Worley, Sarah Gualano, and Alan C.-W. Ko for expert technical assistance. This work was supported by NIH HL51649 (DRS), CA75556 (DRS), and CA48000 (YJL).

REFERENCES

1. WARBURG, O. 1956. On the origin of cancer cells. Science **132:** 309–314.
2. WEBER, G. 1977. Enzymology of cancer cells (first of two parts). New Eng. J. Med. **296:** 486–492.
3. WEBER, G. 1977. Enzymology of cancer cells (second of two parts). New Eng. J. Med. **296:** 541–551.
4. LEHNINGER, A.L. 1976. Biochemistry. New York. Worth Publishers, Inc., pp. 245–441, 467–471, 849–850.

5. GALOFORO, S.S., C.M. BERNS, G. ERDOS, P.M. CORRY & Y.J. LEE. 1996. Hypoglycemia-induced AP-1 transcription factor and basic fibroblast growth factor gene expression in multidrug resistant human breast carcinoma MCF-7/ADR cells. Mol. Cell. Biochem. **155:** 163–171.
6. LEE, Y.J., S.S. GALOFORO, C.M. BERNS, W.P. TONG, H.R. KIM & P.M. CORRY. 1997. Glucose deprivation-induced cytotoxicity in drug resistant human breast carcinoma MCF-7/ADR cells: role of c-myc and bcl-2 in apoptotic cell death. J. Cell Sci. **110:** 681–686.
7. GUPTA, A.K., Y.J. LEE, S.S. GALOFORO, C M. BERNS, A.A. MARTINEZ, P.M. CORRY, X. WU & K.L. GUAN. 1997. Differential effect of glucose deprivation on MAPK activation in drug sensitive human breast carcinoma MCF-7 and multidrug resistant MCF-7/ADR cells. Mol. Cell. Biochem. **170:** 23–30.
8. LIU, X., A.K. GUPTA, P.M. CORRY & Y.J. LEE. 1997. Hypoglycemia-induced c-Jun phosphorylation is mediated by c-Jun N-terminal kinase 1 and Lyn kinase in drug-resistant human breast carcinoma MCF-7/ADR cells. J. Biol. Chem. **272:** 11690–11693.
9. LEE, Y.J., S.S. GALOFORO, C.M. BERNS, J.C. CHEN, B.H. DAVIS, J.E. SIM, P.M. CORRY & D.R. SPITZ. 1998. Glucose deprivation-induced cytotoxicity and alterations in mitogen-activated protein kinase activation are mediated by oxidative stress in multidrug-resistant human breast carcinoma cells. J. Biol. Chem. **273:** 5294–5299.
10. BLACKBURN, R.V., D.R. SPITZ, X. LIU, S.S. GALOFORO, J.E. SIM, L.A. RIDNOUR, J.C. CHEN, B.H. DAVIS, P.M. CORRY & Y.J. LEE. 1999. Metabolic oxidative stress activates signal transduction and gene expression during glucose deprivation in human tumor cells. Free Rad. Biol. Med. **26:** 419–430.
11. STEVENSON, M.A., S.S. POLLOCK, C.N. COLEMAN & S.K. CALDERWOOD. 1994. X-irradiation, phorbol esters, and H_2O_2 stimulate mitogen-activated protein kinase activity in NIH-3T3 cells through the formation of reactive oxygen intermediates. Cancer Res. **54:** 12–15.
12. GUYTON, K.Z., Y. LIU, M. GOROSPE, Q. XU & N.J. HOLBROOK. 1996. Activation of mitogen-activated protein kinase by H_2O_2. J. Biol. Chem. **271:** 4138–4142.
13. NATH, K.A., E.O. NGO, R.P. HEBBEL, A.J. CROATT, B. ZHOU & L.M. NUTTER. 1995. α-Ketoacids scavenge H_2O_2 in vitro and in vivo and reduce menadione-induced DNA injury and cytotoxicity. Am. J. Physiol. (Cell Physiol.) **268:** C227–C236.
14. TUTTLE, S.W., M.E. VARNES, J.B. MITCHELL & J.E. BIAGLOW. 1992. Sensitivity to chemical oxidants and radiation in CHO cell lines deficient in oxidative pentose cycle activity. Int. J. Radiat. Onc. Biol. Phys. **22:** 671–675.
15. AVERILL-BATES, D.A. & E. PRZYBYTKOWSKI. 1994. The role of glucose in cellular defenses against cytotoxicity of hydrogen peroxide in Chinese hamster ovary cells. Arch. Biochem. Biophys. **312:** 52–58.
16. BOVERIS, A. & E. CADENAS. 1982. Production of superoxide radicals and hydrogen peroxide in mitochondria. In Superoxide Dismutase,Vol. II. L.W. Oberley, Ed. :15–30. Boca Raton, FL. CRC Press Inc.
17. XIAO, J., J.E. BIAGLOW, H.J. CHAE-PARK, J. JIN, L. TUEL-AHLGREN, D.E. MYERS, A.L. BURKHARDT, J.B. BOLEN & F.M. UCKUN. 1996. Role of hydroxyl radicals in radiation-induced activation of Lyn tyrosine kinase in human B-cell precursors. Leuk. Lymphoma **22:** 421–430.
18. LANDER, H.M., J.S. OGISTE, K.K. TENG & A. NOVOGRODSKY. 1995. p21ras as a common signaling target of reactive free radicals and cellular redox stress. J. Biol. Chem. **270:** 21195–21198.
19. ABATE, C., L. PATEL, F.J. RAUSCHER III & T. CURRAN. 1990. Redox regulation of Fos and Jun DNA-binding activity in vitro. Science **249:** 1157–1161.
20. NAKAMURA, H., K. NAKAMURA & J. YODOI. 1997. Redox regulation of cellular activation. Annu. Rev. Immunol. **15:** 351–369.
21. ZHENG, M., F. ASLUND & G. STORZ. 1998. Activation of the OxyR transcription factor by reversible disulfide bond fomation. Science **279:** 1718–1721.
22. OBERLEY, L.W., T.D. OBERLEY & G.R. BUETTNER. 1981. Cell division in normal and transformed cells: the possible role of superoxide and hydrogen peroxide. Med. Hypoth. **7:** 21–42.

23. AMES, B.N. 1983. Dietary carcinogens and anticarcinogens: oxygen radicals and degenerative diseases. Science **221:** 1256–1262.
24. CERUTTI, P.A. 1985. Prooxidant states and tumor promotion. Science **227:** 375–381.
25. BOREK, C. & W. TROLL. 1983. Modifiers of free radicals inhibit in vitro the oncogenic actions of x-rays, bleomycin, and the tumor promoter 12-O-tetradecanoylphorbol 13-acetate. Proc. Natl. Acad. Sci. USA **80:** 1304–1307.
26. SZATROWSKI, T.P. & C.F. NATHAN. 1991. Production of large amounts of hydrogen peroxide by human tumor cells. Cancer Res. **51:** 794–798.
27. BUDROE, J.D., T. UMEMURA, K. ANGELOFF & G.M. WILLIAMS. 1992. Dose-response relationships of hepatic acyl-CoA oxidase and catalase activity and liver mitogenesis induced by peroxisome proliferator ciprofibrate in C57BL/6N and BALB/c mice. Toxicol. Appl. Pharmacol. **113:** 192–198.
28. MORAES, E.C., S.M. KEYSE & R.M. TYRRELL. 1990. Mutagenesis by hydrogen peroxide treatment of mammalian cells: a molecular analysis. Carcinogenesis **11:** 283–293.
29. HUNT, C.R., J.E. SIM, S.J. SULLIVAN, T. FEATHERSTONE, W. GOLDEN, C. VON KAPP-HERR, R.A. HOCK, R.A. GOMEZ, A.J. PARSIAN & D.R. SPITZ. 1998. Genomic instability and catalase gene amplification induced by chronic exposure to oxidative stress. Cancer Res. **58:** 3986–3992.
30. BISHOP, J.M. 1987. The molecular genetics of cancer. Science **235:** 305–311.
31. VARMUS, H.E. 1987. Oncogenes and transcriptional control. Science **238:** 1337–1339.
32. WEINBERG, R.A. 1985. The action of oncogenes in the cytoplasm and nucleus. Science **230:** 770-776.
33. BOHMANN, D., T.J. BOS, A. ADMON, T. NISHIMURA, P.K. VOGT & R. TIJAN. 1987. Human proto-oncogene c-jun encodes a DNA binding protein with structural and functional properties of transcription factor AP-1. Science **238:** 1386–1392.
34. SUN, Y. & L.W. OBERLEY. 1996. Redox regulation of transcriptional activators. Free Radic. Biol. Med. **21:** 335–348.
35. CHIDA, K. & P.K. VOGT. 1992. Nuclear translocation of viral Jun but not of cellular Jun is cell cycle dependent. Proc. Natl. Acad. Sci. USA **89:** 4290–4294.
36. OKUNO, H., A. AKAHORI, H. SATO, S. XANTHOUDAKIS, T. CURRAN & H. IBA. 1993. Escape from redox regulation enhances the transforming activity of Fos. Oncogene **8:** 695–701.
37. SZENT-GYÖRGYI, A. 1976. Electronic biology and cancer. Marcel Decker Inc. New York, pp. 34–35.
38. KORSMEYER, S.J., X.-M. YIN, Z.N. OLTVAI, D.J. VEIS-NOVACK & G.P. LINETTE. 1995. Reactive oxygen species and the regulation of cell death by the Bcl-2 gene family. Biochim. Biophys. Acta **1271:** 63–66.
39. VOEHRINGER, D.W., D.J. MCCONKEY, T.J. MCDONNELL, S. BRISBAY & R.E. MEYN. 1998. Bcl-2 expression causes redistribution of glutathione to the nucleus. Proc. Natl. Acad. Sci. USA **95:** 2956–2960.
40. YANG, J., X. LIU, K. BHALLA, C.N. KIM, A. IBRADO, J. CAI, T. PENG, D.P. JONES & X. WANG. 1997. Prevention of apoptosis by Bcl-2: release of cytochrome c from mitochondria blocked. Science **275:** 1129–1132.
41. KLUCK, R.M., E. BOSSY-WETZEL, D.R. GREEN & D.D. NEWMEYER. 1997. The release of cytochrome c from mitochondria: a primary site for Bcl-2 regulation of apoptosis. Science **275:** 1132–1136.
42. POLYAK, K., Y. XIA, J.L. ZWEIER, K.W. KINZLER & B. VOGELSTEIN. 1997. A model for p53-induced apoptosis. Nature **389:** 300–305.

Cytomegalovirus Gene Regulation by Reactive Oxygen Species

Agents in Atherosclerosis

EDITH SPEIR[a]

Cardiology Branch National Heart, Lung, and Blood Institute National Institutes of Health, Bethesda, Maryland 20892-1650, USA

ABSTRACT: Oxidative stress is implicated in the pathogenesis of atherosclerosis, and of viral infections caused by sendai virus, influenza and HIV. Vascular oxidative stress is due to inflammatory and immune responses of vascular cells, and to reperfusion after recanalization of blocked arteries. Because human cytomegalovirus (CMV) may contribute to atherogenesis by several mechanisms, and coronary artery smooth muscle cells (SMC) are permissive for the virus, we examined CMV interactions with SMC. Infection causes generation of intracellular reactive oxygen species (ROS) which activate NF-κB, a cellular transcription factor. NF-κB mediates expression of the CMV promoter and of genes involved in the immune and inflammatory responses. Antioxidants or aspirin inhibit ROS, NF-κB and CMV.

INTRODUCTION

Reactive oxygen species (ROS) were first implicated in cytotoxicity based on the similarity observed between oxygen poisoning and radiation toxicity.[1] Elevated levels of ROS have subsequently been shown to contribute to the development of human atherosclerosis, in particular to ischemia-reperfusion injury and inflammation.[2–4] Cytomegalovirus (CMV), a member of the herpesvirus group, is capable of infecting many tissues and cell types, including coronary and pulmonary arterial smooth muscle cells (SMC). After primary infection, the virus persists in certain tissues in a latent state, during which no viral gene products are expressed. We recently reported the detection by polymerase chain reaction of CMV DNA in coronary atherectomy lesions of patients who had previously undergone successful balloon angioplasty.[5] We found accumulation of the tumor suppressor protein p53 in 38% of atherectomy tissues, and this correlated with the presence of CMV genome in the same specimens. Overexpression of p53 is generally due to mutated p53, which is the most common defect associated with human cancer.[6] In addition, DNA viruses such as adenovirus or papilloma virus were shown to express proteins that specifically bind to and inactivate p53, in order to keep the host cell in an activated state, providing RNA polymerase which these viruses lack and pilfer from the host cell. We showed for the first time

---

[a] Address for correspondence: Edith Speir, Cardiology Branch, National Heart, Lung, and Blood Institute National Institutes of Health, 10/7B15, 10 Center Drive, MSC 1650, Bethesda, MD 20892-1650.
e-mail: speire@nih.gov

that the immediate early CMV protein IE84 binds to and transcriptionally inactivates p53.[5] This could be one of the mechanisms responsible for the excessive SMC proliferation present in atherosclerosis and especially in restenosis. Interestingly, several groups reported recently that oxidative stress induced by diethyl maleate or by ultraviolet radiation inhibited site-specific p53 binding to its cognate DNA,[7,8] which interferes with p53 transcriptional activity, indicating redox regulation of the p53 suppressor function. On the basis of this and other findings, we hypothesized that CMV contributes to the development of restenosis after reactivation of the virus via angioplasty-induced injury. CMV genome also has been detected near arterial plaques of patients with atherosclerosis, and it is possible that periodic reactivation of the virus also predisposes patients to develop atherosclerosis.[9,10]

If CMV does play a role in these diseases, it would be important to identify those changes in the cellular environment that facilitate CMV reactivation and expression of its gene products. Therefore, we focused on whether CMV infection of SMC triggers a cellular response that is conducive to activation-reactivation of the virus and to viral gene expression. We were particularly interested in identifying a response that could activate NF-κB. This pleiotropic transcription factor transactivates the major promoters of several viruses, including HIV and CMV.[11,12] Most important, one of the mechanisms by which HIV gene transcription is enhanced and by which HIV is reactivated from latency is through the generation of ROS, which exert their effects, at least in part, by activation of NF-κB.[11] Several cytokines, upon binding to their receptors, have been shown to activate such a signaling pathway, which ultimately mediates expression of many cellular genes and their products, including those involved in immune and inflammatory responses.[11] Therefore, paradoxically, the same signaling response may both activate latent viruses and constitute an important cellular defense against infecting pathogens. Because CMV has evolved strategies to coopt cellular mechanisms that enhance its own survival,[13] we asked whether CMV infection of SMC generates increased intracellular levels of ROS and, if so, whether such changes in the redox state of the cell activate NF-κB and enhance CMV immediate early (IE) gene transcription and expression. In addition, one of the IE gene products of CMV, IE72, is a potent transactivator of its own promoter, the major immediate early promoter, (MIEP), which has 4 NF-κB binding sites. Although it has been suggested that this action is mediated by NF-κB,[14] the interactions among NF-κB, IE72, and free radicals are largely unknown.

OBJECTIVES

Because CMV may contribute to restenosis and atherosclerosis, and SMC are involved in these disease processes, we examined CMV-SMC interactions. One of the primary mechanisms of host defense against viral invasion is through the production by phagocytes of free radical pulses which destroy infected cells.[15] We hypothesized that the free radical program can also be activated in SMC in response to viral infection. The first series of studies was therefore directed to answering the question: do SMC respond to CMV infection by generating increased levels of ROS? If so, what are the functional consequences, if any, to the redox-sensitive viral and cellular mechanisms, and the response to different types of antioxidants?

EXPERIMENTS AND RESULTS

Human coronary artery SMC and their optimal growth medium were purchased from Clonetics/Biowhittaker. Cells (passages 4–6) were kept at 37°C in an atmosphere of 5% CO_2, 100% humidity, and were infected or transfected during the log phase of their growth. Human CMV, Towne strain, passage 45–50 was propagated in human embryonic lung fibroblasts (HEL 299, American Type Culture Collection) as described before.[16] Viral stock and the CMV immediate early protein expression plasmids (pRcIE72 and pRcIE84), and the major immediate early reporter plasmid (MIEP-CAT) were gifts from Prof. E.S. Huang, UNC, Chapel Hill, NC. The 3X-κB-CAT and 3X-mutκB-CAT plasmids contain 3 copies of a wild-type (TGGGGATTC-CCCA), or 3 copies of a mutated (TGCGGCTTCCCGA) κB DNA binding sequence and were gifts from Prof. A.S. Baldwin, Jr., Chapel Hill, NC. NAC and PDTC were from Janssen Pharmaceuticals. All other reagents were from Sigma Chemical Co.

Effect of CMV Infection on Intracellular Redox State

For the detection of intracellular ROS, we used the lipid-soluble marker dichlorofluorescein diacetate (DCFH-DA, Molecular Probes), that easily enters the cell. Once inside, it is deacetylated by esterases and becomes highly fluorescent upon oxidation.[17] Coronary artery SMC were incubated with 5 µM (final concentration) DCFH-DA in phenol red-free HBSS, (the chambers were sealed with parafilm) for 5 minutes and immediately monitored under the microscope, in a dark room. Fluorescence was monitored and recorded by confocal laser scanning microscopy (Leica TCS4D, Leica Lasertechnik, Heidelberg). Excitation and emission wavelengths were 488 and 520 nm, respectively. Images were collected using a 512×512 pixel format, and were archived for later analysis. The intensity of the fluorescence was quantified with the analysis software provided with the confocal microscopy system. Three different fields were analyzed for each time point of each experiment. By means of computer software, the relative fluorescence was calculated by dividing the total intensity of the fluorescence in the measuring field by the percentage of the area of the field occupied by fluorescent cells. This served to compensate for variations in the number of cells in different measuring fields. SMC were grown in fibronectin-coated 4-well chamber slides for 48–72 hours and pretreated for 1 hour with the agents described below. After removal of the medium/drug, and one wash with Hanks Balanced Salt Solution without phenol red (HBSS, Gibco), cells were infected with 2–5 multiplicities of infection (MOI) of purified CMV in serum-free basal SMC medium for 1 hour. Virus was purified by ultracentrifugation as follows. Supernatant media of infected HEL-299 cells with 100% cytopathic effects (CPE) was centrifuged at 3,000 rpm to remove cell debris. The supernatant (28 ml) was then carefully layered on a 0.5 M sucrose in sterile basal SMC medium, and centrifuged in an SW-28 rotor at 26,000 rpm for 2 hours at 4°C. The supernatant was discarded, and the purified virus pellet was suspended in 1 ml sterile basal medium containing 0.7% dimethylsulfoxide. This stock was divided in 100 ml aliquots and flash-frozen in an alcohol/dry ice bath, and stored in liquid nitrogen repository for future titering and experimental use. Titers typically are $3-5 \times 10^8$ plaque-forming units per milliliter (pfu/ml). Using confocal laser microscopy to detect the intracellularly trapped dye, we found that exogenous H_2O_2 (freshly prepared from a 30% stock,

A
H2O2 (50 nM)
H2O2 (100nM)
H2O2 (200nM)
H2O2 (500nM)
B
Uninfected
Infected 0 min
Infected 15 min
Infected 30 min
Infected 60 min
Infected 120 min
100 µm

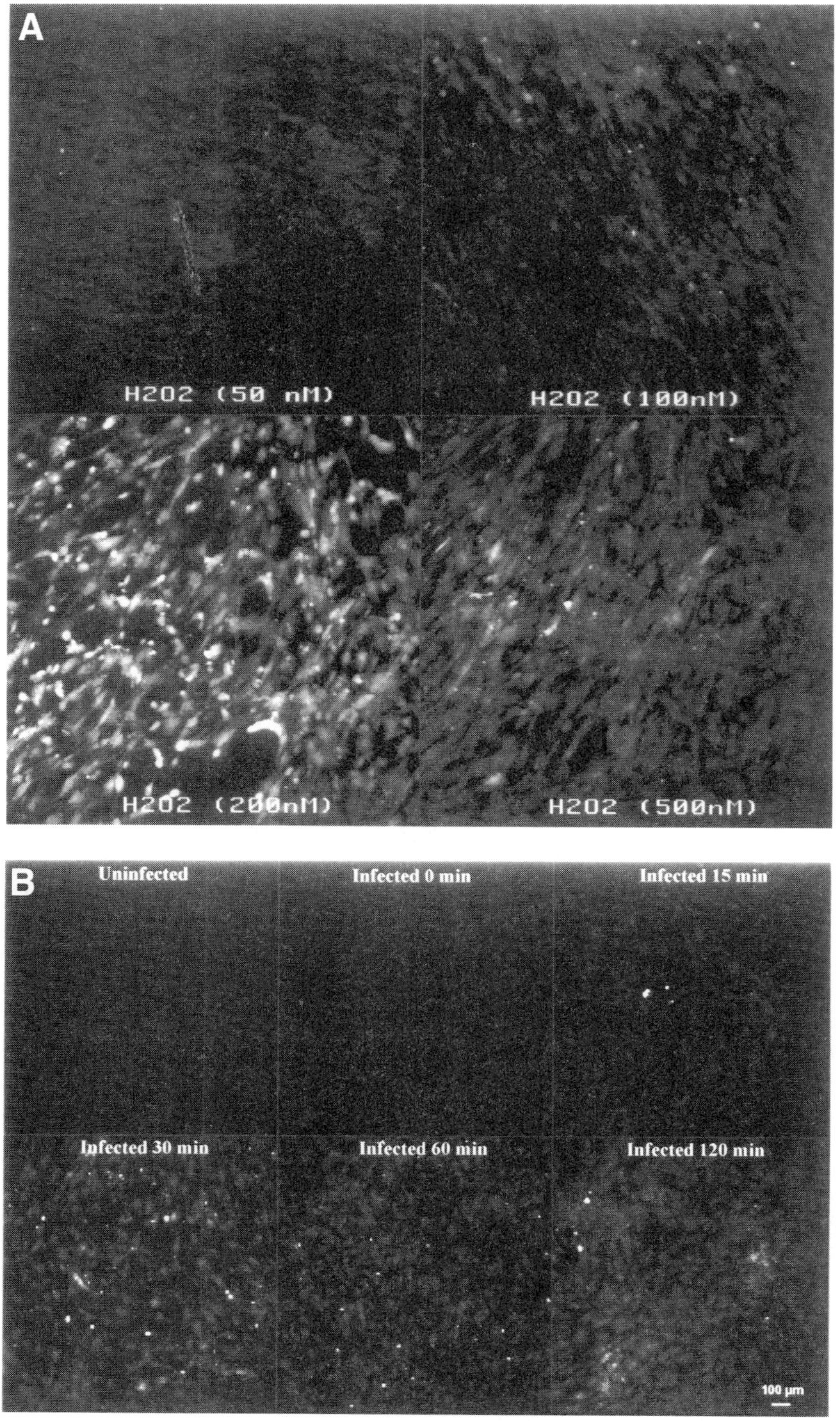

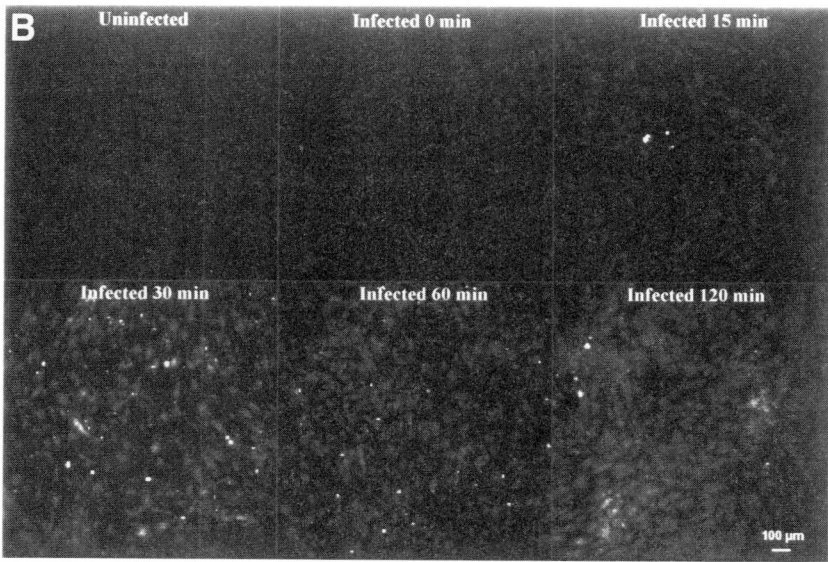

Fisher Scientific) caused a marked increase in intracellular fluorescence at doses ranging from 50–500 nM H_2O_2 (FIG. 1A). We next demonstrated that CMV infection caused an increase in intracellular fluorescence at 30, 60, and 120 minutes after infection (FIG. 1B). That the fluorescence we observed following CMV infection was in fact caused by reactive oxygen species, was confirmed by the finding that the response was suppressed in dose-dependent fashion by the antioxidant N-acetylcysteine, which effectively scavenges ·OH and H_2O_2 but not $O_2^{·-}$ [18] (FIG. 2, bar graph). Relative fluorescence was assessed by computerized measurement of color intensity in 3 fields per experimental well as described above.

Role of Xanthine/Xanthine Oxidase System in the Generation of CMV-Induced ROS

CMV interacts with at least 2 host-cell membrane receptors and induces multiple signaling events before synthesis of IE proteins is detectable.[19] Because SMC of large vessels are an important source of xanthine oxidase,[20] which leads to the production of $O_2^{·-}$ with subsequent generation of H_2O_2 and ·OH, we determined whether this enzyme system plays a role in the CMV/receptor signaling pathway. We found that oxypurinol, a specific inhibitor of xanthine oxidase,[21] dose-dependently inhibited ROS-dependent fluorescence in CMV-infected cells. Oxypurinol (Sigma) was prepared as 100 mM stock solution in DMSO, and added to SMC at a final concentration of 1–10 μM, for 1 hour before CMV infection. Cells were monitored as described with confocal microscopy (not shown).

Effect of CMV Infection on NF-κB Activation

Because CMV infection of fibroblasts activates NF-κB[12] and increases NF-κB binding to DNA,[22] we next determined whether a similar effect is induced by CMV infection of SMC and whether such an effect is mediated by the CMV-induced generation of ROS. Nuclear extracts of HeLa S3 cells of unstimulated or phorbol ester-treated cells were used as negative or as positive controls, respectively. In the same experiment, nuclear extracts of CMV-infected cells were reacted with ^{32}P-labeled oligonucleotide containing an NF-κB binding sequence. As determined by electrophoretic mobility shift assay (EMSA), CMV infection of human coronary SMC caused increased NF-κB binding to its cognate DNA as early as 1 minute after infection and further increased at 30 and 60 minutes after infection (3 μg of protein per well were used). The DNA-protein complexes were resolved on 7% native acrylamide gels. Electrophoresis was performed by running the gels in 5 mM TRIS/38 mM glycine buffer. The upper two bands of the coronary SMC extracts comigrated with the upper two bands of the phorbol ester-treated HeLa extracts, indicating that these bands represent the p65/p50 heterodimers and p50/p50 homodimers of NF-κB[23] (gel

---

FIGURE 1. A: Changes in intracellular redox state of human coronary SMC in response to H_2O_2 Confocal laser scanning microscopy was used to measure changes in the fluorescence induced by the intracellular oxidation of DCFH-DA (5 μM). The photomicrographs are representative of the 4 experiments performed. The fluorescence was maximal after exposure of cells to 200 nM H_2O_2. **B:** Cytomegalovirus infected cells generate intracellular ROS-dependent fluorescence as shown. Uninfected cells are negative.

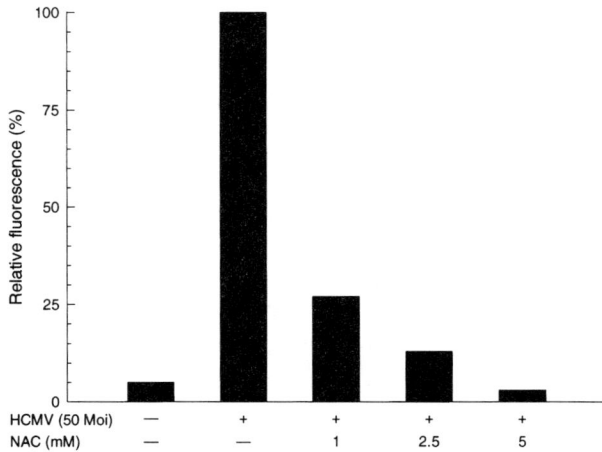

FIGURE 2. NAC pretreatment at 1, 2.5, and 5 mM for 1 hour decreased the fluorescence of CMV-infected cells. Relative fluorescence was quantified with the analysis software provided with the confocal microscope (see text).

not shown). Because increased binding of NF-κB occurs within 1 minute of infection and is abolished by antioxidants, the effect is caused by activation of preexisting NF-κB rather than new protein synthesis, and is mediated by a signaling event triggered by viral interaction with a cell surface receptor, with activation involving ROS.

Effects of H_2O_2, IE72, and Antioxidants on Transcriptional Activity of MIEP-CAT

The MIEP of CMV contains 4 NF-κB binding sites. We have shown increased NF-κB binding in SMC nuclear extracts indicating activation and translocation of NF-κB to the nucleus. To determine a) whether this factor is transcriptionally interactive with the MIEP in this SMC environment, and b) whether the increased ROS levels generated by the SMC response to viral infection enhances viral gene transcription, SMC were transfected a plasmid construct containing the MIEP fused to a CAT reporter gene (MIEP-CAT).[24,25] When H_2O_2 (50 mM) was added to the cells in serum-free medium for 20 hours (40 hours after transfection), MIEP transcriptional activity (CAT expression) increased 10-fold. Pretreatment of cells with NAC inhibited this increase (FIG. 3). These results demonstrate that in SMC, ROS transactivate the CMV MIEP, the first step in the cascade of CMV gene expression that ultimately leads to viral replication.

The immediate early gene product of CMV, IE72, a nuclear phosphoprotein, is expressed within 2–4 hours after infection, and is a potent transactivator of its own promoter, the major immediate early promoter (MIEP), thereby leading to a positive-feedback loop.[26]

To determine whether this activity is redox sensitive, we assessed the effects of antioxidants on the transactivational capacity of IE72. Cotransfection of MIEP-CAT with IE72 for 60 hours increased CAT activity in several cell types; the magnitude of the effect was cell-type specific. The increase was 10-fold in HeLa cells (not shown), 8-fold in primary rat aortic SMC (FIG. 3), and 2-fold in human coronary artery SMC (not shown).

Treatment for 8–10 hours (60–70 hours after cotransfection) with NAC or with the chemically unrelated antioxidant pyrrolidine dithiocarbamate (PDTC) inhibited the activation by IE72. Because of toxicity in human coronary SMC, PDTC was only used in rat SMC. Thus, the ability of IE72 to transactivate its own promoter, presumably through activation of NF-κB, is at least partly ROS dependent. To examine whether transactivation of the CMV MIEP promoter by IE72 is dependent on NF-κB binding sites, as previously suggested,[14] we cotransfected coronary SMC with

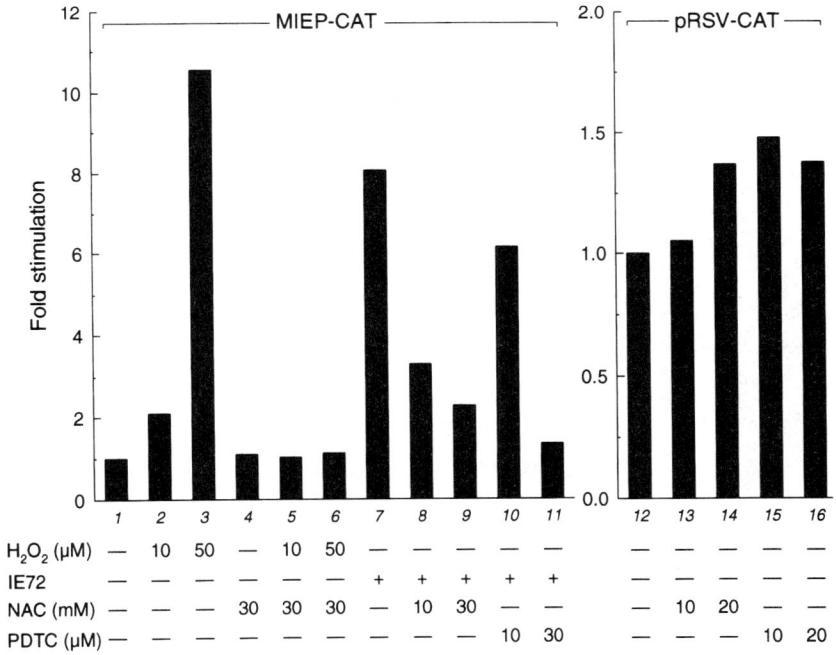

FIGURE 3. Transactivation by H_2O_2 or by IE72 of the MIEP-CAT reporter plasmid and inhibition of transactivation by NAC. Rat aorta SMC were either transfected with only CAT reporter plasmids controlled by the MIEP of CMV (*columns 1–6*) or cotransfected with MIEP-CAT and a vector expressing the IE 72 kD CMV protein (pRSV72, *columns 7–11*). At 40 hours after transfection, cells were treated with NAC or with the antioxidant and metal chelator PDTC for 8 hours. Shown are representative experiments of the CAT activity in SMC treated with or without H_2O_2 and with or without the antioxidants NAC or PDTC. Three independent experiments each were performed. Results from duplicate plates within any experiment varied no more than 15% from the average. Basal CAT activity of the MIEP construct was set to 1. The possibility that this effect was due to inhibition of the RSV promoter was excluded by showing that NAC or PDTC did not inhibit RSV-CAT activity in SMC (*columns 12–16*).

the IE72 expression plasmid and with a CAT reporter gene construct containing a minimal promoter and 3 NF-κB sites in tandem.[27] Cotransfection increased baseline activity 2-fold compared with that measured after cotransfection with the plasmid lacking IE72. This increase was inhibited by NAC. To confirm that the IE72-induced transactivation of the MIEP promoter occurred through NF-κB activation, we tested a construct of the 3XκB-CAT plasmid in which each of the NF-κB sites was mutated (3XmκB-CAT). The mutant construct abolished the transactivation capacity of IE72 in human coronary SMC, indicating that the effect of IE72 was mediated by the functional NF-κB sites (not shown).

Effect of Antioxidant Treatment on Viral Titer

Once we found that ROS transactivated the CMV MIEP and that antioxidants inhibited MIEP transactivation by ROS and by IE72, we determined whether such inhibition impaired the capacity of the virus to replicate and to produce cytopathic effects. Infected SMC (2 MOI) treated with NAC exhibited a concentration-dependent decrease in viral titer (FIG. 4). Treatment with 20 µM PDTC reduced viral titer by 50%. Cell counts (FIG. 4) indicated that NAC-induced decrease in viral titer was not due to cell death. Treatment of infected coronary SMC with 20 mM NAC also decreased the number of cytopathic foci. Thus, the inhibition of CMV gene transcription induced by antioxidants impairs the capacity of the virus to replicate and to exert cytopathic effects.

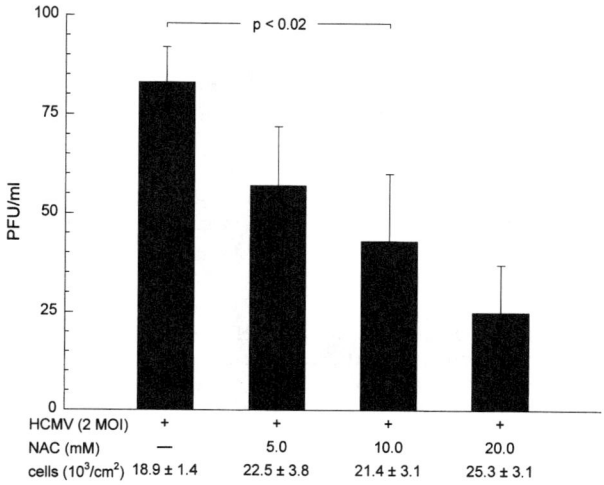

FIGURE 4. NAC treatment decreases viral titer of infected coronary SMC. The effect of NAC on viral titer is concentration-dependent. Shown are virus yields per milliliter of a 10% SMC homogenate and cell counts from parallel cultures at 96 hours after infection (mean of 3 experiments).

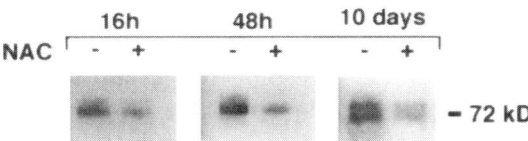

FIGURE 5. Effect of NAC on steady-state expression of IE72 in infected coronary SMC. For Western blot analysis, SMC were infected with 5 MOI of CMV and treated with vehicle or with 20 mM NAC for 16 hours, 48 hours, and at 10 days after infection. IE72 is usually seen as a doublet. The slower band probably is a hyperphosphorylated form of IE72.

Locus of Action of Antioxidants on the Cascade of Viral Gene Expression

The previous experiment demonstrated that antioxidants inhibit the ability of CMV to replicate. To ascertain where in the in the cascade of viral gene expression antioxidants exert their inhibitory effects, we treated CMV-infected coronary SMC with NAC and determined, by immunoblotting, whether these agents blocked the expression of IE72 or whether the effects were exerted further downstream. Immunoblotting of lysates (20 μg of protein per lane) of CMV-infected cells with the monoclonal anti-IE72 antibody (6E1, 1 μg/ml, Vancouver Biotech) demonstrated that NAC treatment reduced steady-state levels of IE72 protein at 16 hours, 48 hours and 10 days after infection (FIG. 5). These experiments demonstrate that antioxidants block expression of IE72. Because IE72 is essential for the expression of subsequent early and late genes of CMV, it is not possible to determine from these experiments whether a reduction of ROS levels also interferes with the expression of these downstream CMV genes independently of its effects on IE72 expression.

SUMMARY

The results of the present investigation indicate that generation of ROS is an essential component of the processes triggered by CMV infection of SMC. Thus, SMC ROS levels increase as early as 15 minutes after infection and persist for at least 2 hours (FIG. 1). Our results also demonstrate that ROS are essential to viral gene transcription, gene expression, and viral replication. One of the primary mechanisms of host defense against viral invasion is through the production of free radical pulses by phagocytes which destroy the infected cell. There is increasing evidence that other cell types produce free radicals, albeit in much lower amounts.[28,29] Because SMC of large vessels are an important source of xanthine oxidase,[20] this enzyme system could be a potential mechanism by which CMV infection might cause ROS generation. Increased xanthine oxidase activity which leads to the production of $O_2^{\cdot-}$ with subsequent generation of H_2O_2 and ·OH, has been implicated in the tissue injury occurring, for example, after experimental myocardial ischemia/reperfusion and, particularly relevant to our investigation, during influenza virus infection in mice.[30] In both of these models, it was proposed that tissue injury is caused by a cellular inflammatory response in which T cells and monocytes or neutrophils, are recruited to the site of injury and mediate release of free radicals through the deaminase-xanthine oxidase system.[31]

We have evidence that multiple enzyme systems are likely contributors. For example, recent reports demonstrated release within 15 minutes after infection of human fibroblasts of arachidonic acid that was blocked by inhibitors of phospholipase A_2 and protein kinase C.[32] Arachidonic acid release provides the substrate for and activates cyclooxygenases and lipoxygenases which produce multiple free radicals. We have recently reported in CMV-infected SMC rapid induction of cyclooxygenase-2 (COX-2), an immediate early inflammatory gene. COX-2 activity generates free radicals and causes release of the immune modulator prostaglandin E2. We further demonstrated that aspirin, via inhibition of ROS, COX-2 and NF-κB, also inhibits CMV gene expression and viral infectivity.[25]

Generation of ROS can be viewed as a protective mechanism of the cell, which can contribute to apoptosis[33] and thereby prevent the infecting virus from replicating and infecting neighboring cells. CMV appears to have adapted to the cellular defense mechanism involving the release of ROS and arachidonic acid, and to subsequent activation of COX-2 and NF-κB. The latter regulates transcription of multiple genes of the immune and inflammatory responses including COX-2 and adhesion molecules. Furthermore, the immediate early viral proteins IE72 and IE84 are transcriptional activators of NF-κB and COX-2.[22,25]

During latent infection, stimuli to host cells that activate NF-κB also lead to activation of the viral promoter and to direct expression of viral IE genes. It is tempting to speculate that the virus has evolved to benefit from the cellular mechanism of ROS generation to activate its own genetic program and to ensure frequent albeit abortive reactivation from latency. Moreover, since CMV DNA is present in the wall of atherosclerotic vessels and in restenosis lesions,[5,9] since ROS are generated in response to injury,[3] and since early atherosclerotic changes in the vessel wall impair the capacity to scavenge free radicals,[34] our finding that ROS are conducive to CMV gene transcription and expression and to viral replication provides additional evidence compatible with the hypothesis that CMV plays a contributory role in the development of restenosis and atherosclerosis.

REFERENCES

1. GERSCHMAN, R., D.L. GILBERT, S.W. NYE, P. DWYER & W.O. FENN. 1954. Oxygen poisoning and x-radiation: a mechanism in common. Science **119:** 623–629.
2. CROSS, C.E., B. HALLIWELL, E.T. BORISH, W.A. PRYOR, B.N. AMES, R.L. SAUL, J.M. MCCORD & D. HARMAN. 1987. Oxygen radicals and human disease. Ann. Intern. Med. **107:** 526–545.
3. GRECH, E.D., N.J.F. DODD, M.J. JACKSON, W.L. MORRISON, E.B. FARAGHER & D. RAMSDALE. 1996. Evidence for free radical generation after primary percutaneous transluminal coronary angioplasty recanalization in acute myocardial infarction. Am. J. Cardiol. **77:** 122–127.
4. AMES, B.N. 1989. Endogenous oxidative DNA damage, aging, and cancer. Free Rad. Res. Commun. **7:** 121–128.
5. SPEIR, E., R. MODALI, E.S. HUANG, M.B. LEON, F. SHAWL, T. FINKEL & S.E. EPSTEIN. 1994. Potential role of human cytomegalovirus and p53 interaction in coronary restenosis. Science **265:** 391–394. Comments.
6. VOGELSTEIN, B. 1990. Cancer. A deadly inheritance. Nature **348:** 681–682.

7. RUSSO, T., N. ZAMBRANO, F. ESPOSITO, R. AMMENDOLA, F. CIMINO, M. FISCELLA, J. JACKMAN, M. O'CONNOR, C.W. ANDERSON & E. APPELLA. 1995. A p53-independent pathway for activation of waf 1/CIP1expression following oxidative stress. J. Biol. Chem. **270:** 29386–29391.
8. HAINAUT, P. & J. MILNER. 1993. Redox modulation of p53 conformation and sequence-specific DNA binding in vitro. Cancer Res. **53:** 4469–4473.
9. MELNICK, J.L., E. ADAM & M.E. DEBAKEY. 1990. Possible role of cytomegalovirus in atherogenesis. JAMA **263:** 2204–2207.
10. BRUGGEMAN, C.A. & M.C. VAN DAM-MIERAS. 1991. The possible role of cytomegalovirus in atherogenesis. Prog. Med. Virol. **38:** 1–26.
11. SCHRECK, R., P. RIEBER & P.A. BAEUERLE. 1991. Reactive oxygen intermediates as apparently widely used messengers in the activation of the NFκB transcription factor and HIV-1. EMBO J. **10:** 2247–2258.
12. KOWALIK, T.F., B. WING, J.S. HASKILL, J.C. AZIZKHAN, A.S. BALDWIN, JR. & E.S. HUANG. 1993. Multiple mechanisms are implicated in the regulation of NF-κB activity during human cytomegalovirus infection. Proc. Natl. Acad. Sci. USA **90:** 1107–1111.
13. GRUNDY, J.E., J.A. MCKEATING, P. WARD, A. SANDERSON & P.A. GRIFFITHS. 1987. Beta-2-microglobulin enhances the infectivity of cytomegalovirus and when bound to the virus enables class I HLA molecules to be used as a virus receptor. J. Gen. Virol. **68:** 793–803.
14. CHERRINGTON, J.M. & E.S. MOCARSKI. 1989. Human cytomegalovirus ie transactivate the alpha promoter-enhancer via an 18-base-pair repeat element. J. Virol. **63:** 1435–1440.
15. SEGAL, A.W. 1989. The electron transport chain of the microbicidal oxidase of phagocytic cells and its involvement in the molecular pathology of chronic granulomatous disease. J. Clin. Invest. **83:** 1785–1793.
16. HUANG, E.S. 1975. Human cytomegalovirus, III: virus-induced DNA polymerase. J. Virol. **16:** 298–310.
17. ZHU, H., G.L. BANNENBERG, P. MOLDEUS & H.G. SHERTZER. 1994. Oxidation pathways for the intracellular probe 2',7'-dichlorofluorescein. Arch. Toxicol. **68:** 582–587.
18. ARUOMA, O.I., B. HALLIWELL, B.M. HOEY & J. BUTLER. 1989. The antioxidant action of N- acetylcysteine: its reaction with hydrogen peroxide, hydroxyl radical, superoxide, and hypochlorous acid. Free Rad. Biol. Med. **6:** 593–597.
19. ALBRECHT, T., I. BOLDOGH, M.P. FONS & T. VALYI-NAGY. 1993. Activation of protooncogenes and cell activation signals in the initiation and progression of human cytomegalovirus infection. *In* Frontiers of Virology 2. Y. Becker, G. Darai & E.S. Huang, Eds. :384–411. New York. Springer Verlag.
20. HELLSTEN-WESTING, Y. 1993. Immunohistochemical localization of xanthine oxidase in human cardiac and skeletal muscle. Histochemistry **100:** 215–222.
21. SPECTOR, T. 1988. Oxypurinol as an inhibitor of xanthine oxidase-catalyzed production of superoxide radical. Biochem. Pharmacol. **37:** 349–352.
22. YUROCHKO, A.D., T.F. KOWALIK, S.M. HUONG & E.S. HUANG. 1995. Human cytomegalovirus upregulates NF-κB activity by transactivating the NF-κB p105/p50 and p65 promoters. J. Virol. **69:** 5391–5400.
23. BAEUERLE, P.A. & D. BALTIMORE. 1989. A 65-kD subunit of active NF-κB is required for inhibition of NF-κB by I-κB. Genes Dev. **3:** 1689–1698.
24. SPEIR, E., T. SHIBUTANI, Z.X. YU, V. FERRANS & S.E. EPSTEIN. 1996. Role of reactive oxygen intermediates in cytomegalovirus gene expression and in the response of human smooth muscle cells to viral infection. Circ. Res. **79:** 1143–1152.
25. SPEIR, E., Z.X. YU, V. FERRANS & E.S. HUANG. 1998. Aspirin attenuates cytomegalovirus infectivity and gene expression mediated by cyclooxygenase-2 in coronary artery smooth muscle cells. Circ. Res. **83:** 210–216.
26. STENBERG, R.M., J. FORTNEY, S.W. BARLOW, B.P. MAGRANE, J.A. NELSON & P. GHAZAL. 1990. Promoter-specific transactivation and repression by human cytomegalovirus immediate-early proteins involves common and unique protein domains. J. Virol. **64:** 1556–1565.

27. FINCO, T.S. & A.S. BALDWIN, JR. 1993. Kappa B site-dependent induction of gene expression by diverse inducers of nuclear factor B requires raf-1. J. Biol. Chem. **268:** 17676–17679.
28. PETERHANS, E. 1979. Sendai virus stimulates chemiluminescence in mouse spleen cells. Biochem. Biophys. Res. Commun. **91:** 383–392.
29. MEIER, B., H.H. RADEKE, S. SELLE, M. YOUNES, H. SIES, K. RESCH & G.G. HABERMEHL. 1989. Human fibroblasts release reactive oxygen species in response to interleukin-1 or tumor necrosis factor-alpha. Biochem. J. **263:** 539–545.
30. AKAIKE, T., M. ANDO, T. ODA, T. DOI, S. IJIRI, S. ARAKI & H. MAEDA. 1990. Dependence on O_2-generation by xanthine oxidase of pathogenesis of influenza virus infection in mice. J. Clin. Invest. **85:** 739–745.
31. GRISHAM, M.B., L.A. HERNANDEZ & D.N. GRANGER. 1986. Xanthine oxidase and neutrophil infiltration in intestinal ischemia. Am. J. Physiol. **251:** G5657–G574.
32. ABUBAKAR, S., I. BOLDOGH & T. ALBRECHT. 1990. Human cytomegalovirus stimulates arachidonic acid metabolism through pathways that are affected by inhibitors of phospholipase A2 and protein kinase C. Biochem. Biophys. Res. Commun. **166:** 953–959.
33. JACOBSON, M.D. 1996. Reactive oxygen species and programmed cell death. Trends Biochem. Sci. **21:** 83–86.
34. OHARA, Y., T.E. PETERSON & D.G. HARRISON. 1993. Hypercholestrolemia increases endothelial superoxide anion production. J. Clin. Invest. **91:** 2546–2551.

Reactive Species in Sickle Cell Disease

MUTAY ASLAN,[a–c] DENYSE THORNLEY-BROWN,[b–d]
AND BRUCE A. FREEMAN[a–c,e]

[a]*Department of Anesthesiology,* [b]*Center for Free Radical Biology,* [c]*Comprehensive Sickle Cell Disease Center,* [d]*Department of Medicine, University of Alabama at Birmingham, Birmingham, Alabama 35233, USA*

> ABSTRACT: The red cell is a relatively abundant locus of both free radical generation and reaction. Erythrocytes have a high content of unsaturated membrane lipids, a rich oxygen supply and are densely packed with redox-active hemoglobin residues. In response, red cells have a highly evolved and well-integrated network of oxidant defense mechanisms that lend an ability to withstand oxidative stress. In the case of congenital hemoglobin mutations that underlie sickle cell disease, they become very susceptible to free radical-mediated injury by virtue of enhanced endogenous rates of production of reactive species and impairment of tissue free radical defense mechanisms. In sickle cell disease, a combination of these susceptibility factors are hypothesized to lead to an overall impairment of vascular function, in large part due to loss of "bioactive" nitric oxide via the free radical-mediated consumption of this vasoactive molecule.

RED CELL PRODUCTION OF REACTIVE OXYGEN SPECIES

Red cell hemoglobin is a quantitatively significant source of superoxide generation in biological systems. There is an electron transfer in the bonding interaction between the heme iron and oxygen in oxygenated hemoglobin.[1] When hemoglobin deoxygenates, the heme iron normally remains in the Fe(II) ferrous state. Alterations in this exchange, wherein hemoglobin auto-oxidizes, results in the formation of methemoglobin and superoxide.[2] There is a normal "physiologic" rate of red cell methemoglobin formation that provides a continual source of superoxide production that in turn generates hydrogen peroxide and oxygen as byproducts of dismutation.[3] In hemoglobin preparations obtained from red cells of sickle cell patients (HbS), autoxidation of oxygenated hemoglobin is 1.7 times faster than oxygenated HbA.[6] Also, sickle red cells have been reported to generate ~2-fold greater extents of superoxide, hydrogen peroxide, hydroxyl radical and lipid oxidation products compared with HbA-containing red cells.[7]

---

[e]Address for correspondence: Bruce A. Freeman, Ph. D., Department of Anesthesiology, 946 Tinsley Harrison Tower, 619 South 19th Street, University of Alabama at Birmingham, Birmingham, AL 35233-6810. Voice: 205-934-4234; fax: 205-934-7437.
  e-mail: bruce.freeman@ccc.uab.edu

ABERRANT IRON HOMEOSTASIS IN SICKLE RED CELLS

Oxidative damage to hemoglobin has been shown to cause irreversible hemoglobin denaturation, precipitation and propensity for further autoxidation.[8] Aggregation of precipitated hemoglobin gives rise to Heinz bodies that can migrate to the cell surface.[9,10] This decompartmentalization of cellular iron can then lead to changes in both membrane structure and function.

Abnormal association of both heme and nonheme iron with sickle red cell membranes is well documented.[11–14] The suppression of oxidative membrane lipid damage after incubation with desferrioxamine indicates that the iron associated with the red cell membrane was in the free Fe(III) state.[13,14] Iron chelators are unable to prevent or reverse iron accumulation in the sickle red cell membrane, revealing that Fe(III) associates with the red cell membrane with extremely high avidity.[15]

MECHANISM AND CONSEQUENCES OF IRON-MEDIATED MEMBRANE DAMAGE

Fe(III) can initiate lipid peroxidation reactions when a reducing agent is present. Initiation of lipid peroxidation *in vitro* and *in vivo* also can occur via production of hydroxyl radical.[16] In an iron catalyzed Haber-Weiss reaction, superoxide, ascorbate, thiols and other reductants can reduce Fe(III) to Fe(II), which in turn reduces hydrogen peroxide to hydroxyl radical.[3] In the presence of a superoxide generating system, hemoglobin-associated heme iron thus promotes hydroxyl radical formation in a dose dependent fashion.[17] Studies with methemoglobin exposed to hydrogen peroxide also support the concept that excess hydrogen peroxide decomposes heme, induces iron release that then in turn reacts with hydrogen peroxide to form hydroxyl or perferryl radical species.[18,19] In the absence of radical chain propagation terminating reductants such as α-tocopherol, lipid peroxidation can become a chain reaction that initiates further peroxidation.

The plasma membrane is a critical site for free radical or oxidative reactions, with unsaturated esterified fatty acids and transmembrane proteins containing oxidizable amino acids being susceptible.[3] The oxidation of unsaturated membrane lipids can result in bilayer fluidity changes, increased ion permeability, covalent crosslinking of protein and lipid membrane constituents, inactivation of membrane enzymes and receptors, polypeptide strand scission, DNA damage and, in nucleated cells, mutagenesis.[20] Malondialdehyde (MDA) and hydroxynonenal, products of lipid peroxidation, can additionally react with proteins, nucleic acids and lipids to further exert toxic effects.[3]

Membrane lipid peroxidation was determined in sickle erythrocytes by the assessment of thiobarbituric acid-reactive products. Sickle red cells showed both an increased endogenous extent of oxidized lipid and a greater tendency for further lipid peroxidation, when compared with HbA-containing erythrocytes.[13,21,22]

HOMOCYSTEINE AND B VITAMIN LEVELS IN SICKLE CELL DISEASE — IMPLICATIONS FOR OXIDATIVE REACTIONS

Homocysteine is a naturally occurring sulfur-containing amino acid that is produced during the metabolism of methionine. Dietary methionine is converted to S-adenosyl methionine (SAM), which is then demethylated to S-adenosyl homocysteine (SAH). SAH is hydrolyzed to homocysteine and adenosine. Homocysteine can be remethylated to methionine by vitamin B_{12}-dependent methionine synthase. This reaction also requires methylenetetrahydrofolate (MTHF) as a cofactor. The majority of homocysteine enters a pyridoxal-5-phosphate (vitamin B_{12})-dependent catabolic pathway where it is converted to cystathonine.[23,24] Human plasma contains both reduced and oxidized species of homocysteine, with 98–99% of total plasma homocysteine in the oxidized form.[23] The sulfhydryl group of homocysteine is labile and will readily autoxidize. Thus, Cu(II) addition to homocysteine will cause oxidation of the thiol and consumption of oxygen in a dose-dependent fashion, inferring the production of reactive oxygen species. Homocysteine oxidation can also be catalyzed by normal human serum. The molar ratio of homocysteine oxidized to oxygen consumed was ~ 4. Addition of catalase decreased the oxygen consumption by nearly half, indicating that hydrogen peroxide was formed during the reaction, probably following dismutation of the primary product of autoxidation, superoxide.[25] Homocysteine exhibits a significant prooxidative potential in the presence of either copper or iron which, in the presence of target cells, can lead to red cell hemolysis and Na^+/K^+ ATPase inhibition.[26]

Since B vitamins are a central requirement for homocysteine metabolism, dietary insufficiency or malabsorbption of vitamin B_{12}, vitamin B_6 or folate can give rise to hyperhomocysteinemia.[23,24,27,28] Importantly, plasma vitamin B_{12}, vitamin B_6 and folic acid levels are lower in patients with sickle cell anemia when compared to normal controls.[29–31] Recently, serum homocysteine and red cell folate levels were measured in the serum of ~100 patients with sickle cell disease.[32] Homocysteine and folate levels were found to be inversely correlated. Also, the incidence of cardiovascular disease (stroke), long associated with hyperhomocysteinemia, was followed and revealed that plasma homocysteine levels for the stroke group were significantly higher than those without stroke.

ALTERED DEFENSE MECHANISMS AGAINST OXIDATIVE STRESS IN SICKLE CELL DISEASE

The enzymatic antioxidant system that protects aerobic organisms from free radical damage include cytoplasmic, mitochondrial and extracellular superoxide dismutases (SOD), glutathione peroxidase (GPx) and catalase. SOD is a metalloprotein that dismutates superoxide to yield hydrogen peroxide and oxygen. At pH 7.4, the rate constant for this enzyme-catalyzed reaction is $\sim 2 \times 10^9 \, M^{-1} \, sec^{-1}$ which is approximately 10^4 times faster than the spontaneous dismutation of superoxide.[33] The hydrogen peroxide that is formed from this reaction and following divalent reduction of oxygen is removed by the heme-containing enzyme catalase, by glutathione peroxidase or by thioredoxin-dependent peroxidases. Catalase catalyzes the dismutation of hydrogen peroxide to water and oxygen or in the presence of alcohols, catalyzes the peroxidatic scavenging of hydrogen peroxide. Glutathione peroxidase catalyzes the reduction of hydrogen peroxide by reduced glutathione (GSH). The resulting glutathione disulfide (GSSG) is reduced by NADPH derived in large part from the pentose phosphate pathway.[3] Glutathione peroxidase has a lower K_m for hydrogen peroxide than catalase;[34] therefore it can scavenge at lower hydrogen peroxide concentrations. Studies of sickle erythrocytes have reported increased levels of superoxide dismutase and decreased levels of glutathione peroxidase and catalase,[21] suggesting that endogenously produced hydrogen peroxide cannot be removed by sickle erythrocytes as efficiently as for normal red blood cells (see FIG. 1 for a summary of this section).

There are also a variety of low molecular weight antioxidants that are important in the cellular defenses against oxidative stress. α-Tocopherol and β-carotene (vitamin A) are lipophilic molecules that scavenge free radicals and thus inhibit lipid

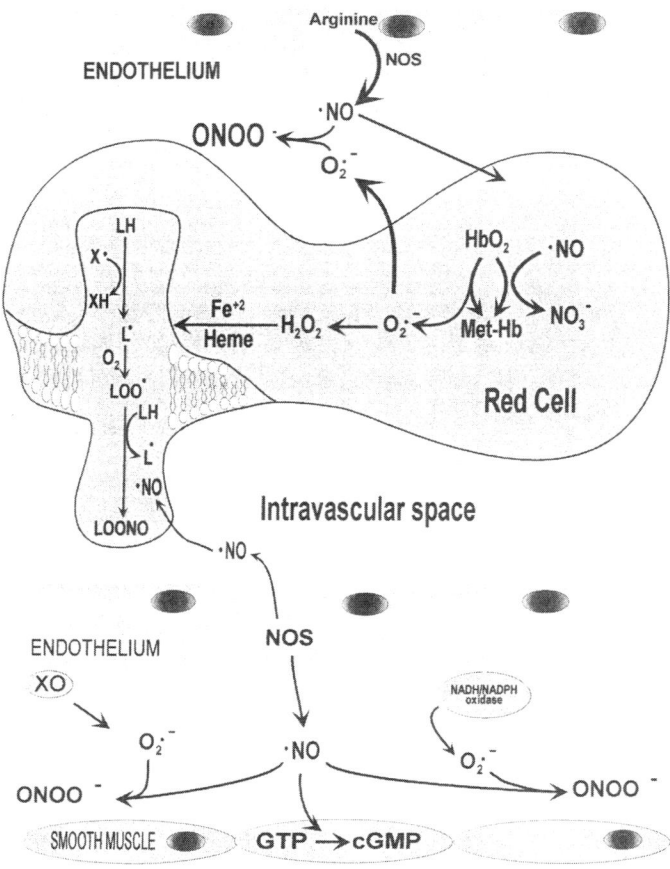

FIGURE 1.

peroxidation.[3] α-Tocopherol and ascorbate can also serve as reducing agents and, during pathological conditions, induce heme release or reduce metals and act as prooxidants. In such cases, the amount of vitamin E that is present and availability of redox-active metal centers determines its prooxidant or antioxidant effect.[35,36] There is ~40% reduction in plasma carotene levels and ~30% reduction in plasma vitamin E levels in sickle cell disease.[37,38] Red blood cell membrane vitamin E levels were also depleted in these patients.[38]

The intracellular thiol reductant that directly scavenges reactive species and serves as a GPx cofactor, GSH, contributes to the reduction of hydrogen peroxide, lipid peroxides and other oxidizing species, thus protecting red cells against oxidative damage.[3] The level of red blood cell GSH was ~50% lower in sickle cell patients when compared with HbA red cells.[39]

NITRIC OXIDE-OXYGEN RADICAL INTERACTIONS IMPAIR VASCULAR FUNCTION

The pathogenesis of sickle cell disease evolves from the polymerization of hemoglobin S (HbS), but the deleterious effects of the disease are related to episodic vascular occlusion.[40] It seems likely that the pathophysiology underlying vasoocclusion is not a simple process. There are a variety of possible contributory abnormalities including hemoglobin polymerization,[41] increased endothelial cell adherence,[42,43] increased blood viscosity, coagulopathy, endothelial dysfunction[44] and altered vascular tone.[45] Local control of vascular tone is mediated in part by the free radical species nitric oxide (NO), that is released by endothelial cell nitric oxide synthase. It is hypothesized herein that defective NO-dependent vascular relaxation and adherence properties may play an important role in the episodic pain, organ blood flow abnormalities and acute chest syndrome occurring in sickle cell disease. Reduced plasma levels of L-arginine,[46,47] increased plasma concentrations of NO metabolites[48] and altered vascular reactivity in response to endogenous angiotensin II [49] in subjects with sickle cell disease support the concept of defective NO action. Since the generation of free radicals (superoxide and lipid peroxyl radicals) are enhanced in sickle cell patients, this can create a reactive milieu that will inactivate NO-mediated vascular relaxation and yield secondary reactive species that may also impair vascular function. To fully grasp this concept, it must first be appreciated that NO reacts at almost diffusion-limited rates with superoxide and lipid peroxyl radicals to yield novel and often reactive products.

NO is a freely diffusible intercellular messenger produced by a variety of mammalian cells including vascular endothelium, neurons, smooth muscle cells, macrophages, neutrophils, platelets and pulmonary epithelium.[50–52] Nitric oxide is synthesized by a family of three nitric oxide synthase (NOS) enzymes that have a C-terminal reductase and an N-terminal oxygenase domain.[50] The synthesis of NO by vascular endothelium regulates the vasodilator tone.[53] Shear stress or receptor activation of vascular endothelium by bradykinin or acetylcholine results in an influx of calcium and activation of calmodulin (CaM).[54] Calmodulin binding to endothelial NOS stimulates enzyme activity by facilitating electron transfer within the reductase domain of NOS.[50] The generation of NO occurs by the oxidative deamination of

L-arginine to L-citrulline in the presence of the cofactors NADPH, flavin adenine dinucleotide (FAD), flavin mononucleotide (FMN), iron protoporphyrin IX and tetrahydrobiopterin (H_4 biopterin).[51] The NO formed diffuses to nearby smooth muscle cells in which it reacts with the ferrous iron in the heme group of guanylate cyclase, resulting in enhanced synthesis of cyclic GMP from guanosine triphosphate.[54] Cyclic GMP then interacts with cGMP-dependent protein kinases, cGMP-regulated cyclic nucleotide phosphodiesterases and c-GMP-regulated ion channels. Cyclic GMP-dependent kinase stimulates Ca^{2+}-ATPase and lowers intracellular calcium levels, thus mediating smooth muscle cell relaxation.[55] Stimulation of guanylate cyclase also inhibits platelet aggregation and adhesion to endothelium.[55,56]

The reaction of NO with metalloproteins, protein sulfhydryls and oxygen-derived free radicals enables it to modulate inflammation and oxidative stress.[51] The reactivity of NO with reactive oxygen species leads to a diversity of both toxic and cytoprotective effects. Nitric oxide is more likely to act as a prooxidant when the concentration of superoxide is greater than or equal to that of NO, via the facile radical reaction of these species to form the potent oxidant peroxynitrite ($ONOO^-$).[57] Physiologic studies have demonstrated that superoxide accelerates the "destruction" of NO and thus inhibits NO-dependent vascular relaxation. This "inactivation" of NO was prevented by SOD.[58,59] Importantly, NO is not lost in this reaction, it reacts with superoxide to give peroxynitrite and its conjugate acid peroxynitrous acid (ONOOH, $pK_a = 6.8$). The rate constant for this reaction is 2×10^{10} M^{-1} sec^{-1}, faster than the SOD-catalyzed enzymatic dismutation of superoxide.[60] Peroxynitrite is a potent oxidant with a half-life of about 1.6 sec at neutral pH.[51] Peroxynitrous acid reacts by two pathways, with the first pathway yielding nitrate (NO_3^-) without forming strong oxidant intermediates. The second pathway forms hydroxyl radical and nitrogen dioxide ($NO_2 \cdot$), a potent oxidant that can initiate fatty acid oxidation and nitration of aromatic amino acids.[61] Peroxynitrite reactivity is also influenced by CO_2, with the formation of a reactive nitrosoperoxocarbonate ($ONOOCO_2^-$) intermediate. Consequently, CO_2 stimulates both peroxynitrite decay and enhances peroxynitrite-mediated nitration of molecules by nearly 2-fold.[62]

The prooxidant versus antioxidant effects of NO on lipid and lipoprotein oxidation depends on the relative local concentrations of reactive oxygen species and on the rate of NO synthesis. In the presence of superoxide, NO serves to enhance lipid peroxidation by a peroxynitrite-mediated and metal independent mechanism. Analysis of membrane and lipoprotein preparations reveals that NO alone serves as a potent terminator of radical chain propagation reactions and can yield novel nitrogen-containing lipid oxidation products.[63,64] Nitric oxide, by virtue of its lipophilicity and high reactivity for lipid peroxyl radical species (LOO·), will also prevent vitamin E depletion during lipid peroxidation. In support of this concept, it has been shown that the reaction between NO and chain propagating peroxyl radical species is three times faster than for vitamin E.[65]

The reaction of NO with the metal centers in proteins is important for its biological activity.[51,54] It has been shown that NO exhibits an antioxidant effect against metal catalyzed lipid oxidation.[51] A recent study examined the effects of NO on excess cellular heme by exposing endothelial cells to heme and an NO donor. Nitrosylation of the cellular heme prevented the increase of catalytic iron in concert with an inhibition of heme-induced heme oxygenase and ferritin synthesis. It was suggested that NO acted as a buffer against excess heme by reducing its redox activity and slowing the release of free iron.[66] Nitric oxide has also been shown to prevent oxidative injury associated with oxoferryl-hemoglobin formation via the reduction of ferryl hemoglobin to methemoglobin.[67] A principal fate of NO in the vasculature is its reaction with hemoglobin. Studies have shown that NO binds very rapidly to deoxyhemoglobin forming a stable hemoglobin (Fe^{+2})-NO complex. Nitric oxide also reacts with and converts oxygenated hemoglobin to methemoglobin and nitrate. It was recently concluded that the major metabolic pathway for endogenously formed NO in healthy human subjects was uptake into the red blood cells and conversion to nitrate and methemoglobin.[68,69] It has been proposed that S-nitrosylation of hemoglobin (SNO-HB) is also important in the regulation of NO metabolism, with the arterial-venous difference in SNO-Hb concentrations suggesting that hemoglobin also acts as an NO donor in the systemic circulation.[70] This concept awaits further confirmation.

Nitric oxide may also modulate the adverse effects of homocysteine. Studies have shown that NO released from endothelial cells in the presence of homocysteine can lead to the formation of S-nitrosohomocysteine, a species that does not support hydrogen peroxide generation and can serve as a vasodilator and platelet inhibitor. These data suggest that the toxic effects of homocysteine may be both a cause and a result of impaired NO activity.[71–73]

Reactive oxygen intermediates have been shown to activate the redox-sensitive transcription factor NFκB. NFκB controls the expression of a variety of genes involved in inflammatory and immune responses,[74] with vascular cell adhesion molecule-1 (VCAM-1) and intercellular adhesion molecule-1 (ICAM-1) being key examples of these genes.[75] Both decreased NO levels and accelerated formation of reactive oxygen species increases the expression of VCAM-1 and ICAM-1,[74] suggesting that NO may function as an important cellular mediator of redox-sensitive gene expression. AP-1 is another transcription factor that can be activated by prooxidant conditions such as exposure to hydrogen peroxide and UV irradiation. AP-1 is composed of jun and fos gene products which form homodimeric (Jun/Jun) or heterodimeric (Jun/Fos) complexes.[76] Induction of fos and jun genes occur rapidly, thus these genes are referred to as early response genes that can be induced by oxidant exposure.[74] Erythrocytes from sickle cell patients display enhanced binding to endothelial cells via interactions between erythrocyte α4β1-integrin and endothelial cell VCAM-1.[43,77] Interestingly, exposure of human endothelial cell cultures to previously sickled erythrocytes resulted in a 4- to 8-fold induction of transcription of the gene encoding the vasoconstrictive mediator endothelin (ET-1).[78] ET-1 mRNA expression and protein levels also increased in the endothelial cells exposed to plasma obtained from sickle cell patients during acute chest syndrome, with plasma obtained from routine visits did not alter ET-1 mRNA expression or protein levels.[79] This induction of the ET-1 gene in endothelial cell cultures is likely to occur via redox sensitive AP-1 or NFκB transcriptional factor activation by red cell or plasma-derived oxidation products.

POTENTIAL NOVEL THERAPIES FOR SICKLE CELL DISEASE BASED ON OXIDATIVE MECHANISMS

Improved understanding of the abnormal oxidative processes that occur in sickle cell disease has led to new insights into the mechanism of action of some currently accepted therapies and suggest new therapies for this disease. Hydroxyurea, which is widely used in sickle cell disease, may have beneficial effects on vascular functions that were not appreciated when the drug was first introduced. Other therapies, including arginine and angiotensin converting enzyme inhibitors may increase nitric oxide levels. The use of agents aimed at lowering the plasma concentration of homocysteine and scavenging free radicals may also be of benefit to patients with sickle cell disease.

HYDROXYUREA IN SICKLE CELL DISEASE

In a multicenter, randomized placebo controlled trial, patients with sickle cell disease who received hydroxyurea had lower rates of vaso-occlusive crises, increased time between crises, a lower incidence of acute chest syndrome and a reduced need for blood transfusions.[80] The mechanism underlying these benefits was initially thought to be in large part related to an increase in the proportion of cells containing fetal hemoglobin and an increase in the erythrocyte mean corpuscular volume, thereby reducing the potential for hemoglobin polymerization.[80] Recent studies have suggested that another beneficial effect of hydroxyurea may be the generation of NO. *In vitro* studies have shown that NO is formed when hydroxyurea is incubated with hydrogen peroxide and a heme-containing protein such as myoglobin or hemoglobin.[81] Administration of hydroxyurea to rats led to a dose dependent increase in nitrosyl hemoglobin, consistent with increased nitric oxide generation.[82] Both oxy- and deoxyhemoglobin S have been shown to react with hydroxyurea to form a methemoglobin hydroxyurea complex. The conversion of deoxyhemoglobin S to methemoglobin is four times faster than that of oxyhemoglobin S, suggesting that one of the beneficial effects of this drug may be a reduction in the concentration of deoxyhemoglobin and therefore a reduction of hemoglobin polymerization. While Stolze and Nohl have demonstrated that hydroxyurea can cause an increase in lipid peroxidation and of free radical formation,[83] no serious adverse effects of this drug were observed in the multicenter trial.[80]

USE OF ANGIOTENSIN CONVERTING ENZYME INHIBITORS IN SICKLE CELL DISEASE

Angiotensin converting enzyme inhibitors inhibit the formation of angiotensin II. A second mechanism of action is decreasing bradykinin metabolism, thereby increasing bradykinin levels. Bradykinin simulates the release of nitric oxide and inhibits the secretion of endothelin from endothelial cells, resulting in vasodilation. In human studies, angiotensin converting enzyme inhibitors have been shown to augment the vasodilatory response to bradykinin in forearm[90] and coronary[91] beds in a nitric oxide mediated manner. Angiotensin converting enzyme inhibitors reduce microalbuminuria[92] and proteinuria[93] in patients with sickle cell anemia, but to date, there have been no published data examining the effect of these agents on nitric oxide metabolism, the frequency of vaso-occlusive crises or other clinical manifestations of the disease. Patients with sickle cell disease have altered potassium metabolism, however, which may limit the usefulness of these agents.[94]

ARGININE SUPPLEMENTATION IN SICKLE CELL DISEASE

The observation that patients with sickle cell anemia have significantly reduced concentrations of arginine in their plasma and urine[84] suggests that this group of patients may have increased utilization of this amino acid and therefore increased nutritional requirements. Arginine is an important metabolic intermediate in several biological reactions including the synthesis of nitric oxide, creatine, agmatine, orotic acid and urea.[85] It also has major endocrine effects, stimulating the secretion of insulin, glucagon, growth hormone, prolactin and corticosteroids and inhibiting the release of somatostatin.[85] To date, there are no published data on the effects of acute or chronic arginine administration to subjects with sickle cell disease; however, the effects of pharmacological doses of arginine both in humans and animals suggest that it may be beneficial to sickle cell patients. In healthy subjects, intravenous arginine increases cerebral and limb blood flow,[86,87] and decreases platelet aggregation and blood viscosity.[87] In addition, arginine, whether given orally or intravenously, increases the glomerular filtration rate[88] and oral arginine increases renal plasma flow.[88] Chronic administration of arginine to hypercholesterolemic rabbits results in decreased surface area of atherosclerotic plaque and improved nitric oxide-dependent vascular function,[89] consistent with improved availability of nitric oxide. Taken together, these observations would suggest the need for clinical studies examining the effects of arginine in patients with sickle cell disease.

STRATEGIES AIMED AT LOWERING THE PLASMA CONCENTRATION OF HOMOCYSTEINE

Hyperhomocysteinemia has been associated with venous and arterial thrombosis,[95] and patients with sickle cell disease have significantly elevated plasma homocysteine concentration.[96] Acquired elevations of plasma homocysteine have been associated with deficiencies of folate, vitamin B_{12} and pyridoxine, with supplementation with these agents shown to lower plasma homocysteine concentrations.

Folate deficiency has long been recognized as a complication of sickle cell disease,[97,98] however, the routine use of folate supplements by many adults with sickle cell disease has made folate deficiency a relatively rare problem. Adults with sickle cell disease have elevated plasma homocysteine concentrations, despite red blood cell folate levels within the normal range.[99] In a recent study, children with sickle cell disease who were not taking folic acid supplements had slight, but statistically significantly elevated plasma homocysteine concentrations.[100] With folic acid treatment, their homocysteine concentrations decreased to values that were lower than controls.[100] Plasma homocysteine concentrations did not change with the administration of vitamin B_{12} and pyridoxine.[100] Since oral supplementation with folate has been shown to lower the plasma homocysteine concentration even in the absence of folate deficiency,[101] it is conceivable that higher doses of folate than are currently used will be needed in order to control homocysteine levels in adult patients with sickle cell disease.

Few studies have attempted to quantify vitamin B_{12} requirements in patients with sickle cell disease. Plasma vitamin B_{12} concentrations were 27% lower in a group of Nigerian subjects with sickle cell disease compared to controls.[102] In a study of 85 Saudi Arabian patients with severe sickle cell disease, 43.5% of the subjects had plasma vitamin B_{12} levels that were below normal, with mean concentrations of 84.3 ± 28.7 pg/ml.[103] These subjects reportedly experienced significant symptomatic improvement following vitamin B_{12} supplementation.[103] Among African American women, the age of onset of pernicious anemia may be earlier than among other ethnic groups and that there is a higher prevalence of antibodies to intrinsic factor.[104] Subtle neurologic abnormalities related to cobalamin deficiency may be difficult to detect in patients with sickle cell disease who have had a stroke. Measuring methylmalonic acid levels in the urine or blood may be a more sensitive technique to evaluate vitamin B_{12} status in these patients.[105]

There are few published data regarding the nutritional requirements of pyridoxine in patients with sickle cell disease. Flores *et al.* found no difference in plasma pyridoxal-5-phosphate levels in patients with sickle cell disease compared to black controls,[106] although the mean concentration in both of these groups were significantly lower than that of white children. Theoretically, supplementation with pyridoxine could be beneficial in sickle cell disease not only for its potential effects on plasma homocysteine concentrations, but also because pyridoxal-5-phosphate forms a Schiff base with the amino terminal end of the hemoglobin β chain, thus potentially inhibiting hemoglobin polymerization.[108] However, *in vitro* experiments have shown that there is a slight increase in blood viscosity following the addition of pyridoxal-5-phosphate.[107] Administration of an oral dose of 100 mg of pyridoxine 4 times daily in 2 patients with sickle cell disease (one with HbSS and the other with ß-thalassemia) showed no change in clinical course.[107] In a study of 5 patients given 50 mg pyridoxine twice daily for 2 months, there was a significant increase in plasma and red cell pyridoxal phosphate concentration.[109] One subject with severe disease had a reduction in emergency room visits and hospitalization days, but, as the other patients in the study had very mild disease prior to receiving pyridoxal phosphate, clinical improvement was difficult to assess.[109] The relationship of pyridoxine supplementation and plasma homocysteine concentration in patients with sickle cell disease has not been addressed.

LOW MOLECULAR WEIGHT ANTIOXIDANTS

The erythrocyte α-tocopherol concentration in patients with sickle cell disease is paradoxically increased,[110,111] while the plasma concentration is normal[110] or low.[111–117] These observations suggest that there may be increased utilization of α-tocopherol by sickle erythrocytes and raise the question of whether there is a relationship between α-tocopherol status and the clinical severity of sickle cell disease. Ndombi *et al.* observed an inverse correlation between α-tocopherol concentrations and the number of irreversibly sickled cells.[114] Phillips and Tanguay, in a retrospective study, observed a small ($r = -0.38$), but significant ($p < 0.04$), correlation between clinical events and α-tocopherol levels and a significant ($r = -0.42$) correlation between clinical events and α-tocopherol levels.[116] They estimated that differences in α-tocopherol levels accounted for 15% of the variability of the severity of sickle cell disease. There are two small prospective studies examining the effect of oral supplementation with vitamin E on clinical outcomes in sickle cell disease. In one study, Vitamin E supplementation in 6 sickle cell patients led to an increase in α-tocopherol concentration from 0.7 ± 0.2 mg/g lipid to 2.3 ± 0.3 mg/g lipid and a significant reduction in irreversibly sickled cells (from $25 \pm 3\%$ to $11 \pm 1\%$.[116] No clinical outcomes were measured. In another study, 13 patients with sickle cell disease received α-tocopherol along with ascorbic acid, zinc and soybean oil. After 8 months, there a 37.5% reduction in the number of irreversibly sickled cells[117] was observed, but there were no effects in hemoglobin concentrations or number of vaso-occlusive or aplastic crises.[117]

Several studies have shown that plasma[112,118] and leukocyte[119] ascorbic acid levels are reduced in patients with sickle cell disease, although other studies show no significant difference in vitamin C concentrations between subjects with sickle cell disease and controls.[110,111,120,121] It is possible that the discrepancies between studies may reflect differences in intake between study populations. To date, there are no published data looking at the effect of ascorbic acid administration on clinical or hematological outcomes.

SUMMARY

Current evidence points towards the critical roles that reactive oxygen species can play in impairing vascular function of sickle cell patients, with a significant component of the pathogenic actions of reactive oxygen species being directed towards the impairment of the salutary vasorelaxant and anti-inflammatory actions of nitric oxide. This encourages further clinical studies that focus on reducing tissue production and reactions of oxygen radicals and enhancing the positive actions of endogenously-produced or exogenously administered nitric oxide.

REFERENCE

1. WALLACE, W.J. & W.S. CAUGHEY. 1979. Superoxide as a participant in the chemistry of oxyhemoglobin. *In* Biochemical and Clinical Aspects of Oxygen. Academic Press, pp. 69–86.
2. CARRELL, W.R., C.C. WINTERBOURN & E.A. RACHMILEWITZ. 1975. Activated oxygen and haemolysis. Br. J. Haematol. **30:** 259–264.
3. FREEMAN, B.A. & J.D. CRAPO. 1982. Biology of disease: free radicals and tissue injury. Lab. Invest. **47:** 412–425.
4. HEBBEL, R.P. 1990. The sickle erythrocyte in double jeopardy: autoxidation and iron decompartmentalization. Semin. Hematol. **27:** 51–59.

5. HEBBEL, R.P. 1996. Oxidative processes in sickle cell disease. *In* Nitric Oxide and Radicals in the Pulmonary Vasculature. E.K. Weir, S.L. Archer, & J.T. Reeves, Eds. Armonk, NY. Futura. **8:** 135–198.
6. HEBBEL, R.P., W.T. MORGAN, J.W. EATON & E.B. HEDLUND. 1988. Accelerated autoxidation and heme loss due to instability of sickle hemoglobin. Proc. Natl. Acad. Sci. USA **83:** 237–241.
7. HEBBEL, R.P., J.W. EATON, M. BALASINGAM & M.H. STEINBERG. 1982. Spontaneous oxygen radical generation by sickle erythrocytes. Clin. Invest. **70:** 1253–1259.
8. CHUI D. & B. LUBIN. 1989. Oxidative hemoglobin denaturation and rbc destruction: the effect of heme on red cell membranes. Semin. Hematol. **26:** 128–135.
9. WINTERBOURN, C.C. & R.W. CARRELL. 1974. Studies of hemoglobin denaturation and Heinz body formation in the unstable hemoglobins. J. Clin. Invest. **54:** 678–689.
10. JACOB, H.S. 1970. Mechanism of Heinz body formation and attachment to red cell membrane. Semin. Hematol. **7:** 341–354.
11. EMBURY, S.H., R.P. HEBBEL, N. MOHANDAS & M.H. STEINBERG. 1996. *In* Sickle Cell Disease. Philadelphia, New York. Lippincott-Raven.
12. KUROSS, S.A., B.H. RANK & R.P. HEBBEL. 1988. Heme in sickle erythrocyte inside-out membranes: possible role in thiol oxidation. Blood **71:** 876–882.
13. EVANS, C.R., S.C. OMORPHOS & E. BAYSAL. 1986. Sickle cell membranes and oxidative damage. Biochem. J. **237:** 265–269.
14. KUROSS, S.A. & R.P. HEBBEL. 1988. Nonheme iron in sickle erythrocytes membranes: association with phospholipids and potential role in lipid peroxidation. Blood **72:** 1278–1285.
15. SHALEV, O. & R.P. HEBBEL. 1996. Extremely high avidity association of Fe(III) with the sickle red cell membrane. Blood **88:** 349–352.
16. BRAUGHLER, M.J., L.A. DUNCAN & R.L. CHASE. 1986. The involvement of iron in lipid peroxidation. J. Biol. Chem. **261:** 10282–10289.
17. SADRZADEH, S.M.H., G. ERNST, S.S. PANTERI, P.E. HALLAWAY & J.W. EATON. 1984. Hemoglobin. J. Biol. Chem. **259:** 14354–14356.
18. PUPPO, A. & B. HALLIWELL. 1988. Formation of hydroxyl radicals from hydrogen peroxide in the presence of iron. Biochem. J. **249:** 185–190.
19. GUTTERIDGE, J.M.C. 1986. Iron promoters of the Fenton reaction and lipid peroxidation can be released from hemoglobin by peroxides. FEBS **201:** 291–295.
20. BUETTNER, G.R. 1993. The pecking order of free radicals and antioxidants: lipid peroxidation, α-tocopherol, and ascorbate. Arch. Biochem. Biophys. **300:** 535–543.
21. DAS, S.K. & R.C. NAIR. 1980. Superoxide dismutase, glutathione peroxidase, catalase and lipid peroxidation of normal and sickled erythrocytes. Br. J. Haematol. **44:** 87–92.
22. JAIN, S.K. & S.B. SHOHET. 1984. A novel phospholipid in irreversibly sickled cells: evidence for in vivo peroxidative membrane damage in sickle cell disease. Blood **63:** 362–367.
23. JOCOBSEN, D.W. 1998. Homocysteine and vitamins in cardiovascular disease. Chemistry **44:** 1833–1843.
24. MINER, S.E.S., J. EVROVSKI & D.E.C. COLE. 1997. Clinical chemistry and molecular biology of homocysteine metabolism: an update. Clin. Biochem. **30:** 189–201.
25. STARKE, G. & J.M. HARLAN. 1986. Endothelial cell injury due to copper catalyzed hydrogen peroxide generation from homocysteine. J. Biol. Chem. **77:** 1370–1376.
26. PREIBISH, G., C. KUFFNER & E.F. ELSTNER. 1993. Model reactions on the prooxidative activity of homocysteine. Z. Naturforsh. **48c:** 58–62.
27. ROBINSON, K., A. GUPTA, V. DENNIS *et al.* 1996. Hyperhomocysteinemia confers an independent increased risk of atherosclerosis in the end stage renal disease and is closely linked to plasma folate and pyridoxine concentrations. Circulation **94:** 2743–2748.
28. ROBINSON, K., E.L. MAYER, D.P. MILLER, *et al.* 1995. Hyperhomocysteinemia and low pyridoxal phosphate. Circulation **92:** 2825–2830.
29. NATTA, C.L. & R.D. REYNOLDS. 1984. Apparent vitamin B_6 deficiency in sickle cell anemia. Am. J. Clin. Nutr. **40:** 235–239.

30. MOMEN, A.K. 1995. Diminished vitamin B_{12} levels in patients with severe sickle cell disease. J. Int. Med. **237:** 551–555.
31. LINDENBAUM, J. & F.A. KLIPSTEIN. 1963. Folic acid deficiency in sickle cell anemia. N. Engl. J. Med. **269:** 875–882.
32. HOUSTON, P.E., S. RANA & S. SEKHSARIA *et al.* 1997. Homocysteine in sickle cell disease: Relationship to stroke. Am. J. Med. **103:** 192–196.
33. FRIDOVICH, I. 1975. Superoxide dismutases. Annu. Rev. Biochem. **44:** 174–157.
34. COHEN, G & P. HOCHSTEIN. 1963. Glutathione peroxidase. The primary agent for the elimination of H_2O_2 in erythrocytes. Biochemistry **2:** 1420–1430.
35. FREI, B., L. ENGLAND & B.N. AMES. 1989. Ascorbate is an outstanding antioxidant in human blood plasma. Proc. Natl. Acad Sci. USA **86:** 6377–6381.
36. YAMAMOTO, K. & E. NIKI. 1998. Interaction of α-tocopherol with iron: antioxidant and prooxidant effects of α-tocopherol in the oxidation of lipids in aqueous dispersions in the presence of iron. Biochim. Biophys. Acta. **958:** 19–28.
37. JAIN, S.K., J.D. ROSS, J. DUETT, J.J. HERBST. 1990. Low plasma pre-albumin and carotenoid levels in sickle cell disease patients. Am. J. Med. Sci. **299:** 13–15.
38. CHUI, D. & B. LUBIN. 1979. Abnormal vitamin E and Glutathione peroxidase levels in sickle cell anemia. J. Clin. Med. **94:** 542–548.
39. TATUM, V.L. & C.K. CHOW. 1996. Status and susceptibility of sickle erythrocytes to oxidative and osmotic stress. Free Rad. Res. **25:** 133–139.
40. JOHNSON, C.S. & R.B. FRANCIS. 1991. Vascular occlusion in sickle cell disease: current concepts and unanswered questions. Blood **77:** 1405–1414.
41. BUNN, H. 1997. Pathogenesis and treatment of sickle cell disease. New Eng. J. Med. **337:** 762–769.
42. HEBBEL, R.P., O. YAMADA, C.F. MOLDOW & H.S. JACOB. 1980. Abnormal adherence of sickle erythrocytes to cultured vascular endothelium. J. Clin. Invest. **65:** 154–160.
43. GEE, B.E. & O.S. PLATT. 1995. Sickle reticulocytes adhere to VCAM-1. Blood **85:** 268–274.
44. KAUL D.K., M.E. FABRY, F. COSTANTINI, E.M. RUBI & R.L. NAGEL. 1995. In vivo demonstration of red cell-endothelial interaction, sickling and altered microvascular response to oxygen in the sickle transgenic mouse. J. Clin. Invest. **96:** 2845–2853.
45. MOSSERI, M., A.N.B. PANDETI, K. WENC, J.M. ISNER & R. WEINSTEIN. Am. Heart J. **126:** 338–346.
46. ENWONWU, C.O. 1989. Increased metabolic demand for arginine in sickle cell anaemia. Med. Sci. Res. **17:** 997–998.
47. ENWONWU, C.O., X. XU & E. TURNER. 1990 Nitrogen metabolism in sickle cell anemia. free amino acids in plasma and urine. Am. J. Med. Sci. **300:** 366–371.
48. REES, D.C., P.D. CERVI, A. O'DRISCOLL, M. HAMILTON, N.E. PARKER & J.B. PORTE. 1995. The metabolites of nitric oxide in sickle-cell disease. Br. J. Haematol. **91:** 834–837.
49. HATCH, F.E., L.R. CROWE, D.E. MILES, J.P. YOUNG & M.E. PORTNER. 1989. Altered vascular reactivity in sickle hemoglobinopathy. Am. J. Hypertens. **2:** 2–8.
50. MAYER, B. & B. HEMMENS. 1997. Biosynthesis and action of nitric oxide in mammalian cells. Trends Biochem. Sci. **22:** 477–481.
51. RUBBO, H., V. DARLEY-USMAR & B.A. FREEMAN. 1996. Nitric oxide regulation of tissue free radical injury. Chem. Res. Toxicol. **9:** 809–820.
52. MONCADA, S. & E.A. HIGGS. 1991. Endogenous nitric oxide: physiology, pathology and clinical relevance. Eur. J. Clin. Inv. **21:** 361–374.
53. VANE, J.R., E. ANGGARD & R.M. BOTTING. 1990. Regulatory functions of the vascular endothelium. N. Engl. J. Med. **323:** 27–36.
54. MONCADA, S. & E.A. HIGGS. 1993. The L-arginine-nitric oxide pathway. New Engl. J. Med. **329:** 2002–2012.
55. CORNWELL, T.L. & T.M. LINCOLN. 1993. Intracellular cyclic GMP receptor proteins. FASEB J. **7:** 328–338.
56. STAMLER, J.S., D.J. SINGEL & J. LOSCALZO. 1992. Biochemistry of nitric oxide and its redox forms. Science **258:** 1898–1902.
57. HUIE, R.E. & S. PADJAMA. 1993. Reactions of NO with $O^{\cdot -}$. Free Rad. Res. Commun. **18:** 195–199.

58. GRYGLEWSKI, R.J., R.M.J. PALMER & S. MONCADA. 1986. Superoxide anion is involved in the breakdown of endothelium-derived vascular relaxing factor. Letters to Nature **320:** 454–456.
59. VANHOUTTE, P.M. & G.M. RUBANYI. 1986. Superoxide anions and hyperoxia inactivate endothelium-derived relaxing factor. Am. J. Physiol. **250:** H822–H827.
60. GUTIERREZ, H.H., B. NIEVES, P. CHUMLEY, A. RIVERA & B.A. FREEMAN 1996. Nitric oxide regulation of superoxide-dependent lung injury: oxidant-protective actions of endogenously produced and exogenously administered nitric oxide. Free Rad. Biol. Med. **21:** 43–45.
61. BECKMAN, J.S., T.W. BECKMAN, J. CHEN, P.A. MARSHALL & B.A. FREEMAN. 1990. Apparent hydroxyl radical production by peroxynitrite: implications for injury from nitric oxide and superoxide. Proc. Natl. Acad. Sci. USA **87:** 1620–1624.
62. RADI, R., A. DENICOLA & B.A. FREEMAN. 1999. Peroxynitrite reactions with carbon dioxide-bicarbonate. Methods Enzmol. **301:** 353–367.
63. RUBBO, H., R. RADI, M. TRUJILLO, R. TELLERI, B. KALYARAMAN, S. BARNES, M. KIRK & B.A. FREEMAN. 1994. Nitric oxide regulation of superoxide and peroxynitrite-dependent lipid peroxidation. J. Biol. Chem. **269:** 26066–26075.
64. RADI, R., J.S. BECKMAN, K.M. BUSH & B.A. FREEMAN. 1991. Peroxynitrite-induced membrane lipid peroxidation: the cytotoxic potential of superoxide and nitric oxide. Arch. Biochem. Biophys. **288:** 481–487.
65. O'DONNELL, V.B, P.H. CHUMLEY, N. HOGG, A. BLOODSWORTH, V.M. DARLEY-USMAR & B.A. FREEMAN. 1997. Nitric oxide inhibition of lipid peroxidation: kinetics of reaction with lipid peroxyl radicals and comparison with α-tocopherol. Biochemistry **36:** 15216–15223.
66. JUCKETT, M., Y. ZHENG, H. YUAN, T. PASTOR, W. ANTHOLINE, M. WEBER & G. VERCELOTTI. 1998. Heme and the endothelium. J. Biol. Chem. **273:** 23388–23397.
67. GURBONOV, N.V., A.N. OSIPOV, B.W. DAY, B.Z. RIVERA, V.E. KAGAN & N.M. ELSAYED. 1995. Reduction of ferrylmyoglobin and ferrylhemoglobin by nitric oxide: A protective mechanism against ferryl hemoprotein-induced oxidations. Biochemistry **34:** 6689–6699.
68. JIA, L., C. BONAVENTURA, J. BONAVENTURA & J.S. STAMLER. 1996. S-nitrosohaemoglobin: a dynamic activity of blood involved in vascular control. Nature **380:** 221–226.
69. WENNMALM, A., G. BENTHIN, A. EDLUND, L. JUNGERSTEN, N.K. JENSEN, S. LUNDIN, U.N. WESTFELT, A.S. PETERSSON & F. WAAGSTEIN. 1993. Metabolism and excretion of nitric oxide in humans. Circ. Res. **73:** 1121–1127.
70. WENNMALM, A. 1995. Metabolism and excretion of nitric oxide in man: basal studies and clinical applications. Mediators in the cardiovascular system regional ischemia. Basel. Birkhauser Verlag. **35:** 129–139.
71. STAMLER, J.S. & J. LOSCALZO. 1992. Endothelium-derived relaxing factor modulates the atherothrombogenic effects of homocysteine. J. Cardiovasc. Pharmacol. **20:** 202–204.
72. STAMLER, J.S., J.A. OSBORNE, O. JARAKI, L.E. RABBANI, M. MULLINS, D. SINGEL & J. LOSCALZO. 1993. Adverse vascular effects of homocysteine are modulated by endothelium-derived relaxing factor and related oxides of Nitrogen. J. Clin. Inv. **61:** 308–318.
73. UPCHURCH, G.R., D.N. WELSH, A.J. FABIAN, A. PIGAZZIA, J.F. KEANEY & J. LOSCALZO. 1977. Stimulation of endothelial nitric oxide production by homocysteine. Atherosclerosis **132:** 177–185.
74. HEININGER, Y.J. 1998. Early response genes and transcriptional factors: AP-1 and NF-κB. Oxygen'98 Sunrise Free Radical School. November 20–23. Washington, DC.
75. KHAN, B.V., D.G. HARRISON, M.T. OLBRYCH, R.W. ALEXANDRE & R.M. MEDFORD. 1996. Nitric oxide regulates vascular cell adhesion molecule-1 gene expression and redox-sensitive transcriptional events in human vascular endothelial cells. Proc. Natl. Acad. Sci. USA **93:** 9114–9119.
76. SCHENK, H., M. KLEINE, W. ERDBRUGER, W. DROGE & K.S. OSTHOFF. 1994. Distinct effects of thioredoxin and antioxidants on the activation of transcription factors NF-κB and AP-1. Proc. Natl. Acad. Sci. USA **91:** 1672–1676.

77. SWERLICK, R.A., J.R. ECKMAN, A. KUMAR, M. JEITLER & T.M. WICK 1993. α4β1-integrin expression on sickle reticulocytes: vascular cell adhesion molecule-1 dependent binding to endothelium. Blood **82:** 1891–1899.
78. PHELAN, M., S.P. PERRINE, M. BRAUER & D.V. FALLER. 1995. Sickle erythrocytes, after sickling, regulate the expression of the endothelin-1 gene and protein in human endothelial cells in culture. J. Clin. Invest. **96:** 1145–1151.
79. HAMMERMAN, S.I., S. KOUREMBANAS, T.J. CONCA, M. TUCCI, M. BRAUER & H.W. FARBER. 1997. Endothelin-1 production during the acute chest syndrome in sickle cell disease. Am. J. Respir. Crit. Care Med. **156:** 280–285.
80. CHARACHE, S., M.L. TERRIN, R.D. MOORE, G.J. DOVER, F.B. BARTON, S.V. ECKERT, R.P. MCMAHON & D.R. BONDS. 1995. Effect of hydroxyurea on the frequency of painful crises in sickle cell anemia. N. Engl. J. Med. **332:** 1317–1322.
81. PACELLI, R., J. TAIRA, J.A. COOK, D.A. WINK & M.C. KRISHNA. 1996. Hydroxyurea reacts with heme proteins to generate nitric oxide. The Lancet **347:** 900.
82. JIANG, J., S.J. JORDAN, D.P. BARR, M.R. GUNTHER, H. MAEDA & R.P. MASON. 1997. In vivo production of nitric oxide in rats after administration of hydroxyurea. Mol. Pharmacol. **52:** 1081–1086.
83. STOLZE, K. & H. NOHL. 1995. Reactions of reducing xenobiotics with oxymyoglobin. Formation of memyoglobin, ferryl myoglobin and free radicals: an electro spin resonance and chemiluminescence study. Biochem. Pharmacol. **49:** 1261–1267.
84. ENWONWU, C.O., X.X. XU & E. TURNER. 1990. Nitrogen metabolism in sickle cell anemia: free amino acids in plasma and urine. Am. J. Med. Sci. **300:** 366–371.
85. REYES, A.A., I.E. KARL & S. KLAHR. 1994. Role of arginine in health and in renal disease. Am. J. Physiol. **36:** F331–F346.
86. REUTENS, D.C., M.D. MCHUGH, P-J. TOUSSAIN, A.C. EVANS, A. GJEDDE, E. MEYER & D.J. STEWART. 1997. L-argentine infusion increases basal but not activated cerebral blood flow in humans. J. Cereb. Blood Flow Metab. **17:** 309–315.
87. GUGLIANA, D., R. MARFELLA, G. VERRAZZO, R. ACAMPORA, L. COPPOLA, D. COZZOLINO & F. D'ONOFRIO. 1997. The vascular effects of L-argentine in humans. J. Clin. Invest. **99:** 433–438.
88. SMOYER, W.E., B.H. BROUHARD, D.K. RASSIN & L. LAGRONE. 1991. Enhanced GFR response to oral versus intravenous argentine administration in normal adults. J. Lab. Clin. Med. **118:** 166–175.
89. COOKE, J.P., A.H. SINGER, P. TSAO, P. ZERA, R.A. ROWAN & M.E. BILLINGHAM. 1992. Antiatherogenic effect of L-arginine in the hypercholesterolemic rabbit. J. Clin. Invest. **90:** 1168–1172.
90. HAEFELI, W.E., L. LINDER & T.F. LUSCHER. 1997. Quinaprilat induces arterial vasodilation mediated by nitric oxide in humans. Hypertension **30:** 912–917.
91. KUGA, T., M. MOHRI, K. EGASHIRA, Y. HIRAKAWA, T. TAGAWA, H. SHIMOKAWA & A. TAKESHITA. 1997. Bradykinin-induced vasodilation of human coronary arteries in vivo: role of nitric oxide and angiotensin-converting enzyme. J. Am. Col. Cardiol. **30:** 1008–1012.
92. AOKI, R.T. & S.T.O. SAAD. 1995. Enalapril reduces the albuminuria of patients with sickle cell disease. Am. J. Med. **98:** 432–435.
93. FALK, R.J., J. SCHEINMAN, G. PHILLIPS, E. ORRINGER, A. JOHNSON & C. JENNETTE. 1992. Prevalence and pathologic features of sickle cell nephropathyand response to inhibition of antiotensin-converting enzyme, N. Engl. J. Med. **326:** 910–915.
94. DEFRONZO, R.A., P.A. TAUFIELD, H. BLACK, P. MCPHEDRAN & C.R. COOKE. 1997. Impaired renal tubular potassium secretion in sickle cell disease. Ann. Int. Med. **90:** 310–316.
95. CLARKE, R., L. DALY, K. ROBINSON, E. NAUGHTEN, S. CAHALANE, B. FOWLER & I. GRAHAM. 1991. Hyperhomocysteinemia: an independent risk factor for vascular disease. N. Engl. J. Med. **324:** 1149–1155.
96. HOUSTON, P.E., S. RANA, S. SEKHASARIA, E. PERLIN, K.S. KIM & O.L. CASTRO. 1997. Homocysteine in sickle cell disease: Relationship stroke. Am. J. Med. **103:** 192-196.
97. LINDENBAUM, J. & F.A. KLIPSTEIN. 1963. Folic acid deficiency in sickle-cell anemia. N. Engl. J. Med. **269:** 875–882.

98. PIERCE, L.E. & C.E. RATH. 1962. Evidence for folic acid deficiency in the genesis of anemic sickle cell crisis. Blood **20:** 19–32.
99. HOUSTON, P.E., S. RANA, S. SEKHSARIA, E. PERLIN, K.S. KIM & O.L. CASTRO. 1997. Homocysteine in sickle cell disease: relationship to stroke. Am. J. Med. **103:** 192–196.
100. VAN DER DIJS, F.P.L., J.B. SCHNOG, D.A.J. BROUWER, H.J.R. VELVIS, G.A. VAN DEN BERG, A.J. BAKKER, A.J. DUITS, F.D. MUSKIET & F.A.J. MUSKIET. 1998. Elevated homocysteine levels indicate suboptimal folate status in pediatric sickle cell patients. Am. J. Hematol. **59:** 192–198.
101. BRATTSTROM, L.E., B. ISRAELSSON, J.O. JEPPSON & B.L. HULTBERG. 1988. Folic acid—an innocuous means to reduce plasma homocysteine. Scand. J. Clin. Lab. Invest. **48:** 215–221.
102. OSIFO, B.O.A., A. ADEYOKUNNO, Y. PARMENTIER, P. GERARD & J.P. NICOLAS. 1983. Abnormalities of serum transcobalamins in sickle cell disease (HbSS) in black Africa. Scand. J. Haematol. **30:** 135–140.
103. AL-MOMEN, A.K. 1995. Diminished vitamin B12 levels in patients with severe sickle cell disease. J. Int. Med. **237:** 551–555.
104. SINOW, R.M., C.S. JOHNSON, D.S. KARNAZE, M.E. SIEGEL & R. CARMEL. 1987. Unsuspected pernicious anemia in a patient with sickle cell disease receiving routine folate supplementation. Arch. Intern. Med. **147:** 1828–1829.
105. LINDENBAUM, J., D.G. SAVAGE, S.P. STABLER & R.H. ALLEN. 1990. Diagnosis of cobalamin deficiency: II. Relative sensitivities of serum cobalamin, methylmalonic acid, and total homocysteine concentrations. Am. J. Hematol. **34:** 99–107.
106. FLORES, L., R. PAIS, I. BUCHANA, D. ARNELLE, V.M. CAMP, M. KUTNER, B.A. FARAJ, J. ECKMAN & A. RAGAB. 1988. Pyridoxal 5′-phosphate levels in children with sickle cell disease. Am. J. Ped. Hematol. Oncol. **10:** 236–240.
107. BEUTLER, E., N.V. PANIKER & C.J. WEST. 1972. Pyridoxine administration in sickle cell disease: An unsuccessful attempt to influence the properties of sickle hemoglobin. Biochem. Med. **6:** 139–143.
108. KARK, J.A., P.G. TARASSOFF & R. BONGIOVANNI. 1983. Pyridoxal phosphate as an antisickling agent in vitro. J. Clin. Invest. **71:** 1224–1229.
109. NATTA, C.L. & R.D. REYNOLDS. 1984. Apparent vitamin B6 deficiency in sickle cell anemia. Am. J. Clin. Nutr. **40:** 235–239.
110. TATUM, V.L. & C.K. CHOW. 1996. Antioxidant status and susceptibility of sickle erythrocytes to oxidative and osmotic stress. Free Rad. Res. **25:** 133–139.
111. TANGNEY, C.C., G. PHILLIPS, R.A. BELL, P. FERNANDES, R. HOPKINS & S.M. WU. 1989. Selected indices of micronutrient status in adult patients with sickle cell anemia (SCA). Am. J. Hematol. **32:** 161–166.
112. ESSIEN, E.U. 1995. Plasma levels of retinol, ascorbic acid and alpha-tocopherol in sickle cell anaemia. Central African J. Med. **41:** 48–50.
113. SESS, D., M.A. CARBONNEAU, M.J. THOMAS, M.F. DUMON. E. PEUCHANT, A. PERROMAT, M. LEBRAS & M. CLERC. 1992. First observations on the main plasma parameters of oxidative stress in homozygous sickle cell disease. Bulletin de la Societe de Pathologie Exotique **85:** 174–179.
114. NDOMBI, I.O. & S.N. KINOTI. 1990. Serum vitamin E and the sickling status in children with sickle cell anaemia. East African Med. J. **67:** 720–725.
115. NATTA, C. & L. MACHLIN. 1979. Plasma levels of tocopherol in sickle cell anemia subjects. Am. J. Clin. Nutr. **32:** 1359–1362.
116. PHILLIPS, G. & C.C. TANGNEY. 1992. Relationship of plasma alpha tocopherol to index of clinical severity in individuals with sickle cell anemia. Am. J. Hematol. **41:** 227–231.
117. NATTA, C., L. MACHLIN & M. BRIN. 1980. A decrease in irreversibly sickled erythrocytes in sickle cell anemia patients given vitamin E. Am. J. Clin. Nutr. **33:** 968–971.
118. MUSKIET, F.A., F.D. MUSKIET, G. MEIBORG & J.B. SCHERMER. 1991. Supplementation of patients with homozygous sickle cell disease. Am. J. Clin. Nutr. **54:** 736–744.
119. AKINKUGBE, F.M. & S.I. ETTE. 1983. Ascorbic acid in sickle cell disease: results of a pilot therapeutic trial. E. African Med. J. **60:** 683–687.

120. CHIU, D., E. VICHINSKY, S.L. HO, T. LIU & B.H. LUBIN. 1990. Vitamin C deficiency in patients with sickle cell anemia. Am. J. Ped. Hematol. Oncol. **12:** 262–267.
121. DELEKAN, D.A., D.I. THURHAM & A.D. ADEKILE. 1989. Reduced antioxidant capacity in paediatric patients with homozygous sickle cell disease. Eur. J. Clin. Nutr. **43:** 609–614.

Dopamine Stimulates Astrocytic C6-D2L Cells via Tyrosine Kinase and p38 MAPK Activation

YONGQUAN LUO[a] AND GEORGE S. ROTH

Molecular Physiology and Genetics Section, Gerontology Research Center, NIA, Baltimore, Maryland 21224, USA

Dopamine (DA), a neurotransmitter of the central nervous system, plays a fundamental role in the control of a variety of physiological functions including locomotor activity, learning, reward behavior, and hormone synthesis and release.[1] However, overflow of DA in the brain has been suggested to participate in certain neurodegenerative processes, which include ischemia,[2] hypoxia,[3] and neurotoxicities induced by excitatory amino acids[4] and methamphetamine.[5] For example, the striatal DA concentration rapidly reaches concentrations as high as 0.2 mM after ligation of the cervical artery in the gerbil ischemic model.[6] Decrease in the endogenous DA concentration by chemical lesion of the nigrostriatal dopaminergic pathway attenuates ischemic insult to the striatum.[7] Moreover, direct administration of DA into striata results in apoptosis and neurodegeneration.[8,9,32] Using *in vitro* neonatal striatal cell cultures, we have demonstrated that DA induces apoptosis through an oxidation-JNK-c-Jun activation pathway.[10]

In an *in vivo* animal model, injection of DA into rat striata also results in activation of astrocytes as indicated by an increase of glial fibrillary acidic protein (GFAP) immunocytochemical staining and astrocyte proliferation determined by [^{3}H]R05-4864 binding to peripheral benzodiazepine receptors on astrocytes.[11] Astrocytes are the intimate partners of neurons and play important roles in both physiological and pathological conditions. Physiologically, astrocytes participate in the maintenance of the microenvironment of neurons, including sequestration and metabolism of various neurotransmitters, and production of proinflammatory and immunomodulatory cytokines and neuropeptides.[12] Reactive astrocytes are also observed in certain DA-related pathological conditions, such as Parkinson's disease (PD), Huntington's disease (HD), and hypoxia-induced ischemia.[13–16] However, the mechanism by which DA activates astrocytes remains unclear.

The C6-D2L cell line is a rat C6 glioma cell line stably expressing recombinant D2L receptors,[17] which are also expressed in cultured striatal astrocytes.[18] The D2 receptor density in C6-D2L cells is about 188 fmol/mg protein and is close to that in striatal membranes (~400 fmol/mg protein).[19,20] The stably expressed D2 receptors in the C6-D2L cells are functionally coupled to pertussis toxin-sensitive G proteins, resulting in inhibition of adenylate cyclase[21] and stimulation of both extracellular signal regulated kinases (ERKs) and c-Jun N-terminal kinase (JNK).[20] This C6-D2L cell line serves as a good *in vitro* cell culture model for studies of DA-induced glial

---

[a]Address for correspondence: Yongquan Luo, Ph.D., Molecular Physiology and Genetics Section, Gerontology Research Center, NIA, 4E02, 5600 Nathan Shock Drive, Baltimore, MD 21224. Voice: 410-558-8507; fax: 410-558-8323.
e-mail: luoyq@helix.nih.gov

response. In the present study, we have used this cell line to initiate a study to explore the potential molecular events through which DA activates astrocytic C6-D2L cells.

First, we have examined the effect of μM levels of DA on mitogenesis in C6-D2L cells using a [^{3}H]thymidine incorporation assay. As shown in FIGURE 1A, DA stimulated [^{3}H]thymidine incorporation in a concentration-dependent manner from 0.1 to 500 μM DA. The mitogenic activity at 500 μM DA was about 3.2 ± 0.2 times higher than that of the control group. This DA-induced mitogenesis could be completely inhibited in the presence of aphidicoline (20 μg/ml, data not shown), a potent DNA polymerase α inhibitor, suggesting that [^{3}H]thymidine incorporation is caused by DNA synthesis and is not due to DNA repair. Cell cycle analysis by flow cytometry indicated that DA (200 μM) increased the 14% of the cells in the S phase with a decreased percentage of cells in G_1 phase (data not shown). This DA-stimulated mitogenesis requires activation of D2 receptors since it can be selectively inhibited by pretreatment of the cells with (+)-butaclamol (10 μM), a potent D2 receptor antagonist, but not by the same concentration of (−)-butaclamol, an inactive form of (+)-butaclamol. Application of the D2 receptor agonist, quinpirole (μM levels), also promoted cell cycling as indicated by an increase in about 10% of cells in S phase with a decrease in G1 phase. Furthermore, we did not observe any mitogenic effect within a DA concentration range of 0–500 μM in wild type C6 cells without expression of D2L receptors (FIG. 1A). Thus, it appears that the mitogenesis induced by μM levels of DA require DA D2 receptor activation.

Next, we have explored the idea that DA-stimulated mitogenesis might be mediated by an intracellular redox-tyrosine kinase pathway. Reactive oxygen species (ROS), including $O\cdot_2$, and H_2O_2, have recently been suggested to be a second messenger for mediation of cell signaling that regulates proliferation and DNA synthesis.[22] Using anti-phosphotyrosine immunoblotting, we have found that μM levels of DA rapidly stimulated protein tyrosine phosphorylation within 5 min, being maximal between 15 and 30 min, then decreased by 1 h (data not shown). The DA-induced protein tyrosine phosphorylation could be blocked by administration of genistein (10–100 μM, data not shown), a tyrosine kinase inhibitor. This protein tyrosine phosphorylation required D2 receptor activation since it could be blocked by preincubation of the cells for 30 min with the specific D2 receptor antagonist, (+)-butaclamol (FIG. 1B). Application of μM levels of quinpirole, a D2 receptor agonist, also resulted in a great increase in the amount of tyrosine-phosphorylated proteins (data not shown). We could not observe a significant changes in tyrosine-phosphorylated proteins in wild-type C6 cells treated with DA concentrations from 10–500 μM (data not shown). Furthermore, the DA-induced protein tyrosine phosphorylation could be blocked by preincubation of the cells with either 1 μM diphenylene iodonium (DPI) or 20 mM N-acetyl cysteine (NAC, FIG. 1B). DPI is an inhibitor of flavonoid-containing oxidases.[23,24] These oxidases can catalyze the formation of superoxide from oxygen by using NADH/NADPH as an electron donor. The enzyme activity has also been shown to be regulated by a variety of membrane receptors, such as intrinsic tyrosine kinase-containing receptors and G protein-coupled receptors.[22,25] NAC is an antioxidant that can scavenge free radicals. These reagents that blocked the protein tyrosine phosphorylation also inhibited the DA-induced mitogenesis (FIG. 1C). Thus, these observations suggest that an intracellular redox-tyrosine kinase pathway is required for DA-stimulated mitogenesis.

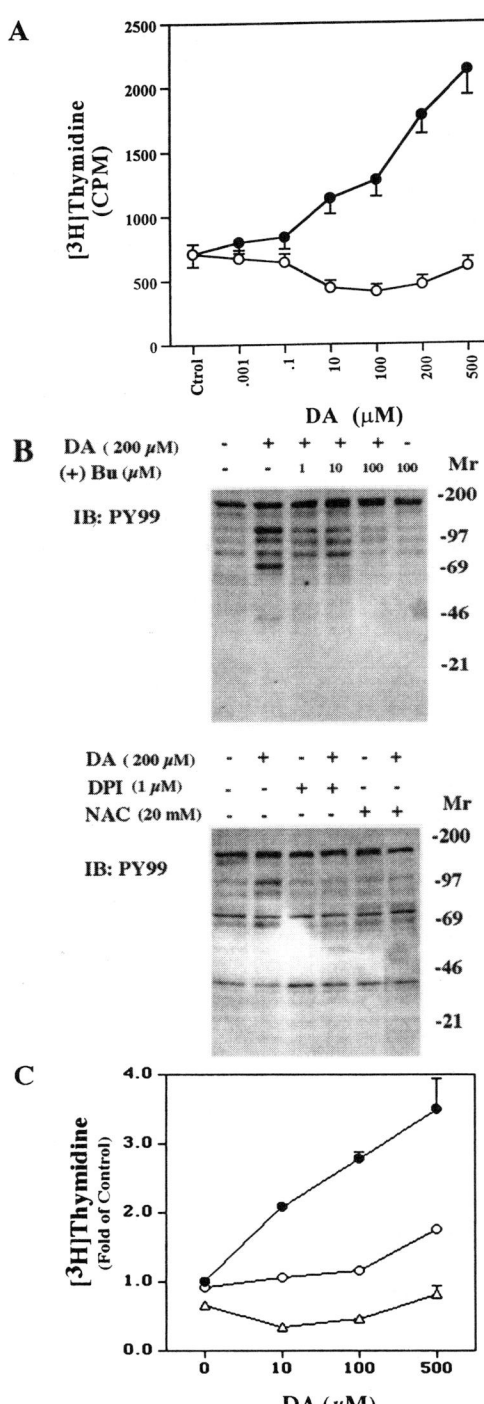

FIGURE 1. DA promotes DNA synthesis via a D2 receptor mediated redox-tyrosine kinase pathway. **A:** Concentration-dependent studies of [^{3}H]thymidine incorporation into DNA of both C6 (empty circle) and C6-D2L (*filled circle*) cells. C6 or C6-D2L cells (1 × 10^5/ml, 0.2 ml/well) were plated in a 96 well plate and grown for 24 h. The cells were rinsed, starved for 8 h, and then stimulated with DA in the serum-free medium containing 0.025% ascorbic acid for 23 h. After that, the cells were pulsed with 2 μCi [^{3}H]thymidine per well for an additional 6 h. The incorporated [^{3}H]thymidine was extracted with a mixture of 0.1% sodium dodecyl sulfate (SDS) and 0.01N NaOH and was counted with a Beckman liquid scintillation counter and expressed as CPM/well. **B:** Activation of redox-tyrosine kinase cascade by DA D2 receptors. C6-D2L cells were treated with the indicated blocking reagents or vehicle (control) for 30 min and then stimulated with 200 μM DA (+) for 5 min. The cell lysates were prepared and protein tyrosine phosphorylation was detected by PY99 antibody. **C:** Effect of inhibitors for redox-tyrosine kinase pathway on DA-induced mitogenesis. C6-D2L cells were plated, treated with either blocking reagents as indicated or vehicle for 30 min, and then stimulated with DA. Mitogenesis was measured by [^{3}H]thymidine incorporation into DNA. *Filled circle*: DA alone; *Empty circle*: DA + 20 mM NAC; *Triangle*: DA + 1 μM DPI. All of the above experiments were performed at least three times and similar results were obtained.

In addition to the above redox-tyrosine kinase pathway, we have also examined other mitogenic signaling by using specific kinase inhibitors. Our data suggest that ERK, PI3 kinase, P38 MAPK and PKC are not be involved in DA-induced DNA synthesis.[33]

Application of μM levels of DA also significantly increased the expression of GFAP, a sensitive parameter serving as another criterion of reactive astrocytes.[26] C6-D2L cells responded to DA (200 μM) treatment with an increase in GFAP expression within the time from 0–48 h. The amount of GFAP expression, which was detected by monoclonal anti-GFAP immunoblotting, was increased about 2.1 ± 0.1 fold after 48 h 200 μM DA incubation. Concentration-dependent studies carried out at 24 h showed that maximal GFAP stimulation occurred at 200 μM. Unlike the DA-stimulated mitogenesis (FIG. 1A), GFAP production was decreased when the cells were exposed to 500 μM DA for 24 h (FIG. 2A). Further examination using the same paradigm as that for mitogenesis suggested that DA-induced GFAP expression was not regulated by the D2 receptor mediated redox-tyrosine pathway since it could not be influenced by the D2 receptor antagonist, (+)-butaclamol, tyrosine kinase inhibitor, genistein and MAPK pathway inhibitor, PD 098059. However, the GFAP expression stimulated by DA could be inhibited by pretreatment of the cells with SB 203580, a selective p38 MAPK inhibitor[27,28] (FIG. 2B), suggesting that activation of p38 MAPK might be required for the regulation of GFAP production. Indeed, application of DA increased the p38 MAPK phosphorylation, a required step for p38 MAPK activity.[29] DA increased p38 MAPK phosphorylation in a time- and concentration-dependent manner (FIG. 2C). DA (100 μM) rapidly stimulated p38 MAPK phosphorylation within 5 min, reached a maximum at 15 min, then gradually decreased to basal level by about 1 hour. The p38 MAPK phosphorylation was greatly increased when the cells were treated with DA concentrations ranging from 10–200 μM for 15 min. The D2 receptor antagonist, (+)-butaclamol, had no effect on DA-stimulated phosphorylation of p38 MAPK (FIG. 2C). Thus, these data suggest that DA-stimulated GFAP expression may be regulated by a D2 receptor-independent p38 MAPK cascade.

As mentioned above, an increase in striatal DA availability is believed to play a role in some neurodegenerative processes, such as ischemia,[2] hypoxia,[3] and neurotoxicity induced by excitatory amino acids[4] and methamphetamine.[5] Direct intrastriatal injection of DA in rats results in neuronal apoptosis and gliosis.[8,9,11] In this communication, we extend these findings by exploring potential molecular events in DA-stimulated astrocytosis using *in vitro* cell cultures. We do indeed observe that μM levels of DA activate C6-D2L glioma cells, indicated by the stimulation of both mitogenesis and GFAP expression. Furthermore, we provide evidence that astrocytic mitogenesis requires activation of D2 receptors, flavinoid-containing oxidases and protein tyrosine kinases; while stimulation of GFAP expression requires D2 receptor-independent activation of the p38 MAPK cascade. Interestingly, in an *in vivo* rat model, reactive astrocytes in the brain responded differently to brain injury. For example, some astrocytes show proliferation as indicated only by [^{3}H]thymidine incorporation, some exhibit hypertrophy labeled only by GFAP, and some are labeled by both GFAP and [^{3}H]thymidine.[30,31] This coincidence suggests that the two separate pathways examined here may operate in the *in vivo* situation.

A

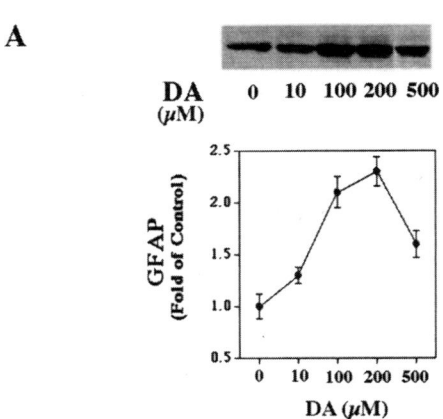

B

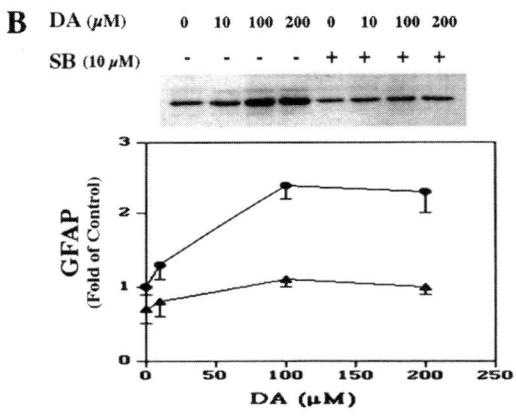

C

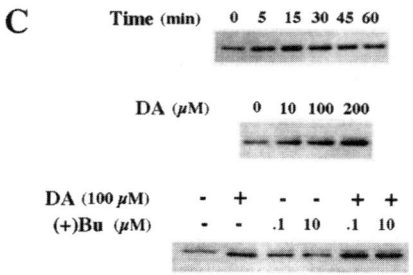

FIGURE 2. DA stimulates GFAP expression with involvement of p38 MAPK activation. C6-D2L cells were plated, seeded, starved with DMEM plus 0.025% ascorbic acid for 6–8 h and then stimulated with DA for 24 h. The cell lysates were prepared with a lysis buffer containing 0.1% SDS and electrophoresed on 8–16% SDS gels. GFAP protein was detected by immunoblotting with monoclonal anti-GFAP (1:2000). **A:** Concentration-dependent stimulation of GFAP expression by DA. **B:** Effect of SB 203580, a specific p38 MAPK inhibitor, on DA-stimulated GFAP expression. SB 203580 (10 μM) was added 30 min before DA stimulation. In both panels **A** and **B**, the top row is a representative anti-GFAP immunoblot; whereas the bottom graph shows immunoreactive GFAP changes ($\pm$ SE) from three experiments. **C:** DA stimulates phosphorylation of p38 MAPK. C6-D2L cells were starved with serum-free DMEM but containing 0.025% ascorbic acid for 8 h, then stimulated with DA. The cell lysates were prepared and immunoblotted with anti-phospho-specific p38 MAPK (1:1000). *Upper row*: Time-course studies for 100 μM DA stimulation; *Middle row*: Concentration-dependent studies. The cells were treated with various concentrations of DA for 15 min. The phosphorylation of p38 MAPK was detected by anti-phospho-specific p38 immunoblotting. *Bottom row*: (+) Butaclamol effects. C6-D2L cells were incubated with (+) butaclamol at the indicated concentration for 30 min and then stimulated with 100 μM DA for 15 min. The phosphorylation of p38 MAPK was immunodetected. Each row shows an anti-phospho-p38 MAPK immunoblot representative of at least three experiments.

ACKNOWLEDGMENTS

We thank Dr. K. A. Neve for providing C6-D2L cells.

REFERENCES

1. PICETTI, R., A. SAIARDI, T A. SAMAD, Y. BOZZI, J.H. BAIK & E. BORRELLI. 1997. Dopamine D2 receptors in signal transduction and behavior. Crit. Rev. Neurobiol. **11:** 121–142.
2. BUISSON, A., J. CALLEBERT, E. MATHIEU, M. PLOTKINE & R. BOULU. 1992. Striatal protection induced by lesioning the substantia nigra of rats subjected to focal ischemia. J. Neurochem. **59:** 1153–1157.
3. AKIYAMA, Y., K. KOSHIMURA, T. OHUE, K. LEE, S. MIWA, S. YAMAGATA & H. KIKUCHI. 1991. Effects of hypoxia on the activity of the dopaminergic neuron system in the rat striatum as studied by in vivo brain microdialysis. J. Neurochem. **57:** 997–1002.
4. FILLOUX, F. & J. WAMSLEY. 1991. Dopaminergic modulation of excitotoxicity in rat striatum: evidence from nigrostriatal lesions. Synapse **8:** 281–288.
5. O'DELL, S., F. WEIHMULLER & J. MARSHALL. 1991. Multiple methamphetamine injections induce marked increases in extracellular striatal dopamine which correlate with subsequent neurotoxicity. Brain Res. **564:** 256–260.
6. SLIVKA, A., T.S. BRANNAN, J. WEINBERGER, P.J. KNOTT & G. COHEN. 1988. Increase in extracellular dopamine in the striatum during cerebral ischemia: a study utilizing cerebral microdialysis. J. Neurochem. **50:** 1714–1718.
7. GLOBUS, M. Y.-T., M.D. GINSBERG, W.D. DEITRICH, R. BUSTO & P. SHEINBERG. 1987. Substantia nigra lesion protects against ischemic damage in the striatum. Neurosci. lett. **80:** 251–256.
8. HASTINGS, T., D. LEWIS & M. ZIGMOND. 1996. Role of oxidation in the neurotoxic effects of intrastriatal dopamine injections. Proc. Natl. Acad. Sci. USA **93:** 1956–1961.
9. HATTORI, A., Y. LUO, H. UMEGAKI, J. MUNOZ & G. S. ROTH. 1998. Intrastriatal injection of dopamine results in DNA damage and apoptosis in rats. NeuroReport **9:** 2569–2572.
10. LUO, Y., H. UMEGAKI, X. WANG, R. ABE & G. S. ROTH. 1998. Dopamine induces apoptosis through an oxidation-involved SAPK/JNK activation pathway. J. Biol. Chem. **273:** 3756–3764.
11. FILLOUX, F. & J. T. TOWNSEND. 1993. Pre- and postsynaptic neurotoxic effects of dopamine demonstrated by intrastriatal injection. Exp. Neurol. **119:** 79–88.
12. WILKIN, G., D. MARRIOTT & A. CHOLEWINSKI. 1990. Astrocyte heterogeneity. Trends Neurosci. **13:** 43.
13. BEACH, T.G., R. WALKER & E. G. MCGEER. 1989. Patterns of gliosis in Alzheimer's disease and aging cerebrum. Glia **2:** 420–436.
14. KUSHNER, P.D., D.T. STEPHENSON & S. WRIGHT. 1991. Reactive astrogliosis is widespread in the subcortical white matter of amyotrophic lateral sclerosis brain. J. Neuropathol. Exp. Neurol. **50:** 263–277.
15. LANGSTON, J.W. 1985. MPTP and Parkinson's disease. Trends Neurosci. **8:** 79.
16. CONWAY, E.L., A.L. GUNDLACH & J.A. CRAVEN. 1998. Temporal changes in glial fibrillary acidic protein messenger RNA and [3H]PK11195 binding in relation to imidazoline-I2-receptor and alpha 2-adrenoceptor binding in the hippocampus following transient global forebrain ischaemia in the rat. Neuroscience **82:** 805–817.
17. NEVE, K., R. HENNINGSEN, J. BUNZOW & O. CIVELLI. 1989. Functional characterization of a rat dopamine D-2 receptor cDNA expressed in a mammalian cell line. Mol. Pharmacol. **36:** 446–451.

18. BAL, A., T. BACHELOT, M. SAVASTA, M. MANIER, J.M. VERNA, A.L. BENABID & C. FEUERSTEIN. 1994. Evidence for dopamine D2 receptor mRNA expression by striatal astrocytes in culture: in situ hybridization and polymerase chain reaction studies. Brain Res. Mol. Brain Res. **23:** 204–212.
19. NEVE, K.A., R.L. NEVE, S. FIDEL, A. JANOWSKY & G.A. HIGGINS. 1991. Increased abundance of alternatively spliced forms of D2 dopamine receptor mRNA after denervation. Proc. Natl. Acad. Sci. U S A **88:** 2802–2806.
20. LUO, Y., G.C. KOKKONEN, X. WANG, K. NEVE & G.R. ROTH. 1998. D2 dopamine receptors stimulate mitogenesis through PTX-sensitive G proteins and Ras involved ERK and SAP/JNK pathways in rat C6-D2L glioma cells. J. Neurochem. **71:** 980–990.
21. WATTS, V.J. & K.A. NEVE. 1996. Sensitization of endogenous and recombinant adenylate cyclase by activation of D2 dopamine receptors Mol. Pharmacol. **50:** 966–976.
22. IRANI, K., Y. XIA, J. ZWEIER, S. SOLLOTT, C. DER, E. FEARON, M. SUNDARESAN, T. FINKEL & P. GOLDSCHMIDT-CLERMONT. 1997. Mitogenic signaling mediated by oxidants in Ras-transformed fibroblasts. Science **275:** 1649–1652.
23. RUPPERSBERG, J.P., M. STOCKER, O. PONGS, S.H. HEINEMANN, R. FRANK & M. KOENEN 1991 Regulation of fast inactivation of cloned mammalian IK(A) channels by cysteine oxidation. Nature **352:** 711–714.
24. LO, Y.Y. C., J.M.S. WONG & T.F. CRUZ. 1996. Reactive oxygen species mediate cytokine activation of c-Jun NH_2-terminal kinases. J. Biol. Chem. **271:** 15703–15707.
25. KRIEGER-BRAUER, H.I., P.K. MEDDA & H. KATHER. 1997. Insulin-induced activation of NADPH-dependent H_2O_2 generation in human adipocyte plasma membranes is mediated by Ga_{i2}. J. Biol. Chem. **272:** 10135–10143.
26. MAJOR, D.E., J.P. KESSLAK, C.W. COTMAN, C.E. FINCH & J.R. DAY. 1997. Life-long dietary restriction attenuates age-related increases in hippocampal glial fibrillary acidic protein mRNA. Neurobiol. Aging **18:** 523–526.
27. CUENDA, A., J. ROUSE, Y.N. DOZA, R. MEIER, P. COHEN, T.F. GALLAGHER, P.R. YOUNG & J.C. LEE. 1995. SB 203580 is a specific inhibitor of a MAP kinase homologue which is stimulated by cellular stresses and interleukin-1. FEBS Lett. **364:** 229–233.
28. LEE, J., J. LAYDON, P. MCDONNELL, T. GALLAGHER, S. KUMAR, D. GREEN, D. MCNULTY, M. BLUMENTHAL, J. HEYS, S. LANDVATTER, J. STRICKLER, M. MCLAUGHLIN, I. SIEMENS, S. FISHER, G. LIVI, J. WHITE, J. ADAMS & P. YOUNG. 1994. A protein kinase involved in the regulation of inflammatory cytokine biosynthesis Nature **372:** 739–754.
29. RAINGEAUD, J., S. GUPTA, J. ROGERS, M. DICKENS, J. HAN, R. ULEVITCH & R. DAVIS. 1995. Pro-inflammatory cytokines and environmental stress cause p38 mitogen-activated protein kinase activation by dual phosphorylation on tyrosine and threonine. J. Biol. Chem. **270:** 7420-7426.
30. LATOV, N., G. NILAVER, E.A. ZIMMERMAN, W.G. JOHNSON, A.J. SILVERMAN, R. DEFENDINI & L. COTE. 1979. A study combining immunoperoxidase technique for glial fibrillary acidic protein and radioautography of tritiated thymidine. Dev. Biol. **72:** 381–384.
31. TAKAMIYA, Y., S. KOHSAKA, M. TOYA, M. OTANI & Y. TSUKADE. 1988. Immunohistochemical studies on the proliferation of reactive astrocytes and the expression of cytoskeletal proteins following brain injury in rats. Dev. Brain Res. **38:** 201–210.
32. LUO, Y., A. HATTORI, J. MUNOZ, Z.-H. QIN & G.S. ROTH. 1999. Intrastriatal dopamine injection induces apoptosis through oxidation-involved activation of transcription factors AP-1 and NF-κB in rats. Mol. Pharmacol. **56:** 254–264.
33. LUO, Y., G.C. KOKKONEN, A. HATTORI, F.J. CHREST & G.S. ROTH. 1999. Dopamine stimulates redox-tyrosine kinase signaling and p38 MAPK in activation of astrocytic C6-D2L cells.. Brain Res. **850:** 21–38.

Selenium Biochemistry

Mammalian Selenoenzymes

THRESSA C. STADTMAN

Laboratory of Biochemistry, NHLBI, National Institutes of Health,
Bethesda, Maryland 20892, USA

Within the past ten years selenium biochemistry has attracted an increasing number of investigators interested both in the beneficial and the toxic effects of the element in biological systems. In 1957 two independent research groups showed that the trace element selenium is an important nutrient for animals. Klaus Schwartz at the National Institutes of Health in Bethesda isolated a selenium-containing factor that prevented rats fed a Torula yeast based diet from developing liver necrosis.[1] Investigators at the Lederle Laboratories in Pearl River, NY, found that exudative diathesis in poultry was prevented by addition of selenium to the diet.[2] Predating these discoveries by three years was a report[3] that *Escherichia coli* required selenium and molybdenum for synthesis of active formate dehydrogenase, but this finding attracted little attention at a time when the revelance of discoveries made in bacterial physiology to mammalian physiology was not widely appreciated.

A possible biochemical explanation of the beneficial effects of selenium in animals came several years later when the trace element was discovered to be an essential component of an important antioxidant enzyme, glutathione peroxidase.[4,5] At the same time a low molecular weight component, protein A, of the Clostridial glycine reductase complex was shown to be a selenoprotein[6] and the selenium-containing moiety was identified as selenocysteine.[7] The techniques developed for alkylation of the reduced selenocysteine and identification of the carboxymethyl and carboxyethyl derivatives of the selenoamino acid in acid hydrolysates of Clostridial selenoprotein A facilitated identification of selenocysteine in glutathione peroxidase and, later, in other selenoproteins.[8]

Continued study of glutathione peroxidase has revealed the presence of four isoenzyme forms in various organs and tissues.[9] The tetrameric enzyme originally isolated from bovine erythrocytes[5] contains one selenocysteine residue per subunit and is now termed cytosolic GPX or GPX-1. Reaction of the ionized selenol in this enzyme with a variety of peroxide substrates converts it to R-SeOH. The selenol is regenerated by successive reactions with GSH. Another tetrameric isoenzyme form, GPX-2, that occurs in plasma is a glycoprotein. GSH levels in plasma are low and therefore the thioredoxin-thioredoxin reductase system serves to regenerate the reduced form of GPX-2 after oxidation by a peroxide. A gastrointestinal glutathione peroxidase, termed GPX-3, another tetrameric isoenzyme, depends on GSH for regeneration. Phospholipid-hydroperoxide glutathione peroxidase (GPX-4), specialized for reaction with fatty acid and cholesterol hydroperoxides, is a monomeric 20 kD protein that contains a single selenocysteine residue. Characterization of this enzyme from *Schistosoma mansoni*, a parasitic platyhelminth known as a blood fluke,[10] provides one of the few known examples of GPX-4 occurrence in species other than mammals.

The discovery in 1991 that Type I iodothyronine 5′-deiodinase is a selenocysteine-containing enzyme[11] was of especial interest at the time because this pointed to a biological role of selenium in developmental processes. Prior to this the only known role of selenium in mammalian physiology was its occurrence as an essential component of an antioxidant enzyme. Subsequent studies have established that other members of the deiodinase family also are selenoproteins. Investigations on the mechanism of selenocysteine insertion into the 5′-deiodinases[12,13] have shown that the eukaryotic process involves use of a stem-loop structure in the 3′ non-translated region of the m-RNA to designate selenocysteine insertion at the UGA codon, instead of chain termination, whereas in prokaryotes the stem-loop is located in the open reading frame immediately downstream from UGA.[14] Similar structures have been identified in the 3′ non-translated regions of other mammalian selenoproteins.[13]

Current interest in the antioxidant properties of selenium has increased greatly with the discovery by Dr. Takashi Tamura in my laboratory that mammalian thioredoxin reductase(TrxR) is a selenocysteine-containing enzyme.[15] This selenoamino acid, located in the sequence -Cys-Secys-Gly at the C-terminus of each subunit[16,17] corresponds to a TGA codon in the c-DNA that originally was interpreted as termination.[18] Thus, in addition to protein bound FAD and the Cys-59, Cys-64 disulfide center that undergo alternate reduction and oxidation, a third potential redox center consisting of Cys-497 and Secys-498 occurs in mammalian thioredoxin reductases. Some evidence indicating that the selenocysteine residue has an essential role in catalysis came from studies showing that TrxR activity in cultured human cells depended on selenium supplementation of media[19] and proteolytic cleavage at the C-terminus of purified TrxR caused loss of enzyme activity.[20] Enzyme preparations isolated from HeLa cells grown under high oxygen saturation conditions proved to be mixtures of low activity, low selenium content enzyme forms and fully active enzyme forms that could be separated by heparin affinity chromatography.[21] The fully active enzyme was not retained on the column whereas the low activity form that comprised 30% to as much as 50% of the enzyme population bound to the heparin column. This enzyme contained 0.5 instead of 1 equivalent of selenium per subunit and was only 50% as active catalytically. The correspondence of low catalytic activity and low selenium content of numerous enzyme preparations isolated from HeLa cells cultured under higher than optimal oxygen levels suggested that oxidative damage to sensitive ionized selenol groups could be involved. In control experiments it was shown that fully active enzyme, when reduced by treatment with tris-(2-carboxyethyl)phosphine, was converted to an extremely oxygen sensitive form.[21] Exposure of this reduced enzyme to air resulted in extensive selenium elimination and concomitant loss of catalytic activity whereas no losses occurred anaerobically under argon. Supplementation of the reduced enzyme with mM levels of pyridine nucleotide, either NADPH or NADP$^+$, induced a form of the enzyme that was stable under aerobic conditions indicating that the labile selenol now was protected from oxidative damage.

The importance of SeCys-498 for catalytic activity of TrxR also was demonstrated by specific alkylation of the ionized selenol group in the reduced enzyme with bromoacetate at low pH. Under these conditions there was quantitative conversion of Secys-498 to CM-Secys when only 1.1 alkyl group was introduced per enzyme subunit and this was sufficient to cause complete inhibition of enzyme activity on

DTNB as substrate.[21] Inhibition of enzyme activity by specific alkylation of the selenocysteine residue indicates participation of the C-terminal redox center in the overall oxidation-reduction process.

Taken together these properties, characteristic of a very reactive selenol group, provide an explanation of the previously observed unusual range of substrates that could be reduced by mammalian TrxR[22] in contrast to the more limited substrate specificity of the enzymes from yeast and prokaryotes that lack selenium. The smaller molecular weight yeast and bacterial thioredoxin reductases, dimers of 35,000 Mr subunits, contain the corresponding FAD and disulfide redox centers but lack the third -Cys-Secys- redox center present in the C-terminal peptide extension of mammalian TrxR.

Recent discovery of a mitochondrial isoenzyme form of mammalian TrxR in rat liver is of especial interest.[23] This enzyme contains the same C-terminal -Cys-Secys-Gly sequence and additionally possesses a leader N-terminal sequence that targets the enzyme to the mitochondrial membrane. Further detailed chemical and kinetic studies on these enzymes together with structural analyses should provide information as to the mechanism of electron transfer between the various redox centers and final acceptor substrates.

REFERENCES

1. SCHWARZ, K. & C.M. FOLTZ. 1957. Selenium as an integral part of Factor 3 against dietary necrotic liver degeneration. J. Am. Chem. Soc. **79:** 3292–3293.
2. PATTERSON, E.L., R. MILSTREY & E.L.R. STOKSTAD. 1957. Effect of selenium in preventing exudative diathesis in chicks. Proc. Soc. Exp. Biol. Med. **95:** 617–620.
3. PINSET, J. 1954. The need for selenite and molybdate in the formation of formate dehydrogenases by members of the Coliaerogenes group of bacteria. Biochem. J. **57:** 10–16.
4. ROTRUCK. J.T., A.L. POPE, H.E. GANTHER, A.B. SWANSON, D.G.F. HAFEMAN & W.G. HOEKSTRA. 1973. Selenium: biochemical role as a component of glutathione peroxidase. Science **179:** 588–590.
5. FLOHE, L., W.A. GUNZLER & H.H. SHOCK. 1973. Glutathione peroxidase. A selenoenzyme. FEBS Lett. **32:** 132–134.
6. TURNER, D.C. & T.C. STADTMAN. 1973. Purification of protein components of the Clostridial glycine reductase system and characterization of protein A as a selenoprotein. Arch. Biochem. Biophys. **154:** 366–381.
7. CONE, J.E., R. MARTIN DEL RIO, J.N. DAVIS & T.C. STADTMAN. 1976. Chemical characterization of the selenoprotein component of Clostridial glycine reductase: identification of selenocysteine as the organoselenium moiety. Proc. Natl. Acad. Sci. USA **73:** 2659–2663.
8. STADTMAN, T.C. 1996. Selenocysteine. Ann. Rev. Biochem. **65:** 83–100.
9. URSINI, F., M. MAIORINO, R. BRIGELIUS-FLOHE, K.D. AUMANN, A. ROVERI, D. SCHOMBURG & L. FLOHE. 1995. Diversity of glutathione peroxidases. Methods Enzymol. **252:** 38–53.
10. MAIORINO, M., C. ROCHE, M. KIESS, K. KOENIG, D. GAWLIK, M. MATTTHES, E. NALDINI, R. PIERCE & L. FLOHE. 1996. A selenium-containing phospholipid-hydroperoxide glutathione peroxidase in *Schistosoma mansoni*. Eur. J. Biochem. **238:** 838–844.
11. BERRY, M.J., L. BANU & P.R. LARSEN. 1991. Type I iodothyronine deiodinase is a selenocysteine-containing enzyme. Nature **349:** 438–440.
12. BERRY, M.J., L. BANU, Y. CHEN, S.J.MANDEL, J.D. KIEFFER, J.W. HARNEY & P.R. LARSEN. 1991. Recognition of UGA as a selenocysteine codon in Type I deiodinase requires sequences in the 3' untranslated region. Nature **353:** 273–276.

13. BERRY, M.J., L. BANU, J.W. HARNEY & P.R. LARSEN. 1993. Functional characterization of the eukaryotic SECIS elements which direct selenocysteine insertion at UGA codons. EMBO J. **12:** 3315–3322.
14. HEIDER, J., C. BARON & A. BOCK. 1992. Coding from a distance: dissection of the mRNA determinants required for the incorporation of selenocysteine into protein. EMBO J. **11:** 3759–3766.
15. TAMURA, T. & T.C. STADTMAN. 1996. A new selenoprotein from human lung adenocarcinoma cells: purification, properties, and thioredoxin reductase activity. Proc. Natl. Acad. Sci. USA **93:** 1006–1011.
16. GLADYSHEV, V.N., K.-T. JEANG & T.C. STADTMAN. 1996. Selenocysteine, identified as the penultimate C-terminal residue in human T cell thioredoxin reductase, corresponds to TGA in the human placental gene. Proc. Natl. Acad. Sci. USA **93:** 6146–6151.
17. LIU, S.-Y. & T.C. STADTMAN. 1997. Heparin-binding properties of selenium-containing thioredoxin reductase from HeLa cells and human lung adenocarcinoma cells. Proc. Natl. Acad. Sci. USA **94:** 6138–6141.
18. GASDASKA, P.Y., J.R. GASDASKA, S. COCHRAN & G. POWIS. 1995. Cloning and sequencing of a human thioredoxin reductase. FEBS Lett. **373:** 5–9.
19. MARCOCCI, L., L. FLOHE & L. PACKER. 1997. Evidence for a functional relevance of the selenocysteine residue in mammalian thioredoxin reductase. BioFactors **6:** 351–358.
20. GROMER, S., J. WISSING, D. BEHNE, K. ASHMAN, R.H.SCHIRMER, L. FLOHE & K. BECKER. 1998. A hypothesis on the catalytic mechanism of the selenoenzyme thioredoxin reductase. Biochem. J. Lett. **332:** 591–592.
21. GORLATOV, S.N. & T.C. STADTMAN. 1998. Human thioredoxin reductase from HeLa cells: selective alkylation of selenocysteine in the protein inhibits enzyme activity and reduction with NADPH influences affinity to heparin. Proc. Natl. Acad. Sci. USA **95:** 8520–8525.
22. HOLMGREN, A. & M. BJORNSTEDT. 1995. Thioredoxin and thioredoxin reductase. Methods Enzymol. **252B:** 199–208.
23. LEE, S.-R., J.-R. KIM, K.-S. KWON, H.W. YOON, R.L. LEVINE, A. GINSBURG & S.G. RHEE. 1999. Molecular cloning and characterization of a mitochondrial selenocysteine-containing thioredoxin reductase from rat liver. J. Biol. Chem. **274:** 4722–4734.

Cytoprotective Properties of Nisoldipine and Amlodipine against Oxidative Endothelial Cell Injury

I. TONG MAK,[a] JINGYUN ZHANG, AND WILLIAM B. WEGLICKI

Department of Physiology & Experimental Medicine, The George Washington University Medical Center, Washington, D.C. 20037, USA

INTRODUCTION

Oxygen free radical-mediated lipid peroxidation might play a major role in the pathogenesis of atherosclerosis.[1] The endothelial cell is a critical target of oxidative injury and may represent an early event of atherosclerosis.[2] Several calcium channel blockers were reported to provide antiatherogenic effects in experimental and clinical studies[3,4] but were unrelated to their Ca-blocking activities. We have reported that certain lipophilic Ca-blockers exhibit antioxidant activities.[5–8] In this study, the ability of two dihydropyridine Ca-blockers, nisoldipine and amlodipine, to protect against oxidative endothelial cell injury was assessed and compared with vitamin E (Trolox).

METHODS

Bovine aortic endothelial cells (AG 8132A) were cultured in Dulbecco's modified Eagle's medium supplemented with 0.01 M HEPES and 10% fetal bovine serum as described.[5–8] After the endothelial cells grew to confluence in 6-well plates, the medium was removed and replaced with 1.0 ml/well of the balanced salt solution containing 10 mM glucose, 125 mM NaCl, 1.2 mM $MgCl_2$, 10 mM potassium phosphate, pH 7.2. The attached cells were preincubated with each drug (3–20 µM) for 45 minutes before exposure to the oxy-radical generating system consisting of 1.7 mM dihydroxyfumarate (DHF) and 10 µM $FeSO_4$ for 30–60 minutes. At the end of incubation, the cells in all wells were retrieved and cell disruption was achieved by brief sonication.[7] Total glutathione levels (GSH+ 1/2 GSSG) were determined by the "cyclic method," which combines the colorimetric reaction of 5,5'-dithiobis-(2-nitrobenzoic acid) with the enzymatic specificity of glutathione reductase as described before.[5–7] Accumulation of lipid peroxidation products was measured by the TBARS assay.[7] Cell survival 24 hours later was used as an overall functional endpoint of cell injury and recovery. At the end of the free radical exposure, the cells in 24-well plates were gently rinsed with 1 ml/well of PBS and received fresh growth medium. Cell viability/survival at 24 hours was determined by the MTT method as described.[8]

---

[a]Address for correspondence: Dr. I.T. Mak, 2300 Eye St. NW, Ross Hall, Room 452, Washington, DC 20037.Voice: 202-994-2865; fax: 202-994-3553.
e-mail: itmak@gwu.edu

RESULTS

We have previously shown that changes in endothelial glutathione status can represent a sensitive indicator of cellular oxidative stress.[5–7] As shown in FIGURE 1, the cells exposed to free radicals in the absence of drug exhibited a 55% loss of total glutathione. Both nisoldipine and amlodipine dose-dependently (3–10 µM) attenuated the loss of glutathione. At 10 µM, both drugs were able to restore the glutathione level to about 73% of control and the efficacy was comparable to that of Trolox (water soluble vitamin E analog). In data not shown, drug levels at 20 µM did not further improve the glutathione level. With identical experimental settings as above, the accumulation of TBAR substances was determined. The samples exposed to free radicals alone displayed about 10-fold higher level of TBARS than the vehicle controls (3.04 ± 0.36 vs. 0.32 ± 0.03 nmol MDA equivalents/well). Both Ca-blockers and Trolox provided concentration-dependent (3–20 µM) inhibition (20–60%) of the TBARS formation (TABLE 1). It appeared that the order of activity was: nisoldipine ≥ Trolox > amlodipine. In separate experiments, endothelial cells (in 24 well-plates) exposed to oxygen radicals for 30 minutes resulted in 43% loss of cell viability which was determined 24 hours later (FIG. 2). Pretreatment of the cells with

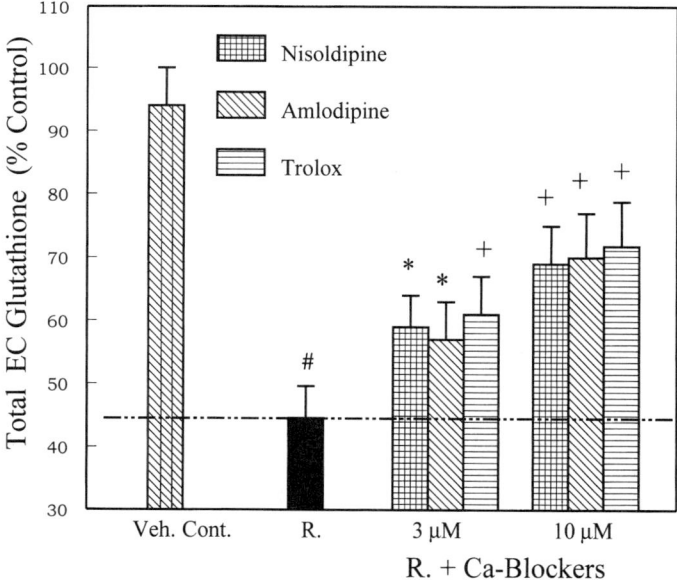

FIGURE 1. Comparative protective effects of nisoldipine, amlodipine and Trolox against oxy-radical induced loss of endothelial GSH. Confluent endothelial cells in 6-well plates were pre-incubated with or without drugs for 45 min before addition of the oxy-radical components (DHF/Fe). After 60 min of incubation, each well was processed for total glutathione determination. Values are mean ± SD of at least 4 experiments; # $p < 0.001$ vs. vehicle (0.25% alcohol to dissolve the drugs) control with no radicals, *$p < 0.05$, +$p < 0.01$ vs. free radical (R.) alone.

TABLE 1. Nisoldipine and amlodipine inhibit endothelial cell lipid peroxidation

Drug Concentration	Percent inhibition of TBARS formation compared to free radical controls		
	Nisoldipine	Amlodipine	Trolox
3 µM	22 ± 4.3	22 ± 4.0	19.7 ± 5.2
10 µM	48 ± 6.8	28 ± 7	44 ± 7.2
20 mM	63 ± 5.9	39.5 ± 10	52 ± 7.3

Experimental conditions were as described for FIGURE 1. The TBARS value for the free radical alone was 3.04 ± 0.36 nmol/well whereas that for the vehicle control was 0.32 ± 0.03 nmol/well. All values are means ± SD from at least 4 separated experiments.

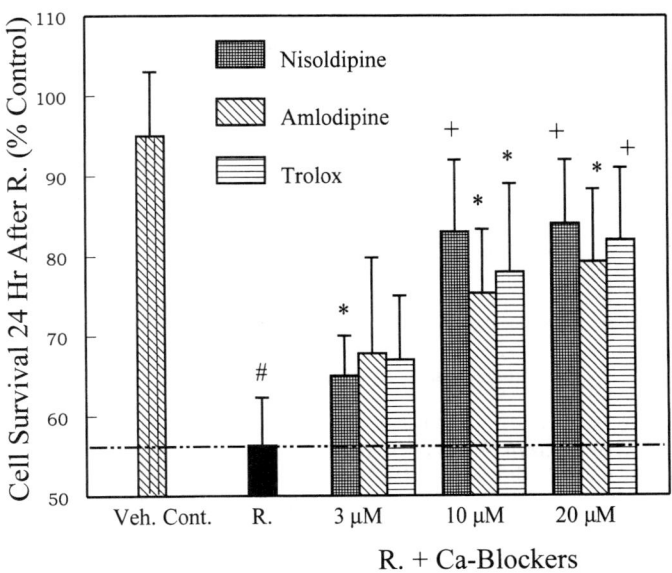

FIGURE 2. Protective effects of nisoldipine, amlodipine and Trolox against oxy-radical-impaired cell survival. Endothelial cells grown in 24-well plates were pre-incubated with or without the Ca-blockers for 45 min before addition of the oxy-radical components. After 30 min of incubation, all wells were replaced with fresh growth medium; 24 hours later, cell viability was determined by the MTT assay. Other conditions were as described for FIGURE 1; $^{\#}p < 0.001$ vs. vehicle control, $^{*}p < 0.05$, $^{+}p < 0.01$ vs. free radical (R.) alone.

3–20 µM of each agents provided varying degrees of significant protection against the loss of cell viability/survival (FIG. 2). At 20 µM, all three agents were able to preserve the cell viability up to ~85% of control. Again, the order of efficacy appeared to be nisodipine > Trolox > amlodipine.

CONCLUSIONS

Both nisoldipine and amlodipine displayed antioxidant and cytoprotective activities against free radical-mediated endothelial injury. The efficacy of nisoldipine was comparable to or slightly better than that of vitamin E. Since calcium was omitted in the incubation system, we suggest that the antiperoxidative properties of the two dihydropyridine Ca-blockers conferred their cytoprotective effects. These properties may contribute to their reported beneficial effects in experimental and clinical studies of atherosclerosis.[3,4]

ACKNOWLEDGMENTS

This study was supported in part by NIH grant RO1-HL-36418. The authors wish to thank Alex Murphy for his excellent technical assistance.

REFERENCES

1. Ross, R. 1986. The pathogenesis of atherosclerosis: an update. N. Engl. J. Med. **314:** 488–500
2. Henning, B. & C.K. Chow. 1988. Lipid peroxidation and endothelial cell injury: Implications in atherosclerosis. Free Rad. Biol. Med. **4:** 99–106.
3. Weinstein, D.B. & J.G. Heider. 1988. Antiatherogenic properties of calcium channel blockers. Am. J. Med. **84** (Suppl. 3B): 102–108.
4. Keogh, A.M. & J.S. Schroeder. 1990. Calcium antagonists and atherosclerosis. Reperfusion **2:** 5–8.
5. Mak, I.T. *et al.* 1992. Antioxidant effects of calcium channel blockers against free radical injury in endothelial cells. Circ. Res. **70:** 1099–1103.
6. Mak, I.T. *et al.* 1992. Antioxidant properties of active and inactive isomers of nicardipine in cardiac membranes, endothelial cells, and perfused rat hearts. Coronary Artery Dis. **3:** 1095–1103.
7. Mak, I.T. & W.B. Weglicki. 1994. Antioxidant activity of calcium channel blocking drugs. Methods Enzymol. **234:** 620–630.
8. Mak, I.T. *et al.* 1995. Protective effects of calcium channel blockers against free radical-impaired endothelial cell proliferation. Biochem. Pharmacol. **50:** 1531–1534.

The Origin of Dinitrosyl-Iron Complex in Endothelial Cells

ANDREI M. KOMAROV,[a] I. TONG MAK, AND WILLIAM B. WEGLICKI

Department of Physiology & Experimental Medicine, The George Washington University Medical Center, Washington, DC 20037, USA

Dinitrosyl-iron complex $(RS)_2Fe(NO)_2$ with characteristic electron paramagnetic resonance (EPR) signal is found in a variety of tissues and cells during inflammation.[1,2] However, the identity of the non-heme iron center(s) involved and the role of such complexes in inflammatory cell response are still unknown. Iron-sulfur clusters of enzymes, such as aconitase, have been proposed as targets of nitric oxide (·NO) and the source of dinitrosyl-iron complexes.[1] Additionally, iron nonspecifically chelated by protein vicinal thiols has been proposed as the source of such complexes.[1] FIGURE 1 demonstrates characteristic EPR signal of the model cysteine (Cys) dinitrosyl-iron complex: $(Cys)_2Fe(NO)_2$ at 77K ($g_\perp = 2.04$ and $g_\| = 2.01$).[3] In this report, we have tested the ability of nitric oxide donors S-nitroso-N-acetyl-penicillamine (SNAP), nitroprusside (NP), peroxynitrite donor 3-morpholinosydnonimine (SIN-1), purified peroxynitrite (^-OONO) and nitrite to form the EPR detectable dinitrosyl-iron complex in cultured bovine endothelial cells (EC). Our results suggest that in exposed cells nitric oxide alone, but not peroxynitrite alone, can produce the dinitrosyl-iron complex and that use of ·NO and ^-OONO simultaneously reduces the amount of dinitrosyl-iron complex formed in cells.

SNAP and SIN-1 were purchased from Cayman Chemical Company (Ann Arbor, MI), sodium nitroprusside was from Sigma (St. Louis, MO). SNAP was dissolved in ethanol and exact concentration was determined at 330 nm (E_{330} = 711 M^{-1} cm^{-1}). Peroxynitrite was synthesized and purified essentially according to the procedure of Beckman *et al.* and stored in NaOH solution at −80 °C until use. The exact concentration was determined at 302 nm ($E_{302} = 1,670$ M^{-1} cm^{-1}). Bovine pulmonary arterial endothelial cells were cultured, then harvested, pelleted, washed twice with the incubation buffer at 37°C and finally resuspended in the buffer containing 11 mM glucose, 120 mM NaCl, 1.2 mM $MgSO_4$, 10 mM KH_2PO_4, pH 7.2 at 8×10^6 cells/ml. EC suspended in the buffer were exposed to ·NO donors, peroxynitrite, or NO_2^- (final concentration up to 1 mM) for 3 h at 37°C. The cells were pelleted again and frozen in liquid nitrogen (8×10^6 cells per sample). Dinitrosyl-iron complex formed in cells ($g_\perp = 2.04$ and $g_\| = 2.01$) was determined by EPR at 77 K.

In endothelial cells treated with 1 mM SNAP the yield of dinitrosyl-iron complex was 135 ± 29 pmol/10^6 cells (FIG. 1). Nitroprusside also produced dinitrosyl-iron complex in EC in a dose-dependent fashion (data not shown). However,

---

[a]Address for correspondence: Department of Physiology & Experimental Medicine, The George Washington University Medical Center, Ross Hall, Room 451A, 2300 Eye Street, NW, Washington, DC 20037. Voice: 202-994-2301;fax: 202-994-3553.
e-mail: amkoma@gwis2.circ.gwu.edu

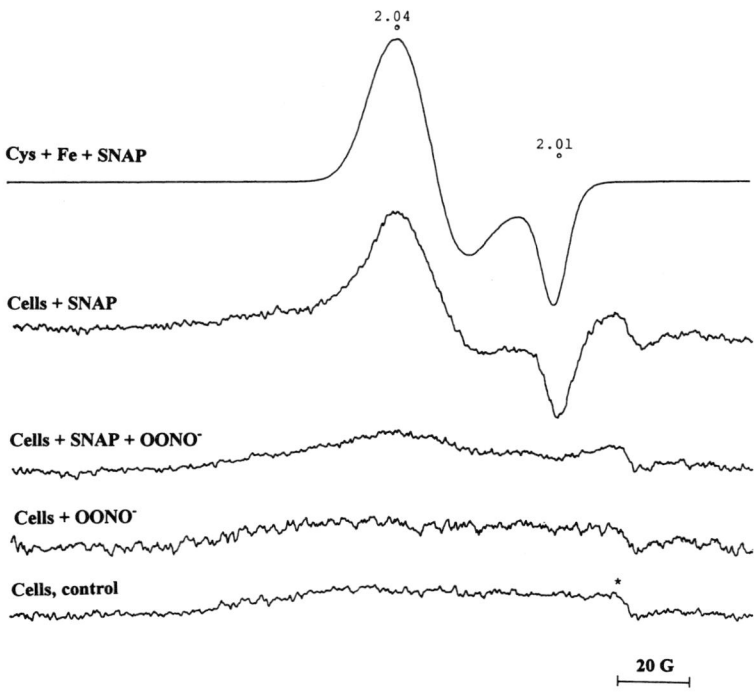

FIGURE 1. EPR signal (77 K) of dinitrosyl-iron complex formed: in solution of cysteine, iron (Cys to iron ratio of 20 to 1, prepared under argon)[3] and 1 mM SNAP (Cys + Fe + SNAP); in suspension of endothelial cells (EC) incubated with 1 mM SNAP (Cells + SNAP); EC incubated with 1 mM SNAP plus 1 mM peroxynitrite (Cells + SNAP + OONO$^-$); EC incubated with 1 mM peroxynitrite (Cells $^+$ OONO$^-$), or EC incubated in plain buffer (Cells, control). The position of $g_\perp = 2.04$ and $g_\parallel = 2.01$ are shown (°). The position of the free radical signal (*) associated with cells is also shown. The instrument gain was 100-fold higher for spectra obtained from cell suspensions compared to the top spectrum, which was obtained after single 200 gauss (G) field scan. Other spectra were an average of four 200 s scans. EPR conditions were: microwave frequency, 9.40 GHz; microwave power, 5 mW; center field, 3,292 G; scan rate 60 G/min; time constant, 0.5 s; modulation amplitude, 6.3 G; modulation frequency 100 KHz.

peroxynitrite donor SIN-1 and peroxynitrite (1 mM final concentration, added in several aliquots during incubation) did not produce such complexes in EC (FIG. 1). Furthermore, co-incubation of SNAP (1 mM) with SIN-1 or peroxynitrite (1 mM) yielded significantly less dinitrosyl-iron complex in EC than SNAP alone (FIG. 1). Nitrite is a decomposition product of SNAP, NP, SIN-1, $^-$OONO and it did not produce any EPR signal upon incubation with EC (data not shown). We have also observed that SIN-1 and peroxynitrite, but not ·NO donors depleted cellular glutathione (> 90%) (FIG. 2). Previously, we have described the same effect of peroxynitrite and SIN-1 on red blood cell glutathione.[4] Peroxynitrite is a well-known

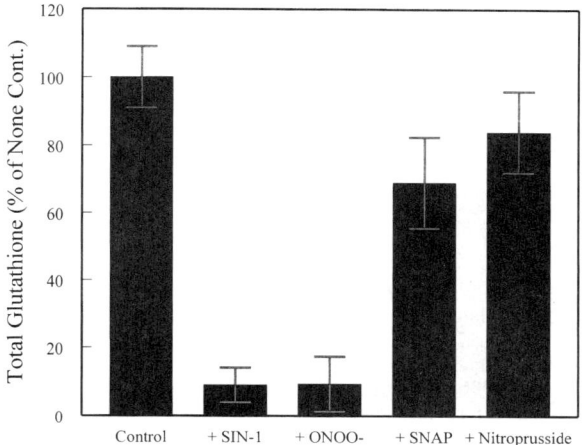

FIGURE 2. Effects of ·NO and peroxynitrite donors on endothelial cell glutathione levels. Endothelial cells were suspended in the buffer and incubated with ·NO donors or peroxynitrite (1 mM, final concentration) for 3 h at 37°C. Control cells were incubated in a plain buffer. At the end of incubating period, the samples were transferred to ice, acidified by adding 5-sulfosalicylic acid (5%) and ruptured by brief sonication. Total glutathione: (GSH) + 1/2 oxidized glutathione (GSSG) was assayed by the 'recycling method', involving GSSG reductase-catalyzed recycling of glutathione. The amount of glutathione in control cells was designated as 100%. Note, that glutathione was used in this study as a representative of intracellular thiols and these data do not imply that intracellular dinitrosyl-iron complexes are formed by glutathione.

thiol oxidant.[5] It may destroy preexisting cellular complexes between thiols and iron by oxidizing thiol groups (2 RSH to RSSR) and iron (Fe^{2+} to Fe^{3+}). This will prevent further formation of dinitrosyl-iron complexes by nitrosothiols (including nitrosothiols derived from peroxynitrite) or other NO donors. Thiol-dependent dinitrosyl-iron complex formation in EC is a marker of cell exposure to ·NO (or nitosothiols), but not to peroxynitrite. Furthermore, peroxynitrite prevents dinitrosyl-iron complex formation from other ·NO sources. Our observation also may explain why in septic shock animals there are no dinitrosyl-iron complexes in tissues which exhibit a high level of nitric oxide and peroxynitrite production.[6,7] Previously, dinitrosyl-iron complex has been proposed as:

1. a candidate for endothelium-derived relaxation factor (EDRF)[1,3];
2. as a reservoir for ·NO, which can transfer NO^+ group to thiols to form nitrosothiols;[1,3] and
3. as a signaling agent capable of oxidizing critical cysteine thiol residues in enzymes.[8]

Peroxynitrite would presumably interfere with or influence any of the above mentioned functions of dinitrosy-iron complex.

ACKNOWLEDGMENTS

This research was supported by Faculty Research Enhancement Fund Award from GWUMC and in part by NIH Grant RO1-HL36418. We also thank Mr. Alex Murphy for excellent technical support and Ms. Patricia Boehme for helping to prepare the manuscript.

REFERENCES

1. HENRY, Y.A. *et al.* 1997. Nitric Oxide Research from Chemistry to Biology: EPR Spectroscopy of Nitrosylated Compounds. New York. Chapman & Hall.
2. YEE, E.L. *et al.* 1996. Effect of nitric oxide on heme metabolism in pulmonary artery endothelial cells. Am. J. Physiol. **271:** L512–L518.
3. VANIN, A.F. *et al.* 1996. Physical properties of dinitrosyl iron complexes with thiol-containing ligands in relation with their vasodilator activity. Biochim. Biophys. Acta **1295:** 5–12.
4. MAK, I.T. *et al.* 1996. Enhanced NO production during Mg deficiency and its role in mediating red blood cell glutathione loss. Am. J. Physiol. **271:** C385–C390.
5. RADI, R. *et al.* 1991. Peroxynitrite oxidation of sulfhydryls. The cytotoxic potential of superoxide and nitric oxide. J. Biol. Chem. **266:** 4244–4250.
6. KOMAROV, A.M. *et al.* 1997. Iron potentiates nitric oxide scavenging by dithiocarbamates in tissue of septic shock mice. Biochim. Biophys. Acta **1361:** 229–234.
7. GALEGAN, M.Y. *et al.* 1997. Reaction of dinitrosyl complexes of non-heme iron with diethyldithiocarbamate in the blood of anaesthetized rats: its specific manifestation at physicochemical and physiological levels. Biofizika **42:** 687–693.
8. BECKER, K. *et al.* 1998. Enzyme inactivation through sulfhydryl oxidation by physiologic NO-carriers. Nature Struct. Biol. **5:** 267–271.

Neutrophil-Endothelial Cell Interactions

Inverse Correlation between Nitric Oxide and Superoxide Anions[a]

JAYASREE NATH[b] AND SANTHANAM KAUSALYA

Department of Respiratory Research, Division of Medical Casualty Research, Walter Reed Army Institute of Research, Silver Spring, Maryland 20910-7500, USA

INTRODUCTION

Nitric oxide (·NO), a free-radical gas produced by many different cell types, has diverse effects in biological systems, which may be either beneficial or detrimental, depending on the amounts of ·NO generated and also on the nature of the immediate microenvironment where ·NO is synthesized and released.[1-4] Interaction of neutrophils (PMN) with endothelial cells (EC) is of central importance in the regulation of acute inflammatory response. PMN-mediated cell, tissue or organ injury forms the basis of many pathophysiologic conditions, and the vascular endothelium is an important target of oxidant-induced injuries in a variety of inflammatory conditions.[5] Although an excess production of ·NO in the presence of superoxide anions ($·O_2^-$) is known to have cytotoxic effects,[6] ·NO is also implicated in the protection of injury to target cells or tissues, during cellular events that are associated with production of reactive oxygen intermediates (ROI). The interactive role of ROI and ·NO is particularly relevant in the context of inflammatory phagocytes, which generate large amounts of ROI during diapedesis and infiltration to sites of infection, inflammation and/or injury.[3,4]

In our present *in vitro* study, we attempted to delineate the interactive role of ·NO and $·O_2^-$ in PMN-mediated EC injury, by direct quantitation of ·NO and $·O_2^-$ in PMN-EC co-cultures. We utilized freshly isolated human peripheral blood PMN and cryopreserved human umbilical vein endothelial cells (HUVEC, from Clonetics) to test the interactive role of ·NO and $·O_2^-$ during PMN-EC interactions under a variety of experimental conditions. The details of these studies have been recently published.[7] Our results demonstrate an inverse correlation between ·NO and $·O_2^-$ during PMN-EC interactions and also indicate a protective role of ·NO in PMN-mediated EC injury.

---

[a]The opinions or assertions contained herein are the private views of the authors, and are not to be construed as official or reflecting the views of the Department of the Army or the Department of Defense.

[b]Address for correspondence: Dr. Jayasree Nath, Department of Respiratory Research, Division of Medical Casualty Research, Walter Reed Army Institute of Research, 503 Robert Grant Street, Rm. 1N-85, Silver Spring, MD 20910-7500. Voice: 301-319-9772; fax: 301-319-9706.
e-mail: jayasree.nath@na.amedd.army.mil

RESULTS AND CONCLUSION

In studies reported elsewhere, we have shown significant decrease in PMN adhesion to EC in the presence of ·NO-generating agents.[7] These results prompted us to measure ·NO levels in PMN-EC co-culture supernatants by a sensitive fluorimetric assay, the details of which have been published elsewhere.[7,8] For studies involving PMN-EC interactions, we have routinely stimulated the PMN with the synthetic chemotactic peptide fmet-leu-phe (fMLF) and primed the EC with pro-inflammatory cytokines IFN-γ and/or TNF-α. When necessary, PMN were primed with bacterial lipopolysaccharide (LPS), before stimulation with fMLF. Other details of these studies have been published.[7] As shown in TABLE 1, when ·NO levels were quantitated in PMN-EC co-cultures, there was a significant decrease in the ·NO levels (measured as nitrite) in PMN-EC co-culture supernatants, as compared to EC cultured alone under identical experimental conditions. This suggested that PMN-derived product(s) were responsible for the observed decrease in ·NO levels in PMN-EC co-culture supernatants. Interestingly, the presence of superoxide dismutase (SOD), a scavenger of $·O_2^-$, substantially restored the ·NO levels in the PMN-EC co-culture supernatants (TABLE 1, bottom line). Furthermore, an inverse correlation between $·O_2^-$ and ·NO was also evident in the presence of SOD, when almost a twofold increase in the measurable ·NO levels in the supernatants from stimulated PMNs cultured alone, was

TABLE 1. NO levels in PMN-EC co-cultures

Cells	Conditions	Nitrite (in pmoles)
EC	Control	355.50 ± 26.56
	LPS	402.93 ± 22.36
	IFNγ + TNFα	775.62 ± 35.98
	IFNγ + TNFα + LPS	800.79 ± 41.69
	IFNγ + TNFα + SOD	762.38 ± 28.62
PMN	Control	58.01 ± 10.33
	LPS	50.66 ± 21.24
	IFNγ + TNFα	56.45 ± 12.43
	IFNγ + TNFα + LPS + SOD	105.66 ± 16.34
PMN + EC	Control	339.64 ± 27.32
	LPS	372.88 ± 19.87
	IFNγ + TNFα + LPS	422.45 ± 35.76*
	IFNγ + TNFα + LPS + SOD	672.99 ± 42.98

EC were treated with cytokines TNFα (20 ng/ml) and IFNγ (20 ng/ml) for 18 h prior to co-culture with PMN. PMN were incubated with LPS (1 μg/ml) or with cytokines, for 1 h before the co-culture. NO levels were measured in the supernatants after 1 h co-culture. PMN and EC cultured alone served as controls. Each condition was measured in triplicate and the data represent mean ± SE of four independent experiments. *Significant decrease in NO levels as compared to the parallel condition (IFNγ + TNFα + LPS) where EC were cultured alone, in the absence of PMN ($p < 0.01$; t-test, unpaired). (Courtesy: *J. Leukocyte Biology*).

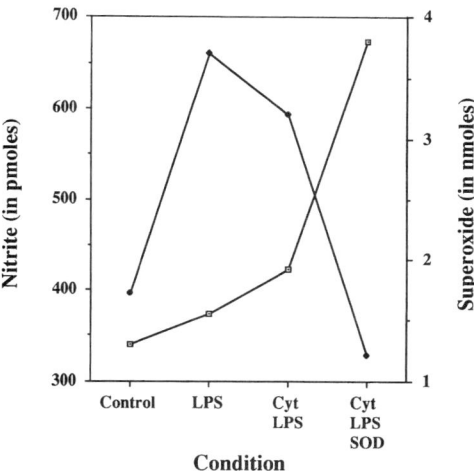

FIGURE 1. Inverse correlation between ·NO and $·O_2^-$ levels in PMN-EC co-culture. □, nitrite, ◆, $·O_2^-$. EC and PMN were co-cultured at a cell ratio of 1:10, and incubated for 1 h at 37°C. $·O_2^-$ was measured spectrophotometrically and ·NO as nitrite was measured fluorimetrically as described in the text. Each condition was measured in triplicate and results are representative of two independent experiments. Cyt, IFNγ + TNFα. (Courtesy: *J. Leukocyte Biology*.)

observed. These results clearly suggested an active role of PMN-derived $·O_2^-$ in inactivation of ·NO in PMN-EC co-cultures. The inverse correlation between ·NO and $·O_2^-$ levels was further substantiated, when ·NO and $·O_2^-$ levels were directly quantitated, in parallel, in PMN-EC co-cultures. These results are shown in FIGURE 1. In order to delineate the interactive role of ·NO and $·O_2^-$ in PMN-mediated EC injury, we also assessed PMN-mediated killing of EC by cytofluorimetry. As shown in FIGURE 2, a strong correlation between PMN-derived $·O_2^-$ generation and PMN-mediated EC cytotoxicity, was observed. In addition, a protective role of ·NO in PMN-mediated killing of EC was noted in the presence of a ·NO donor, S-nitroso-N-penicillamine (SNAP), and also in the presence of L-arginine, the physiological substrate for nitric oxide synthase, the primary enzyme responsible for generation of ·NO in biological systems. As is also evident from FIGURE 2, the protective role of ·NO was most pronounced when the EC were primed with a combination of the

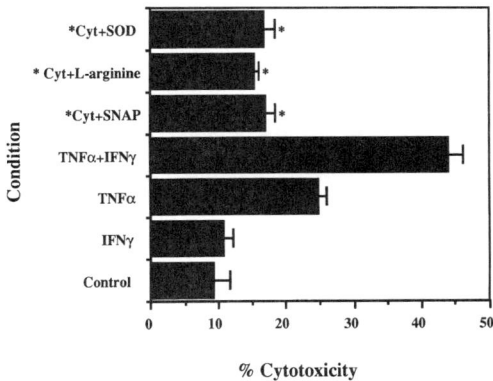

FIGURE 2. Cytotoxicity of EC by activated PMN. EC were labeled with calcien-AM and co-cultured with PMN at a cell ratio of 1:50 and incubated for 3 h at 37°C. Cytotoxicity was measured as dye release in the supernatants by fluorimetry. Each condition was measured in triplicate and data represent mean ± SE of four independent experiments. *Significant decrease as compared to the cytotoxicity of cytokine (IFNγ + TNFα)-treated EC ($P < 0.01$; t-test, unpaired). Cyt, IFNγ + TNFα. (Courtesy: *J. Leukocyte Biology*.)

pro-inflammatory cytokines, IFN-γ and TNF-α, which causes a marked increase in ·NO generation by EC (TABLE 1). Further details of these studies have been published.[7]

The results of our PMN-EC co-culture studies suggest that by generating ·NO, the EC inhibit PMN adhesion, thereby preventing their own damage. However, with the release of large amounts of ROI by activated PMN, ·NO is rapidly inactivated, which results in an increased PMN-EC adhesion that leads to cytotoxicity of EC. Thus, the complex interactions between EC-derived ·NO and PMN-derived ROI appear to play a central role in the regulation of inflammatory responses that lead to cell and tissue injury.

ACKNOWLEDGMENT

S.K. was a recipient of the National Research Council Associateship.

REFERENCES

1. NATHAN, C. 1992. Nitric oxide as a secretory product of mammalian cells. FASEB J. **6:** 3051–3064.
2. MONCADA, S. & E.A. HIGGS. 1993. The L-arginine nitric oxide pathway. N. Engl. J. Med. **329:** 2002–2012.
3. RUBANYI, G.M., E.H. HO, E.H. CANTOR, W.C. LUMMA & L.H. PARKER-BOTELHO. 1991. Cytoprotective function of nitric oxide: inactivation of superoxide radicals produced by human leukocytes. Biochem. Biophys. Res. Commun. **181:** 1392–1397.
4. WINK, D.A., I. HANBAUER, M.C. KRISHNA, W. DEGRAFF, J. GAMSON & J.B. MITCHELL. 1993. Nitric oxide protects against cellular damage and cytotoxicity from reactive oxygen species. Proc. Natl. Acad. Sci. USA. **90:** 9813–9817.
5. ANDERSON, B.O., J.M. BROWN & A.H. HARKEN. 1991. Mechanisms of neutrophil-mediated tissue injury. J. Surg. Res. **51:** 170–179.
6. RADI, R., J.S. BECKMAN, K. BUSH & B.A. FREEMAN. 1991. Peroxynitrite oxidation of sulfhydryls. The cytotoxic potential of superoxide and nitric oxide. J. Biol. Chem. **266:** 4244–4250.
7. KAUSALYA, S. & J. NATH. 1998. Interactive role of nitric oxide and superoxide anion in neutrophil-mediated endothelial cell injury. J. Leuk. Biol. **64:** 185–191.
8. NATH, J. & A. POWLEDGE. 1997. Modulation of human neutrophil inflammatory responses by nitric oxide: studies in unprimed and LPS-primed cells. J. Leuk. Biol. **62:** 805–816.

Daniel L. Gilbert: Explorer of Life

CLAIRE GILBERT[a,b] AND RAYMOND L. GILBERT[c]

[a]Springfield Hospital Center, 6655 Sykesville Road, Sykesville, Maryland 21784, USA

[c]Department of Administration, Information Services Division, State of Indiana, Indianapolis, Indiana 46204, USA

ABSTRACT: From the time Dan was 12 years old to the present, he always was the explorer of life. He was very fortunate in having an opportunity to see how life exists in several human societies around the world. His many professional foreign travels are presented. Dan went to a 1959 meeting in Argentina, to England in 1963, and to Chile in 1963. After he married Claire, she accompanied him on his 44 day around the world trip in 1965 and on all his later foreign trips. In 1966, there was a Latin American trip on the way to doing research at a laboratory in Chile. Next on his travels was being a consultant in Venezuela in 1969 and attending a meeting in Russia in 1972. His son Raymond joined in trips to France in 1977, to Australia in 1983, to a 1990 trip to a Peruvian meeting, and finally joined the 1991 meeting to a meeting in Japan, with a side trip to China where Dan gave lectures. The last 2 foreign trips for Dan with Claire, but not with Raymond, were the meetings in 1994 in Buenos Aires, Argentina, and the 1998 meeting, which was held in both Israel and Jordan.

INTRODUCTION

When Daniel first met Claire, whom he married in 1964, he told her about a life changing experience he had one day when he was 12 years old; he was visiting a science museum with his class and looking at a skull. A sudden storm blew up, and he looked out the window to see a man on the street struck by lightning. The interface of humans and nature, and the very nature of life itself, seemed to him mysterious and also the most important topic anyone could study. He decided then to devote his life to exploring life at its most basic level, although he wasn't quite sure how to do that. His own chapter, elsewhere in this volume, describes what happened next in his research career.

He was born on July 2, 1925, in Brooklyn, New York, the younger of his parents' two sons. His older brother, Morton, later became a dentist. His family lived on Long Island except for 1 year when he lived in Denver as an infant. When he was 6, he moved to Morristown, New Jersey, where Daniel completed high school in 1943.

When his father bought a small pool table for 10 year old Morton, he asked if he could play also. Morton agreed. At the time, 6 year old Dan, who is right handed, didn't realize that Morton was left handed. However, being stubborn, he practiced almost daily for a year before he could even hit a ball with the cue stick. Finally, he could beat his brother in pool. His education was interrupted, like so many of his generation, by service in the military in World War II. While he was in the Army, an

[b]e-mail: dangil@helix.nih.gov

Army buddy said to him, "Dan, I thought you are right handed, but you play pool left handed." It was only then that he realized that Morton was left handed.

He was in the Battle of the Bulge, and received the Purple Heart and Combat Infantryman Badge. He returned home to his parents in New Jersey. Aided by Public Law 16, the GI Bill for disabled veterans, he commuted to a local institution, Drew University, where he earned his A.B. in Biology in 1948. He attended the State University of Iowa to earn his M.S. degree in Physiology and published his first research. He went on to the University of Rochester for his Ph.D. in Physiology, which he completed in 1955. He spent one additional year at Rochester as an Instructor, and then moved to Albany Medical College, where he was an Instructor, then Assistant Professor, of Physiology 1956–60. He then became Assistant, then Associate Professor, at Jefferson Medical School in Philadelphia from 1960 until he was asked to join the National Institutes of Health, where he has been a Research Physiologist since 1962.

He accepted the offer because he was interested in the antioxidant barrier to the toxic effects of oxygen, which can literally burn up the cellular constituents. Oxygen easily diffuses across the cellular membrane. However, the cellular membrane is a barrier to most other chemicals. Thus, there are two types of barriers which retain the energy difference between the inside of the living cell and the environment of the cell; an antioxidant barrier and the cell membrane. Now he had a chance to study the other barrier, the cellular membrane. One of his important papers was the one which showed that diphenylhydantoin (Dilantin or phenyltoin) inhibits the sodium current.[1]

Dan's research has been presented in conferences which were key points in his career. In 1959, he attended the first of many International Physiological Congresses; this one was in Buenos Aires, Argentina. He used this opportunity to travel, as he was always eager to explore. He visited San Juan, Puerto Rico; Caracas, Venezuela; Rio de Janeiro, Brazil; Asuncion, Paraguay; and La Paz, Bolivia. Peru was the site of his visit to a high altitude research lab in Morococha, to the jungle village of Iquitos on the Amazon, to Cuzco with its remnants of Inca buildings, and to Machu Picchu, the lost city of the Incas in the mountains outside Cuzco. His next international conference was in London, in 1963; on this occasion, he saw Paris, Rome, and Naples. He stood at the edge of the Vesuvius crater and almost got locked into the Herculaneum ruins. In the fall of 1963, he went to Chile via Iguassu Falls, Brazil; Asuncion, Paraguay; and Buenos Aires, Argentina; for the first of two research trips there to Montemar, a marine biology station 4 miles north of Viña del Mar. He flew down to Punta Arenas, the southernmost city in Chile, and hired a private plane to look for Cape Horn; they hit bad weather and had to turn back before running out of fuel.

MARRIAGE: A NEW PARTNER

His life changed radically when he married Claire Plunguian on Sunday, July 26, 1964. She was then a graduate student in Romance Languages at Johns Hopkins; she would go on to complete her doctorate there, and later her master's in Social Work at the University of Maryland. She was very willing to share his travels. Their first trip, the day after the wedding, was to Woods Hole, on Cape Cod, Massachusetts, where he did research at the Marine Biological Laboratory on the squid giant axon and synapse each summer from 1962 to 1988. That morning, as his new bride waited to be introduced, his lab chief, Kenneth S. Cole, "Kacy," glared at Dan, demanding to know why he was late to work, and pointing out he was wasting time by not working on some fine large squid in the lab tank. This was Kacy's style of humor, which Claire eventually got used to. She decided that when Dan was separated from her on axon duty, she was a "squidow." That year also featured a trip to Brown University, in Providence, Rhode Island, where Dan gave the Bowditch Lecture at the American Physiological Society meeting.

His first foreign trip with Claire was to attend the 1965 International Physiological Congress in Tokyo, Japan. On the way to Tokyo, they stopped in San Francisco to see Dan's relatives and then stopped in Hawaii, seeing Honolulu, on Oahu, and the Kilauea Crater and Mauna Kea, both on the big Island of Hawaii. They picked a side tour to Bangkok, but then realized this was halfway around the world, and there was no point flying back from Japan. So they went around the world in 44 days, stopping in Japan, Hong Kong, Bangkok, Nepal, India, Lebanon, Syria, Egypt, Jordan, Israel, Greece, Austria, Denmark, France, and England before heading home. Highlights included seeing temples in Nepal, with the backdrop of the Himalayas; seeing a totally deserted Taj Mahal at sunrise, with a rosy glow shining through the marble; seeing Dan walk a couple of miles in the desert to Aqarquf, a ziggurat, outside of Baghdad, Iraq (soldiers congratulated him for doing this walk); visiting the tombs in Luxor and realizing they were looking at the original paint, over 3,000 years old; seeing Claire's twin cousins in Israel; strolling among the Cretan ruins, after meeting friendly natives who confessed they had never traveled 20 miles down the road to see Phaestos, an important Cretan archeological site; and a romantic day in Paris where Claire got to practice her French. Dan took many stereo pictures of the trip including one of both Claire and Dan in front of a Roman lion in Baalbek, Lebanon. He took the right and left images and reversed them; this produced a headache, since other clues showed them in front of the lion, and not behind. As they sorted through all the photos, they decided this was the honeymoon they didn't have time for a year earlier.

After just a couple of months, they were off again, to another three-month research stint on the coast in Chile. This time they went first to Houston, Texas, so Claire could show off Rice University, where she received her B.A. degree; then came Mexico City and neighboring Teotihuacán, with its two big Pyramids of the Sun and Moon. After that, they left for Guatemala to see the active Pacaya Volcano. Dan was always interested in volcanoes to see how new land is born. On the way to Ecuador, they stopped off in Panama to see the Panama Canal. They saw the celebration in Quito of the New Year, when children burned dummies representing Viejo, the old year, at midnight; they went through the town asking for money for the Vedova, the widow. They took a trip on New Year's Day, 1966, to see the Colorado Indians living in the jungle. To get there from Quito, they had to travel on a mountain road where Dan noticed many signs for "Peligro," which he assumed was a big town up ahead. When he mentioned this, Claire told him that meant "danger." On the way back, it started to rain. Dan insisted that he be on the right side, so he could get pictures of any rocks coming down the mountain. Just then, a huge boulder as big as the car they were riding in, came roaring down the mountain. Dan froze! The boulder hit the road, just after they were there, and bounced off the road to continue its journey down the mountain. Some smaller rockslides collided with the car; only the car was dented, not the passengers. The next stop was Peru, where they visited Pachacamac, a pre-Inca temple on the coastal desert and then visited Cuzco and Machu Picchu; they continued their travels back to Lima and then on to Santiago and Montemar, Chile. While Dan worked hard with the giant squid there, Claire flew south to see the Lake District in Chile and Argentina, and then Punta Arenas; from there, she rode out to a ranch so she could tour the Torres Del Paine, snow capped peaks that rise from an alpine meadow where guanacos graze. They saw a whale factory in Quintay on the coast. They were able to see penguins on an island off the coast, an animal that always fascinated Dan. On the way home, they stopped in Buenos Aires, Argentina. The next stop was Brazil; Iguassu Falls, Rio de Janeiro, Brasila, Manaus, where they saw the famous Opera House, and then flew to Belem on the coast. When they missed the plane back to the USA, they had to fly a thousand miles back to Rio in order to fly to New York.

They visited Montreal in 1967, to see the World's Fair and some Canadian relatives of Claire's; they returned to this city several times after that, stopping to see Quebec City, Toronto, and Ontario as well. In 1969, Dan was invited to be a consultant in Venezuela; that turned into a trip that included Merida and the Mayan ruins of the Yucatan in Mexico, El Salvador, Costa Rica, and the walled city of Cartagena and Bogota with its stunning Gold Museum in Columbia, Puerto Rico with its rain forest, Curacao, and Aruba. Claire had to return early to the USA; Dan then saw Angel Falls and some of the jungle areas of Venezuela.

The next foreign trip was to Russia in 1972, for the International Biophysical Congress. By this time, Claire and Dan had a son, Raymond, born in 1971; he was too young to travel, so he stayed home in Bethesda with his sitter. Because of a clerical mistake on Dan's visa, which didn't match his passport number, there was a delay at the airport clearing customs in Moscow. Russia was then still a closed society, and scientists feared listening devices in hotel rooms; they preferred to chat as they strolled along the street. Official receptions at the meeting really divided the hosts from their foreign guests; while the latter made a fuss over the caviar, the Russians went straight for the vodka and polished it off. A highlight was the visit to the Hermitage in the city now called St. Petersburg, and the Summer Palace, an hour's ride away up the coast in a modern ship. Russia seemed subdued, and people in Moscow were the most guarded, although there were a few chances to talk to ordinary people off the tourist paths. The visa error led to a scary confrontation upon leaving Russia, when Dan was asked to explain it; Dan said, "It was just a clerical error." The Russian official replied, "I believe you," then leaned out over his counter towards Dan's ear and whispered, "I think." Dan pointed out his passport picture and asked, "Can't you see? That's me." The official, with a wink, replied, "Ah... funny things happen here," but finally let Dan pass through. It was a relief to be on the plane, flying to beautiful Helsinki, Finland. They explored Finland's Lapland above the Arctic Circle, and were impressed by the scenery and friendliness of Sweden, Norway, and Denmark.

RAYMOND: A PARTNER IN EXPLORATION

When Raymond was 6, he got to travel with his parents for the 1977 Physiological Congress in Paris, France. The family drove around in a rented car to see Versailles, the castles on the Loire, and cathedrals all over. They visited the Normandy area where Dan landed with the Army in 1944, and made friends with a friendly countess in a nearby chateau who remembered the Americans coming. They visited Grenoble, near the Alps, where Claire had studied for two years as a Fulbright Fellow. A real highlight was driving through the French Riviera, to see Cannes and Nice, and then driving through the Mont Blanc tunnel to take the cable car up to the top of a popular ski run. Dan got to revisit the scene of the Battle of the Bulge, and finally drive across the bridge into Germany; he met some other veterans of that war who looked much older than Dan. The family was able to visit sights in Belgium, Luxembourg, Monaco, Switzerland, and Italy as well. In Paris they took a boat ride on the Seine and saw the Mona Lisa and so much more at the Louvre. During the week long meeting in Paris, Claire and Raymond had some opportunities to explore on their own. One day they waited for Dan before going up the Eiffel tower; he was so busy that by the time the three got together, they had closed the top floor of the tower. Alas, the second floor was all they could get to.

In 1979, Dan took Raymond on a different type of exploration. In celebration of Raymond's birthday, he took Raymond to the NASA Goddard Space Flight Center in Greenbelt, Maryland. At that time, they were able to witness the first live pictures from Voyager of its fly-by with Jupiter. This was a rather dramatic way to inspire the spirit of exploration in his son, who later majored in Astronomy at the University of Maryland.

The next big family trip was to attend the 1983 Physiological Congress in Sydney, Australia; Raymond was 12 by then, and a seasoned traveler. On the way to Australia, the family visited Hawaii, with its active volcano, and Fiji, where they saw firewalking. Following this Sydney Congress, Dan was invited to a meeting on the History of Respiratory Physiology in Leura, Blue Mountains, Australia. That trip featured visits to Alice Springs to see Ayres Rock, which Raymond succeeded in climbing; off shore trips to see the coral reefs; visits to Melbourne, a very cosmopolitan city with an avant garde theater show; and Canberra, where the tour of the Supreme Court showed they were still wearing wigs like the British. On the way back, they visited Auckland, the hot springs in Rota Rua, and the glacier at Mt. Cook in New Zealand before heading home.

Dan was a featured speaker at a 1990 meeting in Cuzco, Peru, where he presented his research on the 1590 publication by Father Acosta on high altitude sickness, one of the first observations of the effects of a low oxygen atmosphere. Raymond got to see Machu Picchu this time, and Claire got to visit a barren men's prison, which was very revealing about modern Peruvian culture. Dan had helped to determine that the mountain with the modern name "Tullujuto," between Lima and Cuzco, was actually the ancient mountain of Pariacaca mentioned by Father Acosta. A real highlight of this trip was chartering a small private plane to fly the three Gilberts over this mountain, a very distinctly beautiful one with twin snow covered peaks the local pilots call "raspadilla" or snow cone. Claire managed to take many photos of this stunning

FIGURE 1. Dan. a black and white picture (original in color), pictured on the weekend Science magazine cover of *La Nacion*, a daily Argentinean newspaper, published in Buenos Aires on Saturday, April 23, 1994.

sight, in spite of throwing up during the entire flight due to a bad dinner the night before, giving a whole new meaning to altitude sickness.

When Dan was invited to a meeting in Tokyo in 1991, Raymond was persuaded to skip several days of college to come along. Beijing was a thrill, with two host families who served as guides through the Forbidden City and the Great Wall. Flying 1000 miles inland to Xian, land of the terra cotta warriors, was a bit tricky. Dan got on a plane first, said goodby to his family, and then was quite surprised when Raymond and Claire managed to get on a stand-by flight and show up to greet him later that night. Dinner in Xian allowed the family to get up close and personal with one of the local Communist leaders there. Then it was on to Tokyo for the meeting; this big city had changed since last seen in 1965, when the Ginza was the real center; now there was a new area, Shinjuku, to explore.

In 1994, Dan was a featured speaker at a conference in Buenos Aires (FIG. 1), which highlighted research he had done with Rebeca Gerschman, an Argentinean scientist, on Oxygen Toxicity. All members of the conference were given a copy of their pioneering work on Oxygen Toxicity, published in Science in 1954. Claire walked her feet off exploring this elegant city; a highlight was a symphony performance at the Opera House, where the Gilberts were given free tickets by the Public Relations Director. Side trips included visiting the Moreno glacier, while enjoying the rustic charms of nearby Calafate, and the frontier town of Ushuaia, on Tierra del Fuego. A private boat was chartered to visit Puerto Williams, on a Chilean island across the Darwin Strait; there was much fuss about passports and visas in little shacks for crossing back and forth across the international boundaries that day. A tour of the strait got them close to dolphins and sea lions as well as to penguins and other Antarctic birds.

The latest foreign trip was to Israel, in 1998, where Dan was invited to speak at a conference in Jerusalem. This was a chance to reunite with Claire's cousins, and to tour Tel Aviv, Jerusalem, and the Golan Heights with them. The conference was partly held in Amman, Jordan, where they saw some Roman ruins and some very ancient carvings as well as modern mansions. Dan and Claire decided to take a side trip to Turkey, and saw some of the glories of Istanbul's mosques and monuments. They enjoyed a boat ride to admire the estates on both the Asian and European sides of the Bosphorus. Dan did this trip in a wheelchair, as he was recovering from his surgery in December, 1997, to remove his fourth meningioma brain tumor.

DOMESTIC TRAVELS AND OTHER HIGHLIGHTS

Domestic travels have been a treat as well. Dan and Claire have explored California each time they can, seeing San Francisco, Los Angeles, San Diego, Yosemite, Lake Tahoe, Palm Springs, the Napa wine country, and points in between. Western travels include Las Vegas, the Grand Canyon, Yellowstone, Jackson Hole, Montana, and Denver, where side trips included Pike's Peak and the Royal Gorge. Raymond joined his parents for a trip to South Dakota, to see land jointly owned with cousins in a family trust; there actually were deer and antelope playing on the land once homesteaded by Dan's uncle. There have been several trips to Florida, for Miami meetings and to see relatives, that have led to exploring Miami Beach, Ft. Lauderdale, and Sarasota; animal parks and mansions and museums were all enjoyed. Florida trips included a trip down to Key West, seeing the Hemingway and Truman houses, and jugglers on the pier. Another memorable trip was a 3 day cruise in total luxury from Miami to the Bahamas, complete with swimming on a private island with catered barbecue. From the Cape Cod summers, there have been side trips to New Hampshire, Maine, Boston, Plymouth Rock, and to Provincetown, with whale watches; there were boat tours of Hyannis and summer stock theaters in Falmouth and Dennis. Going back to New Jersey to see family members often has included side trips to New York City. Several times, the Gilberts traveled back to Maryland from Woods Hole by taking the ferry from Connecticut to Long Island. To celebrate their anniversary one year, Claire and Dan drove to Dan's old home town, Rockville Center, and then went to a rock concert by Santana at Jones Beach, where he recalled going in 1961. At that time, there were no rock concerts, but only water shows with graceful diving and swimming.

There were several lawn parties after they moved into their present home in Bethesda in 1973. One occasion was recorded on videotape. At a special lawn party to celebrate 25 years of marriage, they asked Raymond to perform the ceremony all over again, much more informally. At the original ceremony, Claire never got a chance to say "I do." With Raymond in charge, she got to say that as loud as she wanted.

Summers in Woods Hole were always a treat. Dan and Claire bought a cottage there at the end of her first summer, in 1964, and have coped with its aging electric and plumbing problems ever since. This house was the scene of an annual lawn party with some special theme, usually around the first week in July, when both the national holiday and Dan's birthday were celebrated. Many friends from the Marine Biological Laboratory (MBL) and Woods Hole came. One memorable party combined celebrations for Kacy's 75th birthday and for Dan's 50th. In 1969, Claire was determined to make Dan wear an orange shirt she bought him; she decided that the theme was going to be an Orange Party; she was shocked when some guests showed up in green, and some came with newspaper articles pointing out that the chosen date, July 12, was William of Orange Day, which annually sparks controversies and even riots in Ireland.

They were usually faithful in attending the Friday Night Lectures at the MBL, and the wine and cheese receptions afterwards; Claire caught up on some of the biology education she missed in college. Claire was active in the MBL Club for several years, where she edited a newsletter and helped to run the Thursday night movies. Raymond from an early age could visit the laboratory and learn to work the computers. From age 7, Raymond got a great science education by attending the Children's School of Science in Woods Hole, taking 6 week courses in topics such as Marine Invertebrates, and enjoyed digging about in sand and marshes to get specimens to study. Claire enjoyed serving on the Board of this school, working with some dedicated and talented parents to keep up its great reputation.

No social occasion in Woods Hole, it seemed, was complete without a recounting of boat stories, mostly that of intrepid sailors racing their way around the rocks of Woods Hole harbors. Dan never agreed to a sailboat, but bought a used motorboat in 1969 as a graduation present for Claire when she finished her Ph.D. This was actually a 14 foot row boat, with a 10 horsepower motor attached, and some oars to rescue passengers when the motor died, which it did frequently. It was named the Clan (CLaire plus dAN), and was used to explore the Elizabeth Islands off the Cape. A favorite destination was Hadley Harbor, off Naushon, and Bull Island in that harbor. It was a big adventure when with Raymond and with an extra tank of gas, all three went completely around Naushon in this little boat. When Raymond captained the boat with Dan one day, he didn't stop steering until he had taken the two of them all around the outermost Elizabeth Island, Cuttyhunk, about 25 miles away from Woods Hole; that was a remarkable, and probably dangerous, trip in such a small boat. On one trip to Naushon, the boat was leaking badly. A passing sailor in a small dinghy agreed to help. It turned out he was the owner of a large impressive yacht; he came back with a Corbel champagne cork, which plugged the hole very nicely. Raymond had to stretch to keep his toe tight against the cork as the three Gilberts headed back through the Woods Hole passage to the home dock in Eel Pond, next to the MBL.

Another boating adventure, although not in this boat, was for all three Gilberts and some neighbors to go out to Cuttyhunk Island, and then on to Penikese Island to visit some goats the neighbors had donated. This island, the original home of the Marine Biological Laboratory, was being used as a juvenile rehabilitation camp for those who had some legal troubles.

The final Gilbert trip in the boat "Clan," just before it was sold to good friends in 1988, was planned for Dan and Raymond to see if they could go through the Woods Hole passage, come around on the Buzzards Bay side, and go all the way up the coast to the Cape Cod Canal. It was a bright and sunny day when they started, but after several hours they noticed some very dark threatening clouds. Inevitably, the rains came, with thunder and lightning. They got down in the boat, and somehow steered toward a beach, where they pulled the boat out of the water and sought refuge on a porch of the nearest house. Claire was very worried about them, and was much relieved to be reunited with them later that night.

CONCLUSION

Dan and Claire are very proud of Raymond (FIG. 2). When Raymond was in high school, he was already collecting awards: the election to the National Honor Society of Secondary Schools in 1987, the tutorial award in 1988, the Outstanding Academic Award in 1988, and a citation for participating in mathematics competition from the American Regions Mathematics League. In college at the University of Maryland at College Park he received a B.S. in physics and in mathematics with honors. While he was an undergraduate, he was elected to the Golden Key National Honor Society and was the president of the UMCP Sigma Pi Sigma chapter of the National Physics Honor Society. He did graduate work at Indiana University for 2 years, but had to drop out of school due to health problems, now resolved. He is now employed by the State of Indiana as a computer systems analyst in Indianapolis. His favorite recreation is playing bridge in tournaments each week, and he has been a Life Master in bridge since 1997.

There is always another type of traveling to do when Dan and Claire keep busy at home in Bethesda with domestic projects, such as remodeling the kitchen or getting the cherry trees mulched. Clair has been a social worker and supervisor since 1991

FIGURE 2. Left to right Raymond, Claire, and Dan. Picture taken on September 25, 1999 by Raymond using a self timer.

at Springfield Hospital Center, a Maryland state psychiatric hospital. Dan continues to collaborate with colleagues on publishing and research projects, and searches the web for resources. In spite of the difficulties caused by his six operations from 1980 to March 2000 to remove his meningioma brain tumor, which has kept growing back, he has remained active in his scientific searches for the ultimate keys to how life originated, and how it is sustained at a very basic level. His curiosity for new horizons has kept him on his true career, that of the Explorer of Life.

REFERENCE

1. LIPICKY, R.J., D.L. GILBERT & I.M. STILLMAN. 1972. Diphenylhydantoin inhibition of sodium conductance in squid giant axon. Proc. Natl. Acad. Sci. USA **69:** 1758–1760.

Index of Contributors

Akman, S.A., 88–102
Aquilla, E.M., 61–68
Aslan, M., 375
Axelrod, J., 274–282

Blumberg, J.B., 159–167
Boveris, A., 121–135

Cadenas, E., 121–135
Carreras, M.C., 121–135
Chen, R.-C., 255–261
Chernyshev, O.N., 292
Chevion, M., 308–325
Chevion, S., 308–325
Chiueh, C.C., ix–xi, 238–254, 255–261
Chock, P.B., 168–181
Clerch, L.B., 103–111
Coffin, D., 61–68
Cohen, G., 112–120
Colton, C.A., 292–307
Costa, L.E., 121–135

Day, R.M., 159–167
Demple, B., 69–87

Ehrenstein, G., 283–291
Espey M.G., 209–221

Feelisch, M., 209–221
Floyd, R.A., 222–237
Freeman, B.A., 375–391
Fukuto, J., 209–221

Galdzicki, Z., 283–291
Galoforo, S.S., 349–362
Gilbert, C., 415–424
Gilbert, D.L., 1–14, 292–307
Gilbert, R.L., 415–424
González-Flecha, B., 69–87

Grimaldi, M., 274–282
Grisham, M.B., 209–221
Grünblatt, E., 262–273
Gutteridge, J.M.C., 136–147

Halliwell, B., 136–147
Hampson, A.J., 274–282
Hanbauer, I., 182–190
Haugaard, N., 148–158
Hensley, K., 222–237

Jovanovic, S.V., 326–335

Kang, S.-O., 168–181
Kausalya, S., 411–414
Komarov, A.M., 407–410
Krishna, M.C., 28–43
Kuppusamy, P., 28–43

Landauer, M.R., 44–60
Lange, G.D., 283–291
Lee, Y.J., 349–362
Levine, R.L., 191–208
Lolic, M., 274–282
Luo, Y., 392–398

Mak, I.T., 403–406, 407–410
Mandel, S., 262–273
McKinney, L.C., 61–68
Mišík, V., 335–348
Miranda, K.M., 209–221
Mitchell, J.B., 28–43

Nath, J., 411–414

O'Connor, T.R., 88–102

Poderoso, J.J., 121–135

Rauhala, P., 238–254
Ridnour, L.A., 349–362

Riesz, P., 335–348
Rodriguez, H., 88–102
Rosenthal, R., 274–282
Roth, G.S., 392–398
Russo, A., 28–43

Scortegagna, M., 182–190
Shi, S.S., 159–167
Sim, J.E., 349–362
Simic, M.G., 326–335
Speir, E., 363–374
Spitz, D.R., 349–362
Stadtman, E.R., 191–208
Stadtman, T., 399–402
Suzuki, Y.J., 159–167

Thornley-Brown, D., 375–391

Vitek, M.P., 209–221, 292–307
Vodovotz, Y., 61–68

Weglicki, W.B., 403–406, 407–410
Weiss, J.F., 44–60
West, J.B., 15–27
Wink, D., 274–282
Wink, D.A., 61–68, 209–221
Wu, R.-M., 255–261

Yim, M.B., 168–181
Youdim, M.B.H., 262–273

Zhang, J., 403–406